AF479335

A History of Mathematics in the United States and Canada

Volume 1: 1492–1900

AMS/MAA | SPECTRUM

VOL **94**

A History of Mathematics in the United States and Canada

Volume 1: 1492–1900

David E. Zitarelli

MAA PRESS

An Imprint of the

AMERICAN MATHEMATICAL SOCIETY

Providence, Rhode Island

2010 *Mathematics Subject Classification.* Primary 01A60, 01A70, 01A72, 01A73.

Photographs on the cover are (clockwise from top of the star) Benjamin Peirce, E.H. Moore, Felix Klein, Christine Ladd Franklin, and Kelly Miller.

For additional information and updates on this book, visit
www.ams.org/bookpages/spec-94

Library of Congress Cataloging-in-Publication Data

Names: Zitarelli, David E., author. | American Mathematical Society.
Title: A history of mathematics in the United States and Canada / David E. Zitarelli.
Description: Providence, Rhode Island : MAA Press, an imprint of the American Mathematical Society,
   [2019]– | Series: Spectrum ; volume 94 | Includes index. Contents: volume 1. 1492–1900
Identifiers: LCCN 2018024243 | ISBN 9781470448295 (alk. paper)
Subjects: LCSH: Mathematics–United States–History. | Mathematics–Canada–History. | AMS: History and
   biography – History of mathematics and mathematicians – General histories, source books. msc | History
   and biography – History of mathematics and mathematicians – 15th and 16th centuries, Renaissance.
   msc | History and biography – History of mathematics and mathematicians – 17th century. msc | History
   and biography – History of mathematics and mathematicians – 18th century. msc | History and biography
   – History of mathematics and mathematicians – 19th century. msc | History and biography – History of
   mathematics and mathematicians – Biographies, obituaries, personalia, bibliographies. msc | History
   and biography – History of mathematics and mathematicians – Schools of mathematics. msc | History
   and biography – History of mathematics and mathematicians – Universities. msc
Classification: LCC QA27.U5 Z58 2019 | DDC 510.973–dc23
LC record available at https://lccn.loc.gov/2018024243

To my wife, Anita,
for 50 + N years,
where N > 0

Contents

Contents ix

Foreword from the Editor

In the late spring of 2018, David Zitarelli submitted the files for this book to AMS for publication. On December 2, as work was underway on them, David died. The book that David submitted covered the years 1492–1930. At nearly 800 pages it was considerably longer that the draft I reviewed and approved several years ago. In consultation with David's family and in order to make the book a more manageable size, I decided to truncate the manuscript at 1900 and move the remaining content into the forthcoming second volume. That volume is currently being assembled from David's files. It will cover the years 1901–1941 and should appear in approximately two years.

The manuscript David submitted needed, in addition to truncation, a significant amount of cleaning up. Jennifer Wright Sharp, the AMS production editor assigned to the book, has done yeoman's work removing redundancies, resolving contradictions, chasing down missing references, and indexing. This book would not exist without her contributions.

Over the years of development of this project David became a friend, and his death came as a shock. It has been an honor to bring this project, his passion and obsession for the last fifteen years of his life, to completion. In this book, in addition to his deep scholarship, you see much of David's personality. He loved people and this history is very people-centered; you can hear David's admiration and affection for his mathematical cast of thousands in his prose. To illustrate the admiration and affection that David inspired in others, we have reprinted the obituary written by his son Paul on the following pages. We thank the Zitarelli family for permission to include it and for all their assistance with this project.

Stephen F. Kennedy
MAA Press Acquisitions

David Zitarelli (1941–2018)

The day of December 2 was perfectly normal for David Zitarelli. He jogged four miles, which brought his 2018 total to 1,110 miles and his lifetime total to 57,366 miles. (We know this because he tracked his mileage daily, beginning in 1973.) He wrote in his journal (the "Z Notes"), the same one he has been keeping daily since 1977, first on yellow notepads, then on a series of ever-more-complex word processors. He ate a rotisserie chicken for dinner with his wife, spent time babysitting his two Minneapolis grandkids, then FaceTimed with his two Seattle grandkids. He read a few pages of his Jill Lepore book, slipped his bookmark into page 422, and went to sleep. And then he died. His heart, whose genetic insufficiencies he had tried like hell to mitigate over many years via near-daily exercise, finally gave out. It was a heart that many of us knew—family, friends, acquaintances, colleagues, baristas, dental hygienists—because he wore it proudly on his sleeve. For better or worse (mostly better), all of us knew where we stood with Dr. Z, a gentile mensch if ever there was one.

He was born on August 12, 1941, in Chester, Pennsylvania, and grew up in Holmes, PA. He was a proud graduate of Ridley High School (Class of 1959), then went on to earn his Bachelor's (1963) and Master's (1965) degrees at Temple University, followed by a PhD in Mathematics at Penn State. The same year he secured his PhD (1970), he began a teaching career at Temple that would span 42 years before his retirement as Full Professor in 2012. During that tenure, he was awarded multiple teaching honors—most notably, a Great Teacher Award in 2005—but was perhaps most proud of the legions of students that he inspired, persuaded, and cajoled into mathematics-related careers. His love of teaching combined with his love of sports in another important aspect of his life—softball coaching—where he nurtured and inspired legions of young women (including his daughter) to strive for athletic excellence.

Zitarelli loved words as much as he loved numbers. Beginning in 1984, he enjoyed a fruitful, decade-long collaboration with Dr. Raymond Coughlin, authoring five successful mathematics textbooks. He went on to write two more textbooks (one alone; one with Dr. David Hill) before turning his attention to his late-career passion: the history of mathematics. In 2001 he published *EPADEL: A Semisesquicentennial History, 1926–2000*, which charts the history of Eastern Pennsylvania and Delaware Section of the Mathematical Association of America, and then in 2004 he turned to the large work that would dominate the last fifteen years of his life: *A History of Mathematics in the United States and Canada*. He completed Volume 1, covering the years 1492–1930, in 2018, and it will be published in 2019 by MAA Press, an imprint of the American Mathematical Society. He was hard at work on Volume 2 (1930–2000) at the time of his death, and may well have finished it sooner had it not been for the welcome distractions of four grandchildren, all of whom arrived between 2009 and 2016.

He and his wife Anita (whom he married in 1966 and fiercely loved for all of their 52 years of marriage) split their retirement time between stints in Minneapolis (with daughter Nicole Danielsen, son-in-law Ryan Danielsen, and grandchildren Oliver and Zoey) and Bainbridge Island, WA (with son Paul Zitarelli, daughter-in-law Kelli Larsen, and grandchildren Lenna and Solomon). Despite initial sadness about leaving his "Philly roots," Zitarelli flourished in the upper Midwest and Pacific Northwest, making easy friends in both places and cherishing the chance to participate in the daily lives of his children and grandchildren.

Was Zitarelli perfect? Of course not. He could exaggerate (he would doubtless revise this to "embellish") the truth of a matter in the pursuit of "a good story." Sometimes, if he was losing in a game of one-on-one basketball, he would knee his own family member in the back of the leg to secure a rebound and then flatly deny a foul. No, he was not perfect. But he was very very good, and he was ours for a time, and we loved him.

Paul Zitarelli

Preface

In 2001 I received an email message from Michael Schein, a prize-winning Caltech graduate who was then a 19-year-old, first-year graduate student at Harvard. Michael has taught at Bar-Ilan University in Israel since receiving his PhD in 2006 with a dissertation supervised by Richard Taylor. I have never met this algebraist, but knew his family well. His father, Boris, a Russian mentor for my PhD dissertation, had stayed at my family's house (along with Michael's mother and sister) after immigrating to the United States, where Michael was born shortly thereafter. The younger Schein commented on my paper that had recently appeared in the *American Mathematical Monthly*:

> Thank you very much for sending me a reprint of your article on the development of American mathematics in the first half of the twentieth century. I had wondered how it had grown so quickly from such humble origins and had assumed this was mostly due to the large number of foreign mathematicians who arrived in the thirties. I was very interested to learn that American mathematics had already produced its own infrastructure by then.

Up to that time, I had taught a course on the history of mathematics in America every other spring to complement the more standard general history course I taught annually. Since no book was appropriate for the American course, I used the Parshall and Rowe classic, *The Emergence of the American Mathematical Research Community*, and began to supplement it with my own notes. Michael's message inspired me to expand those notes into book form, totally unbeknownst to him.

The notable mathematician Irving Kaplansky also felt the need for a modern history of mathematics when he nominated Hyman Bass for President of the American Mathematical Society in 1999: "The 'Stone' age was at its peak. (I hope that many *young* mathematicians will be reading this, and I realize they may not have a clue as to what I am talking about.)"[1] I feel that Kaplansky's parenthetical expression could have omitted the adjective "young," because all individuals versed in the first year or two of undergraduate mathematics would benefit from knowing the history of the subject.

Volume 1 covers the period 1492–1930. I hope to complete Volume II covering 1930–2000 in the near future. Overall, the two volumes are aimed at filling a gaping hole in the development of mathematics communities in the United States and Canada. Collectively, I refer to these two countries as "America," matching the sense in which both professional organizations of mathematicians use it—the American Mathematical Society (**AMS**) and the Mathematical Association of America (**MAA**). Indeed, there is no professional organization in the US for historians of mathematics; instead, most such historians who live in the US belong to the Canadian Society for the History of Mathematics.

Periodization of any historical study is an artificial construct, yet it provides a structure that enhances an understanding of the subject area whose history is being examined. Mathematics is similar: the axiomatic approach provides an external structure to an area that enhances an understanding of the area. The account here is mainly chronological, arranged by periods, called parts. Part I covers the Colonial Era up to 1800, including the establishment of various colleges and universities, as well as original investigations by a limited number of scientists. Part II examines two periods: The first (up to 1838) highlights initial quests to form a mathematical community via organizations and periodicals. The remaining years (up to 1876) were dominated by Benjamin Peirce, who engaged in legitimate research into various mathematical arenas. This latter period also witnessed the growth in the number of individuals who were interested in pursuing higher degrees in mathematics when J.J. Sylvester was hired by Johns Hopkins University to inaugurate the first graduate program. Part III, 1876–1900, saw revolutionary growth in American mathematics in both numbers and quality of contributions to the field.

Overall, my approach presents a bird's-eye view that intertwines descriptions of leading (and secondary) figures in the field, along with thematic developments in the subject. Each of the four parts is followed by a "Transition" section into the next part, based on a topic that reflects the continuous nature of development rather than a discrete break. The one exception might be the "Transition 1876" section, because that watershed year saw a revolutionary leap (a jump discontinuity) in research within the American mathematical community.

I find that the human factor lends life and vitality to any subject, but it is particularly central for mathematics. Consequently, I devoted a lot of time to finding material even about secondary and tertiary figures. Hopefully, this human facet enhances the focus of the book. Major figures are not restricted to white males. Native Americans, African Americans, Chinese Americans, and women appear in the context of the subject when they occur naturally, but sometimes entire sections are devoted to their contributions. Along these lines, it is my fervent hope to make household names out of leading figures like Benjamin Peirce, E.H. Moore, Oswald Veblen, George Birkhoff, R.L. Moore, and Marston Morse, but also Caleb Cheeshahteamuck, Isaac Greenwood, Mary Winston, Anna Pell Wheeler, Li-fu Chiang, Elwood Cox, and William Claytor.

There are two ways in which I have deviated from many standard works on the history of mathematics. For one, I mainly use first names rather than initials. Thus, I cite George Birkhoff instead of G.D. Birkhoff. However, some names are so entrenched with initials—E.H. Moore, R.L. Moore—that it seemed unwise to depart from standard usage. The other deviation refers to dates: I use August 12, 2018, not the more logical 12 August 2018, mainly because it is the form used more often in the US.

In addition to the subject of mathematics and the American mathematicians who took part in developing it, the two volumes take into account external social and political factors that have also impacted advances in, and in turn have been influenced by, the subject significantly—such as changes in education, government support of basic research, and wars.

Nonetheless, this labor of love presents a history of *mathematics*—so it is the development of the subject that guides my presentation. Thus, it might sometimes be necessary for the reader to skip over unfamiliar topics; the work has been formatted to accommodate such omissions. Indeed, I have not always been confident in some of the fields within mathematics that I have described, so I was compelled sometimes to lean on colleagues for help. I am grateful for having been grounded broadly by Herstein's algebra, Kelley's topology, Rudin's real and complex analysis, and Feller's probability, yet writing about mathematics in the twentieth century presents a formidable task, one I hope I can capture accurately. Although a reader with little formal training in mathematics can gain an overall impression of the development of the subject in America over the past four centuries and an acquaintance with the major figures in that development, the higher the level of attainment in mathematics, the more the reader will understand the development.

David E. Zitarelli
Minneapolis
February 2018

Acknowledgments

It takes a village to complete a project of this size and depth. My village consisted of family members, administrators and students at Temple University, friends, MAA Press editors, and AMS production personnel.

None of this would have been possible without the support of my wife, Anita, who has assumed all of life's nonmathematical necessities for me over the past ten years or so. I have also benefited from the love and encouragement from our children Paul and Nicole, and their families (spouses Kelli and Ryan, and their two children each).

I owe a debt of gratitude to many of the students in the biannual course I taught at Temple starting in 1996, "The history of mathematics in America," for providing help ridding early versions of salient mistakes. The 2004 class was the first where I wrote my own notes to supplement the Parshall–Rowe book. The website, https://davidzitarelli.wordpress.com/hoam/ contains many of the term papers written by the twenty students in that class. Endnote 26 in Chapter 4 reads, "Most of my material on the *Cambridge Miscellany* is based on ... archival research carried out by Gino Pagano, an undergraduate in my course." That paper is available on my website under "HoAM", but was originally developed by another student in the class, Neil Lampton.

I am indebted to Honors Program director Ruth Ost for arranging for the course to be offered in the Department of History in 2012. This class, based entirely on Volume 1, convinced me that the material was appropriate for nonscience majors, as well as the usual suspects. Heartfelt appreciation is extended to former dean of Temple's College of Science, the chemist Hai-Lung Dai, whose knowledge of the history of science led to my promotion to full professor based on my publications in the history of mathematics.

Since retiring to Minneapolis in June 2012, I have found a group of friends who meet at the Dunn Brothers coffee shop in Linden Hills and have been keen on this project. They come from various walks of life—librarian, history teacher, city manager, lawyer, realtor, engineer, and detective. Three proofread the manuscript and gleefully pointed out unwelcome gremlins—Mark Bernhardson, Linda Clark, and Jim Hendricks.

MAA Press has been enthusiastic about this project since its inception. I thank acquisition editor Steve Kennedy for unwavering backing and the Spectrum Series committee, headed by editor Jim Tattersall, for their close reading of, and vast improvements to, an early version. Finally, I thank AMS Publisher Sergei Gelfand for turning Word files into the attractive product you are now holding in your hands (either the paper- or e-version), and AMS acquisitions assistant Christine Thivierge for her help in locating sources for photos and obtaining proper permissions for publishing them.

Permissions and credits

The AMS is grateful to the following individuals and organizations for providing photographs and for permission to reproduce copyrighted material. While every effort has been made to trace and acknowledge copyright holders, the publishers would like to apologize for any omissions and will be pleased to incorporate missing acknowledgments in any future edition.

Dover Publications, New York
  Source for inequalities "<" and ">"; p. 15; from Thomas Harriot, *A Briefe and True Report of the New-found Land of Virginia* (reprint of 1588 original), Courtesy of Dover Publications

Peabody Essex Museum, Salem, Massachusetts
  Portrait of Nathaniel Bowditch by Charles Osgood; p. 144; Courtesy of the Peabody Essex Museum

University of Pennsylvania, Philadelphia, Pennsylvania
  Portrait of Reverend John Ewing; p. 103; Courtesy of the Archives of the University of Pennsylvania
  Portrait of Robert Patterson; p. 106; Courtesy of the Archives of the University of Pennsylvania

University of Rhode Island, Kingston, Rhode Island
  Photo of Anne Lucy Bosworth Focke; p. 418; Courtesy of University Archives, University of Rhode Island Library

The AMS gratefully acknowledges the kindness of the following individuals in granting the following permissions:

Kelli Larsen
  Photo of David E. Zitarelli; back cover; Courtesy of Kelli Larsen

David E. Zitarelli
  Cover page of memorial article on Alexander Fisher; p. 133; author's personal collection
  Cover page of Jeremiah Day's *Algebra*; p. 134; author's personal collection

Paul Zitarelli
  Obituary, David E. Zitarelli (1941–2018); pp. xii–xiii; Courtesy of Paul Zitarelli

The American Mathematical Society holds the copyright to the following photographs:
 Photo of George William (G.W.) Hill; p. 200
 Photo of Henry Burchard Fine; p. 275
 Photo of William Fogg Osgood; p. 279
 Photo of Maxime Bôcher; p. 281
 Photo of Henry Seely White; p. 283
 Photo of Edward Burr Van Vleck; p. 284
 Photo of Thomas Scott Fiske; p. 289
 Photo of John Howard Van Amringe; p. 291
 Photo of John Emory McClintock; p. 292
 Photo of Robert Simpson Woodward; p. 293
 Photo of Eliakim Hastings (E.H.) Moore; p. 319
 Photo of Leonard Eugene Dickson; p. 349
 Photo of Gilbert Ames Bliss; p. 352
 Photo of Virgil Snyder; p. 393

The Mathematical Association of America holds the copyright to the following photographs:
 Photo of George Abram Miller; p. 384
 Photo of Earle Raymond Hedrick; p. 420
 Photo of William De Weese Cairns; p. 426

The following photographs are in the public domain:
 Portrait of Thomas Harriot; p. 13
 Portrait of Ezra Stiles; p. 49
 Portrait of Nehemiah Strong; p. 50
 Portrait of John Witherspoon; p. 57
 Portrait of James Alexander; p. 58
 Woodcut of Cadwallader Colden; p. 59
 Portrait of David Rittenhouse; p. 68
 Woodcut of Benjamin Banneker; p. 83
 Title page of Baltimore edition of Banneker's 1792 almanac; p. 86
 Portrait of Hugh Williamson; p. 104
 Portrait of Ferdinand Hassler; p. 114
 Portrait of Alden Partridge; p. 115
 Portrait of Sylvanus Thayer; p. 116
 Photo of Claudius Crozet; p. 117
 Portrait of Robert Adrain; p. 123
 Photo of Benjamin Silliman; p. 132
 Portrait of Jeremiah Day; p. 135
 Photo of William Chauvenet; p. 147
 Photo of Francis Amasa Walker; p. 159
 Photo of James Dunwoody Brownson DeBow; p. 160
 Photo of Joseph Lovering; p. 172
 Photo of Benjamin Peirce; p. 176
 Photo of Josiah Willard Gibbs; p. 201
 Photo of Christine Ladd Franklin; p. 208

Photo of Artemas Martin; p. 218
Photo of Hubert A. Newton; p. 221
Photo of Daniel Coit Gilman; p. 240
Photo of James Joseph (J.J.) Sylvester; p. 245
Photo of William Edward Story; p. 247
Photo of George Bruce Halsted; p. 249
Photo of Charles Sanders Peirce; p. 250
Photo of Washington Irving Stringham; p. 253
Photo of Kelly Miller; p. 272
Photo of Felix Klein; p. 273
Photo of Frank Nelson Cole; p. 276 (Fair Use)
Photo of Mary Frances Winston Newson; p. 286
Photo of Amasa Leland Stanford; p. 297
Photo of Jonas Gilman Clark; p. 299
Photo of Oskar Bolza; p. 302
Photo of William Rainey Harper; p. 317
Photo of Heinrich Maschke; p. 320
Photo of Looking West from Peristyle, Court of Honor and Grand Basin of the
    1893 World's Columbian Exposition; p. 326
Photo of William Holding Echols, Jr.; p. 334
Photo of Clarence Abiathar Waldo; p. 343
Photo of James Loudon; p. 363
Photo of Sophus Lie; p. 372
Photo of Edgar Odell Lovett; p. 375
Photo of Hans Frederick Blichfeld at ICM, Zürich, 1932; p. 378
Photo of Ezra Otis Kendall; p. 387
Photo of Winifred Edgerton Merrill; p. 399
Photo of Charlotte Angas Scott; p. 405 (Fair Use)
Photo of Katharine Coman; p. 413
Photo of Charles Max Mason; p. 423
Photo of Wallie Abraham Hurwitz; p. 424

Introduction to Volume I

A brief look at three individuals who contributed vitally to mathematics in America provides a window into the goals and coverage of this work. Those with a mathematical background might want to guess each person's identity. Answers appear at the end of this Introduction, which also includes 13 questions for anyone desiring a prereading quiz of knowledge of mathematics in the United States and Canada. (Answers to this list are scattered throughout this volume.)

Before presenting the three individuals, it is important to point out that many accounts of mathematicians and the mathematics involved have been expanded with files stored on my personal website in the section on this book. Thus, to read more about Thomas Harriot's mathematical exploits, the text directs the reader to "The online file 'Web01-Harriot'." To access this file, go to

https://davidzitarelli.wordpress.com/

The reader will then have to select "Read More" under this book title, and then the volume and the part where the file is mentioned in the text. In this case, click on the book title, Volume 1, and then Part I. The pdf file "Web01-Harriot" is listed on the resulting menu. There is also a second section on the website for my courses at Temple University, and within that section the reader will find many papers written by my students filled with additional information.

The first individual (1906–1978) was the only logician ever invited to deliver an address at an International Congress of Mathematicians, a quadrennial event that was held in Cambridge, MA, in 1950. He was not a native-born American, but an Austrian who achieved his most startling results in the early 1930s, a couple of years before emigrating to the US via the trans-Siberian railroad and a trans-Pacific cruise. During World War II, while walking along the coast of Maine in typical mathematical style—deliberate cadence, hands folded behind his back, totally absorbed in abstruse thought—local residents suspected him of being a German spy. Eventually, he adjusted to life in the US and became a naturalized citizen. His fundamental contributions continued apace by publishing a paper in the *American Mathematical Monthly* and delivering an AMS lecture despite suffering from severe paranoia and hypochondria. This first individual represents the initial period in Volume 2, the 1930s.

Who was he?

The next individual comes from an entirely different period, one that ended about 65 years earlier. Since 1876, an historic year for research mathematics in the US, the daily workings of mathematicians bear many similarities with conditions today, yet the present use of the term "mathematician," meaning someone who publishes original research, would have been inappropriate before then. Therefore, our second person

(1809–1882) represents a time when mathematics was carried out by general scientists. This scientist was by far the best in mathematics the US had produced by then, some achievements even being known in Europe. A particular incident tells the story.

When this person hopped aboard a carriage bound from Cambridge to Boston in 1847, he gleefully boasted to a Harvard colleague, "Gauss says I am right!" In his hand, he held a letter from the Göttingen giant stating that his criticism of the work of Urbain Jean Joseph Leverrier was correct. The previous year that French astronomer became the first person ever to calculate the trajectory of a planet (Neptune) based on the perturbations of another planet (Uranus). Yet the unknown American carried out his own investigation, with the healthy skepticism characteristic of scientists then and now, and concluded that the subsequent sighting of Neptune was "a happy accident." The fact that he stated his conclusion formally at a meeting of the American Academy of Arts and Sciences (called the "American Academy" in this volume to distinguish it from the AAAS, the American Academy for the Advancement of Science) would not have been significant because few Europeans had any interest in developments across the Atlantic. However, our individual allowed his comments to be published in the leading astronomical journal, the *Astronomische Nachrichten*. Not surprisingly, this caused a public dispute between the two men, so when our individual received confirmation from Gauss, he gained even more confidence that his work was correct. Today, there is a myth that mathematicians do their best work before age 40, yet our second person's most substantial contribution, the book *Linear Associative Algebras*, appeared more than 20 years later when he was 61. Even though this work contained important results from the first legitimate mathematical research carried out by an American and was based on lectures delivered at the prestigious National Academy of Sciences, it had to be published privately, thus severely limiting its impact. Consequently, this American scientific giant gained little recognition, at home or abroad. In fact, his major work did not gain public appreciation until it appeared posthumously in the *American Journal of Mathematics*.

The history of mathematics boasts of the Brothers Bernoulli. Our mathematician just described comes from what we might call the Peirce Pride. Which Peirce was he? How do you pronounce his family name?

Many modern mathematicians will be able to identify our first figure, and some can hazard a guess at the second, but very few (even amongst historians) will have heard of the third. Yet this person's advances were vital to the growth of mathematics in the US during colonial days, and at least one of his contributions continues to this day. This third mathematician was educated at Harvard in the 1720s before sailing to England to pursue further studies in divinity. While abroad, his knowledge of mathematics impressed a philanthropist so much that the patron endowed the Hollis Professorship at Harvard that continues to this day, even though that donor never set foot in the New World. This third entry was the first Hollis Professor, an appointment he celebrated by delivering lectures on algebra, a subject virtually unknown in America at the time. Unfortunately, however, he had a drinking problem, one that caused a thinking problem—on the part of the Harvard Corporation, which ended up expelling him from his endowed professorship in the College.

Who was this virtually unknown inebriate?

There you have it—three central characters representing three different periods in American mathematics. Yet not one of these leading figures flourished in the revolutionary quarter-century, 1876–1900. Before 1876 all contributors to American mathematics were isolated individuals who worked with only a modicum of support. I refer to them as "rugged individualists," a term that includes Christine Ladd.

Conditions changed dramatically in 1876 with the opening of the Johns Hopkins University, whose mission required faculty members to publish original research and to train future researchers, all the while carrying out their teaching assignments. The next 25 years witnessed incipient signs of four vital elements in the life of modern mathematicians mentioned in the brief accounts of our three mathematicians:

- professional organizations (such as the AMS, MAA, AAAS, and NAS);
- journals (such as the *Transactions of the AMS* and the *American Mathematical Monthly*);
- specialized meetings (such as international congresses and national gatherings); and
- universities that support research.

Additional Questions

1. What trigonometry tables accompanied Christopher Columbus on his voyage to the Americas?
2. Who was the first legitimate mathematician to set foot in the US?
3. How many years separated the landing at Plymouth Rock in 1620 from the founding of the first college in the US? Who were the professors? What mathematics did they teach?
4. How many years separated the landing at Jamestown in 1607 from the establishment of the second college in the US? Who were the professors? What mathematics did they teach?
5. What scoundrel landed the first endowed position in mathematics? What mathematical subject did he advance at Harvard?
6. A very popular biography of Benjamin Franklin (by Walter Isaacson) declared that the famous statesman had no mathematical talent. List the ways that this assertion is false.
7. How was the format and wording of the Declaration of Independence influenced by Thomas Jefferson's knowledge of geometry?
8. The first journal devoted entirely to mathematics appeared in 1804. Who was the editor? Who became its most important contributor? How long did the journal survive?
9. Who translated Laplace's *Mécanique Céleste* with annotations aimed as an aid for English-speaking readers?
10. What undergraduate student helped with the Laplace translation, single-handedly brought out the fifth and final volume of this monumental work, and subsequently became America's first research mathematician?
11. What university revolutionized higher education in America by putting research and the training of graduate students on an equal footing with teaching?
12. Where did most American students go for PhDs in mathematics during 1880–1900? Who was the leading mentor?

13. Which university was more successful in developing a graduate program in mathematics in the 1890s, Clark or Chicago? Why?

Answers:

1) Kurt Gödel
2) Benjamin Peirce (pronounced "purse")
3) Isaac Greenwood

Part I

Colonial Era
and Period of Confederation,
1492–1800

Introduction to Part I

Part I describes the development of mathematics in the United States and Canada up to 1800. The first settlers in New England and Nouvelle France were not just uneducated farmers and merchants—several individuals possessed impressive mathematical abilities. To reinforce this, some examples cited here show how material from the history of mathematics can be used effectively in classrooms today, ranging from beginning algebra to calculus, and even extending to numerical analysis.

Part I consists of two chapters. The first begins with the voyages of Christopher Columbus and Giovanni da Verrazzano. What role did mathematics play in these passages across the Atlantic Ocean? Few people associate mathematics with Columbus's historic journey, but a moment's reflection raises the question: Just how did explorers get from point A to point B before the advent of GPS? This chapter will show that fifteenth- and sixteenth-century pioneers relied on a combination of astronomy and navigational trigonometry to guide them.

Those two centuries saw numerous boats reach these shores, but mathematics played little role beyond supplying improvements in both observational and theoretical navigation. This includes a 1585 voyage to the Atlantic coast under the command of Sir Walter Raleigh, when the first world-class mathematician set foot on American soil. Thomas Harriot was only 25 years old when he crossed the Atlantic as a scientist charged with describing the land and the people he encountered in the New World. (In general, I shall use the word "scientist" for someone who worked in the mathematical and physical sciences, even though the term was not adopted until William Whewell coined it in 1833. Until then the terms "natural philosopher," "scholar," and "savant" were used.) Harriot was accompanied by an artist who sketched pictures to illustrate his descriptions, resulting in a format like *National Geographic* magazine. Harriot's presence provides an opportunity to discuss the evolution of algebra, an area where he made original contributions.

Matters changed dramatically for America right from the start of the seventeenth century, when some colonial settlements began to prosper. Think of Jamestown in 1607, Quebec City in 1608, and Plymouth in 1620. Now, one could hardly expect Virginia settlers, French pioneers, or New England pilgrims to be too concerned with education, let alone have time for leisurely pursuits like mathematical investigations. Besides, these settlers came to the New World in search of a better life or to escape religious persecution, so they were very practical minded. Yet this hardy bunch included several with college educations from England, Scotland, and France who harbored hopes of offering the same opportunities for their offspring. These early adventurers established grammar schools from the beginning. However, given limitations on free time and a sparsely situated population, one could hardly expect them to found a college in these new areas. Yet, that is exactly what happened. Cambridge College, as it was initially called, was launched in 1636, just 16 years after the first pilgrims set foot on Plymouth Rock, and six years after the Massachusetts Bay Colony was established. This is astounding. Three years later, the name was changed to Harvard College, the name of the undergraduate part of Harvard University to this day.

Farther north, along the Saint Lawrence River, the Jesuits launched a college one year before Harvard was founded. However, the Collège de Québec did not last as long as its younger counterpart, closing in 1761 shortly before the final French and Indian War.

Moreover, even before these two institutions of higher learning were established, a movement was launched in the Colony of Virginia to found a college. Unfortunately, clashes with Native Americans delayed that undertaking until 1660, when it arose for the second time, only to be derailed by external events once again. Consequently, it was 1693 before the Royal College of William and Mary was established, thus becoming the third institution of higher learning in America. Yale, the next college founded in the 13 original colonies, got its start in 1701, but it took 45 more years before another institution of higher learning joined these four.

Who were the professors at these initial colleges? Was mathematics part of the curricula? What mathematical topics were covered?

Despite this rather auspicious beginning, harsh and inhospitable conditions in the colonies permitted little time for independent mathematical pursuits. So while seventeenth-century Europe witnessed the most radical advances in the 4000-year history of the subject, weary colonists were unable to keep abreast of such endeavors, let alone partake of their discovery. Only a few small farmers, artisans, and merchants needed mathematics beyond the ability to add and subtract numbers and to perform some multiplication; division was rarely needed. Even trade was mostly carried out by bartering. While those hardy individuals who fled Europe were eking out a living, the Old World produced revolutionary developments in combining algebra with geometry (led by René Descartes and Pierre de Fermat), number theory (Fermat again and the priest Mersenne), calculating devices (Napier and Pascal), probability (Fermat yet again and Pascal again), projective geometry (Desargues), calculus (Newton and Leibniz), and astronomy (Kepler, Galileo, and Halley).

Throughout the eighteenth century, economic progress in the New World created a need for mathematical tools to solve problems in three areas: surveying, astronomy, and navigation. This in turn created a demand for elementary mathematical notions to be taught in all educational institutions, thus this account will examine the rise of various schools, from public and private grammar schools to colleges. Here the emphasis will be on developments that took place at the Collège de Québec, Harvard, William and Mary, and Yale, the four initial institutions of higher learning that preceded the spate of colleges founded between 1746 and 1776.

Along the way certain individuals of legendary misbehavior are introduced. For instance, the first mathematics professor in America lasted less than a year because of scandalous behavior at William and Mary. Similarly, the first Hollis Professor of Mathematics at Harvard was evicted from his endowed chair for similar offenses. And the Harvard tutor, who would have seemed to be the obvious successor, was bypassed due to his own shameful acts. This trio womanized, boozed, and generally did not act collegially. Nonetheless, while most colonial people were upstanding citizens, these few rascals hint at the wide range of behaviors of mathematicians throughout the ages. Not all scientific advances in America took place in the colleges, however. To reinforce this, Chapter 1 contains examinations of almanacs for predicting future events, surveys conducted to establish formal boundaries, and the founding of America's first scientific society.

Chapter 2 is concerned with the period 1750–1800. The British defeat of the French in the Seven Years War put an end to higher education in Canada for half a century beyond 1761. Naturally, the American Revolutionary War put an abrupt halt to this progress in the US, though it did produce a country that would ultimately occupy a territory far larger than Great Britain. Virtually no mathematical activity occurred over the decade 1776–1786. However, in the days leading up to the cry for independence, two small groups of scientists formed professional organizations that merged into one that published the first general science journal in the country—the American Philosophical Society (**APS**).

Two Harvard professors, Isaac Greenwood and John Winthrop, contributed appreciably to mathematics and astronomy, respectively. Chapter 2 also describes the mathematics contained in some of the published works of Greenwood, David Rittenhouse, Benjamin Franklin, and Benjamin Banneker. Although these four brilliant scientists—plus Winthrop—were born and bred in America, only Rittenhouse, and perhaps Greenwood, might be considered mathematicians by today's standards. The lives and accomplishments of two African American scientists are also examined. While Benjamin Banneker established a reputation as a gifted astronomer and mathematician, Thomas Fuller was known only for lightning-quick mental calculations. Both men, however, were celebrated by abolitionists as examples of inherent talent in African Americans.

Accounts of the accomplishments of these men raise a vital question: What is a mathematician? Chapter 2 opens with an answer to this question. The chapter also discusses cryptology, which was only slightly used during the American Revolutionary War, including a clever code designed and used by Benjamin Franklin.

Chapter 2 ends with the Period of Confederation, 1783–1800. For mathematics the two leading developments were the formation of a second scientific organization (the American Academy of Arts and Sciences) and the publication of its journal, as well as the publication of an important textbook by an American author (Nicholas Pike) for an American audience.

1
Beginnings

This chapter begins with a brief discussion of the slight use of mathematics in early voyages to America. The coverage then jumps ahead to 1585, when the first world-class mathematician set foot on American soil. The instruction of mathematics in American schools is described next, first at Harvard (also referred to as "the College") and later at William and Mary as well as Yale. The leading college, Harvard, produced the first American-born mathematician, Isaac Greenwood, and his more illustrious successor, John Winthrop. Yale produced Thomas Clap. Along the way, this chapter introduces the first American mathematics professor (Hugh Jones), authors of scientific almanacs, and the founding of other American colleges.

Columbus

The story of how Columbus encountered America is of interest mathematically, and it provides a springboard for developments that took place before the American Revolutionary War. As a youngster **Christopher Columbus** (1451–1506) traveled from his native Genoa, Italy, to Lisbon, Portugal, so he could learn to speak both Portuguese and Castilian (the languages of seamen), learn to read Latin (for geography), and study mathematics and astronomy (for navigation). This enabled him to read a Latin translation of Strabo's *Geographica* (1469), where the circumference of the earth was listed as roughly 18,000 statute miles. Columbus discovered an error (though not the right one) and deduced that the circumference was 16,500 statute miles. Columbus was not a very good mathematics student, but he was very lucky. Moreover, he found a patron to finance his voyage, departing from the Canary Islands on September 6, 1492. On October 7, the ships reached the desired mark with no land in sight. Expecting land nearby, Columbus changed his course to follow flocks of birds flying by. What serendipity! It has been said that Columbus didn't know where he was going when he started out, didn't know where he was when he got there, and didn't know where he had been after he got back.

Where did he land? There seems to be no agreement on the exact island where Columbus first set foot but, as historian of mathematics Fred Rickey wrote, "Columbus did more than encounter America. He discovered a way to sail there—and back."[1]

It seems that years congruent to 76 (mod 100) are witness to pivotal events in the political and mathematical histories of the United States. It is well known that the Revolutionary War began with the signing of the Declaration of Independence in 1776. It is not as well known that while the country was celebrating its centennial in Philadelphia in 1876, an equally important revolution in higher education was taking place in Baltimore. This latter development is described in Chapter 3. For now, let's go back to 1476, when the German mathematician **Johannes Müller** (1436–1476) died. Using the regal *nom de plume* Regiomontanus, he had published three classic works. One was the book *De Triangulis omnimodis*, the first textbook on trigonometry ever written. Another, *Epitoma*, became the most important astronomical text published in the fifteenth century. But of relevance here is Regiomontanus's *Canon Sinuum*, which provided the geometry of both plane and spherical triangle. It is known that Christopher Columbus carried the seven-place table of sines—which was printed as a supplement to *Canon Sinuum*—with him on his famous voyages of discovery.[2] So this, then, was the first mathematics book to reach America.

First mathematician to visit the Colonies

Sir Walter Raleigh sent Thomas Harriot on one of his explorations to the New World. An Elizabethan scholar and scientist who contributed to almost all endeavors of his time, Harriot was arguably the most distinguished English mathematician before Isaac Newton. He was surely the first mathematician to land in the United States.

Thomas Harriot (1560–1621) was born in or near Oxford.[3] He entered Oxford University's Oriel College in 1576 and graduated four years later. Upon graduation, he moved to London, though it is unclear what a 20-year-old person with largely theoretical skills in mathematics could do. Yet he developed an interest in overseas explorations, for within a few years he became expert in a branch of applied mathematics—instrumental navigation—that brought him in contact with **Sir Walter Raleigh** (1554–1618; also spelled Ralegh). In addition Harriot had cultivated a reputation as a teacher of navigation and geography. By 1583 he had moved into Raleigh's household, where he tested navigational instruments the famed explorer had bought for him and instructed Raleigh's seamen in methods for using them.

By the time Raleigh's first reconnaissance voyage set out for America in 1584, Harriot had greatly refined the technique of using various instruments to take latitude sights from both the Sun and the Pole Star. Moreover, he revised existing tables based on his own astronomical observations, and, accordingly, he made physical modifications to his instruments. Within a short time, he determined theoretically how to construct the Mercator projection and thus provide true directional sailing. One problem eluded his grasp, however—the determination of longitude—a challenge that eluded the grasp of scientists for another 200 years. Nonetheless, Harriot exemplified problem-solving at its very best!

In 1585 Harriot became the first mathematician to land in America when he sailed on one of Sir Walter Raleigh's vessels in the official capacity of surveyor.[4] In preparation for that trip he acquired a working knowledge of several studies: botany, horticulture, geology, chemistry, metallurgy, and linguistics. He put his knowledge of linguistics to use by teaching English to, and learning Algonquin from, a Native American named Manteo who had been brought back to England from the earlier voyage. This enabled

Figure 1.1. Thomas Harriot

Harriot to communicate with the Native Americans when his boat, *Tiger*, arrived on the shores of modern North Carolina (called Virginia then).

A settlement of 108 men was established on Roanoke Island, which served as their home base for ten months until their precipitous departure in June 1586. The artist John White accompanied Harriot on all his investigations. While Harriot maintained a journal of what he saw and heard, White sketched the inhabitants, their villages, plants, and living creatures. During the winter the men charted the shores from what is now South Carolina to Chesapeake Bay. By April 1586 the two had finished the scaling down of survey sheets and created a map that White painted; this was the first surveyed map of any part of North America. It was so accurate that a satellite photo of the area taken in 1984 showed a remarkable similarity, if not congruence, of areas despite alterations to the shoreline in the intervening four centuries.

With Native American relations souring and colonists' unfulfilled expectations for gold leading to frustration, the settlement was abandoned and the crew returned to England in July 1586. Unfortunately, many of the papers, charts, and specimens prepared by Harriot and White were tragically tipped into the ocean while they were being moved onto the ship that transported them back to England. Consequently, their grand plan for a great encyclopedia, an illustrated work intended to provide the first detailed history of the colony, was thwarted. Nonetheless, two years later, in 1588, they published their findings in Thomas Harriot's book *A Briefe and True Report of the New-found Land of Virginia*, a thin, oversized source that is still available today and makes interesting reading for anyone interested in the Native Americans he encountered.[5] This became the only publication in his lifetime, although he left behind numerous mathematical manuscripts that were published posthumously.

Harriot holds interest for several reasons, even though he did not work on mathematics in the New World. For one thing, there was no such thing as a scientific periodical at the time, let alone a mathematics journal. The only medium for disseminating one's work to the public was the book, and that was a rare commodity. Today Harriot is regarded as one of the world's first scientists to adopt symbols in place of sets of words, a notational advance that deserves to be known as one of the most important in the history of mankind.

Harriot had hoped to return to America, especially to a tract of land that he had discovered near Chesapeake Bay, but Queen Elizabeth never gave him permission. Worse, his patron, the Duke of Northumberland, fell out of favor with the queen and was summarily executed. Nonetheless, he was granted an annual pension by Henry Percy, the Ninth Earl of Northumberland, that provided sufficient funds for securing the comfort and leisure which Harriot's scientific pursuits demanded. Unfortunately for Harriot, and especially regrettably for the earl, Percy was imprisoned in 1605 for the Gunpowder Plot. Harriot too was interrogated and briefly imprisoned, but soon released. Such major distractions, plus an increasingly painful cancerous growth on his nose, took a toll on his work in mathematics after that. Harriot died at age 61. The cause of his death might have been predicted beforehand—he died of cancer in the left nostril of his nose, a condition brought about by the habit of smoking he had acquired while visiting America.

The online file "Web01-Harriot" (see p. 1) provides a brief account of Harriot's mathematical accomplishments within the history of algebra and mathematical notation. Harriot's main contributions to mathematics were in the field of what we call algebra today, which was termed the "analytic art" at the time. The modern inequality symbols "<" and ">" first appeared in *Artis analyticae Praxis*, so Harriot is frequently given credit for introducing them into mathematics. They perhaps derive from the marking ⋈ on the shoulder of a Native American drawn by the artist John White who accompanied Harriot on his historic voyage (see Figure 1.2). From a mathematical viewpoint, "the significance of the inequality signs lies in the fact that this is the first time that such signs were used and accorded the same status as the equality sign."[6]

The book *Artis analyticae Praxis* was put together from Harriot's manuscript papers and published in 1631, ten years after his death. Even if he had attempted to disseminate his mathematical knowledge in America, none of the early settlers had the leisure time to read him (and few had the requisite education to understand his works anyway).

Harriot contributed to scientific fields other than mathematics. His astronomical observations improved upon existing tables; he engaged in a notable correspondence with Kepler on this matter, and his sketches of the Moon predate Galileo's drawings. In addition, Harriot's study of the rainbow led him to state the sine law of refraction 20 years before Willebrord Snell did so in 1621. He also contributed to ballistics and optics. Oxford historian Robert Fox wrote that Harriot "was a towering figure in his day, and he remains for us a distinguished exemplar of ... an Elizabethan man of science."[7]

Harriot's journey to America in 1585 represents the only event of any mathematical consequence in the century after Columbus encountered the New World. But real progress occurred at the beginning of the next century.

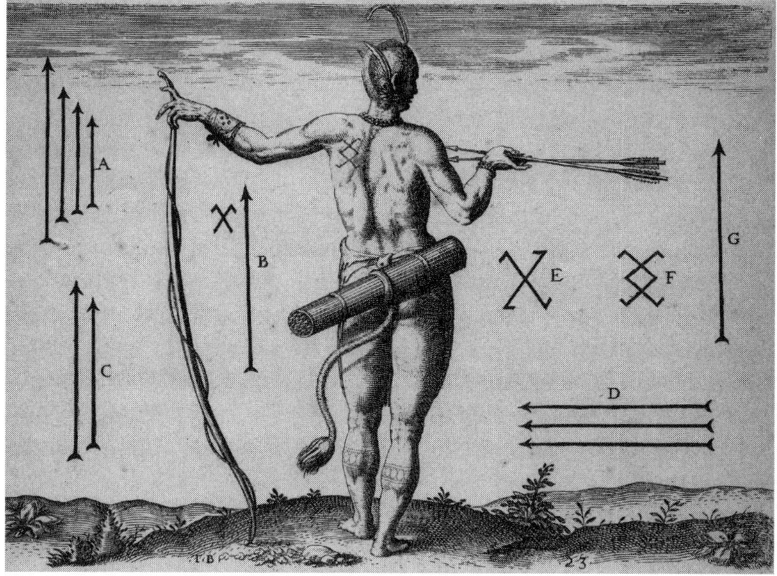

Figure 1.2. Source for inequality signs "<" and ">"

Seventeenth century

It might appear that scant attention was paid to education in the US and Canada throughout most of the seventeenth century, so progress in mathematics, as measured by individual investigations and original contributions, was exceedingly slow. However, under the surface lurked impressive gains in education by a people whose main concern was with daily existence. Here I describe the teaching of mathematics in grammar schools and secondary schools in New England before turning our attention to the first college established in the US.

Secondary schools. During the early part of the seventeenth century, only densely populated settlements established schools, mainly those near Jamestown and Plymouth, the first two areas settled in 1607 and 1620, respectively. Some of those grammar schools did not offer instruction in mathematics, while those that did restricted their attention to counting and three of the four operations on integers (division was excluded). As late as 1750 a town in New Hampshire voted "to hire a school-master for six months in ye summer season to teach ye children to read and writing."[8] As late as 1799 one school administrator bemoaned, "Until within a few years no studies have been permitted in the day school but spelling, reading, and writing."[9] It seems that the "third R" was still missing from some curricula up to the beginning of the nineteenth century.

In general, grammar schools were established for the purpose of educating boys beginning at ages six to eight. Although the Puritans remained loyal to England, they were forbidden from enrolling in colleges there. However, there was a strongly felt need for all boys to know how to "cipher" for business purposes, that is, how to perform simple computations involving measures (pounds and ounces, inches and yards, etc.) and financial transactions (using various nondecimal currencies). Girls were never taught the subject. Even in Philadelphia in the first half of the eighteenth century, two of the most celebrated schools in the colonies replaced instruction in arithmetic with sewing for girls, who were otherwise taught the same material as boys in reading and writing.

A few of the grammar schools evolved into secondary schools, some of which retain their names today. In the secondary schools, ciphering was taught by drilling students in the manipulation of integers because books were rare. Fractions were rarely taught. Slate boards were not introduced for school use until after the Revolutionary War, and blackboards not until the nineteenth century (and whiteboards at the end of the twentieth century). Paper too was costly in the Colonial Era, so birch bark was sometimes used in schools to teach children to write and calculate. Most instruction consisted of teachers dictating lessons, with students listening attentively, perhaps copying some notes, and then solving problems on birch bark. In such instances it was not unusual for up to a dozen students to be lined up at the teacher's desk awaiting evaluation of their solutions.

Who were the teachers? Most were drawn from the clergy, but the best were college students or college graduates using the position as a stepping stone to a better future. A shining example is former US President John Adams, who wrote of his experiences in 1755: "Sometimes paper, sometimes his penknife, now birch, now arithmetic, now a ferule, then ABC, then scolding, then flattering, then thwacking, calls for the pedagogue's attention."[10]

An outstanding teacher from these times was **Ezekiel Cheever** (1614–1708), called the "father of Connecticut School-masters, the Pioneer, and Patriarch of elementary classical culture in New England."[11] Cheever serves as an example of those who emigrated from England (he was born in London and emigrated in 1637) and whose lot in life rose beyond those of their ancestors in the old country—his grandfather was a yeoman and his father a skinner. Shortly after arriving in Boston, he moved to New Haven, where he was a schoolmaster 1638–1650; he was also deacon of the first church of New Haven during six of these years. Initially, Cheever was the head of the Free Schoole but soon it expanded to the Free Grammar School. Generally, colonial grammar schools provided instruction in Greek and Latin. They were endowed by either a grant of land or a bequest of money, and they were open to all white males, regardless of the family's economic standing. They later evolved into academies, not public schools.

Ezekiel Cheever moved to Ipswich, MA, in November 1650 to take charge of that town's grammar school which was supported by donations from publicly spirited individuals. Eleven years later he moved to Charlestown, MA, to become schoolmaster of that town's Free School. In 1670 Cheever returned to Boston to become headmaster of the Free School, which was renamed the Latin School in 1790. Boston Latin School can thus claim distinction as being the oldest public secondary school in the US. He remained at the school for the rest of his life, and under his guidance it became the principal classical school throughout the colonies.

Altogether, Ezekiel Cheever taught for seventy years, right up to the time of his death at age 94. He contributed little to the literature, but his *A Short Introduction to the Latin Tongue* was probably the earliest American school book. It was written when he was in New Haven, hence before 1650. The book was so popular that its twentieth edition appeared in 1790.

Most early settlers were too busy adapting to life in the New World to engage in higher education. Even those who came to America with a university education knew mathematics only up to the rule of three and *pons asinorum*. (The rule of three refers to problems that reduce to solving an equation of the form $ab = cd$, where all but one of the four variables is known. *Pons asinorum* refers to the inability to transcend Proposition I.5 in the first book of Euclid's *Elements:* in an isosceles triangle the base angles are equal.)

Harvard College

The founding of a college so early in the history of the US is somewhat surprising given the challenging conditions the Pilgrims faced. Nonetheless, some of the college-educated leaders of the Massachusetts Bay Colony sought to afford their sons the same opportunity they had experienced at Cambridge, Oxford, or Edinburgh. To achieve that goal, in October 1636 the Great and General Court of the Colony "agreed to give 400£ towards a schoale or colledge."[12] Classes at what would ultimately become Harvard University began in July 1638 with nine students and one master in a single-frame house. (Today a typical Harvard freshman class consists of 1600 students.)

When the Englishman and Puritan minister **John Harvard** (1607–1638) died at age 30, having emigrated from England only the year before, he bequeathed half his estate (780£) plus his entire library (260 books) to the fledgling institution. In appreciation, the school's name was changed from the "college at Newtowne" to Harvard College in 1639. Thus the first college in the US dates its founding to 1636, just 16 years after the first Pilgrims landed at Plymouth, a rather short period of time. Although the population of all of New England at the time was only 4000 and Boston consisted of no more than 30 houses, almost 100 men were graduates of Oxford, Cambridge, and Scottish universities. Nonetheless, Harvard remained a relatively small, provincial institution until the presidency of Charles W. Eliot, who by the turn of the twentieth century transformed it into the modern university of today. To put enrollment figures in perspective, only 465 students graduated in all of the seventeenth century.

Radcliffe College was established in 1894, during Eliot's presidency. The idea to found a women's college associated with the all-male Harvard had floated since 1878, and the following year saw the formation of the "Harvard Annex," which was chartered in 1894. The school was named after Ann Radcliffe (Lady Mowlson; 1576–1661), who had created Harvard's first scholarship fund in 1643. The year 1643 set a precedent for collegiate fundraising.

Even though classes at Harvard began in July 1638, the real founding of higher education in the colonies began with an unusual man, **Henry Dunster** (1609–1659), who became the first president of Harvard in 1640. Dunster studied oriental languages at Cambridge in England, where he earned a bachelor's degree in 1630 and a master's degree four years later. He then taught at Magdalene College, Cambridge, achieving a reputation as a Hebrew scholar, especially regarding biblical studies. He immigrated to Boston in 1640, one year after Nathaniel Eaton was dismissed as master of Harvard College. Dunster chose the title "president" in place of "master." A Congregationalist minister, he modeled Harvard after Cambridge University, and ended up playing an important role in inaugurating mathematics instruction. There are no known images of Dunster.

Harvard College did not obtain its charter from the Massachusetts Bay Colony until 1650. The charter established the President and Fellows of the College, which thereby became the first corporation in the Western Hemisphere. Known today as the Harvard Corporation, this body still governs the university, a remarkable testament to Dunster's foresight; in fact, its membership was increased from the original seven to thirteen only in 2010.

Most colleges and universities in the US and Canada scramble for funds, and Harvard was no exception in the early Colonial Era. Although the Massachusetts Bay Colony's grant was generous, as was John Harvard's bequest, the need for additional funds was evident, so in 1643 the College launched an appeal called "New England's First Fruits." The pamphlet announcing the appeal reveals that Dunster read mathematics and astronomy to third-year students, with arithmetic and geometry taking up the first three quarters of the year. Later in Dunster's tenure he taught Harvard students plane surveying and a smattering of navigation.

The curriculum was entirely prescribed, and students followed a regular schedule. Each day was devoted to one or two subjects, so that in the final year of the program, for instance, mathematics and astronomy were read at 10 a.m. on Monday and Tuesday, Greek on Wednesday, Latin on Thursday, and rhetoric on Friday. Here "rhetoric" refers to public speaking, especially disputations carried out in Latin. However, to show the relative emphasis placed on mathematics, ten hours a week were devoted to philosophy, seven to Greek, six to rhetoric, four to Oriental languages, and just two to mathematics.

Initially, Harvard's degree program lasted three years, so Dunster was instructing third-year seniors. In 1655 the College switched to a four-year degree program, thus setting the stage for every college and university that followed. All Harvard students, called "scholars" at the time, took the same courses in six liberal arts (grammar, logic, rhetoric, arithmetic, geometry, and astronomy) and three philosophies (metaphysics, ethics, and natural science) as well as Greek, Hebrew, and ancient history. Final exams were given at the end of the four years when, over a six-day period, any graduate with a master's degree or any member of the Board of Overseers could quiz the students orally. All examinations were oral. This tradition continued until the 1870s; even then, most faculty members opposed the switch to written examinations.

At the end of a week of festive graduation ceremonies, the valedictorian delivered an oration in Greek, the salutatorian one in Latin, and the third-ranking student an address in Hebrew. Today a Harvard graduate, not necessarily the salutatorian, still delivers a commencement address in Latin.

Most Harvard students owned few if any mathematics textbooks; arithmetic and geometry books were a rarity in the colonies. Therefore, some teachers wrote manuscript textbooks based on original English sources. In the classroom they would mainly dictate the material from these works. For instance, in 1649 President Dunster based his geometry course on Euclid's *Elements* and from a book by Petrus Ramus. Moreover, he taught in English, even though Latin was the official language of the College, so in his copies of these texts he translated each proposition from Latin to English in the margin "with purpose to ripen it on fuller thought."[13] In 1652, Dunster managed to purchase a 1570 copy of Billingsley's *Elements of Geometrie*, the first edition of Euclid ever to appear in English. Students were expected to make their own copies of these manuscripts, and in this way mathematics was handed down from one generation to the next.

Henry Dunster was succeeded as Harvard president by **Charles Chauncy** (1592–1672) in 1654. Chauncy was a Cambridge graduate who emigrated from England when he ran afoul of the Anglican Church. He landed in Plymouth 18 years after the original Pilgrims came ashore. In 1641 he settled in Scituate, MA, where he was a minister until he accepted the Harvard presidency. However, Chauncy did not share Dunster's vigor for mathematics, so instruction in the subject declined to the point that Thomas Brattle, who attended 1672–1676, never received instruction in geometry during his student days.

Only a few areas of human endeavor in the early colonial settlements had a need for mathematics. One was surveying. Even Thomas Harriot came to the country as a surveyor. The demand for skilled surveyors increased during 1625–1675, when the Colony of Virginia initiated a program of enticing middle-class Englishmen to settle there with an offer of 50 acres of ungranted land for each person the settler brought to the New World. The operative word here is *ungranted*. A surveyor had to lay out the new property to ensure that it was disjoint from all contiguous properties, and then to record the deed at the county court. (Since the clerk was "seated," the court house became known as the county seat.) Surveying required knowledge of geometry and trigonometry beyond those simple computations needed for calculating measures and making financial transactions, resulting in the need for additional mathematics in the schools.

Indian College. Harvard's founding charter from 1650 dedicated the college to "the advancement and education of youth in all manners of good literature, arts, and sciences." But it went even further in an attempt to obtain funding for the College. Part of that document read:

> Whereas through the good hand of God many well devoted persons have beene and dayly are moved and stirred up to give and bestowe sundry guiftes legacies lands and Revennewes for the advancement of all good literature artes and Sciences at Harvard Colledge in Cambridge in the County of Middlesex and to the maintenance of the President and Fellowes and for all accommodacons of Buildings and all other necessary provisions that may conduce to the education of English & Indian youth of this country in knowledge and godliness.

Promoting the education of "Indian youth" carried with it a critical ulterior motive. Namely, it sought to raise funds from the Society for the Propagation of the Gospel in New England, which aimed to convert the Native American population to Christianity. By agreeing to support this purpose with donations to Harvard College in exchange for the free tuition, room, and board of Native American students, this London-based society expected its graduates to proselytize back in their home communities upon graduation.

Native American students had attended Harvard before 1650, although none completed requirements for a degree. But in 1655 Harvard established a formal Indian College to teach the English language and Protestantism to Native Americans. The next year, funding from the society enabled the College to construct a two-story brick building in what is now Harvard Yard. Because no Native American students attended at first, the building was used for English students until the first Native Americans enrolled in 1661. This building also housed a printing press that two years later published the first Bible in any language printed in British North America. One of the "first fruits of the vine," James Printer, did much of the translation and typesetting for John Eliot's edition of the Bible. Written in the native Algonquin Indian language, it became known as the "Indian Bible."

These Native American males had to satisfy the same entrance requirements as their colonial cohorts: recite and translate from Cicero and Virgil in Latin, and Isocrates, Xenophon, and the New Testament in Greek. They too would have already attended a grammar school, where they were expected to master Latin. (How many Harvard students today would have qualified for admission under these requirements?)

The entering class of eight included the first two Native Americans to enroll in the Indian College and live in the building. Neither was able to achieve what the Society for the Propagation of the Gospel envisioned. Tragically, Joel Iacoomes "was murdered by seafaring marauders on the way to Commencement,"[14] so he did not receive his degree. His classmate and fellow Wampanoag tribe member, Caleb Cheeshahteamuck, became the first Native American to graduate from college in 1665, but he died of pneumonia just a few months later and so was never able to carry out the society's hope for proselytizing to the tribes on his native Martha's Vineyard. He had been educated at Daniel Weld's school in Roxbury and then Elijah Corlett's elite grammar school in Cambridge before entering Harvard. Nothing is known about either of their performances in arithmetic and geometry, but a recent book, *Caleb's Crossing*, paints a very engaging portrait of student life at Harvard in the seventeenth century.[15]

Harvard's original outreach to Native American students essentially ended in 1675 with the onset of King Philip's War with the Massachusetts Bay Colony. It seems that no Native American students enrolled after that, and the building was razed in 1693. Of related interest, when Tiffany Smalley received her Harvard BA in 2011, she became the first member of the Wampanoag nation from Martha's Vineyard to graduate since Caleb Cheeshahteamuck earned his degree 346 years earlier; she resided in Dunster House.

Almanacs

The Harvard curriculum indicates that mathematics was born as a handmaiden for astronomy. Apparently, Harriot's advances had no influence on this side of the Atlantic at all, as algebra remained an unknown subject. Moreover, there was little change in the curriculum until the beginning of the eighteenth century, mainly because the course of study was designed for students seeking to enter the ministry through Cambridge or Oxford. Therefore, the subjects that monopolized the curriculum were philosophy, linguistics, and theology; mathematics was nearly excluded.

The main admission requirement was the ability to speak, read, and write Latin. There was no mathematics prerequisite, so by the time students became seniors, they had not studied mathematics for at least two years. Some excelled anyway, beginning with **Samuel Danforth** (1626–1674), certainly one of the "ripest fruits" in the class of 1643. Danforth came to the New World with his father when he was eight years old, his mother having died five years earlier, so he received his entire education in the colonies. By 1650, he had achieved local fame as an astronomer of distinction, probably due to calculations he carried out for almanacs in 1646, 1647, and 1648. All three were printed in Cambridge, England. In 1651 a New England writer gushed, "[Danforth] hath not only studied divinity, but also astronomy; he put forth many almanacs."[16]

The major mathematical activity in America during the seventeenth century was the writing of almanacs, which was regarded as proof of profound erudition. Four almanac authors were associated with Harvard and attained a fairly high degree of sophistication. Samuel Danforth has already been mentioned. Two others were **Urian Oakes** (1631–1681), who served as Harvard's acting president 1675–1680 and was full president for the final year of his life, and **John Sherman** (1613–1685), who published *An Almanack of Coelestial Motions* in at least 1674, 1676, and 1677.[17] (The online file "Web01-Almanacs" provides additional material on Danforth, Oakes, and Sherman.)

The best of the seventeenth-century almanac authors, and an outstanding colonial scientist, was **Thomas Brattle** (1658–1713), whose work *An Almanack of Coelestial Motions of the Sun and Planets, with Their Principal Aspects, for the Year of the Christian æra 1678* was printed in Cambridge, MA. Unlike Oakes and Sherman, Brattle was born in Boston. He spent his entire life in the area, graduating from Harvard in 1676, and he later amassed a considerable fortune that enabled him to make several generous gifts to Harvard and to serve as treasurer 1693–1713. Brattle used a telescope (which had been given to Harvard in 1672) to observe a famous comet of 1680. He sent his observations to the first Astronomer Royal at Greenwich, who in turn forwarded them to Isaac Newton. That famous Englishman used the observations in his *Principia Mathematica* to prove that comets travel in paths determined by the law of gravity. Newton praised "the observer in New England"[18] amongst some European astronomers who also aided his investigations.

Thomas Brattle became the first American scientist to have scientific work published abroad, with articles from 1705 and 1707 appearing in the *Philosophical Transactions* of the Royal Society of London. The first work was done jointly with the astronomer James Hodgson in London, showing that collaboration was carried out effectively 300 years before the advent of electronic mail. In a letter that summarizes the plight up to 1876 of American scientists with an abiding interest in mathematics, Brattle wrote to a friend that he was working "here alone by myself, without a meet help in respect to my studies."[19] A well-known street in Cambridge, MA, commemorates the Brattle family; it almost runs into Chauncy Street.

Clergymen played a prominent role at Harvard, and in general religion was another area with an expressed need for mathematics in the seventeenth century, though the level of mathematics needed by ministers and priests for establishing the dates of certain holidays and major saint days was considerably less than what astronomers and surveyors required. In conjunction with the needs in surveying and astronomy, this in turn led to the founding of several grammar schools, especially by the Jesuits. This order played a pivotal role in higher education in Canada when French Jesuits opened the Collège de Québec in 1635, one year before the founding of Harvard. Thirty years later the French-born Martin Boutet was appointed "professor matheseos," thereby becoming the first mathematics professor in America. The course of study at the Collège de Québec lasted five years and was in the hands of two priests, with help from six brothers. Mathematics was taught in the last two years in the curriculum as an aid for navigation, surveying, and cartography during this time, but it was doubtless limited to elementary and commercial arithmetic. Nonetheless, a book on the history of Jesuit education reports "a pitiful contrast between the intellectual culture in the newly acquired Canada and the uncultured backwardness of the older English colonies."[20]

Royal College of William and Mary

The first permanent settlement in the US was established in 1607 at Jamestown by the Virginia Company of London. Another successful settlement took place in 1620 when a group of Anglicans and Separatists landed at Plymouth, MA; they later became known as Pilgrims. Therefore it seems appropriate that the first two American colleges would be founded near these sites. Harvard College was established only 16 years after the initial landing at Plymouth Harbor. Its early founding might suggest that the US prized higher education right from the start, but the College was far inferior to Oxford, Cambridge, and Edinburgh, even though these prestigious universities recognized degrees from the American version at once. Harvard enjoyed a 57-year head start on the competition, as the second US institution of higher learning, the Royal College of William and Mary, was not founded until 1693 and the third, Yale, in 1701. Incidentally, the first college in all of North America was Real y Pontifica Universidad de México, which was founded in 1551 and opened two years later.

Why did it take so much longer for a southern colony to establish an institution of higher learning? Were Jamestown's inhabitants more involved with mere survival? Was education undervalued there?

The answers to all three questions emanate from the main difference between the two groups of settlers. New Englanders were mainly farmers, with a small percentage having earned a college degree, yet many had a burning desire for their sons to enjoy a better life through education. Those afforded the time and expense of being able to prepare their sons for Harvard, either in public schools or by private tutors, benefited from the College's founding in 1636.

Many of the settlers in the Colony of Virginia, on the other hand, were plantation owners and successful merchants. A very high percentage of inhabitants in the colony were Englishmen, "and as men from the English and Scotch universities were continually arriving, the need of a home institution was not as acutely felt during the [seventeenth] century."[21] Instead, these wealthy colonists sent their sons abroad to study at Cambridge, Oxford, or Edinburgh "to an extent never dreamed of in the northern colonies. The ocean was, in fact, a connecting bridge to the shipping people and merchants who really settled Virginia."[22] A sense of adventure and commercial enterprise minimized perceived risks associated with trans-Atlantic travel for these men, who often made multiple crossings in order to obtain additional land grants. Extended families supported young students too. For instance, a will from 1671 provided funds so that a magistrate's two nephews "should be sent to school in London and afterwards returned to Virginia."[23]

The desire for a college education caused a concomitant need for preparatory education. Essentially the Colony of Virginia initiated a system of compulsory education in 1646 when a legislative act empowered the eight counties to provide schooling even for children of "such parents whose poverty extends not to give them good breeding."[24] The colony consisted of twenty parishes, each with a minister who provided instruction in the school along with ministerial duties. Income was derived from taxes assessed for this purpose. At this early date lessons were provided only in reading and writing; there is no evidence of any formal instruction in arithmetic.

It is noteworthy that the act provided instruction without regard to gender or race, often with the pupils being indentured as apprentices. As an example, in 1690 a girl named Rebecca was apprenticed until age 21 to a married couple who were required to provide "a Compleat yeares schooling … to bee taught … within the aforesaid term of … Apprenticeship."[25] Here a complete year included summertime, so this provision amounted to three years of instruction. By contrast, in the seventeenth century generally schools in New England were open only for two months in the summer and two months in the winter.

The clause stating that education should be provided regardless of race did not always apply to slaves. It did, however, extend to children with one white parent and one black parent. For instance, in 1716 George Petsworth, "a molattoe boy of the age of 2 years," was apprenticed to Ralph Bevis, who was required to provide "3 years' schooling, and carefully to Instruct him afterwards that he may read well in any part of the Bible, also to Instruct and Learn him ye s^d molattoe boy such Lawful way or ways that he may be able after his Indenture time expired to gitt his own Liveing."[26]

One particularly successful free school was established in the 1634 will of the childless planter Benjamin Syms (b. 1590) for children in two adjoining parishes. Syms bequeathed 200 acres of land in Elizabeth County and eight cows whose milk and ultimate sale provided funds for a school building that was erected by 1647. The same year Syms wrote his will, another successful Virginian, Thomas Eaton, bequeathed 250 acres for a nearby school called the Eaton Charity School. Sixty years later, in 1695, a county register recorded that "a negroe Joan belonging to Eaton's free school … be free from paying Levyes."[27] After roughly 125 years of successes, both the Syms and Eaton free schools experienced financial difficulties during and after the Revolutionary War. However, the depths to which they fell by 1800 were bemoaned by the inhabitants of coastal Virginia, leading the new state's general assembly to incorporate them as the Hampton Academy in 1805. State funding became more critical over the next 46 years, resulting in the former academy evolving into the public Hampton High School. Noted historian and William and Mary president, Lyon G. Tyler, concluded, "Hampton, the oldest existing English town in the United States, has the oldest free school."[28]

Various smaller schools dotted the Colony of Virginia landscape up to the Revolutionary War, some being unrelated to public parish schools. Two were established for teaching Native American children. A fashionable boarding school for girls was established in Williamsburg about 1760.

Wealthy Virginia colonists, who did not send children to England or Scotland for their education, sent them to private schools or provided tutors at home. There was a fine line between these two means for providing preparatory education because private tutors generally taught other children in addition to those residing in the house. Young ministers often came over from abroad to serve as tutors or as teachers. Some educated but poor colonists were also employed as teachers by indenture.

Advertisements placed in the *Virginia Gazette* indicate ways in which individuals or schools sought teachers or tutors. For instance, Theophilus Field advertised for "Any single man capable of teaching Greek, Latin, and the Mathematicks, who can be well recommended."[29] An unnamed person required "A tutor for a private family, who among other things thoroughly understands mathematics."[30] The Norborne Parish was looking to hire "A school-master well qualified to teach writing and Arithmetic"[31] to a class of 15 to 20 students. On a higher level, a "publick School" at Cabin Point, VA, sought to hire "a Master properly qualified to teach English, writing and Arithmetick. This school will consist of nearly Thirty scholars."[32]

Third college in America. There were plans to found a college in present day Williamsburg even before the Pilgrims landed at Plymouth. As early as 1617 King James decreed that funds should be collected for a college to be located in Henrico, near Richmond, VA. At about the same time, money was raised to start "a collegiate or free school"[33] to be situated inland at City Point, VA, and named the East India School in honor of its benefactors from the East India Company. Two years later the First Colonial Assembly at Jamestown discussed the possibility of establishing a college. Over the next few years those ideas began to gain credence when appropriately skilled individuals were sent to the colony—a manager for the lands as well as a rector, master, and usher for the college. However, a massacre by Native Americans in 1622 killed 350 settlers and thus scuttled any idea about starting a college.

The idea of founding an institution of higher learning gained traction a generation later when, in 1660, the Colonial General Assembly passed an act founding "a college and free schoole."[34] However, a different kind of uprising conspired to douse this second set of flames when Virginia settlers became so upset with the actions of the colony's governor, William Berkeley, a proponent of the college, that popular rebellions broke out in several parishes. The best known, Bacon's Rebellion, was the first uprising among discontented colonists against a king's appointed official in the colonies. These rebellions became so violent that the settlement at Jamestown, which served as the capital of the colony 1607–1698, was burned to the ground, including the state house.

In 1690, thirty years after the Virginia legislature passed an act to found a college, the General Assembly again proposed the idea. This time the legislature instructed Dr. James Blair to "endeav^r to procure from their Ma^{tys} an ample charter for a *Free Schoole and colledge*, wherein shall be taught the Latin, Greek; and Hebrew tongues, together wth Philosophy, Mathematicks, and Divinity."[35] James Blair was a Scottish clergyman brought to Virginia to marshal resources for the proposed college. To do so, he first gained the support of the new governor's council and a Convention of Clergy. Both groups accepted his proposals and, moreover, recommended them to the General Assembly, which approved them in 1691. That assembly appointed Blair as an agent in England, where he sailed with the intent of securing a charter from King William and Queen Mary. This time Blair gained the support of influential merchants, the Bishop of London, and Archbishop of Canterbury before appealing to the rulers. Once again his approach was successful, and on February 2, 1693, King William issued a charter for Their Majesties' Royal College of William and Mary.

Initially the Royal College of William and Mary was to consist of three divisions. The lowest was the Grammar School, which was essentially a "free," meaning "public," secondary school. Above that would be the Philosophical School, offering a four-year program, and the Divinity School, providing another three years of biblical training. William and Mary was founded as an Anglican school, so all students, including Native Americans, had to be members of the Church of England and all professors had to declare adherence to the Thirty-Nine Articles of the Anglican faith. The first building, later named for English mathematician Christopher Wren, was completed in 1695.

The charter for William and Mary provided for five professorships—two in divinity and one each in mathematics, Greek and Latin, and moral philosophy. By contrast, Harvard still had no mathematics professor at the time. But this was only in theory. In practice, only the Grammar School was operational until about 1712. Classes were taught by the first president, James Blair, as well as a grammar master, a writing master, and an usher.

The first professor—of mathematics and natural philosophy—was appointed in 1712, which marked the real beginning of the institution as a college. This professor taught "physicks, metaphysics, and mathematics."[36] A second professor, of moral philosophy, was soon added to teach rhetoric, logic, and ethics. Their salaries consisted of a fixed amount plus an added amount for each student except those on scholarship. Professors were also entitled to apartments in the college building. Curiously, the president was allowed to be married, but the professors were not, though this requirement was not always enforced.

At the end of four years each student took an oral examination in front of the president, professors, and any ministers proficient in Latin and Greek. If the student's performance was satisfactory, he was awarded a Bachelor of Arts degree and, three years later, a Master of Arts degree. (I used the masculine pronoun deliberately because William and Mary did not admit women until 1918.) Passing the examination also enabled the student to proceed to the Divinity School, which prepared these young men to become ministers in the Church of England.

By 1729 the foundation for the college was complete. There were six professors, all graduates of Edinburgh, Oxford, or Cambridge. In 1779 the Grammar School and Divinity School were abolished and replaced by three schools: modern languages, constitutional and court law, and medicine. Enrollment increased from 29 in 1704 to 60 in 1737 and then to 115 in 1754. Records show that 75 of the 115 students boarded at the college—52 paying students, 15 on scholarship, and eight Native Americans; the other 40 lived in town.

A second college in the US can trace its roots to the Colony of Virginia. In 1749 the Edinburgh graduate Robert Alexander founded the Augusta Academy that was run by the Princeton graduate John Brown for the next 20 years. It was renamed the Liberty Hall Academy in 1776 in a burst of revolutionary pride, but it did not become a college until awarding its first bachelor's degree nine years later. Its name was changed to Washington Academy when George Washington donated a substantial amount of money in 1796, and it was chartered as Washington College 17 years after that. Another famous general, Robert E. Lee, took over the presidency of the college after the Civil War and, upon his death in 1870, the name was changed to its present Washington and Lee University.

Some of America's first professors of mathematics were very colorful figures—if not rogues. Three of these scalawags will be introduced shortly, two at Harvard and the other one at William and Mary.

First mathematics professor. Harvard's instruction in mathematics was carried out by tutors after Henry Dunster was removed as president in 1654, so it was the second college in the US that appointed the country's first professor of mathematics and philosophy. That particular individual turned out to be a scoundrel, with misbehavior curtailing his tenure in less than a year.

The distinction of being the first professor of mathematics in the US rests with the **Reverend Tanaquil Lefevre** (dates unknown). Toward the early part of the eighteenth century, authorities at William and Mary sought to appoint someone versed in mathematics. They kept in close touch with the college chancellor, the Lord Bishop of London, who carried out his duties from abroad. Upon his recommendation, Tanaquil Lefevre was appointed professor of mathematics and philosophy on April 25, 1711, with a salary of £80 a year. Lefevre was probably chosen for his excellent background and education. In fact, he published a work on algebra in 1714 with his father, who spelled his name Tanaguy Lefebre.

Many settlers of French descent arrived in Jamestown about 1700. Lefèvre is one of the most common French surnames, meaning "metal worker," and it was spelled variously as Lefebure, Lefebvre, Lefeubre, Lefeuvre, and Fèvre. This produced some confusion; several internet sites state that Isaac Lefevre was the first professor of mathematics and physics at William and Mary. The authors of such assertions were apparently not acquainted with an article in the *American Mathematical Monthly* that settled the matter almost 70 years ago in favor of *Tanaquil* Lefevre.[37]

William and Mary extended the offer to Lefevre with the hope that he would bring scholarship and honor to the college. Instead, he brought shame. In a letter addressed to the Lord Bishop of London the following May, an authority wrote:[38]

> I gave your lordship an account of Mr. Lefevre's admission into the college upon your lordship's recommendation and aim to acquaint you now that after a tryal of three quarters of a year he appeared so negligent in all of the posts of duty and guilty of some other very great irregularities, that the governors of the college could no longer bear with him, and were obliged to remove him from office.

What were these "irregularities," and what might have caused them?[39]

> I am apt to believe that most of his irregularities were owing to an idle hussy he brought over with him, because since she left him (I got her a passage back to England last February) he has left off that scandalous custom of drinking and appears quite another man, being now settled at a gentleman's house for teaching his son, and has a competent salary enough to keep him from being any more burdensome to your lordship or his other friends.

Consequently, Lefevre achieved no fame in mathematics—and professorships in our field got off to a rocky start. Similar misconduct occurred with the first mathematics professors at Harvard and at West Point, but for now we remain with William and Mary.

Hugh Jones. William and Mary authorities were in no hurry to name a successor to Lefevre, taking five years—yes, five *years*—before appointing the **Reverend Hugh Jones** (1692–1760) in 1717. Yet it was another ten years after that when Harvard appointed its first mathematics professor. Reverend Jones was a university graduate, but it is unclear whether he attended Cambridge or Oxford.[40] He was regarded by William and Mary leaders as possessing outstanding culture and scholarship as well as virtuosity. He seems to have emigrated from England in 1716 shortly after receiving his MA degree. An aristocratic, loyal Hanoverian and zealous churchman, Jones wrote the first English grammar in America. He is also known for his book published in

Table 1.1. Professors of mathematics at William and Mary, 1711–1805

Name	Appointment	Name	Appointment
Tanaquil Lefevre	1711	William Small	1758
Hugh Jones	1717	Richard Graham	1763
Alexander Irvine	1728	John Carom	1766
Joshua Fry	1732	Thomas Gwatkin	1769
John Graeme	1737	James Madison	1773
Richard Graham	1749	Robert Andrews	1784

1724, *Present State of Virginia*, which served as "an invaluable source of material for subsequent historians."[41] This was the first historical work in America written by a professor in a college.

Jones attained some recognition in mathematics for a work arguing that the base-8 number system is superior to our decimal system for arithmetical computations. The 47-page manuscript, titled "The Reasons, Rules, and Uses of Octave Computation or Natural Arithmetic," is shelved in the British Museum. The purpose, Jones wrote, was to devise this system because even common arithmetic had become "mysterious to Women and Youths and often troublesome to the best Artists."[42] Written about 1745, it espouses that the octonary system was the most practical numerical system for dealing with coins, measures, and weights. Besides applications to arithmetic, geometry, and natural philosophy, the manuscript also presents utilitarian uses in land surveying, grain storage, and ship displacement. Nonetheless, the author was doubtful that his proposal would be adopted, for "there seems no Probability that this will be soon, if ever, complied with."[43]

Hugh Jones may also have published the book *Accidence to the Mathematicks*, but no printed copy is extant. Because his experience with colonists had taught him that they are "for the most part only desirous of learning what is absolutely necessary in the shortest and best method,"[44] he composed manuscripts on algebra, geometry, surveying of land, and navigation, but none of these were printed in book form. Overall, then, it seems that Hugh Jones was educated in basic mathematics and attempted to convey the subject to his young students, but the low level of the material is what we typically associate today with the Colonial Era.

Reverend Jones held the professorship until 1722, when he was only 30 years old. For unknown reasons, he left the post and returned to England. Although he sailed back to the colonies in 1724, he became a minister in churches in Virginia and Maryland for his remaining time. However, he was appointed the chief mathematician for the determination of the boundaries delineating Maryland, Pennsylvania, Delaware, and present-day West Virginia (then part of the Colony of Virginia), an historic project later known as the Mason–Dixon Line.

Table 1.1 lists the initial year of appointment of the eleven professors who held the mathematics and philosophy chair at William and Mary up 1784.[45] Note that Richard Graham served two different terms. I provide brief sketches of most of these William and Mary mathematics professors; the online file "Web01-W&M" supplies more details.

It is advisable to notice the varying degrees of loyalty among these first mathematics professors toward the English monarchy, which was an important issue at this royal college. Little is known about **Alexander Irvine** (d. 1732) except that he was educated in Edinburgh, sailed to Philadelphia in 1727, and was appointed professor of mathematics and surveying the next year. Apparently, he was the surveyor who determined the boundary between the colonies of Virginia and North Carolina. **John Graeme** tested George Washington as a surveyor. **Joshua Fry** (1699–1754) graduated from Oxford in 1718 and came to William and Mary as professor of natural philosophy and mathematics upon the death of Irvine in 1732. Fry served on the commission with Irvine to determine the Virginia–North Carolina border. In 1752 he collaborated with Thomas Jefferson's father on an influential map of Virginia.

Richard Graham (b. 1720) entered Queen's College at Oxford in 1737, received his bachelor of arts degree in 1742 and then a master of arts in 1746. This qualified him to be appointed professor of natural philosophy and mathematics at William and Mary in 1749. For unknown reasons he was removed from his position in 1758, yet three years later he returned to William and Mary as chair of moral philosophy. Graham resumed his professorship of natural philosophy and mathematics in early 1764 but sailed back to England to become a fellow at Oxford two years later.

William Small (1734–1775) served between Graham's two terms. Born in Scotland, Small received a master's degree in 1755 from Aberdeen College. Three years later he was appointed professor of natural philosophy and mathematics at William and Mary. It is believed that he was the first professor in America to adopt the lecture method of teaching in place of the custom of reading from a manuscript that had been favored up to that time. Small's most famous student from 1760 to 1762 was Thomas Jefferson, who penned an endearing tribute to his teacher.[46] Sadly, Small passed away at age 40 from malaria he had contracted during his stay in Virginia.

Thomas Gwatkin (1741–1800) matriculated at Jesus College, Oxford, in 1763 and was ordained four years later by the Chancellor of the Royal College of William and Mary. Gwatkin set sail for Virginia in 1770, a year after being nominated at age 28 for the chair of natural philosophy and mathematics at William and Mary. In early June 1775, he was asked by Richard Henry Lee and Thomas Jefferson to support the proceedings of the Congress of the Colonies. A devout Loyalist, however, he was disgusted "with rebellious colonists and disorderly collegians,"[47] so he refused. From that day forward "he was subjected to a variety of cruel treatment, by which his life was put into imminent danger and which was the cause of his subsequent permanent ill-health."[48] He sought protection from the Governor of Virginia but instead was removed from his professorship. He sailed back to England later in June.

Bishop James Madison (1747–1812) was a William and Mary graduate (1771) who was appointed professor of natural philosophy and mathematics at his *alma mater* in 1773 and was associated with the college during the turbulent revolutionary period. He traveled to England in 1775 to be ordained a priest of the Church of England, but he returned to the Colony of Virginia as instructor at William and Mary shortly before independence was declared. He organized a student militia in 1777 when he assumed the presidency of the college. Two years later he abolished the college's grammar and divinity schools.

About 1770, some prospective students wanted to study only a particular subject, but the time was not yet ripe for colleges to satisfy such interests. In May of that year the six professors wrote, "The president and masters or professors beg leave to represent that the college is not designed to be the sole place of resort for education in the colony."[49] That made it clear that the main purpose of a William and Mary education was to pursue, in descending order, 1) classical studies, 2) natural and moral philosophy, and 3) the sciences. The document added that this three-sided mission for education "cannot be departed from or occasionally altered even for the sake of extraordinary geniuses … who aim at no more than a skill in Vulgare Arithmetic and some practical branch of mathematics to qualify them for an inferior office in life."[50] Although such specialization did not arise in America for another century, a slight change took place at William and Mary in 1779, when Madison adopted a kind of elective system. Madison allowed "irregular" students (called "part-time" today) to study only those subjects they wanted to. Regarding this change, Thomas Jefferson wrote, "At William and Mary students are allowed to attend the schools of their choice, and those branches of science only which will be useful to them in the line of life they propose."[51]

Robert Andrews (c. 1750–1804) is noteworthy for two reasons. First, he was born in America (Pennsylvania), a first at William and Mary, but others preceded him at Harvard and Yale in this regard. Andrews graduated from the University of Pennsylvania (then the College of Philadelphia). Upon graduation, he was a tutor for several years before traveling to London to be ordained in the Church of England. He was appointed professor of moral philosophy during the Revolutionary War (1779). However a historic event occurred in 1784, when President Madison (of William and Mary) was relieved of the duty of teaching mathematics and was made professor of moral philosophy and natural philosophy. The outspoken patriot Robert Andrews was then transferred to the mathematics chair, making him probably the first person in America to teach mathematics only, which he did from 1784 to at least 1789, becoming the second reason why I list Andrews in this chapter.

First Harvard mathematics professor

Harvard created its first professorship of mathematics in 1727, five years after Hugh Jones had departed from William and Mary. I regard that person, Isaac Greenwood (see pp. 31), as the closest thing to a mathematician in America based on the works he produced and the material he taught. Greenwood is not well known, perhaps due to misbehavior that puts him in league with Lefevre, but an examination of his writings shows that he was current with mathematical advances taking place in England, particularly with the writings of Isaac Newton. Moreover, he presented some of this material to his Harvard students and included it in at least one book.

To place Greenwood's contributions in context, I first view texts that were available for instruction in the Colonial Era. Then I introduce Thomas Hollis, the Harvard benefactor who established the first endowed professorship of mathematics in America. (As of March 12, 2018, the Hollis Professor of Mathematics and Natural Philosophy was the physicist Bertrand Halperin. The mathematician **Andrew Gleason** (1921–2008) held the chair from 1969 to 1992, when Halperin succeeded him.)

Textbooks. During the first part of the eighteenth century, printing presses in the colonies churned out numerous books, but few dealt with mathematics, and those that did principally covered arithmetic. Initially even these works were reprints of English

versions. For instance, the twenty-fifth edition of an arithmetic book written by James
Hodder in London in 1661, but revised by Henry Mose after his death, was published in
Boston in 1719 by Benjamin Franklin's uncle, James Franklin. Called merely *Hodder's
Arithmetic*, the American edition contained nothing relating directly to the colonies.
Franklin's autobiography informs us that he learned his arithmetic in 1722 not from
his uncle's book but from a similar work of that time, *Cocker's Arithmetic*. This work
first appeared in 1678 and "attained a record of approximately one hundred English
editions in something over one hundred years."[52] Another European book known to
have an American edition was an English translation from the Dutch of an arithmetic
written by Pieter Venema. Published in New York in 1730, this book is notable because
it contained material on algebra as well as arithmetic.

A textbook imported from England for instruction at both Harvard and Yale was
the third edition of the work *The Young Mathematician's Guide*. Written by John Ward,
"formerly the chief surveyor and now Professor of Mathematics in the City of Chester,"
it was published in London in 1707 and was apparently used at Harvard and Yale soon
thereafter. This 400-page text consists of five parts: arithmetic (pp. 1–142), algebra
(pp. 143–276), the elements of Euclid's geometry (pp. 277–354), conic sections (pp. 355–
390), and "the arithmetic of infinities" (pp. 391–426). A modern reader is advised to
refrain from drawing pre-Cantorian conclusions from the title of the fifth part; it is
concerned with finding solutions, mainly to problems from finance, by progressions
and series. This material does not adumbrate a calculus of infinite quantities.

More than a hundred English books on arithmetic were published in Great Britain
in the early 1700s; I single out one author because of his role in America. The Scotsman
Alexander Malcolm published two hefty treatises on arithmetic in London in 1726
and 1730. It is known that Malcolm came to New York City about 1740 and set up a
school, yet, for unknown reasons, no American publisher printed either of his works.

One other arithmetic book of smaller interest, published in New Haven in 1747,
carried a title that indicates a kind of applied usage that differed from other works on
arithmetic both in its coverage and its intended audience: *Small Tract of Arithmetick,
for the Use of Farmers and Country-People*. I do not know anything about its author,
Jonathan Burnham.

The authoritative source on textbooks in the Colonial Era remains *Bibliography of
Mathematical Works Printed in America through 1850*[53] by the historian of mathematics
Louis Charles Karpinski (1878–1956), who spent the bulk of his professional career
at the University of Michigan. The interested reader is advised to consult this tome for
additional information on mathematics books published in America up to 1850.

Hollis Professorship. The first textbook on mathematics written by an American
contained important advances in the teaching of the subject in the colonies, but it also
has a curious history. The scandalous behavior of the author is equally fascinating.
To set the scene when he appeared on stage, we go back to London in 1721, when
the aristocratic **Thomas Hollis** (1659–1731) became deeply interested in Harvard and
established a professorship of divinity. A successful merchant in London, Hollis had
been a trustee for an uncle's will that bequeathed a donation to Harvard and this seems
to have attracted his attention to the college. Inspired to do likewise, he established an
endowed chair in divinity in 1721 and, moreover, stipulated that a Baptist could not be
forbidden from holding the professorship even though that religion was not favored by

the Church of England. Hollis planned to establish a second endowed chair at Harvard himself in his will, a professorship in mathematics and natural philosophy.

Two years later, in 1723, Isaac Greenwood delivered a parcel of items from Harvard to Hollis. The two men must have kept in contact because in May 1725, Hollis reported that Greenwood had dined with him and had made a lasting impression. He wrote[54] that Greenwood

> … was much admired as a Divine [Minister] … & had several advantageous offers made him [*sic*] of spending the remainder of his life [in London]: but either thro' too great a Fondness for his native Country and Friends, or a much more prevalent Passion to Philosophy, he chose rather to postpone his Interest for that Time. His Genius leading him chiefly to the Mathematics and Philosophy, he applied himself mostly to these.

Hollis added that Greenwood had "made such strange and surprizing Advances" in science that his English tutor, the Copley Medalist John Theophilus Desaguliers, was duly impressed.

Isaac Newton was equally captivated. Hollis wrote that Greenwood had been invited to attend meetings of the Royal Society by Desaguliers and another English scientist, William Derham. According to Hollis, Greenwood "had the Honour of answering several Questions to their Satisfaction propos'd to him by [the Royal Society's] President, the great Sir Isaac Newton."[55]

Six months later, in December 1725, Hollis proposed endowing a chair in mathematics and natural philosophy at Harvard. The two Hollis professorships, combined with the Hancock Professor of Religion established in 1765, were the only other named chairs in America before 1800. Today the Harvard Online Library Information System is named HOLLIS in honor of this early benefactor, who left the university a generous donation, apparatus for scientific investigations, a substantial number of books, and sets of Hebrew and Greek types for printing. (The types were treasured gifts in the days before desktop publishing.)

Isaac Greenwood. The person who induced Thomas Hollis to endow the professorship in mathematics and natural philosophy at Harvard was the scalawag **Isaac Greenwood** (1702–1745). The case can be made that he was America's first true mathematician. I base this assertion on three of his activities: 1) he taught advanced mathematics, 2) he published works on mathematics, and 3) he mentored advanced students in the field. Greenwood was assuredly the first "true fruit" to grow on the mathematics vine at Harvard, using terminology from the 1643 appeal for funding, but ultimately we consider him a "practitioner" (see pp. 55, 118). Born in Boston, where his father was a carpenter skilled in ship construction and repair, he did not do well in school initially; his obituary in the *Boston Gazette* read that "he was Old before he could read letters."[56] That account quickly added, however, that due to "the Advantage of an uncommon Memory, he soon made so prodigious a Progress in his Studies as is scarce credible. At 15 he was thought fit for an Admission into College."[57]

Thus in 1717 Isaac Greenwood enrolled at Harvard, where an older brother had graduated in 1709 and an uncle in 1685. He did not live there, nor did he seem to engage in student activities, but he studied under an excellent tutor, Thomas Robie. (The online file "Web01-Robie" provides further details about Robie.) For now, it is pertinent to note that Robie's emphasis on science and knowledge of Isaac Newton's writings would help to mold Greenwood during his formative years.

A particularly virulent smallpox epidemic struck Boston in 1721, Greenwood's senior year at Harvard. A controversial way of combating the disease then was by inoculation, whereby a person was infected with a strain of it (from a patient who had a mild case of smallpox) through a cut in the skin. Greenwood was inoculated, developed a light case with a few pocks, and soon fully recovered. Convinced of the efficacy of the method, he then wrote a small pamphlet defending a major proponent of this predecessor of vaccination, the prominent Puritan minister Cotton Mather.

It is unknown what Isaac Greenwood did with his time between graduation and his voyage to England in 1723. He arrived in London armed with two important items. One was the packet of documents delivered to Thomas Hollis, thanking him also for the 24-foot telescope he had donated to Harvard the year before. The other was a letter from his pastor Cotton Mather to James Jurin (1684–1750), the English scientist and physician who was conducting research on smallpox vaccines. In mathematics, Jurin is known as an early and outspoken advocate of Newton's approach to calculus over Leibniz's.

Greenwood made a distinct impression on Hollis in the two years between the time he arrived in London in 1723 and when Hollis wrote to Harvard about him two years later. This indicates that, in the interim, Greenwood studied mathematics beyond what Robie had taught at Harvard. In the middle of this period Greenwood was awarded an AM degree from Harvard (1724). The College's graduates could obtain a master's degree three years beyond the bachelor's by paying a nominal fee. Most did.

In June 1725 Hollis informed Harvard authorities that Greenwood would be returning in time for the start of the fall semester armed with apparatus for teaching mathematics. He wrote, "I hope he will prove an usefull instructor in your college ... [he] appears to me sober, and dilligent to acquire knowledge."[58] Now, the word "sober" can mean a person of earnestly thoughtful character. I am unsure what Hollis meant, but correspondence from a year later suggests he intended the word to mean "not drunk."

In any event, Greenwood did not return to the colonies in 1725 as Hollis had expected. That December the two met, and Hollis informed Greenwood of his plan to endow a professorship in science at Harvard and to recommend him for the position. The aspiring mathematician then helped his benefactor draw up a set of guidelines for the endowed chair. However, Greenwood was still in London the next July, 1726, when Hollis received word that Greenwood had suddenly fled for Lisbon on his way back to Boston. It appears that Greenwood's nonmathematical character caused some concern, because even before he sailed back to the colonies, Hollis warned Harvard, "Mr. Greenwood has left us on a sudden without paying his debts or taking leave of Dr. Desaguliers, his Landlord or Tutor."[59] Not to mention that Hollis himself had not been informed of the colonist's departure.

Greenwood arrived in Boston in late October, preached a few times over the next couple of months, and in January 1727 offered the first public lecture course on science in the New World. The newspaper advertisement for the course declared that it would make subscribers "better acquainted with the Principles of Nature, and the wonderful Discoveries of the incomparable Sir Isaac Newton, than by a Years Application to Books."[60] The lectures were assembled and published as a pamphlet.[61]

That spring Hollis considered the appointment of Greenwood that was being proposed by Harvard authorities—curiously based on Hollis's initial recommendation—for the chair he had endowed that was to begin with the fall 1727 semester. Hollis informed the College that "concerning Mr Greenwood … [and] wishing his future cariage may be sober, Religious, dilligent, and becoming his Profession … I think I shall accept him as my Professor."[62] Note once again the use of the word "sober." The Harvard Board immediately appointed Greenwood as "Professor of Mathematicks & Natural & Experimental Philosophy," thus creating the first position in America where a person could make a living by mathematics. It should be kept in mind, however, that "The Hollis chair … was expected to lecture, not engage in research."[63] It would be another 150 years before the idea that professors should engage in original investigations would gain a foothold in American universities.

The inaugural ceremony celebrating this new position was held in February 1728. Greenwood's teaching load was light (this was similar to how leading universities woo top-notch researchers today)—he was required only to lecture on Wednesday afternoons at 2 p.m. Moreover, he was able to restrict attendance to those upperclassmen who first obtained his approval and whose parents paid a fee beforehand. What a wonderful precedent!

Isaac Greenwood was an especially effective teacher whose remarkably lucid lectures inspired several of his students to pursue science as a career. His most successful charge in this regard was John Winthrop, who entered Harvard in 1728 as a 14-year-old prodigy and succeeded Greenwood as Hollis Professor ten years later. Mostly, Greenwood lectured on work by Isaac Newton, some of which his tutor, Thomas Robie, had introduced to him, but most of which he had learned while in England. In fact, his personal copy of *Principia Mathematica* was one of only three in the colonies. Moreover Greenwood actively engaged his students in cutting-edge research by having them participate with him in using the telescope that Hollis had donated to Harvard to make observations that were subsequently transmitted to the Royal Society through James Jurin. As a result of this joint effort, Greenwood published the paper "An account of an aurora borealis on the 22d of October, 1730" in the Royal Society's *Philosophical Transactions*.[64] He had published two other papers in the journal before then.[65]

In addition, Greenwood offered to organize private classes outside the College. In 1727 he advertised private instruction on two different but related topics. One was called "The modern discoveries in astronomy and philosophy." The other was "Sir Isaac Newton's incomparable method of fluxions, or the differential calculus, together with any of the universal methods of investigation used by the moderns; the *Elements* of Euclid and Appollonius [*sic*]." The latter class is intriguing because it indicates that Greenwood was aware of the Leibniz approach to calculus as well as the fluxions of Newton.

At the start of his Harvard tenure, Greenwood's appointment was propitious, as he became the first American to write a mathematics textbook, which he titled *Arithmetic, Vulgar and Decimal.* Curiously, this book was published anonymously in 1729. It took a sleuth, the twentieth-century historian of mathematics Lao G. Simons, to identify the author when she found conclusive evidence in an advertisement for the book in a Boston newspaper from 1729.[66] (A brief biography of Simons is given in the online file "Web01-LaoSimons.") The preface to Greenwood's book states:[67]

> There are many things in the following Treatise of greater Curiosity,
> than Necessity in the Practice of Numbers. Such are the Methods... of
> contracting Decimal Multiplication and Division, the Rules concern-
> ing Circulating Figures, Sir Isaac Newton's Contraction in the Evolu-
> tion of the Square Root, &c.

Although Greenwood lectured on this material at Harvard, it appears that it was
unknown elsewhere; there is no evidence of a reference to it at either of the two other
existing colleges or any founded between 1746 and 1776. Indeed, no second edition
was ever written, and by 1890 only three copies of it were known to exist.

Although this first mathematics book written by someone born in the US was
restricted to arithmetic properties and their applications, Isaac Greenwood lectured on
higher topics to his advanced students at Harvard. Two manuscript notebooks from
students toward the beginning and end of his tenure as Hollis Professor indicate his
breadth of knowledge and show what he had learned from reading the *Treatise on
Algebra* by John Wallis (1616–1703), one of the leading English mathematicians before
Isaac Newton. This was the first time that topics beyond mere arithmetic were available
in the colonies—methods for solving "adfected Quadratic Equations," "the Method of
Converging Series," and "Dr. Halley's Theorems for Solving Equations of all Sorts."
Along the way he cited works by two English mathematicians (William Oughtred and
Joseph Raphson), in addition to Halley, that provide evidence of his wide scholarship.

Notebooks from two illustrious Harvard students show that Greenwood lectured
on algebra throughout his tenure at Harvard. One of the manuscripts was written by
James Diman, a 1730 Harvard graduate who served as college librarian 1735–1737.
The first page of the 129-page notebook contains the inscription "James Diman's Book
1730/31," indicating that the notes were taken toward the beginning of Greenwood's
professorship. The second manuscript was written by Samuel Langdon, a 1740 Harvard graduate who served as president of the college during the revolutionary period 1774–1780. The inscription on the front cover of this 93-page manuscript reads,
"Samuel Langdon's Book, July 25, 1739," indicating that Greenwood lectured on this
material right on up to his dismissal. This conclusion is confirmed by the titles of the
two notebooks, Diman's reading, "Algebra or Universal Mathematics reviewed 1738
with Notes and Additions," and Langdon's reading, "Algebra by Isaac Greenwood, MA
Began July 25, 1739." The two manuscripts resemble each other so closely that their
contents must have been derived from the same source.

As the titles show, the topics were in algebra, a subject so lacking in the colonies
that Greenwood was the only colonist capable of teaching it at that time. Elsewhere,
arithmetic was the mainstay up to 1800. What did the term "algebra" mean for Greenwood? Basically, it meant material that is taught in a two-year high-school algebra
sequence today. For instance, he introduced the notation for powers of a variable x
as x^2, x^3, x^4, signifying the "Square, Cube & Biquadratick of x: so $x^{\frac{1}{2}}$, $x^{\frac{1}{3}}$, $x^{\frac{1}{4}}$ will
signifie y^e Square, Cube & Biquadratick Root of x." From there Greenwood discussed
the meaning of the notation

$$\sqrt[3]{7 + \sqrt[2]{2}},$$

which he expressed as

$$\sqrt[3]{:\ 7 + \sqrt{:\ 2}}.$$

Greenwood also treated higher order equations, providing three distinct methods for solving cubic equations. The third, called the "Method of converging series," is a version of the Newton–Raphson method of successive approximations. This discussion led naturally into "M^r. Raphson's Theorems for Simple Powers," which was used to calculate $\sqrt[4]{90}$. He extended this method to solve equations such as

$$a^4 \pm pa^3 = N$$

and

$$\frac{N - g^4 \mp pg^3}{4g^3 \pm 3pg^2} = x.$$

I state here a passage (in the original old English) that offers the interested reader an opportunity to interpret a result covered in most high schools and which is well known to professional mathematicians. This is the kind of example showing how the history of mathematics can be used in high-school or university courses. The challenge is to understand the original wording in order to determine what rule is being described, and then to express the rule in modern terms.

> To find ye Coefficients in Binomial Powers. Rule: Multiply ye Coefficient into ye Index of ye Power and Divide that Product by ye Number of terms, counting from ye left hand, and ye Quotient will be ye Coefficient or Numerical Figure of ye next successive Quantity.

The online file "Web01-Binomial" gives the solution.

Greenwood solved geometric problems by reducing them to linear and quadratic equations. One such quadratic in the 1740 manuscript of Langdon has complex roots, indicating not only Greenwood's sophistication but also the accomplishments of his very advanced students. As another indication that Greenwood was a product of a time when using algebra to solve geometric problems was in vogue as well as using techniques articulated by René Descartes a century earlier, he presented these algebraic topics amongst methods for solving a series of 24 problems from geometry. Results are taken from Euclid's *Elements* throughout the solutions of these problems, indicating that Greenwood's students had already completed a very good course in geometry.

Greenwood's appointment was propitious initially, both for him and for Harvard. However, it was known that he had fled London and was on the lam from debts incurred from trusting, influential, supportive figures. But it appears that money matters were only the tip of the iceberg. Lurking below was an expansive problem that floated financial affairs above the seemingly calm waters on all sides of Isaac Greenwood. He had a very serious drinking problem, one that, curiously, does not seem to have morphed into a thinking problem for almost a decade. After all, the manuscripts and public lectures cited above attest to his command of advanced topics and recent developments over his first ten years at Harvard.

But ultimately the "demon rum," as the Puritans called it (a term later adopted by twentieth-century prohibitionists), caused a precipitous fall from Greenwood's perch as the top mathematician in America to a sorry, solitary figure. In April 1737, almost ten years after his appointment, Greenwood was called before the board and admonished for dereliction of duties and repeated confrontations with the few other professors at the time. The board reported "he confessed the Charge of intemperance … and cast himself on the Lenity of the Overseers & professed his resolution of reformation."[68]

Like a duck on a lake, Greenwood appeared calm on the surface while paddling furiously below. In his case the paddling amounted to irresponsible behavior in regards to his professional and familial obligations—he had a wife and five children.

By that November the governing corporation had had enough. Greenwood was accused of disregarding repeated warnings, ordered to exhibit "an humble confession" in the public hall, and warned that if he did not remain sober he would be removed from office within five months. How did he react? He got drunk again. To add insult to injury, when summoned before the board, he refused to appear. Yet the overseers continued to coddle their prize professor in spite of a vote on December 7, 1737, to remove him from office. The board then accepted his humble confession and vow to reform. This time, however, a set of members was charged with overseeing the astronomy apparatus on a daily basis and charged with locking the door to the building if he appeared in a drunken state.

Greenwood was able to win his battle against the demon rum for the next four months, for which he was rewarded with a pay raise and renewed control of the Hollis telescope. He promptly celebrated with a bout of uncontrollable intemperance. When dragged before the corporation once again, he was too sick and befuddled to even offer yet another humble confession. This was the final straw. Isaac Greenwood was removed as Hollis Professor in July 1738. He was replaced by John Winthrop, who had benefited enormously from Greenwood as a mentor in the classroom and astronomical observatory. Winthrop, in fact, had carried out observations under Greenwood's guidance. Winthrop would prove to be a sober successor—in both senses of the word.

What happened to Greenwood after his firing?

In the fall of 1738 he established a school of experimental philosophy to teach from Newton's *Principia Mathematica*, modeled along the lines of the one he had witnessed under Desaguliers in London. This position seemed to agree with Greenwood, who exhibited the elusive talent of reaching students of heterogeneous ability. The *Boston Gazette* reported, "He had a happy Talent … of representing the most obscure and difficult Things in such a plain and easy Light, as it could not fail to satisfy the most ignorant, at the same Time that it would please the most learned."[69] Yet the school faltered within two years. Like Benjamin Franklin before him, Greenwood set out from Boston to seek greater opportunity in Philadelphia. Once again he advertised a course of lectures on scientific experimentation. Franklin himself arranged for Greenwood to use astronomical equipment owned by the nascent American Philosophical Society. But the duck's frantic paddling beneath the surface was unable to keep him afloat even amongst an emerging community of scientists in the largest colonial city.

Greenwood returned to Boston by July 1742, when he accepted a position as instructor aboard a Royal Navy man-of-war that sailed to England. His whereabouts for the next three years remain unknown, but after a cruise on a similar ship, he landed in South Carolina, where he died a pitiful, solitary man in October 1745. No autopsy report is available but if it were, it would probably list his death due to cirrhosis of the liver.

Isaac Greenwood was surely America's first mathematician. He was the most learned person in mathematics in the colonies up to 1738, standing head and shoulders above his contemporaries until being surpassed by his very best student, John Winthrop. His public lectures on the work of Isaac Newton were very popular, yet he never became a productive scientist because he was dogged by alcohol. In spite of the demon rum, or perhaps because of it, Greenwood exerted a profound influence on some of his students. Overall, although he was not the first mathematics professor in the US, he was the country's first important one and the most accomplished up to that time.

When were mathematics books first written in the New World? Greenwood's *Arithmetic, Vulgar and Decimal* was the first book on mathematics penned by an American, but it was not the first in North America. That honor goes to Brother Juan Diez, who arrived in Mexico with the conquistador Hernando Cortés in 1519 and wrote *Sumario Compendioso* (*Comprehensive Summary*), the first nonreligious book (and the twenty-fifth overall) to be published in the New World.[70] Its publication date of 1556 is remarkable because it precedes the settlements at Jamestown (1607), Quebec (1608), and Plymouth (1620) by over 50 years. Greenwood's 1729 book was not the first in English either, as several appeared in the first quarter of the eighteenth century before his *Arithmetic*.

Advertisements posted in various cities in the eighteenth century indicate that qualified individuals (and perhaps some charlatans) offered private instruction in mathematics. The earliest extant example is a 1709 advertisement by Owen Harris in Boston offering instruction in surveying, dialing (related to sun dials), navigation, and astronomy. As noted above, an advertisement from 1729 showed Isaac Greenwood's desire to teach calculus via Newton's fluxions. In 1743 Nathan Prince, who had been a tutor at Harvard, advertised the opening of a school offering instruction in the application of mathematics to areas typical for that time, like surveying and navigation, and emerging areas, like gunnery.[71]

Sober successor

The Royal College of William and Mary hired Hugh Jones to succeed Tanaquil Lefevre, and Jones proved to be a counterweight to his predecessor's reprehensible behavior. Harvard did the same after firing Isaac Greenwood for shameful conduct. Whereas it took William and Mary five years to find a suitable replacement, Harvard acted with dispatch. And with prudence.

The natural person for Harvard College to consider was the tutor for mathematics and natural philosophy, **Nathan Prince** (1698–1748), who had held the position since 1723. However, his life bore too many resemblances to Greenwood to make him an appropriate candidate. Like Greenwood, Prince was born in the Massachusetts Bay Colony and educated at Harvard, receiving his BA in 1718. This means his time there overlapped with Greenwood, though I do not know if they knew each other. Both had an older brother who had graduated from the College. Upon graduation Prince taught school for two years, returned to Harvard, received an MA degree in 1721, and preached a year in Rhode Island and Nantucket before returning to Harvard once again. In April 1723 Harvard's president John Leverett appointed him tutor to succeed Thomas Robie.

Even though Prince published only one scientific article (in astronomy), he was known as a remarkable scholar. However, he was also widely known to be hot tempered and unreliable, two traits that might explain why the Harvard Corporation bypassed him as a candidate for the Hollis Professorship in 1738. Obviously, he was disappointed when not offered the post, and the rest of his years at Harvard were marked by myriad complaints from faculty and students as well as accusations of intoxication.

Finally, in February 1742 the Board of Overseers dismissed him after bringing him up on multiple charges:

(1)   intemperance;
(2)   disturbing the peace;
(3)   contemptuous speech toward the president and fellows;
(4)   ridiculing his peers;
(5)   stirring up strife;
(6)   numerous other misdeeds.

The historian of mathematics Florian Cajori described Prince's personal habits as "notoriously irregular."[72] In light of the problems Harvard had with Greenwood, it is not surprising that the College looked elsewhere for a replacement.

Prince seemed to follow Greenwood's example after his removal. First he set up a school in Boston that was equally unsuccessful. Next he taught school in the Colony of Connecticut for several years before accepting a position as schoolmaster on a man-of-war bound for Lisbon. While there in the summer of 1746, he learned that the Society for the Propagation of the Gospel wanted him to be a missionary to Native Americans on the Island of Roatán near Honduras. Prince accepted, and sailed there to assume his duties in June 1748. However, he died just one month later.

What is it with these colonial mathematicians? Lefevre womanized, Greenwood boozed, and Prince was personally disagreeable. What kind of examples did they set? Fortunately, the next mathematics professor served as a much better role model.

John Winthrop IV. Today, **John Winthrop IV** (1714–1779) is known primarily as an astronomer, mainly for discovering a moon of Jupiter. Here I provide a brief account of his career. (The online file "Web01-Winthrop" adds further details.) He was one of the few American scientists from the Colonial Era to make contributions to astronomy using precise spherical trigonometry. A child prodigy who entered Harvard in 1728 at age 13 (just when Greenwood began lecturing at the College), Winthrop received his AB four years later, finishing as class valedictorian. He returned to Cambridge in the fall to continue his studies, which culminated with a master's degree in 1735 based on his disputation that it is not permissible for magistrates to impose hardships on anyone who maintained his own religious views. This was the first public sign of a protest against then-prevailing Puritanical beliefs.

Three years later, in 1738, Harvard sought a notable scientist to succeed Isaac Greenwood as "Hollisian Professor of the Mathematicks and of natural and Experimental Philosophy." The Corporation rejected longtime tutor Nathan Prince because his vices extended even beyond Greenwood's faults. Instead, that August the Corporation elected the 23-year-old John Winthrop, although he still had to pass muster before two committees. Only one person publicly opposed Winthrop's election—the disappointed Nathan Prince. Unlike Greenwood, he held the position a long time, 41 years altogether.

In fact, John Winthrop took the responsibilities of this chair very seriously. In addition to teaching such topics as plane and spherical trigonometry, mensuration of solids, and conic sections, he set about mastering Newton's *Principia Mathematica*, thereby becoming one of the very few Americans able to understand calculus. Harvard records indicate that he lectured on the method of fluxions as early as 1756, thus adopting the terminology of Newton instead of Leibniz. In a document from 1764 Winthrop defined the mathematical areas that the Hollis Professor should teach: geometry, algebra, conic sections, and plane and spherical trigonometry. Although calculus is not included in this list, his inventory contained areas of remarkable scope, including hydrostatics, mechanics, statics, optics, astronomy, geography, navigation, and surveying.

The main mission of a Harvard professor at the time was teaching, so Winthrop had to carve out time for his scientific endeavors from these duties as well as personal responsibilities. In 1746 he married Elizabeth Townsend, a member of the Chauncy family. This couple thus represented a generation of colonists that was inching farther and farther away from Great Britain.

Winthrop also mentored students in the laboratory located on the second floor of Old Harvard, where he "established for the first time America's independence in scientific development, and also gave to Harvard College her early prominence in scientific investigations."[73] The modern laboratory, as we know it today, did not exist then, so students were reduced to observing experiments without actively engaging in them. Equipment was just not generally available, even to the professor of natural philosophy. Yet in 1746 Winthrop gave the first practical demonstration of electricity and magnetism in America, using instruments that were obtained for him from London by none other than Benjamin Franklin. This emerging breed of American scientist felt no nccd to travel abroad for education or for research.

Socially, Winthrop was a product of his era in another way. Although a slave owner, he was aghast at the spectacle of a slave woman burning at the stake for murdering her master. A historian from 50 years ago reported, "When his own slave boy, George, died of the measles, he was mourned as one of the family, not as an unfortunate investment. His successor, Scipio, was watched over like the white children of the family."[74]

Winthrop's wife died suddenly in 1753. To overcome loneliness, he began to visit other colonial scientists with whom he had only corresponded with beforehand. One year later, for instance, he traveled to Yale to meet William Johnson, to Princeton to visit President Burr, and to Philadelphia to meet with Benjamin Franklin and his compatriots in the American Philosophical Society.

Back in Cambridge, in 1756, Winthrop married the widow Hannah Fayerweather Tolman, who was "a geyser of patriotism,"[75] even though her new husband initially supported strong ties to Great Britain. Winthrop opposed a movement to form an American "philosophic society" beyond the one foundering in Philadelphia because "our Country has hardly arrived yet to a state of maturity"[76] like the Royal Society in London. However, during the war scare of 1759, he agreed to the Massachusetts Bay Colony governor's request to update the Admiralty chart of Boston harbor that had been made in 1705. Winthrop also became indignant at the subsequent Stamp Act and Massacre, which compelled him to add his name to the list of candidates for Province Council in 1773. His reply to Benjamin Franklin's congratulatory letter, upon his successful election, evinced an ability in matters political as well as scientific: "If the Ministry are determin'd to inforce these Measures, I dread the Consequences: I verily fear they will turn America into a field of blood."[77] By this time John Winthrop's patriotism matched his wife Hannah's, and his bold public pronouncements on redressing perceived injustices made him a popular figure. In April 1776 Winthrop wrote to John Adams in Philadelphia, where utterances of independence still remained whispered, "Our people are impatiently waiting for the Congress to declare off from Great Britain.

If they should not do it pretty soon, I am not sure but this colony will do it for themselves. Pray, how would such a step be relished by the Congress."[78] Congress declared Independence three months later.

By late 1778 John Winthrop's physical condition worsened and he died the next May, deprived of seeing the colonies' ultimate victory and participating in the critical Constitutional Convention. Forty years earlier, just a year after assuming the Hollis Professorship, he had observed sun spots that led to his first known scientific investigation. The notes he recorded on that occasion were transcribed by the prominent Harvard librarian Frederick G. Kilgour.[79] Winthrop began (the italics are due to Winthrop):[80]

> 1739 April 19th at Boston. Walking on the Common a little before sunset, the air being so hazy that I was able to look on the sun, I plainly saw *with my naked eye* a very large and remarkable spot. Its shape was oblong and the length of it was perpendicular to the horizon. I observed it several minutes till the sun was actually set. ... The next day, Friday, coming back to Cambridge, I looked at the sun with an 8 foot telescope from 6 A.M. till sunset and discovered not only the same spot which I saw before but several others in his disk.

Winthrop's notes afford the modern reader an appreciation of how long such observations took. From our vantage point, the observations themselves are not the essence—rather it is the drawing of a conclusion from the evidence and then the establishment of a proof for its theoretical underpinning.

John Winthrop began his astronomy investigations in 1739 with a telescope that Thomas Hollis himself had obtained from the famous astronomer Edmund Halley. That instrument enabled Winthrop to observe a solar eclipse and to become the first American to view the transit of Mercury. His findings were published in the *Philosophical Transactions* of the Royal Society. Because this scientific periodical became such an important outlet for Winthrop's research, it is discussed before his published scientific work is described.

Royal Society of London. During the 1640s, a group of English scientists, then called natural philosophers, began meeting to discuss the idea of promoting knowledge of the natural world through observation and experiment. A dozen of them, who had assembled in 1660 to hear a lecture by mathematician Christopher Wren, decided to formalize their association by constituting what they called a "Colledge for the Promoting of Physico-Mathematicall Experimentall Learning." (Recall that the first building at William and Mary College was named in honor of Wren.) The group agreed to meet weekly to observe experiments and discuss advances in the field. Three years later the Royal Charter of King Charles II established this group as "The Royal Society of London for Improving Natural Knowledge." The last four words in the title were ultimately dropped.

The Royal Society became the leading scientific organization in the British Empire during the eighteenth century. The original 12 organizers became charter members. John Winthrop's namesake and great-granduncle was one of the original members, who were called Fellows. From that time onward, Fellows had to be elected, the third of whom was John Winthrop. Initially, criteria for membership were not well defined, but in 1731 a rule established that each candidate had to be proposed in writing and the certificate signed by those who supported election.

By 1665, Royal Society members felt the need for an outlet to publish their findings, resulting in the first issue of the *Philosophical Transactions*. This leading scientific journal celebrated its 350th anniversary in 2015, making it the oldest scientific periodical in the world still in continuous publication. One of the early articles in the journal, published in 1670 by John Winthrop, the "Governour of Connecticut in New England," described some features of trees in the New World, "matrices" (a kind of shellfish the Native Americans used for bartering), and some "very strange and very curiously contrived fish" existing in that area.[81] This article was typical of most colonial contributions to the *Philosophical Transactions* in that it described natural phenomena or living creatures in New England and Nova Scotia. Nonetheless, some articles were published by a few of these early Americans containing theoretical material, namely Thomas Brattle (two articles), Thomas Robie (two), and Isaac Greenwood (three).

Initially, scientists who did not live in London could get an article published by mailing it to any member of the Royal Society, who decided whether to read the submission at a weekly meeting. The Secretary of the Society then alone determined which readings qualified for publication in the *Philosophical Transactions*. Volumes of the journal appeared about once every two years. In 1750, however, the Royal Society reorganized its publication procedure by adding other competent peers to work with the Secretary in deciding which submissions to accept for publication, thus establishing the idea of what has become the norm today of (sometimes blind) refereeing.

A paper from 30 years ago from the *Notes and Records of the Royal Society of London* analyzed the 23 colonial authors of papers deemed worthy of publication in the *Philosophical Transactions* during 1753–1775 and provided brief snippets of their careers.[82] The work of four of these scientists in this chapter is mentioned—Cadwallader Colden, Benjamin Franklin, David Rittenhouse, and John Winthrop. Colden and Rittenhouse published one paper each (altogether, 14 of the 23 colonial authors produced one article) while Winthrop led the way with twelve and Franklin followed with nine. We turn now to the leading American scientist before the Revolutionary War.

Winthrop's publications. One biographer wrote, "Winthrop is undoubtedly the most important American pioneer in mathematics and astronomy."[83] Yet only one of Winthrop's published articles evinced mathematics explicitly. As a result, I tend to agree with the Harvard mathematician and historian Julian Coolidge, who wrote, "I can not find that [Winthrop's] interest in pure mathematics was outstanding."[84] Nevertheless, Winthrop used mathematics in novel ways in other articles to carry out calculations allowing him to aim his telescope at the proper position in the sky.

To illustrate this hidden use of mathematics, I examine his first article on the transit of Mercury and eclipse of the moon he observed in 1740. The results of these first authentic astronomy observations in America were included in a letter he sent to the Royal Society on December 30, 1740. The letter was read in November 1743, almost three years later, by secretary C. Mortimer and published in the 1742–1743 issue of the *Philosophical Transactions*. Knowing he was a 26-year-old unknown, Winthrop opened in a very modest way that tied him to two predecessors known to Royal Society Fellows:[85]

> Though I have not the honor to be known to you, I flatter myself
> you will excuse the freedom of this letter, since the design of it is to
> lay before you an observation which, I hope, may be of some use in
> astronomy. In confidence of this, I take the liberty to inform you, that,
> on the 21st of April 1740, I had an opportunity to observe Mercury
> … transiting the sun's disk. Being advertised by the calculations
> of that excellent astronomer, Dr. Halley, that the former part of this
> transit would be visible in our horizon, I was resolved to observe it in
> the best manner I could, with those few instruments I was furnished
> with; which were only those I had received from my predecessor Mr.
> Isaac Greenwood, and are the same that are mentioned by the late Mr.
> Thomas Robie in *Philosophical Transactions* N° 382.

Winthrop revealed his novel approach to the transit of Mercury by writing, "I
chose rather to deduce Mercury's right ascensions and declinations by calculation from
hence, than to observe them immediately in the common way of placing one of the
cross hairs parallel to the equator."[86] He listed the results in tabular form:[87]

	h.	′.	″.
The sun at the horizontal	5.	37.	59.
The sun at the vertical		39.	1.
Mercury at the vertical		39.	16.
Mercury at the horizontal		40.	1.

This table demonstrates the precision—accurate to the second—with which
Winthrop's calculations allowed him to compute in an era that was without clocks
or chronographs. Although Winthrop's computations differ very little from what is
known today, the transit of Mercury was not of great astronomical importance overall, yet by ascertaining the planet's position, knowledge of its orbit was considerably
improved.

An eclipse of the moon holds even less importance, so Winthrop devoted little
attention to it in his first published article. In the concluding few paragraphs he wrote,
"I was in hopes to have made a good observation of the lunar eclipse, which happened
last week. But the Sky … became overcast, which hindered me from making above
One or Two Observations that I could depend upon."[88] He concluded the article by
presenting his observations again in tabular form.

Transits of Mercury continued to fascinate Winthrop over the course of his career,
as attested by two short notes he published in the *Philosophical Transactions* regarding
transits from 1743, the year his first article appeared in that journal, and 1769, by which
time he was recognized as a leading scientist and had been elected a Fellow of the Royal
Society. For unknown reasons he demurred 20 years from the time of the October
1743 transit to the time when he reported on it in June 1763 by a letter to Nathaniel
Bliss, the Astronomer Royal of the United Kingdom. The objective of the note was to
establish that Winthrop's observations "determine the longitude of Cambridge, New
England, with more exactness than any of the observations that have been used for
that purpose."[89] Perhaps the reason for publishing his observations at this later time
was that Winthrop found himself among leading European scholars. He wrote, "The
comparison of this observation with those made in Europe will, I presume, determine
the difference of meridians within a few seconds of time."[90]

Winthrop's observations of the 1769 transit of Mercury were limited by the fact that sunset occurred before the planet had crossed over the sun. Nonetheless, he reported on his calculations from Mercury's entry across the sun in a letter sent that December to Benjamin Franklin in London. Results traveled slowly 250 years ago—Franklin forwarded the letter the following February, it was read at the Royal Society meeting in January 1771, and finally published in the *Philosophical Transactions* later that year. Winthrop stated that the reason for publishing the note was because, "This transit completes three periods of 46 years, since the first observation of [Peter] Gaffendi at Paris, in 1631."[91]

Back in 1742, shortly after Winthrop's first observations of Mercury, he became involved in an application of mathematics that is still current today—weather forecasts. Boston was then a small city of about 25,000 inhabitants surrounded by a fairly prosperous farming community, yet there was no need for systematic weather prognostication. Nonetheless, that year Winthrop began keeping daily records of weather observations. However, lack of proper instruments prevented accurate measurements of even such seemingly innocuous facts as maximal and minimal temperatures during a day. Thus, his daily meteorological observations were generally inaccurate up to 1759, when he obtained a Fahrenheit thermometer. He continued recording temperatures for the next several years but did not posit a theory for predicting weather.

In 1755 a different act of Mother Nature compelled Winthrop to study seismology, an area where he did reach a conclusion. His observations and analysis of an earthquake were published in the very first article in the volume of the *Philosophical Transactions* that appeared in 1758. The opening paragraph in the letter submitted to Royal Society secretary Thomas Birch set the stage for what was to follow:[92]

> I beg leave to lay before you the best account I am able to give of the great earthquake, which shook New England, and the neighboring parts of America, on Tuesday the eighteenth day of November 1755, about a quarter after four in the morning. I deferred writing till this time, in order to obtain the most distinct information of the several particulars relating to it, both here and in the other places where it was felt; especially the extent of it.

After describing his own observations of the movement of a pendulum caused by the rather severe earthquake, Winthrop questioned neighbors for corroboration of the effects of a series of four separate shocks. His introductory statement shows that he sought information from people as far away as New York, Philadelphia, and the Chesapeake Bay in one direction, Halifax, Nova Scotia, in another, the British Fort of Oswego and Lake George, New York, in a third, and even St. Martin's in what was then known as the West Indies.

Winthrop did not advance his theory in this publication; rather he announced it in two public lectures delivered at Harvard. A later analysis concluded, "that the disturbances of the earth-crust were in the form of waves, and transmitted a pendulum-like motion to buildings and objects on the surface." He was the first to apply computation to the phenomena, consequently discovering the analogy between seismic motion and musical vibrations; he also discovered the principle that "the quicker the motion the shorter the wave length of the disturbance."[93]

Winthrop's public pronouncements served another purpose as well. His theological views were consonant with the times, so he made sure to emphasize that his findings were aimed at describing the acts of God. In this way he hoped to combat the idea that earthquakes were a direct expression of the wrath of God. Overall, Winthrop's study of the earthquake thrust him into becoming America's first seismologist.

A particularly notable worldwide scientific event took place in 1761 with the transit of Venus, the first time such a passage had been observed since 1639. Winthrop sensed the value of this episode beforehand and diligently prepared for what would become America's first astronomy expedition. By a special act of the Massachusetts Bay Colony, he arranged for the use of a sloop to transport him, two of his students as assistants, and instruments loaned from Harvard to St. John's, Newfoundland, where conditions for observing the passage of the planet across the sun were maximal. Winthrop and his assistants recorded various aspects of the transit methodically, providing him with invaluable data for analysis once he returned to Cambridge. Gathering all available evidence, he calculated the sun's parallax at its mean distance from the Earth to be 8".68. The computation of a parallax in general involves the principle of triangulation for long, narrow triangles, which was considered advanced mathematics at the time. Today, use of spacecraft telemetry puts the accepted value of solar parallax at 8".794 143.

Once again Winthrop presented his observations in tabular form in the *Philosophical Transactions*.[94] He ended the article with a statement that he was hoping to find more celestial bodies during his time in Newfoundland: "I viewed the Sun with great attention ... in hopes to find a satellite of Venus; but in vain. There were several spots then on the Sun; but none that I saw could be a satellite."[95]

The next transit of Venus occurred eight years later, in 1769. (Occurring in pairs, the ensuing two, with Simon Newcomb the main American observer, took place in 1874 and 1882. The next pair occurred in our time—2004 and 2012.) The British Astronomer Royal, Nevil Maskelyne, urged Winthrop to lead an expedition to the Lake Superior region, which afforded views of the beginning and end of the transit, but ill health forced him to carry out his observations at Cambridge, where only the planet's entry was visible. His findings were published as part of a large report on worldwide observations in the initial volume of the *Transactions of the American Physical Society*.[96] Again Winthrop made a detailed study of his observations but was unable to improve upon his earlier computation of the sun's parallax. An analysis of his conclusions of these transits based on public lectures he delivered was published a century ago.[97]

The results of Winthrop's study of comets bears special mention because of its mathematical structure. Inspired by the first predicted return of Halley's Comet from 1682, he delivered two public lectures on the nature of comets in general that found light of day in a *Philosophical Transactions* paper from 1767. Curiously, this long article was communicated to the Royal Society and was written in Latin.[98] He posed five problems whose solutions were presented in terms of lemmas, corollaries, scholia, and mathematical demonstrations. Overall, Winthrop determined the limits of attraction between a comet and the sun, as well as the laws of motion and direction governing the vapor trail following the head of a comet. In this paper he also explained the curved appearance of the tail of a comet.

The Royal Society was so impressed with Winthrop's work that it elected him a Fellow in 1765, one year after his article on the 1761 transit of Venus appeared. He garnered other honors as well. In 1768 he was awarded the honorary degree of LLD (Doctor of Laws) by the University of Edinburgh, thus becoming the first American to be awarded that degree; Harvard also bestowed it upon him five years later.

Overall, it is difficult to classify Winthrop as a mathematician because, although versed in the subject, it was never his major focus. His reputation as the leading colonial astronomer of the time, on the other hand, is well deserved, with many of his contributions partly due to his skills in mathematics. Like Isaac Greenwood and Nathan Prince before him, he was born and educated in the colonies, reflecting a difference between the settlements in New England and Virginia, where the two initial professors at William and Mary were both born and educated in England.

John Winthrop IV died in 1779 at age 64. The Hollis Professorship had gotten off to a promising start with Isaac Greenwood—in mathematics, if not in citizenship—and had attained even greater heights with John Winthrop, but went into decline with the subsequent appointments of Samuel Williams and Samuel Webber.[99]

Yale

Let us turn our attention now from Harvard to Yale. Like William and Mary, Yale traces its roots back to the 1640s, when colonial clergymen led an effort to establish a college in the New Haven Colony, which had been established in 1638. However, this vision remained unfulfilled until 1701, when the Collegiate School was founded in nearby Killingworth. The third college in the colonies maintained a nomadic existence during its first 15 years at several locations in the cultural wilderness of colonial Connecticut before finally settling in New Haven in 1716, when that town outbid all other communities in both land and money to support the college.

Although mathematics and natural philosophy were included in the curriculum from the beginning, the conservative-minded trustees adhered to the medieval tradition of emphasizing moral philosophy, divinity, and classical languages, particularly Latin. As a result, surveying was taught but without the rudiments of geometry and trigonometry, a practice that reflects the colonies' practical tendencies at the time. Yet some scientifically interested students benefited from their training anyway, especially after a collection of books was donated to Yale in 1715 to form the first holdings of a library.

The person responsible for securing the books was **Jeremiah Dummer** (1681–1739). Dummer was born in Boston; his father was an accomplished portrait artist, silversmith, and engraver. Jeremiah Dummer graduated from Harvard in 1699 and remained there for two more years before sailing to Holland to pursue further studies in theology. First he enrolled at the University of Leiden, earning a degree in 1703. Just ten days later he was examined in Universal Philosophy at the University of Utrecht, where he defended two theses and performed well in two disputations. As a result, Dummer was awarded a doctorate, which was then the highest degree available, so when he sailed back to the colonies, he became "the first Harvard man to return from the Continent with a PhD."[100]

Lacking a community of scholars with similar interests, Dummer sailed to London in 1708. He remained in England for the rest of his life, engaging in controversial politics in his adopted land while concurrently serving as the agent for the colonies of Massachusetts and Connecticut. His biographer, Harvard librarian John Langdon

Sibley, wrote, "Dummer was the greatest colonial agent before Franklin."[101] In 1715 Dummer wrote an essay titled "The defense of the New England charters" that was published as a pamphlet six years later and became instrumental in defeating a proposal to abolish those charters.

For our purposes, Jeremiah Dummer's most important contribution to American mathematics was the collection of some 700–800 books he arranged to be donated to Yale. It is unclear why Dummer acted on behalf of Yale instead of his *alma mater*, Harvard, but he was unrelenting in his zeal to obtain donations for a library. He had appealed to the leading citizens in Old England, and his biggest conquest was undoubtedly **Elihu Yale** (1649–1721), the governor of the British East India Company, who never set foot in the New World yet was convinced to make a gift of nine bales of goods that included 417 books and a portrait and arms of King George I. This generosity induced the Collegiate School's trustees to rename it Yale College in 1718. Dummer also induced important works from two notable scientific figures. He wrote, "Sir Isaac Newton gives the second edition of his *Principia* (which appeared in 1713) ... Doctor Halley sends his edition of *Apollonius*."[102]

One Yale student who benefited from this largess at once was **Samuel Johnson** (1696–1772). Born in Guilford, CT, to parents who traced their American lineage back to 1637, Johnson attended Yale 1712–1716. Upon graduation, he moved with the college to the new location in New Haven, where he was appointed tutor for three years. Thus the Collegiate School underwent a name change while Johnson was a student there. During those three years Johnson initiated curricular changes that marked the first important advances in mathematics and science offerings at his *alma mater*. To prepare himself for the upper-level material he sought to teach, he underwent an intensive self-study of "Euclid, Algebra and Conic Sections"[103] during his first year in this position.

When Samuel Johnson was two years into his tutorship (in 1718), he was able to make extensive use of the library of books donated by Elihu Yale. However, in 1720 Johnson left Yale to become a minister at the Congregational Church in West Haven, and two years after that, influenced by extensive readings of Locke and Newton available in the library, he along with six other Yale graduates publicly expressed doubts about the legitimacy of Congregational ordination. Therefore, they sailed to England to be ordained by a Bishop of the Church of England. When they returned to Connecticut the next year, they established the colony's first Anglican Church in Stratford. Over the next three decades Johnson became identified as a vigorous advocate for Anglican causes.

Columbia

We digress here briefly to link Samuel Johnson with another new college established in the colonies. Several American colleges were founded during the late 1740s, including Princeton and the University of Pennsylvania. A third was Columbia University. Then called King's College, it was envisioned initially as an "Episcopal Colledge" in New York City. Samuel Johnson was elected the first president in November 1753, and classes opened the next July in an unused room in a schoolhouse adjacent to Trinity Church.

At first Johnson was the only faculty member for the eight students. During part of his eight-year tenure, he shared teaching responsibilities with a tutor, his younger son, William Samuel Johnson, in 1755 and then with Leonard Cutting (1756–1762) and Columbia's first professor, **Daniel Treadwell** (1757–1760). The death of Johnson's wife in 1759, the death of Treadwell the next year, plus ongoing political and denominational haggling within Columbia, impelled Johnson to leave his post in 1762. He was replaced by the 26-year-old, Oxford-trained Anglican minister **Myles Cooper** (1737–1781). Johnson's older son, the attorney Samuel William Johnson, became its third president in 1787, holding the post until 1800. Our Samuel Johnson did not live to see this succession, having died in 1772.

Thomas Clap. The courses Samuel Johnson instituted at Yale were continued by the tutors Daniel Brown and Jonathan Edwards up to 1728. However, Elisha Williams, the college's rector (1726–1739), held little interest in mathematics and science (called natural philosophy), so those subjects were marginalized and taught in a desultory manner.

That situation was reversed dramatically within three years of the appointment of a new rector, **Thomas Stephen Clap** (1703–1767), whose 26-year presidency (1740–1766) has been called "the golden age of mathematics and science in Yale's colonial history."[104] Thomas Clap was born and educated in Scituate, MA. His father was a descendant of a family that had immigrated to New England in 1630 and settled in Scituate ten years later. Clap graduated in 1722 from Harvard. His tutor there was **Thomas Robie** (1689–1729). Described by one exuberant biographer as "the most famous New Englander in science in his day,"[105] Robie was a mentor of Clap's contemporary Isaac Greenwood. (As noted earlier, details about Robie are available online at "Web01-Robie.") Robie immersed Clap in a doctrine of natural philosophy that presented the natural causation of Isaac Newton as mutually compatible with the natural causation of Puritan beliefs. One of Clap's essays contained a tribute to Newton that expressed his own views on the role of science and mathematics *vis-à-vis* religion:[106]

> There are many important Truths in natural Philosophy and Mathematics, which, when they come to be fairly proposed, were never doubted of; such as the general Laws of Attraction, the Weight of the Atmosphere, Rules of Fluxions, etc.

Clap carried on a vigorous correspondence with Benjamin Franklin and Cadwallader Colden on the matters described in this quotation.

Thomas Clap made significant changes in the curriculum shortly after assuming the Yale presidency in 1740. He adhered to a philosophy in which "Languages and Mathematics (which are themselves indeed a kind of Language) for these are both of them a necessary Furniture in order to the attainment of any considerable Perfection in the other parts of the Learning."[107] In 1742 Clap instituted a new curriculum described as follows: "In the first year to study principally the tongues, arithmetic, and algebra; the second, logic, rhetoric, and geometry; the third, mathematics, and natural philosophy; and the fourth, ethics and divinity."[108] It is noteworthy that mathematics preceded natural philosophy, the reverse of the Harvard curriculum in which Clap had been educated.

In 1745, to ensure that entering students would be able to handle the new curriculum, Yale instituted a policy that would eventually revolutionize admission policies at institutions of higher learning throughout the country by establishing a mathematical requirement of proficiency in arithmetic. Beyond that, mathematics continued to play a prominent role under President Clap, whose second curricular change, instituted in 1766, his last year in office, offered a course "Mathematics" for freshmen, algebra and trigonometry for sophomores, and "most branches of the mathematics"[109] for juniors. In this curriculum, applications in the senior year included surveying and navigation, the latter based on conic sections and fluxions. This was the earliest mention of conic sections and calculus as part of college courses in the country. The emphasis on navigation reflects the rising importance of the East India Company and its role in England's colonial expansion throughout the world. Beginning in 1770, Yale also established a second professorship, in mathematics, its first since the one in sacred theology had been established 25 years earlier. This one, like those at some other colleges of the time, was combined with philosophy and natural philosophy.

However, by the end of his Yale days, Clap was not so revered by students. He resigned the presidency in 1766 due to a rebellious bunch that almost reduced the college's buildings to rubble because of his imposition of old-style religious values on this younger generation. This is especially surprising in light of his earlier effort to teach Newtonianism within the confines of Puritan beliefs.

Competition between Harvard and Yale heated up over the first half of the eighteenth century—and has continued unabated ever since. In general, the advantage went to Harvard, which gained a slight edge with its curricular offerings. For instance, higher plane curves entered Harvard under Isaac Greenwood in 1735 but not at Yale until four years later, while fluxions were offered at Harvard in 1751 under John Winthrop but not for another seven years at Yale.

In this regard, bachelors' theses provide evidence of these subjects being known before they resided officially in courses. At that time a thesis was a statement the degree candidate was prepared to defend or a question he could answer. (I purposely use the male form here because female students were not accepted at Yale until 1969 and Harvard in 1977.) Undergraduate theses first appeared in Harvard's commencement programs in 1653 but none was in mathematics for another 40 years, yet that one only reflects the low state of affairs at the time because the student answered "Yes" to the question, "Is the quadrature of a circle possible?" (Perhaps the student did not restrict constructions to straightedge and compass.) However, beginning with a 1711 statement on conic sections, three theses from the period 1711–1721 suggest that some students were exposed to algebra, conics, and calculus. In 1719 one Harvard thesis read, "A fluxion is the velocity of an increasing or decreasing flowing quantity," while another from 1721 stated, "Algebra is the art of reasoning with unknown quantities in order to define their relation to known quantities." Moreover, reflecting the influence of Samuel Johnson, a Yale thesis from 1720 deals with a topic related to Fermat's last theorem, showing that there are no nontrivial solutions to the equation $x^4 + y^4 = z^2$. Theses during Clap's presidency ran the gamut of mathematics from simple algebra to complex integral calculus, the latter appearing as fluxions for the first time in 1758.

Samuel Johnson had stayed at Yale as a tutor for three years after graduation. Such positions were generally held by recent graduates of the college who served terms of two or three years before moving on to other professions; tutoring was conceived as an interim career. During his long tenure as Yale president, Thomas Clap appointed 28 tutors, and many of them exhibited proficiency in mathematics and natural philosophy;

Figure 1.3. Ezra Stiles

he was acutely aware that students required such specialized instruction in technical areas. These tutors underwent a special, more intensive course of preparation, including two such notable mathematical figures introduced here briefly because of subsequent accomplishments in the young country. The online file "Web01-YaleTutors" adds more information.

Ezra Stiles (1727–1795) was born in North Haven, CT, into a family that had emigrated from England in 1635. Ezra's father was a 1722 graduate of Yale who was then ordained a pastor in North Haven, then part of New Haven. Ezra Stiles studied theology at Yale and received his bachelor's degree in 1746. Three years later he was singled out by Thomas Clap to be a tutor, but he held this position for only one year. That was an important time in his life because Benjamin Franklin had donated an electric apparatus to Yale that Stiles used to conduct the first experiments in electricity in New England. In 1755 Stiles delivered an oration in honor of Benjamin Franklin even though he was no longer officially connected to Yale. The two colonial scientists had formed a friendship that lasted a lifetime. The next year Stiles moved to Newport, RI, where he remained until 1777, when his congregation had to disperse due to a British attack on the town. His pastoral duties ended when he accepted the presidency of Yale in June 1778, a post he held until his death 17 years later.

Nehemiah Strong (c. 1729–1807) was born in Northampton, MA. He received his AB in 1755 and his AM three years later, both from Yale. Strong was chosen by Thomas Clap to be a tutor 1757–1760. Once ordained, he became pastor of a church, but he ended up losing that post in 1767 due to a very embarrassing incident. During this time he married a woman whose first husband was believed to have perished at sea, but he appeared unexpectedly and claimed his wife, who left Strong for her first husband. He was dismissed from the pastorate due to entanglements involved in the resulting marriage annulment.

Figure 1.4. Nehemiah Strong

Fortunately, Strong's reputation in mathematics earned him an appointment as the first professor of mathematics and natural philosophy at Yale in December of 1770. He held the position throughout the Revolutionary War, though his salary was reduced, not just due to uncertain economics but friction with the governing corporation over his Tory views. Strong was gradually "squeezed out when the Corporation first reduced his salary, then forbade him to lecture, and finally required him to take an oath of allegiance as though he were a closet loyalist."[110] As a result, he resigned his professorship in 1781. He became known for his 1784 book *Astronomy Improved*.

Dartmouth College

The third accomplished mathematician who served as a tutor at Yale under Thomas Clap was Bezaleel Woodward, who shortly thereafter moved to Dartmouth College in remote New Hampshire. The founding of Dartmouth College is described below, followed by a brief overview of Woodward's career.

Dartmouth's first president, the Puritan minister **Eleazar Wheelock** (1711–1779), was a 1733 graduate of Yale. One year later he was installed as the pastor of a Congregational church in Lebanon, CT, where he remained another 34 years until moving to Hanover, NH, to establish a college. During his ministering, Wheelock became so impressed with the learning ability of one of his Native American students, Samson Occom, who himself went on to become an ordained minister, that in 1755 he established Moor's Indian Charity School to train other Native Americans as missionaries. Occom and another minister sailed to England 11 years later to raise funds to establish a trust for the school. The head of that trust was William Legge, the 2nd Earl of Dartmouth. Although funding was sufficient to support the Charity School, Wheelock was unable to attract enough Native American students to attend it. In fact, Wheelock had a bigger plan in mind—to establish a college for the benefit of the sons of colonists. When the Royal Governor of New Hampshire offered him land to build the college and sufficient resources to support it, Wheeler moved his operations from Connecticut to Hanover, NH. Moreover, Governor John Wentworth issued a charter in the name of King George III in December for a college "for the education and instruction of Youth of the Indian Tribes in this Land … and also of English Youth and any others."[111] Cleverly, the inclusion of wording about educating Native Americans enabled the new college to use the Charity School's unspent trust money. The college was named after the 2nd Earl of Dartmouth even though he not only opposed the institution but never donated anything to it—books, laboratory equipment, nor money.

Dartmouth thus became the ninth college in the US and the last one established under colonial rule. Success happened quickly, with four students graduating in 1771, one of whom was Wheelock's son John, who succeeded him as president upon his father's death at age 68. However, those early gains were soon negated by the Revolutionary War, which exacted such a heavy toll on Dartmouth that it took another 125 years before the college attained national renown under the presidency of William Jewell Tucker from 1893 to 1909.

Dartmouth president Eleazar Wheelock had the good fortune to recruit a very able mathematics professor to his outpost in Hanover, NH. **Bezaleel Woodward** (1745–1804) was born in Lebanon, CT, where Wheelock was a minister, into a family that traced its roots back to Dorchester, MA, in 1638 on one side of the family and to Northampton in 1639 on the other. Woodward graduated from Yale in 1764 at age 19 and then became a tutor under Thomas Clap. He then spent a few years in the ministry before being contacted by Eleazar Wheelock, who appointed him tutor at Dartmouth in October 1770. Woodward played an important role at Dartmouth in those early, challenging days:[112]

> [Wheelock's] first associate in instruction, who acted in the capacity
> of tutor, was Mr. Bezaleel Woodward. … The fact that Mr. Wood-
> ward was subsequently, for many years, a highly esteemed professor
> of Mathematics in the college indicates that he was a worthy pupil of
> his distinguished teacher [Clap].

In 1771 the trustees of the college donated an acre of land to Woodward for a home.

The Yale curriculum doubtlessly served as the model for Dartmouth, with Woodward as professor of mathematics and philosophy. Bezaleel Woodward was married to Wheelock's daughter and served the college in many ways over the years, including being its librarian, acting president on two occasions, and treasurer. His salary was small, even when he was promoted to professor of mathematics and natural science in 1782.

Overall, Bezaleel Woodward had a long and illustrious career until his death in August 1804. Yet there is no evidence of any contributions to science, let alone mathematics. So why was he called a mathematician? Similarly, Thomas Clap wrote about his Yale graduates:[113]

> Many of them well understand Surveying, Navigation and the Calcu-
> lation of the Eclipses; and some of them are considerable Proficients
> in Conic Sections and Fluxions.

This statement encompasses the reason that Thomas Clap was regarded by his contemporaries as an accomplished mathematician. He taught applications of mathematics to fields that were critical at the time—surveying, navigation, and astronomy. Nonetheless, he too did not produce any original works in mathematics. By today's standards neither Wheelock nor Clap would be regarded as mathematicians. Chapter 2 begins with a discussion of this issue.

2

Independence

Colonial America sought political liberation from Great Britain when the Thirteen Colonies declared independence in 1776. When King George III dispatched soldiers to American shores, Richard Henry Lee, a Virginia delegate to the Second Continental Congress, introduced a resolution on June 7, 1776, proposing independence for the Thirteen Colonies:

> **Resolved**, That these United Colonies are, and of right ought to be, free and independent States, that they are absolved from all allegiance to the British Crown, and that all political connection between them and the State of Great Britain is, and ought to be, totally dissolved.

Three days later, the Continental Congress appointed a committee to draft a statement of independence for the United Colonies. A 33-year-old delegate from Virginia, Thomas Jefferson, a notoriously poor speaker but acclaimed writer, worked feverishly to craft the proper wording throughout the month of June before finally presenting it on July 2. After two days, during which Congress haggled over revisions and Benjamin Franklin doodled with magic squares, the Declaration of Independence was adopted on the afternoon of July 4. Given Jefferson's sound footing in mathematics at the College of William and Mary, it is not surprising that part of the Declaration reads like an axiomatic system:

> We hold these truths to be self-evident, that all men are created equal, that they are endowed by their Creator with certain unalienable Rights, that among these are Life, Liberty, and the pursuit of Happiness.

In this analogy, self-evident truths are axioms while items such as "Rights," "Liberty," "Happiness," and even "men," are undefined terms.

This chapter begins with the question asked at the end of Chapter 1: What is a mathematician? This thorny issue was not settled until about 1876, after which original contributions to mathematics became the main criterion for being regarded as a mathematician.

The chapter then proceeds, mostly chronologically, from roughly 1750 up to 1800. An early development was the founding of several notable colleges in America between

1746 and 1776, with Princeton emphasized. Two other pre-war advances were the formation of America's first scientific organization and the publication of the first scientific journal. Before the outbreak of the war, groups of scientifically interested men in Philadelphia formed a learned society—the American Philosophical Society (**APS**)—that inaugurated the *Transactions of the APS* in 1771. However, a report from the APS stated:[1]

> The American Revolution struck a heavy blow at the Society. ... Political feelings could no longer be put aside; Loyalists and Quaker pacifists soon ceased entirely from attendance.

Consequently, the second volume of the journal was not published for another 15 years, finally appearing in 1786.

What about Canada? It had been more closely aligned with Great Britain than with the colonies, but it had a mix of French and English speakers. This chapter discusses initial developments in education that took place mainly in the Province of Québec, where Jesuit priests launched a college one year before Harvard was founded. However, the Collège de Québec did not last as long as its younger counterpart, closing in 1761, when the British defeat of the French in the Seven Years' War put an end to higher education in Canada, which did not revive for a half century.

The author of the first paper in the *Transactions of the APS*, David Rittenhouse, made surprisingly advanced contributions to mathematics in subsequent issues of the journal. A section details two of his papers—one on calculus and one on numerical analysis. Another section provides examples of mature mathematical thinking in another leading colonial statesman, Benjamin Franklin, whose political and scientific exploits are better known. Both played vital roles in the APS. Their contributions are appropriate for courses today in finite mathematics, calculus, and numerical analysis.

Throughout US history, warfare has had a symbiotic relationship with the development of mathematics, beginning with the Revolutionary War, as seen here in applications by David Rittenhouse. The war had devastating effects on American colleges up to its conclusion in 1783. For instance, Rutgers (then Queen's College) was closed, and its students were sent packing when British forces occupied its buildings. Even colleges that remained open were affected—conic sections and fluxions entered the curriculum at Yale in 1766, but neither course remained on the books in 1777.[2]

A section on surveying examines the lives and accomplishments of the first African American scientists to warrant inclusion in a history of mathematics in the US. While Benjamin Banneker established a reputation as a gifted astronomer and mathematical practitioner, Thomas Fuller was known only for lightning-quick mental calculations. Both men were celebrated by abolitionists as examples of inherent talent in blacks.

Although cryptology was only slightly used during the Revolutionary War, a section introduces its prehistory, including the death of one traitor caused by decoded messages. The coverage includes a clever code designed and used by Benjamin Franklin.

Although political independence for the US began in 1776–1783, mathematics in America only started to throw off its British yoke during the Period of Confederation, 1783–1800. This period witnessed the first textbooks written by American authors for an American audience, particularly the *Arithmetic* of Nicholas Pike. Another related development at this time was the formation of another scientific organization (the American Academy of Arts and Sciences), along with its journal, the *Memoirs.*

What is a mathematician?

Thomas Clap was ahead of his time in teacher training, offering prospective tutors a strong dose of advanced mathematics when they studied directly with him in their senior year. He also kept a close eye on their teaching when they became tutors. For instance, Ezra Stiles wrote that Clap "always spoke to the Person on whom he set his eyes for Tutor and desired him to adapt his Studies preparatory."[3] A description of Clap as "one of the most influential intellectual leaders of eighteenth-century America"[4] rings true in light of his contributions to the science curriculum, the offering of advanced courses, and the ability to teach advanced subjects. Although mathematics played a prominent role in the curriculum Clap designed at Yale, he did not produce a single work on mathematics. In spite of this fact, he was highly regarded by his contemporaries as the second most accomplished mathematician in the colonies:[5]

> In mathematics and natural philosophy, I have not [*sic*] reason to think that he was equalled [*sic*] by any man in America, except the most learned Professor Winthrop.

In something of an understatement, Samuel Johnson wrote that Clap was "much of a mathematician."[6]

This latter quotation raises the question, What *is* a mathematician? A modern mathematician, after all, would chafe at the notion of someone who did not produce one mathematical work as being labeled in such a way. The lesson here is that the definition of "mathematician" is not fixed—it has varied over time. Albert Einstein was regarded as a mathematician in 1930. Sixty years later a Nobel Prize in Economics was awarded to John Nash, who specialized in mathematics before his untimely death in 2015, yet he has been more generally recognized for his contributions to economics.

Up to the time of the watershed year 1876, a mathematician in America was someone sufficiently steeped in the subject to be able to teach advanced parts of the subject and, moreover, to apply these topics to related fields. Thomas Clap and John Winthrop are shining examples. To avoid confusion, in Chapters 2–4 we will call them "mathematical practitioners."[7] However, I claim that David Rittenhouse is a mathematician by today's standards, because he published papers on mainstream mathematics that were entirely new to him. Other figures defy this easy distinction, such as Isaac Greenwood; even though he presented his own approach to topics novel to American students at the time, they were not original, and so I label him a mathematical practitioner. Generally, the only four individuals (up to 1876) I call mathematicians are Rittenhouse, Nathaniel Bowditch, Robert Adrain, and Benjamin Peirce. All others were mathematical practitioners. (See p. 118 for the definitions I follow.)

Princeton

There were three colleges in the colonies up to 1746, at which time Princeton was founded as the College of New Jersey. There was a demand for higher education due to the emergence of prosperous colonial gentry over the next 25 years, with the founding of Penn, Dartmouth, Columbia, Brown, and Rutgers. However, disruptions caused by the Revolutionary War put a halt to this advance.

This section briefly describes those professors who taught mathematics at Princeton, which originally opened at Elizabethtown, NJ, and moved to other places before settling into its present location. The college's founding was mainly due to the efforts of Jonathan Dickinson, who conducted a classical school in Elizabethtown before becoming the first president. He was aided in founding the college by Aaron Burr (father of the future vice president), who conducted a school in Newark similar to Dickinson's. Upon Dickinson's death in October 1747, Burr took control of the embryonic college and moved it to Newark. Although Burr's tenure too was cut short by untimely death, during the ten years he served as president, he devised the curriculum, drew up the first entrance requirements, enlarged the student body from seven to 70, found a permanent location at Princeton in 1756, and oversaw construction that year for the college's first building, Nassau Hall. Aaron Burr's wife, Esther, survived him by less than a year when she died of smallpox at age 26, leaving behind their two young children, Sarah (age four) and Aaron, Jr. (age two). The younger Burr (1756–1836) graduated from Princeton in 1772 and became the third vice president of the United States, serving under Thomas Jefferson (1801–1805). In 1804, the last full year of Burr's single term in that office, he killed political rival Alexander Hamilton in a duel.

Although Burr *le père* had graduated from Yale when Ezra Stiles was a tutor, he did not teach mathematics, instead leaving instruction in the hands of the college's two tutors. John Witherspoon, who assumed the presidency in 1768, continued this tradition, but changed it by hiring a student who had graduated that year. Upon graduation, **William Churchill Houston** (1740–1788) remained at the College of New Jersey as a tutor. However, Witherspoon considered himself less than an accomplished scholar in mathematics and astronomy, so in 1771 he appointed Houston as the college's first professor of mathematics and natural philosophy, thus leaving the president with responsibility for instruction in moral philosophy, divinity, rhetoric, history, and French. When the Revolutionary War broke out in 1776, the college's only two professors were Witherspoon and Houston, but the college closed when the town of Princeton was overrun by British troops. Houston returned to campus after the war but resigned in 1783 to participate in the Continental Congress that finally produced the country's constitution.

John Witherspoon (1723–1794), similar to later Princeton mathematical practitioner, Walter Minto, was an immigrant from Scotland who arrived with an MA and divinity degrees from the University of Edinburgh. Witherspoon believed firmly in the American cause; he signed the Declaration of Independence and became a leading member of the Continental Congress 1776–1782. He lost a son, James, a 1770 Princeton graduate, in the Revolutionary War. Just two years earlier Witherspoon assumed the presidency of Princeton, a post he held until his death. Although he did not approve of academic learning for its own sake, he felt that education must produce good citizens. Therefore he shifted Princeton's focus from preparing students for the ministry to equipping them for civic leadership. Just three years before his appointment as president, Witherspoon caused a mild stir when, as a 68-year-old widower of two years, he married a 24-year-old widow.

Course listings at Princeton and other emerging colleges varied widely. To put these new institutions in perspective, recall that the only mathematics courses offered at Yale were arithmetic and surveying up to 1726, and surveying was restricted to students in their final year. At Harvard, Isaac Greenwood introduced material beyond arithmetic 1727–1738, teaching topics he had learned during travels to England, and

Figure 2.1. John Witherspoon

John Winthrop offered calculus based on Newton's fluxions in the 1750s. Yet the University of Pennsylvania listed a course on fluxions in 1758 for students in their second year with Colin Maclaurin's *Treatise on Fluxions* named as the recommended text. However, all these newer institutions were chiefly concerned with classical education steeped in languages, especially Greek and Latin; mathematics garnered scant attention. How prepared for college were perspective students in light of the fact that a large number of colonists opposed the idea of public education altogether? Initially, few boys were prepared for even the merest of mathematics courses offered; many sought private instruction to qualify for admission. But it was not until the mid-eighteenth century that colleges adopted mathematics entrance requirements. Yale was the first in 1745, but this change did not gain favor for some time, with Princeton adopting requirements in 1760, but Harvard not until 1807. Generally, the main requirement for admission was knowledge of Latin, both written and spoken, and it was the college president who administered the oral exam and placed accepted students in the four-year curriculum accordingly.

American Philosophical Society

The American Philosophical Society was this country's first learned society, and the only one before the Revolutionary War. Formed officially in January 1769, its roots go back to two previously existing societies, each of which had sought to stimulate the advance of science.[8] Both were located in Philadelphia, whose wealth of naturalists, artists, gardeners, craftsmen, collectors, and generally scientifically literate populace produced the first community of scientists in America.

The first of the two societies began in 1743, when Benjamin Franklin wrote in his *Proposal for Promoting Useful Knowledge among the British Plantations in America*:

> The first Drudgery of Settling new Colonies ... is now pretty well over;
> and there are many in every Province in Circumstances that set them
> at Ease, and afford Leisure to cultivate the finer Arts, and improve the
> common Stock of Knowledge.[9]

The scholarly society he advocated became a reality that year when it became known as the American Philosophical Society (**APS**). The three leaders within the nine-person group were the botanist John Bartram, physician Thomas Bond, and Franklin. This society drew its inspiration from the Royal Society of London (which had elected 19 American members up to 1743) and the Dublin Philosophical Society (which had been founded in 1731 "for improving Husbandry, Manufactures, and other useful Arts").

Two APS members elected in this early stage were regarded as mathematicians, even though I would classify them as surveyors. Because surveying was one of the three primary areas that relied on mathematics in the eighteenth century, those who were particularly adept were regarded as mathematicians. The two elected to the APS before 1746 were James Alexander and Cadwallader Colden, both living in what is now New York State and both involved in setting the boundary between the Provinces of New York and New Jersey.

Both New York and New Jersey have roots in The Netherlands, though France was the first European country to occupy the area of New York. The Dutch stake began when Holland bought the island of Manhattan from Native Americans in 1626. The present state was renamed the Colony of New York by England in 1664. The colony's population grew ninefold from 18,067 in the first colonial census of 1698 to 168,007 in the last of 1771. In the seventeenth century the area of North Jersey was called New Netherlands but was renamed by Charles II the Province of New Jersey after the English Channel Island of Jersey to honor the loyalty of the islanders to his throne. North Jersey was united with South Jersey to form the Colony of New Jersey in 1702. The border between the colonies was in dispute from 1701 to 1765, and both Alexander and Colden were involved in these skirmishes.

Figure 2.2. James Alexander

James Alexander[1] (c. 1691–1756) was born in Scotland but fled to the colonies due to a political uprising. He soon became the surveyor for Perth Amboy, a small settlement in what is now New Jersey. He was then appointed surveyor for both New York and New Jersey, whereby one of his first duties was to determine the boundary between the two colonies. Alexander was also an accomplished barrister, serving as attorney general for these two colonies. Politically he was elected to several terms in the Colonial Assembly and was appointed to the Governor's Council of New York

---

[1]Subsequent chapters will introduce a twentieth-century mathematician of the same name as this James Alexander, who is discussed here.

and New Jersey. As far as I am concerned, however, Alexander's biggest accomplishment was founding the American Philosophical Society in 1743 along with Benjamin Franklin, Thomas Bond, and John Bartram. On education issues, he was a very successful fundraiser for Columbia University (then King's College) when it was founded in 1754.

Like Alexander, **Cadwallader Colden** (1688–1776) was born of Scottish parents, though his mother was visiting Ireland when he was born. After graduating from Edinburgh University in 1705 he remained there to study natural philosophy, including physics, anatomy, chemistry, and botany. This enabled him to set up a medical practice in Philadelphia when one of his aunts invited him to the colonies five years later. He returned to Scotland in 1715 to marry, whereupon the newlyweds landed back in the "City of Brotherly Love" later that year. Colden's study of a yellow fever epidemic in New York City resulted in the publication of a series of essays in 1743 correlating unsanitary living conditions with high rates of disease.

Figure 2.3. Cadwallader Colden

Subsequent efforts to improve New York's sanitation system improved Colden's standing in the scientific community at about the same time he was corresponding with Benjamin Franklin about forming the APS. Colden was considered a mathematician in the Society because of his role as surveyor general for the Colony of New York beginning in 1743. He published an article on an earthquake in November 1755 in the *Philosophical Transactions* of the Royal Society. Fourteen years later he served as governor of New York (1769–1771). He spent much of his spare time studying the work of Isaac Newton and is credited with finding errors in some publications. Colden's daughter Jane became the first female American botanist.

The moving force behind the APS had been the botanist John Bartram (1699–1777), who in 1739 proposed forming a society of the "most ingenious and curious men"[10] in America. Bartram's plan was to purchase a building, sponsor lectures, and underwrite expeditions, but it was too ambitious for the colonies. Yet Franklin's idea to revise and simplify the plan worked four years later. Nonetheless, by 1746 interest languished and the Society was moribund.

The second group that joined to reform the APS consisted of an obscure collection of young men who formed in the wake of American resistance to the Stamp Act. This group gathered for the purpose of self-improvement through conversation and study. This "young Junto," as it called itself, was founded in 1750 but, like the APS, its membership soon languished, and by 1763 it stopped meeting. It was revived three years later with a membership of seven that was driven by three people of diverse backgrounds—importer Charles Thomson, civil servant Edmund Physick, and iron-monger Isaac Paschall. This time the group assumed a more formal name that also de-scribed its aim—the American Society for Promoting and Propagating Useful Knowl-edge. The Society grew quickly and elected officers in 1768. Although the aim was to deal with improved methods of farming, new manufactures, and mineral wealth, one of its founders, Charles Thomson (who later became secretary of the Continental Con-gress) wrote, "But it is not proposed to confine the view of the Society, wholly, to these things, so as to exclude other useful subjects, either in physics, mechanics, astronomy, mathematics, &c."[11] The American Society was very resourceful.

Meanwhile, the APS too reconstituted itself in 1767 with the same name because some of its original members took umbrage at not being invited to join the new one. This newest APS also chose officers. Both organizations emphasized the "useful" arts of business and industry.

Both the APS and the American Society were founded in Philadelphia, so within two years the groups were able to drop their religious differences and merge into one organization on January 2, 1769, under the name "The American Philosophical Soci-ety, held at Philadelphia, for Promoting Useful Knowledge." Benjamin Franklin was chosen as president even though he resided in London as a colonial agent in England at the time. The primary difference between this Society and its two predecessors that enabled it to survive (right on up to this day) was that Philadelphia merchants joined in droves, thus providing a critical mass with sufficient wealth to support such an organi-zation. Besides, the leadership of James Logan, John Bartram, and David Rittenhouse was exemplary.

In 1769 there were 244 members living in America, and another 21 foreign mem-bers who were elected into one of six sections; mathematics was listed in Section 1 along with natural philosophy, astronomy, and geography. The membership consti-tuted the beginning of a critical mass needed to form a community of scientists and, later, mathematicians. Membership classification was changed in 1936 to four classes, with the first being the mathematical and physical sciences; a fifth class was added in 1976 to recognize people who, in the Jeffersonian tradition, had multiple accomplish-ments and broad interests.

Up to 1776 the study of natural philosophy included mathematics and the kinds of investigations now considered scientific and technological. Contributing somewhat to the Society's international fame was its participation in astronomical observations. Attracting the recognition of the scholarly world was the plotting of the 1769 tran-sit of Venus by David Rittenhouse, who used one of his telescopes erected on a plat-form behind the Pennsylvania State House (now Independence Hall). His observa-tions were published in the *Philosophical Transactions* of the Royal Society of London, which elected only two American members during 1776–1800, Rittenhouse and James Bowdoin. The historian of mathematics Todd Timmons wrote, "After Franklin's ex-periments with electricity, possibly the most important event in the physical sciences in eighteenth-century America was the observation of the transit of Venus in 1769."[12] Nonetheless, this event was observational, not theoretical, like most advances in nat-ural history. Besides, American scientists were invited to join in observations of the transit mainly because of their unique locations in the world. The results were useful for the colonies, however, because they provided precise locations of American cities.

Some APS traditions have proved to be enduring. For instance, starting in 1769, its meetings were held on the first and third Friday evenings of each month except during summer, when they were held on the third Friday only. Also, officers were chosen in January. Both traditions continued until 1902, when an annual general meeting was established and meetings were held once a month until 1936, when an annual summer meeting was initiated and monthly gatherings eliminated.

Several papers in an important book[13] on scientific developments during the New Republic provide information on the development of mathematics during this period. One of those papers discusses why no society resembling the APS was formed in New York City, which would become the site of the first truly national organization of mathematics one century later. Historian John Greene wrote:[14]

> In New York, where population, commerce, industry, wealth, and geographic location might have led one to expect the development of an array of institutions for promoting knowledge rivaling or surpassing the achievements of Philadelphia, nothing of the kind occurred.

Greene suggested that the most important circumstance causing this lack of a local community of scientific enthusiasts seems to have been the removal of the state capital to Albany, which divided the scientific resources of the area between two "poles."

From the time of its founding in 1769, the APS felt the need to publish the proceedings of its meetings, so the Society inaugurated its first journal, the *Transactions of the APS*, which appeared for the first time in 1771. The first volume described Society activities, results of elections, and deaths of members, but mostly it reported on scholarly research in various disciplines in the humanities and sciences divided into four sections:

> Section I. Mathematical and Astronomical Papers
> Section II. Essays on Agriculture
> Section III. Miscellaneous Papers
> Section IV. Medical Papers

Authors of papers in Section I formed the first, nascent publication community.

The initial article in the *Transactions of the APS*, and hence the first scientific paper published in America, was titled "A description of an orrery, executed on a new plan," submitted by David Rittenhouse. It was followed by his one-page report on the transit of Venus that had already been published by the Royal Society. To underline the importance of this transit for colonial scientists, the next seven articles were devoted to observations recorded at other locations. William Smith and John Ewing authored two of these articles. Section I in the journal also included accounts of a comet from 1770 and a transit of Mercury, as well as a long paper on the Sun's parallax by William Smith, the provost of the College of Philadelphia (now the University of Pennsylvania).

In the early 1770s the APS sponsored an investigation of a canal linking the Chesapeake Bay to the Delaware River, thus bringing trade from Maryland to the port of Philadelphia. Surveys conducted for this study resulted in the publication of several detailed maps, but the project had to be aborted because of the impending war with Great Britain. This was but one bit of collateral damage inflicted on the development of mathematics in America in the period leading up to the Revolutionary War, as well as the immediate period afterward.

Another bit of collateral damage was the *Transactions of the APS* itself; the second volume was not issued for another 15 years. That 1786 publication was the only other volume to appear before Franklin's death four years later. The papers in Volume II were not divided into sections as in Volume I, but they followed in the same order, beginning with astronomy and ending with medicine. The topics of these investigations reflect colonial strengths in observational astronomy and geography, as well as medicine, but no theoretical work appeared until 1799 (Volume IV), when David Rittenhouse published a short paper on a method of calculating common logarithms. Overall, Rittenhouse dominated the mathematical and astronomical sections of the *Transactions of the APS* through the end of the eighteenth century. Two of his papers examined below dealt with pure mathematics. A third, "To determine the true place of a planet in an elliptical orbit, directly from the mean anomaly, by converging series," provided the most advanced work in applied mathematics in America in the eighteenth century.

The American Revolution took a toll on the pursuit of science generally in America. By 1782 the APS appeared on the verge of extinction, and one member described it as being "in a very languishing state."[15] Once again Benjamin Franklin came to the rescue. After having served as US Minister to the French Court from December 1776 to September 1785, he returned to Philadelphia and set about reinvigorating the APS. He made plans to raise funds for, and then to build, Philosophical Hall, which was constructed 1785–1789. Meetings had been held in various public buildings as well as the homes of officers theretofore. The first stated meeting in Philosophical Hall took place in November 1789. The next four volumes of the *Transactions of the APS* appeared in 1793, 1799, 1802, and 1809. By the time the next issue was published in 1818, numbered Volume I in a new series, mathematics in America had progressed materially.

Canada

Although Canada was explored before the colonies, European settlements did not expand as quickly there as those in the colonies, so mathematics lagged. Nonetheless, there were a few noteworthy developments that took place before 1800. This section first recalls a few facts from Canada's early history that place subsequent material in historical perspective. Of particular note is the strong influence of the French on Canadian mathematics and the important role that Jesuit priests played in this early period. Along the way, four notable characters are introduced—Martin Boutet, Louis Jolliet, Jean-Baptiste-Louis Franquelin, and Joseph-Pierre de Bonnécamps.

About 1000 years ago the Norwegian explorer Leif Erickson landed at Vinland, which is possibly today's aptly named Newfoundland. He and his fellow Norsemen (Vikings) thus arrived in America almost 500 years before Columbus, but their settlement did not survive long. Almost 500 years later, Portuguese explorers claimed Newfoundland and Labrador for their country shortly after Columbus's voyages to the New World, but these settlements too were short lived. In 1524 King Francis I (of France) sponsored Giovanni da Verrazzano to explore the Atlantic Coast from Florida to Newfoundland, resulting in several trading posts being established in modern day Canada, yet these too were short lived. Nevertheless, in 1608 the geographer Samuel de

Champlain founded the first permanent settlement in Nouvelle France (New France), settling in what is now Quebec City. Ultimately, an Iroquoian word for *settlement*, "Kanata," became the name of this new country.

The rise of British interests in North America over the next 150 years resulted in several skirmishes with the French headquartered in Quebec City. The last of the four "French and Indian Wars" ended with a decisive victory by Great Britain in 1763. In the treaty concluding that war, the French ceded all of Nouvelle France to the British.

In the American Revolutionary War 13 years later, New England invaded Quebec City with the hopes of liberating the city and thereby gaining allies in the struggle for independence. However, British troops managed to rebuff this charge in what is called the Battle of Quebec. A major byproduct of this conflict was a split of British North America in two—one English, the other French. Mathematics in the country would reflect this divide into the twentieth century. The authoritative source on the history of mathematics in Canada related the subject's development to the above political history as follows: "Mathematics has an old history in Canada ... and like the country itself it represents two cultures based on and evolving from distinct national traditions."[16]

In 1812, some 30 years after gaining independence, the US engaged in further hostilities with Great Britain. The 32-month military conflict has come to be known as "The War of 1812." When concluded in 1815, there were no changes in lands, but the conflict did resolve many issues between the US and Great Britain that had not been settled earlier. During this war, the US again attempted to gain control of Quebec City in order to enlarge its boundary northward, but this effort too failed. To prevent further incursions, the British built the Citadelle in 1820 and manned it until 1871. It has become a popular tourist attraction today. Canada had been formed back in 1840. The capital city varied until 1867, when Ottawa became its permanent capital.

Collège de Québec. As noted above, settlements along the Saint Lawrence River were established before either Jamestown or Plymouth, with Quebec City the largest settlement in Nouvelle France. Even though the population of this French colony grew slowly, a college was founded in 1635, one year before Harvard. However, the Collège de Québec did not last as long as the younger institution, closing in 1761, shortly before the final French and Indian War (the Seven Years' War).

In the meantime the small Jesuit college provided the only opportunity for higher education in Canada. During the first two years of its existence, the Collège de Québec offered the traditional French five-year curriculum. Generally, ten Jesuit priests lived at the college, with two in charge of all instruction; eight brothers assisted them. One of the two priests taught mathematics which was mostly elementary, as it was at Harvard, but it also included some commercial arithmetic. The aim was to teach applications of mathematics to surveying, cartography, and navigation.

That situation changed dramatically in 1659 when the Jesuits began offering a complete classical course of instruction that extended study for two more years. Similar to Jesuit colleges in France at the time, mathematics teaching was concentrated in these last two years. Despite an increasing population in the settlement and the concomitant rise in importance for the Collège, enrollment remained small. For instance, at this time only a dozen students matriculated, including pupils in the preparatory *petite école*.

The first mathematics professor at the Collège de Québec was the Frenchman **Martin Boutet de Saint-Martin** (c. 1612–1683), who sailed to Canada in 1645 with his wife and two daughters. Except for a return trip to France in 1677 to accept an honor decreed by King Louis XIV, he spent the rest of his life in Quebec City. Virtually nothing is known about his years in France, including his education. Boutet was also an accomplished musician and singer; one biographer wrote, "Mentally a mathematician, Saint-Martin was, emotionally, a musician."[17] His participation in activities at the Notre Dame de Recouvrance led to his appointment in 1651 as cantor and master of the children's choir at this parish church. This position brought with it a house in place of a salary; it was one of only two buildings, both wooden huts, in Quebec City at the time. This home served not only as the domicile for Boutet and his wife, but as boarding for some of the pupils in the Collège.

In 1661 Boutet began offering mathematics courses at the Collège de Québec that were oriented toward two important aspects of life along the Saint Lawrence River: surveying and navigation. The Collège's main mission was training for the priesthood, however, so mathematics instruction was not emphasized. One of Boutet's earliest students was **Louis Jolliet** (c. 1645–1700), the Canadian explorer known for mapping the Mississippi River. Jolliet enrolled at the Collège de Québec at age 11 and in 1662 took the minor orders to become a priest. However, after defending a thesis in philosophy five years later, he left the seminary to pursue fur trading. One year later, after a trip to France, this pursuit led him across Lake Michigan and down the mighty and, at that time, mysterious Mississippi.

Meanwhile, Martin Boutet was establishing a reputation as a particularly effective teacher. About 1666 he was asked to extend his courses beyond the Collège to include the training of pilots for navigating the Saint Lawrence River. As the historians Thomas Archibald and Louis Charbonneau observed, "Besides the chronic shortage of navigators, there was also the need for accurate maps. Once again the natural choice to provide such training was Boutet."[18] As a result, the chief administrator for Nouvelle France established a chair of hydrography and named Boutet its first professor. There, the root word "hydro" refers to the Saint Lawrence River and beyond, in both directions. Boutet became known as "the mathematician" of Québec, a position that extended over the whole French colony.

Due to slow communication with France, the official title of the position was not bestowed upon Boutet until 1678, when he was designated officially a professional engineer. This was the first time he received remuneration for the position. Moreover, since as royal engineer for Nouvelle France he was mandated to "teach hydrography, piloting, and other parts of mathematics,"[19] I regard him as the first mathematics professor in America, predating by more than 40 years the appointments of Tanaquil Lefevre at William & Mary and Isaac Greenwood at Harvard. To the best of my knowledge, Boutet did not become engaged in the disreputable misbehavior of the scalawags who followed him in the US. The chair of hydrography became the most eminent mathematical position in Canada throughout the eighteenth century and well into the nineteenth. Next, hydrographers who held the position from the time of Boutet's death around 1683 until 1763, when the French ceded the colony to the British, are introduced.

In 1687 **Jean-Baptiste-Louis Franquelin** (1652–1718) not only succeeded Boutet but received the title of "Hydrographe du roi à Québec," or Royal Hydrographer.[20]

Franquelin is mainly known today as a cartographer. Like Boutet, he was born in France but sailed to Canada at a young age (in 1671). Initially he was a fur trader, but after three years the Governor General of Nouvelle France, Louis de Buade Frontenac, persuaded him to devote his time to mapmaking, starting with a larger and more accurate version of the map Jolliet had drawn from memory of the Mississippi River. Apparently, Franquelin had studied cartography at one of the French colleges because he arrived in Canada with professional drafting instruments, paints, and brushes. He spent the next 19 years making detailed maps of the New World for the governor to send back to Paris. Perhaps the most impressive map depicted the area from Nouvelle France to the Gulf of Mexico and measured roughly 6 feet by 4-1/2 feet.

The need for a navigation teacher in Quebec became acute after Boutet died. Franquelin and Jolliet were recommended. Franquelin was selected, but he actually opposed the appointment because the salary was too low and his cartography business too lucrative. Nonetheless the governor ordered him to begin teaching navigation at once. He obeyed—of course. Earlier Franquelin had given private lessons in hydrography so he was well suited to the task of a Royal Hydrographer, a position he held 1687–1697. During this decade, however, he spent a number of years in Paris, leaving nobody behind to teach mathematics at the Collège de Québec. King Louis XIV was unconcerned with his absences because he was more interested in detailed maps of Nouvelle France and New England for use in a war against England, Holland, and Spain. In particular, Governor Frontenac was given instructions to attack Albany by land and Manhattan by sea, incursions that never materialized mainly due to the killing of several Canadian colonists by the Iroquois. Moreover, the king arranged for Franquelin's wife and ten of his 13 children to sail to France on one of his majesty's ships, but tragedy struck when the vessel hit a reef, foundered, and all aboard perished except for a few crew members. It appears that Franquelin never returned to Canada after this calamity. It is unknown when and exactly where he died. But in 1697, Jolliet returned from France to Quebec to succeed him.

Louis Jolliet was an accomplished explorer who had studied under Martin Boutet. Jolliet thus became the first Canadian-born Royal Hydrographer. However, he only held the post for three years because in May 1700 he left for Anticosti Island, a sparsely populated isle located at the outlet of the Saint Lawrence River into the Gulf of Saint Lawrence, and he is presumed to have died along the way. His body was never found.

I do not know if Jolliet ever taught mathematics courses during his tenure. In fact, instruction in the subject remained elusive for several years. Franquelin was appointed Royal Hydrographer a second time in 1701 but he never returned to Canada, so the position remained vacant until 1703 when Jean Deshayes was appointed. Little is known about this astronomer except that he died after serving for only three years. Earlier, in 1685, he had sailed to Quebec to observe a lunar eclipse, and upon his death 21 years later, he bequeathed 15 volumes that constituted the first scientific library in Canada, including a copy of l'Hôpital's famous 1696 calculus book.

The Jesuits. Jesuit priests had been offering hydrography courses at the Collège de Québec since at least 1700, so they undoubtedly taught the prerequisite mathematics subjects. In light of the recurring difficulties involved in filling the post of Royal Hydrographer, the governor petitioned the king to grant a Chair in Hydrography to the Jesuits at Québec. The request was sanctioned in 1708, and Jesuit priests held the professorship at the Collège from then until 1759. Generally each priest held the chair for about five years. During winters, the students resided in Quebec and took courses in geometry, trigonometry, physics, and naval theory. In summers they apprenticed as pilots.

The most well-known chair of hydrography was Father **Joseph-Pierre de Bonnécamps** (1707–1790), who taught at the Collège de Québec (1741–1759), stopping only when Nouvelle France fell to the British and the Collège was closed. Bonnécamps became a Jesuit novice in 1727, studied philosophy at the Collège de la Flèche (1729–1732), and theology at the Collège Louis-le-Grand (1739–1742). In the interim he taught at Jesuit colleges in Caen (1732–1736) and Vannes (1736–1739); all four of these institutions were in France. Upon being ordained in 1743, he was sent as professor of hydrography to the Collège de Québec, where over the next 16 years he taught courses in mathematics, hydrography, cartography, and astronomy. He published a memoir on the aurora borealis in a Jesuit scientific periodical in 1746 and reported on precise astronomical observations at present-day Kingston, Ontario, to the Paris Academy of Sciences six years later. In 1749 he also made the first map of the Ohio River country based on one of his expeditions. But ten years later he returned to France after the capture of Quebec, which resulted in a desperate situation in Canada.

The mathematics curriculum during the time that Bonnécamps was at the Collège de Québec consisted of two parts. Pure mathematics included arithmetic, algebra, geometry, and plane trigonometry. Applied topics (called "mixed mathematics") included practical geometry (measurement of length, area, and volume), mechanics, hydrostatics, spherical astronomy, and optics. Although this program sounds impressive, in reality the classical course led to only one end—the priesthood—and therefore mathematics played only a minor role in the curriculum. Moreover, all instruction was dictated, there were no exercises, and practical applications remained at a low level. One of Bonnécamps's students, the military engineer Michel Chartier de Lotbinière, was the architect of Fort Ticonderoga in upstate New York.

Clearly, mathematics had gotten off to a very slow start in Canada. The subject also wallowed under British rule well into the nineteenth century.

Founding Fathers

Generally today the term "founding fathers" of the US refers to political leaders. This section presents Benjamin Franklin and David Rittenhouse as founding fathers of American mathematics in addition to Isaac Greenwood and John Winthrop. Franklin and Rittenhouse were also founding fathers in the traditional sense as well. This section singles them out because their accomplishments should be known more widely. Benjamin Franklin, for instance, is celebrated for political leadership and experimental science, but this section describes his more obscure, but multifaceted, knowledge of a variety of different areas of modern mathematics. David Rittenhouse, on the other hand, is not well known even for his exploits in astronomy, let alone mathematics. This section cites two of his published papers as examples of historical topics that could be used in calculus and numerical analysis courses today.

David Rittenhouse. Arguably the most accomplished colonial scientist among all those considered in this chapter was David Rittenhouse, whose investigations in astronomy and mathematical publications exceed those of John Winthrop. A true polymath, Rittenhouse should be remembered today as the very first person in America to be associated with three different aspects of mathematics that are taken for granted today:

- publishing papers on mathematics *per se* in a scientific journal,
- applying mathematical reasoning to warfare, and
- receiving funding from the government for research.

His inner drive to pursue science and mathematics launched him into the forefront of those fields under conditions symptomatic of most colonial leaders—little formal education and only a small band of supporters. The emphasis here is Rittenhouse's contributions to mathematics, but the account begins with a brief review of his life and career. The online website "Web02-Rittenhouse" adds further information.

David Rittenhouse (1732–1796) was born on a farm near the Germantown section of Philadelphia, which was then a separate settlement.[21] He was generally self-educated, although a biographical piece celebrating the bicentennial of his birth suggests that as a youth he attended school at a Presbyterian church located near the farm.[22] Historian of science Silvio Bedini wrote that Rittenhouse "was fascinated with mathematics from his early years but, with little opportunity for schooling, was largely self-taught from books on elementary arithmetic and geometry and a box of tools inherited from an uncle, David Williams."[23]

Indeed, it was Rittenhouse's mother's brother David Williams who introduced him to mathematics and supplied him with books and mechanical devices. Williams possessed a copy of a translation of Isaac Newton's *Principia Mathematica* in English and tutored his nephew on its contents up to his death in 1744, when the youngster was only 12. Only later was Rittenhouse able to return to this classic volume, becoming, according to one biographer, "a mathematical disciple of Newton."[24] In the meantime Rittenhouse and his uncle were fond of solving mathematical problems posed in Franklin's *Almanacs*, a habit Rittenhouse maintained throughout his teenage years.

A big change in Rittenhouse's career occurred when he constructed a wooden clock at age 17. His scientific studies were aided by books from University of Dublin graduate Thomas Barton, who opened a lending library nearby and later married one of Rittenhouse's sisters. By 1751 Rittenhouse's reputation for building precision clocks had advanced to the point that his mother persuaded his father to provide him with sufficient tools and material to set up a shop at the edge of the farm. The shop was a financial success, with sales of precision mathematical instruments for astronomy and surveying. Moreover, his standing was enhanced by a mechanical model of the universe (called an "orrery") and a surveyor's compass for Colonel George Washington. His reputation for accuracy in surveying led him in 1764 to determine boundaries between various states, notably the dispute between the heirs of William Penn and Lord Baltimore. Mason and Dixon used his intricate calculations three years later when determining the dividing line between Pennsylvania and Maryland.

By 1769 Rittenhouse had built several precision instruments for observing the famous transit of Venus—an astronomical quadrant, an equal altitude instrument, a transit telescope, a zenith sector, and, of course, an accurate clock. The legislature

Figure 2.4. David Rittenhouse

in the Colony of Pennsylvania had allotted 100 pounds for Rittenhouse's scientific investigations, which thus became the first instance of an individual receiving a grant from a governmental agency for research. He carried out his observations from an observatory he had constructed at his farm about 20 miles outside Philadelphia.

Rittenhouse moved to Philadelphia upon the death of his first wife in 1770 because he expected to be appointed to a newly created position in the Land Office. But the legislature was totally absorbed with overarching political events that culminated six years later, so the position was never created. In 1771 he began delivering occasional lectures at what was then the College of Philadelphia (referred to hereafter as "Penn"), but he soon became ensnared in Revolutionary events along with Benjamin Franklin because their partisanship brought authority to the conflict.

Rittenhouse became the first American scientist to lend his mathematical expertise to a war effort in a couple of different ways during the Revolutionary War. For one, in late 1775 the Colonial Committee on Safety (later more accurately called the War Board) charged him with surveying the Delaware River to chart its depth and the nature of its shoreline to determine the best way to fortify the river against British attack and thus afford maximal protection for General Washington's army. Rittenhouse's mechanical skills, combined with increased acquaintance with explosives and ballistics, also enabled him to design and produce rifled cannons. This led him to experiment with telescopic sights by measuring the distances that cannon balls traveled in a manner that Oswald Veblen replicated and extended in World War I almost 150 years later. Like Veblen's advances at the Aberdeen Proving Ground, Rittenhouse's improvements on rifled heavy cannons could not be implemented during the war.

Even before hostilities in the Revolutionary War ended in 1783, however, Rittenhouse had entered academic pursuits. In 1780 he was appointed to the first American professorship of astronomy at Penn, a position he held for only one year. His major contributions to science took place between 1780, when he was 48 years old, until his death 16 years later. During this interval he published at least 20 papers over a wide range of fields—optics, psychology, magnetism, and electricity. One of his works on optics published in the *Transactions of the APS* is particularly noteworthy.[25] In 1785 the famed American author and jurist Francis Hopkinson (1737–1791) posed an optical problem to him. When his solution was read to the APS in February 1786, Rittenhouse "designed a plane transmission grating, made of fine wire fitted into a frame, that resolved a problem in the inflexion of light."[26] A recent book on optics ranks this discovery as one of the most important scientific advances right on up to today:[27]

> Two centuries after the discovery of gratings by D. Rittenhouse, gratings and more complicated periodic structures have become common not only in spectroscopy but also in numerous domains of physics, such as acoustics, solid state physics, nonlinear optics, X-ray instrumentation, optical communications, information processing. Moreover, gratings appear in the common life in CD players, as safety features on credit cards and banknotes, as well as in a variety of display and advertising applications.

Overall, Rittenhouse published only two papers on mathematics proper, but not until the 1790s, when he was in his 60s. Both appeared in the *Transactions of the APS* and are historically important. One included results on sine functions that translate into integrals and display a recursion formula for integrating the powers of the sine based on geometrical proofs, infinite series, and induction. Rittenhouse's other paper stated an algorithm for computing common logarithms, and his illustrative example shows a modern numerical analyst at work. Both works are examined for their inherent worth and because they provide examples for the use of the history of mathematics in the classroom today. Another paper, published posthumously, dealt with astronomy but made conspicuous use of convergent series; it is not explored here.

Calculus. David Rittenhouse read a paper at an APS meeting that was published as a short note devoted to mathematics *per se* in the Society's *Transactions of the APS* in 1793.[28] It is a marvel and still can be used as an example of the use of an original source in a calculus course. Titled, "Dr. Rittenhouse, to Mr. Patterson, relative to a method of finding the sum of the several powers of the sines, etc.," in modern terms it is equivalent to evaluating

$$\int_0^{\pi/2} \sin^n x \, dx \quad \text{for } n = 1, 2, \ldots.$$

Historians frequently exhort students to "Read the masters!"—and Rittenhouse was a master.

Like most modern authors of mathematics publications, Rittenhouse was terse and provided no clue to underlying scaffolding. Yet he provided enough evidence that, with over 200 years hindsight, his aim becomes clear. Moreover, his approach marks him as a disciple of Isaac Newton. The title of the paper belies the fact that the motivation for this study came from determining the times of vibration of a pendulum.

Yet Rittenhouse felt that the resulting problem was of interest in its own right. He wrote:

> I was induced to attempt the means of doing this solely by its usefulness, but in prosecuting the enquiry I found much of that pleasing regularity, the discovery of which the geometrician often thinks a sufficient reward for his labours.

This statement shows that Rittenhouse was a mathematician at heart, drawn to a problem because of the inherent beauty of the pattern in its solution. Earlier in his life, when his brother-in-law presented a set of mathematical problems to him back in 1771, Rittenhouse replied, "You cannot conceive how much I despise this kind of juggle, where no use is proposed."[29] As the quotation from the 1793 paper attests, he had changed his mind 20 years later.

Rittenhouse stated that he was able to prove that, in the first quadrant, the sum of the sines was equal to 1 and the sum of the squares of sines was $\pi/4$. In terms of definite integrals, as areas, these two sums become

$$\int_0^{\pi/2} \sin x \, dx = 1 \tag{2.1}$$

and

$$\int_0^{\pi/2} \sin^2 x \, dx = \pi/4. \tag{2.2}$$

I have taken the liberty of writing the result of the latter integral as $\pi/4$, whereas Rittenhouse expressed it as "1/2 multiplied by the arch of 90^2." This expression refers to the part of the circumference of a circle in the first quadrant, which is $\pi/2$.

Rittenhouse did not supply proofs of these two results but, in strict Newtonian tradition, he announced that he was in possession of geometric demonstrations. Next he listed the results of

$$\int_0^{\pi/2} \sin^n x \, dx$$

for $n = 3, \ldots, 8$. They are recorded in Table 2.1.

Rittenhouse bemoaned his inability to find geometric proofs of these cases: "I have not been able strictly to demonstrate any more than the two first cases. The others were investigated by the method of infinite series," meaning that he was able to evaluate the four definite integrals

$$\int_0^{\pi/2} \sin^n x \, dx$$

Table 2.1

n	Sum
3	2/3
4	$3\pi/16$
5	8/15
6	$5\pi/32$
7	16/35
8	$35\pi/256$

for $n = 3, 4, 5, 6$ using infinite series, an approach also discovered by Newton but regarded as less desirable than a strictly geometric proof. Calculus students today use trigonometric identities to evaluate these integrals. Once again, Rittenhouse omitted proofs, thus resulting in a particularly tantalizing circumstance for the modern reader who cries out for the method (or methods) used.

Next Rittenhouse stated that he obtained the results in Table 2.1 for $n = 7$ and $n = 8$ "By the Law of Continuation." This "law" probably means that he observed a pattern for $n = 1$ to $n = 6$ and extended it by induction to the next two cases. Based on today's methods for evaluating $\int \sin^n x\, dx$ by trigonometric methods, it is known that these cases must be taken in pairs because the trigonometric identities differ according to the parity of n.

Drawing from this evidence, Rittenhouse stated a rule for evaluating these sums in a manner reminiscent of directions for working with unit fractions as stated in the *Rhind Papyrus* from ancient Egypt some 3,600 years earlier:

> Make a fraction whose denominator is the index of the given power, and its numerator the same index, diminished by unity...; by this fraction multiply the sum of the next but one lower power, and we have the form of the given power.

Putting this rule into modern terms, Rittenhouse had thus discovered the recursion formula,

$$\int_0^{\pi/2} \sin^n x\, dx = \frac{n-1}{n} \int_0^{\pi/2} \sin^{n-2} x\, dx,$$

based on the generating integrals (2.1) and (2.2).

Yet apparently, Rittenhouse still yearned for geometric proofs. The first part of the title of the paper containing these results is relevant here: "Dr. Rittenhouse, to Mr. Patterson." This is in reference to Penn professor Robert Patterson. Rittenhouse concluded the paper with a plea for Patterson to supply proofs for all cases, writing, "Should your leisure permit you to give any attention to this subject I shall be glad to see you furnish a demonization for the 3^d, or any subsequent case abovementioned."

No response from Patterson is known. However, in 1816 **Eugenius "Owen" Nulty**, sent a letter to the Penn professor's son, Robert Maskell Patterson. Nulty was a member of the APS who was then at Dickinson College but was later a "calculator" for the US Coast Survey. The younger Patterson published Nulty's letter in the first volume of the new series of the *Transactions of the APS* started two years later. Nulty stated, "As this subject has not been resumed in any of the subsequent volumes of the transactions, I have thought the following investigation would not be uninteresting to you."[30] He then provided a complete treatment using standard integral notation, thus providing a glimpse of how quickly mathematics developed over the previous 20 years. Apparently, neither Rittenhouse nor Nulty was aware that the English mathematician John Wallis had evaluated these sums by 1655.

Numerical analysis. Despite ill health, Rittenhouse read his other paper on mathematics to the APS in August 1795, when he was president of the Society. However, it was only published posthumously in the 1799 volume of the *Transactions of the APS*.[31] The title describes its content, "Method of raising [calculating] the common logarithm of any number immediately." Rittenhouse's aim was to describe a procedure for evaluating $\log N$ for any number N correct to any desired degree of accuracy. The subject of logarithms might seem to reside on a lower level than calculus, but the description in the title belies the depth of his analysis and the height of the mathematics used to obtain the algorithm he discovered. Moreover, his use of the word "immediately" in the title is particularly misleading.

Rittenhouse was as terse as ever. Once again he began with a brief description, stated a rule, and then demonstrated it without providing a clue about the underlying reasoning. First, he defined the (common) logarithm of a number in terms of the characteristic (which he called the *index*) and the mantissa. Then he stated the rule I will call Rittenhouse's algorithm:

> If the number be greater than 10, divide it by the highest power of 10 that will leave the quotient not less than 1. The index of that power is the first figure, or index of the logarithm. Divide 10 by the quotient so found raised to the highest power that will leave the new quotient not less than unity. Divide the last divisor by the last quotient raised to its proper power, and proceed in this manner until a sufficient number of divisions are made, which will be when the quotient approaches nearly to unity. Make a compound fraction, taking the successive indexes of the powers you divide by for denominators and unity for numerators. Reduce this compound fraction to a simple one, and that by division to a decimal fraction, which together with the index first found (if any) will be the logarithm required.

To demonstrate this rule, Rittenhouse calculated $\log 99$ to nine decimal places. The interested reader might want to consult an article by the late Jesuit priest, **Father Frederick Anthony Homann** (1929–2011), who illustrated the Rittenhouse algorithm with $\log 20$ because the computations are much more manageable.[32] Instead, I supply details for the computations used in determining $\log 99$ by describing the first two iterations to illustrate the algorithm and to emphasize Rittenhouse's professional standing as a numerical analyst. The reader might want to refer to the two pages of computations from the original paper; those pages, as well as details of Rittenhouse's solution, are available online at "Web02-RittNumAnal."

To begin, $99 = 9.9 \times 10^1$ so the index (characteristic) is 1 and so $\log 99 \approx 1.\cdots$. I introduce modern notation to illustrate the iterative procedure that Rittenhouse adopted to approximate the mantissa 9.9 to nine decimal places using continued fractions. Set $Q_{-1} = 10$ (for the base of the common logarithm) and $Q_0 = 9.9$ (for the mantissa). Then define the quotients

$$Q_{k+1} = \frac{Q_{k-1}}{Q_k^{n_k}} \quad \text{for } k = 0, 1, 2, \ldots,$$

where n_k is the largest integer for which $Q_{k+1} \geq 1$. The mantissa for $\log 99$ will be the continued fraction $[n_0, n_1, n_2, \ldots]$.

By definition $Q_1 = \dfrac{Q_{-1}}{Q_0^{n_0}} = \dfrac{10}{9.9^{n_0}}$. Clearly, $n_0 = 1$ is the largest integer for which $Q_1 \geq 1$. Therefore, $Q_1 = \dfrac{10}{9.9}$. Now the calculations become tedious because the next step is to find the largest integer n_1 for which $Q_2 \geq 1$, where

$$Q_2 = \frac{Q_0}{Q_1^{n_1}} = \frac{9.9}{\left(\dfrac{10}{9.9}\right)^{n_1}}.$$

This was easy for n_0. But it turns out that $n_1 = 228$. This yields the continued fraction form of the second Rittenhouse approximation,

$$R_2 = \cfrac{1}{1 + \cfrac{1}{228}} = \frac{228}{229} = 0.995633188.$$

This result is accurate to five decimal places because $\log 99 \approx 1.995\,635$.

How did Rittenhouse conclude that $n_1 = 228$? His computations show a master numerical analyst at work. First he set $a = \dfrac{10}{9.9}$ and calculated a sequence of values for a^2, a^4, a^8, ... by squaring successive terms. The aim was to find the first term greater than the numerator 9.9 in Q_2. He stopped at $a^{128} \approx 3.6$ because he knew that $3.6^2 > 9.9$. Therefore, $128 \le n_1 \le 256$. The next step was to find appropriate powers of a^n for $n \ge 128$ whose sum remained less than the numerator 9.9. Rittenhouse found that $a^{128} \times a^{64} = a^{192} < 9.9$ and $a^{128} \times a^{64} \times a^{32} = a^{224} < 9.9$. He did not display that $a^{128} \times a^{64} \times a^{32} \times a^{16} > 9.9$ and $a^{128} \times a^{64} \times a^{32} \times a^8 > 9.9$ because mathematics papers do not supply such computations. Ultimately, he recorded $a^{128} \times a^{64} \times a^{32} \times a^4 = a^{228} = 9.889\,521 < 9.9$, leading to the conclusion that $n_1 = 228$.

It is especially impressive that all of these calculations were carried out by hand. Yet, how did Rittenhouse not become discouraged after the first several iterations of a^n? We will never know the answer because, like most modern mathematics papers, the author published only the finished product, not the underlying scaffolding.

Subsequently, Rittenhouse carried out similar calculations to obtain $n_2 = 9$, $n_3 = 2$, and $n_4 = 75$, producing the continued fraction approximation $[1,\ 228,\ 9,\ 2,\ 75\,]$. (For details, see the online file "Web02-RittNumAnal.") This means that the fifth Rittenhouse approximation for the mantissa of $\log 99$ is

$$R_5 = \cfrac{1}{1 + \cfrac{1}{228 + \cfrac{1}{9 + \cfrac{1}{2 + \frac{1}{75}}}}} = \frac{327\,103}{328\,537} = 0.995\,635\,194.$$

Thus, correct to nine decimal places,

$$\log 99 = 1.995\,635\,194.$$

Rittenhouse's closing statement, "3 [is] too much in the 10th [place]," reflects the style of modern numerical analysts to obtain bounds on approximations.

David Rittenhouse did not supply the reason why he desired logarithms to such accuracy, but logarithms were of great use to colonial surveyors and astronomers. Although he knew much of the mathematics carried out in England in the eighteenth century, he was apparently unaware that his algorithm had already appeared in a 1717 article by Brook Taylor in the *Philosophical Transactions* of the Royal Society of London.[33] As Rittenhouse's papers on the sums of powers of sines showed, he had a masterful command of series and thus it is surprising that he missed Taylor's work. Yet in 1954 the numerical analyst **Daniel Shanks** (1917–1996) discovered the algorithm independently and published it in the journal *Mathematical Tables and Aids to Computation* without knowing the work of either predecessor. Shanks began, "The method of calculating logarithms given in this paper is quite unlike anything previously known to the author and seems worth recording because of its mathematical beauty and its adaptability to high speed computing machines. ... This algorithm is based directly upon ... arithmetic continued fractions."[34] Shanks illustrated the algorithm by computing log 2.

Our colonial hero David Rittenhouse was one of only two Americans elected to the Royal Society between the Revolutionary War and 1800. (James Bowdoin was the other.) Rittenhouse would surely have been pleased to know of Shanks's independent discovery 150 years later. Based on these two papers, written while he was in his sixties, I regard him as the most accomplished colonial mathematician.

Benjamin Franklin. Attention now turns to one of the most complex characters in the history of America, **Benjamin Franklin** (1706–1790), who became a leading political figure in the colonies and then in the US after retiring at age 42 from a very successful, 20-year career as a printer. "*Words* may shew a man's Wit, but *actions* his Meaning," he wrote, an aphorism he followed by devoting the next 40 years to scientific study and public service. Like several notable politicos of the time, Franklin was known to dabble in mathematics, especially with magic squares. But the word "dabble" connotes amateurism. And indeed, that was how Franklin came to be viewed, even in highly respected circles. For instance, the renowned Stanford computer scientist Donald Knuth summed up such a view in his influential book, *The Art of Computer Programming*, where he opined that Franklin "was a *polymath* who excelled at everything *except* mathematics."[35] Knuth's charge appeared in the first edition of his book, but a more recent, detailed analysis of Franklin's writings by Villanova mathematician Paul Pasles compelled the venerable computer scientist to change his mind and vow to correct his assertion in the next edition.[36]

Generally Franklin is known for two accomplishments in mathematics—*Poor Richard's Almanack* and magic circles. Each item is discussed below, but in order to emphasize the mature mathematical mind he possessed and the variety of ways in which he applied it, this section next considers his contributions to four related fields:

- postulate theory,
- demography,
- social science, and
- applied mathematics.

Historians of mathematics generally list two areas in which American (pure) mathematicians first specialized around 1900—foundations and group theory. Those involved with foundations up to the 1920s have been labeled "postulate theorists" because of their work on axiomatics.[37] This field has a long and glorious history, going back to classical Greek times, when the geometer Euclid wrote his book *The Elements*. The content within this work, impressive as it is, does not alone account for its enduring influence. Rather, it is the structure of the enterprise. That structure, now known as deduction, refers to any system that begins with undefined terms, adds definitions of new terms and concepts in terms of them, makes a set of assumptions about these terms and concepts (called "postulates" by Euclid, "axioms" today), and finally uses logic to prove results about the terms and concepts based only on the assumptions and previously proved results. Franklin applied two aspects of this structure to the Declaration of Independence. Recent historical evidence has suggested that Thomas Jefferson began the initial draft of the document by writing, "We hold these truths to be sacred and undeniable." Although Jefferson too was versed in Euclidean geometry, it was Franklin, along with John Adams, who urged that "sacred and undeniable" be replaced with "self-evident," the wording Euclid adopted for his postulates. The Declaration next asserted that "all men are created equal." What does the term *men* mean? Is it generic for *people*, thus including women? Does it include African Americans? No, "men" remained an undefined term, and hence caused considerable difficulty over the next 200 years. Finally, there is the word "equal." Euclid went to great lengths in the first part of *The Elements* to define his use of this term in different settings, ranging from the generally accepted usage of being identical, to being congruent, and to being similar. So from this, I follow Pasles in concluding that Benjamin Franklin was a postulate theorist, perhaps the country's first, as reflected by the wording in and the structure of the Declaration of Independence.

Now consider Franklin as population statistician. In his 1755 essay, "Observations concerning the increase of mankind and the peopling of countries," he applied reasoning with exponentials to conclude that the population of the colonies would double every 20 years. Later he revised this estimate to 25 years based on additional information gleaned from further analysis. Isn't that typical in mathematical modeling? When the US census was first taken in 1790, it was revealed that the population had doubled every 23 years, so Franklin was not far off in his estimation (though by 1790 he was no longer able to refine his model). In 1770, 15 years after his essay and while serving as an ambassador of the colonies in England, he used his knowledge of the relationship between the linear increase in the British population and the exponential growth in the colonies to warn England about the custom of imposing onerous taxes on beleaguered colonists. He stated in unequivocal, humorous, and yet amazingly prescient terms, "the time will quickly come when the majority of subjects will be in America ... Britain will be taxed by an American parliament."[38] As is known now, though Franklin could not have known six years before the historic Declaration, the branch of American government dealing with taxes is the senate, not the parliament. However, one can be justified in heralding him as the founder of modern demography. In fact, he also used a very creative twist to estimate populations. In his *Almanack* he challenged his reader in these words: "Total of souls in 1737, 47369; ditto in 1745, 61403; increase 14034. *Query*, at this rate of increase, in what number of years will that province double its inhabitants?"[39] Here Franklin was approximating the population of New Jersey using numbers he obtained in a particularly clever and useful way. Living in Philadelphia, he knew that cemeteries there buried 2100 people over a seven-year period, which, given the total population, enabled him to conclude that the death rate is roughly one in 35. So he extrapolated his thinking to obtain appropriate figures for the neighboring colony.

I also view Franklin as a mathematical social scientist because of the way he used statistics to argue against the institution of slavery. It is true that at one point Franklin did own slaves, but later in life he argued against what he called "the notion of hereditary order."[40] The idea here is that when one person has a positive quality, the probability of passing that quality on to an offspring is $\frac{1}{2}$. Similarly, the probability of passing it to the next generation is $\frac{1}{2} \cdot \frac{1}{2} = \frac{1}{4}$. Continuing this line of thought, the probability of passing it to the third generation is $1/8$, to the fourth generation is $1/16$, etc. Thus, using mathematical reasoning, descendants have little right to claim benefits from great ancestors, and so slaves should not be passed from generation to generation as prized possessions. "*Words* may shew a man's Wit, but *actions* his Meaning." As confirmation of his changed viewpoint, the former small-time slave-owner became president of the Pennsylvania Abolition Society.

As well, Franklin serves as an example of an applied mathematical practitioner who solved real-world problems even when society might not know a problem exists. For instance, while serving in France, Franklin adopted mathematical reasoning to convince the French to adopt Daylight Saving Time, which made it easier to transport this notion across the ocean upon his return from Europe. How did he convince the French? With finances. He argued that Daylight Saving Time would afford one hour a day in which candles did *not* have to be lit. That translates into seven hours a week for 183 days (half a year). Multiply that by the half-pound of wax that was burned by the candle, for each of the 100,000 families living in Paris at the time, and you derive great savings. "A penny saved is a penny earned." The French adopted the proposal at once. His reasoning resounds with only one idea—the multiplication principle.

The account now returns to the two major areas in which Franklin gained his reputation in mathematics over the years, beginning with his famous almanac. *Poor Richard's Almanack* has been an American classic for almost 300 years even though its contents are very similar to many others from that time. Its sayings, however, are arguably wiser and wittier than the competition. Almanacs were popular in the country from the start; when the first printing press in America was constructed in 1639, it produced three works—the Massachusetts Freeman's Oath, a bible, and an almanac. Almanacs frequently appeared throughout the remainder of the seventeenth century, written by such notable figures as Samuel Danforth, John Sherman, Urian Oakes, and Thomas Brattle. Generally, they were the provenance of mathematicians and astronomers, but when produced by others, their astronomical calculations were often performed by local schoolmasters. Therefore, when Benjamin Franklin opened the New Printing Office in Philadelphia in 1728 with Hugh Meredith, he enlisted Thomas Godfrey to prepare calculations for the 1730 edition of the *Pennsylvania Almanack*. However, two years later, Godfrey sold his calculations to another firm, causing Franklin to hastily prepare his own. Written in 1732, it appeared the next year under the title *Poor Richard, 1733. An Almanack for the Year of Christ 1733*, and it was an instant success, selling upwards of 10,000 copies. It seems to me that the most important parts of the almanacs were the tables predicting solar and lunar eclipses (with comments about whether each would be visible in Philadelphia, and if so, when) and the tables giving information for each day (such as locations of the moon and planets, the length of the day, and the rising and setting of the sun and moon). Even Franklin's notation seems mathematical in spirit: he wrote 7 14 5 to denote that the sun rose at 7:14 a.m. and set at 14:46 p.m. (14 minutes before 5). While sunrise and sunset were given for each day, the moon appeared every fifth day, so he advised the reader to add 1-1/2 hours per day to obtain intervening times. This also serves as the type of interpolation problem that students used for logarithm and trigonometric tables before the advent of personal computers. The 24-page almanac, a handy pocket-size 3 × 6 inches, concluded with the distances of various points along the highway between two historically important cities, Philadelphia and Williamsburg.

Subsequent issues of *Poor Richard's Almanack* contained additional inklings into the workings of the mind of this closet mathematician. For instance, he wrote that the 22 letters "may be joined 5852616738497664000 ways."[41] Franklin expert Paul Pasles puzzled over this conclusion for a long time before discovering Franklin's method and uncovering an error in his computations. Pasles argues convincingly that Franklin was seeking the number of ways to arrange 23 distinct letters so that no letter repeats and no letter is omitted. Why 23 letters instead of 22? Because the English alphabet then consisted of 23 letters—J, W, and U were not included. Yet even with this analysis and the use of a computer, one does not get the same number of ways. Here Pasles admitted, "It took me some time to figure out what Poor Richard meant by [this] example." He added, "Someone, likely an inattentive apprentice [in the printing shop], must have miscopied one digit and outright deleted four others, including the one in front. For if these digits are restored:

$$\_5852616738\_\_497664000\_$$
$$\uparrow \quad \uparrow \quad \uparrow\uparrow \quad\quad \uparrow$$
$$2 \quad 0 \quad 8\,8 \quad\quad 0$$

then the result becomes 2585201673888497664000000."[42]

Sometimes Franklin has the reputation of being a 100% practical person, yet the question of permuting letters in the alphabet seems to serve no real purpose. Nor do perfect numbers, though they made their way into two issues of *Poor Richard*. Franklin noted that the first three perfect numbers are 6, 28, and 496, and then, in his inimitable style, in two consecutive almanacs, he challenged readers to calculate the next one: "Let the curious reader, fond of mathematical questions, find the fourth."[43] Incidentally, perfect numbers were not new at the time; they had been studied as far back as Euclid in about 300 B.C.

Magic squares grabbed Franklin's attention from the 1730s through the 1770s, yet it was not the oldest magic square that piqued his intellectual curiosity. According to ancient Chinese tradition, from perhaps as long ago as 2800 B.C., a turtle carried a special square from the river Lo to a man:

4	9	2
3	5	7
8	1	6

This is called a "magic square" because each of the three rows, three columns, and two diagonals adds up to the same number, in this case 15. About 250 B.C. a Chinese book called *Nine Chapters of the Mathematical Art* devoted one entire section to constructing them. But this work remained unknown in the West, so Franklin could not possibly have been influenced by it. What, then, was the source of his inspiration? It is not known. Indeed, there is nothing in the massive *Papers of Benjamin Franklin* (36 volumes, and counting) to suggest what inspired him. It is known that he achieved amazingly impressive feats not only with magic squares, but magic circles as well. Curiously, none of these figures ever appeared in *Poor Richard*. And while the source of this pastime remains unknown, Franklin's autobiography pursues the topic when he found time in his early years as clerk of the Pennsylvania Assembly:[44]

> I was at length tired with sitting there to hear debates in which I could take no part, and which were often so unentertaining, that I was induced to amuse myself with making magic squares, or circles, or any thing [sic] to avoid weariness.

The 8 × 8 magic square below is Franklin's "lost square," whose existence had remained elusive until located and identified by Pasles, who wrote, "Surely this is

[Franklin's] greatest numerical discovery."[45] This magic square contains properties that transcend the mere sums of rows, columns, and diagonals (with a "magic sum" of 260; for instance, the bent rows in **bold** forming a V at the top, and the bent rows in *italics* on the left and continuing onto the right have the same magic sum.) This example provides some idea of the profundity of Franklin's inventiveness.

2	57	6	61	8	63	*4*	**59**
7	**64**	3	60	1	58	**5**	*62*
49	10	**53**	14	55	**16**	51	12
56	*15*	52	**11**	**50**	9	54	13
42	*17*	46	21	48	23	44	19
47	24	43	20	41	18	45	22
25	34	29	38	31	40	*27*	*36*
32	39	28	35	26	33	*30*	37

In a letter to his friend Peter Collinson, Franklin demonstrated a 16×16 magic square he called the "most magically magical of any magic square ever made by any magician."[46] In another letter to Collinson he produced a magic circle.

Yet a different kind of influential contribution Franklin gave to mathematics was his essay "On the usefulness of mathematics," published in an issue of the *Pennsylvania Gazette* from 1735. However, in spite of extolling the virtues of mathematics, he opined that the only value of science was its practical side: "Nor is it of much importance to us to know the manner in which nature executes her laws. 'Tis enough if we know the laws themselves."[47] Clearly, he changed his opinion later in life.

Overall, this account of Franklin shows a remarkable ability and knowledge in pure and applied mathematics. I agree with Paul Pasles, who concluded:

> [Franklin] was adept at ... systematic and creative ways of thinking about numbers, arrangements, and relationships that characterize mathematical thought.

Indeed, Benjamin Franklin was a mathematical enthusiast of first rank.

David Rittenhouse and Benjamin Franklin were part of an extended community of mathematically able scientists in Philadelphia during the eighteenth century, including **James Logan** (1674–1751) and **Thomas Godfrey** (1704–1749). Logan was conversant with the work of Isaac Newton (on algebra and calculus) and the Dutch scientist Christiaan Huygens (on dioptrics). Godfrey, introduced in connection to *Poor Richard*, was known as a maker of astronomical instruments, including a quadrant that was a predecessor of the sextant.[48]

Even political figures participated actively in mathematics at this time. At a 1998 AMS conference on the History of Mathematics in America, held in Philadelphia, the late British historian of mathematics John Fauvel (1947–2001) delivered a lecture on the role of **Thomas Jefferson** (1743–1826) in popularizing mathematics with the clever title, "The declaration of American independence and other mathematical texts of the eighteenth century." A more recent essay on Jefferson's contributions to mathematics indicated that it was Jefferson who early saw the need for international collaboration in mathematics education, which would benefit a young country like the United States until it produced more indigenous teachers of high quality. An excellent source on Jefferson by the late historian of science, I. Bernard Cohen, captures Jefferson's mastery of mathematics and indicates the level of teaching at his *alma mater*, the Royal College of William and Mary:[49]

> At first [Jefferson] had simply worked on this part of the design [of a plow] by trial and error, but in correspondence with Robert Patterson, professor of mathematics at the University of Pennsylvania in Philadelphia, he learned that this was a problem requiring the calculus for its solution. Patterson told Jefferson that this very problem of a shape of least resistance had been explored in William Emerson's *Doctrine of Fluxions*. ... It was then that Jefferson told Patterson that he had actually learned the calculus of fluxions as a student by reading Emerson's textbook. Jefferson made use of Emerson's work and so the problem of designing a plow became in good part a Newtonian exercise in applied mathematics.

A recent discovery (2016) by Frank Swetz revealed:[50]

> [Jefferson] seemed fascinated with the geometric and structural features of an octagon ... [He] appreciated the functional aspects of the space it enclosed. [His] retirement retreat, "Plantation House," was laid out in the shape of an octagon.

While such public figures as Franklin and Jefferson may have contributed little directly to mathematics, they certainly appreciated the subject's importance and took pride in their ability to apply it. The tradition of American leaders playing a role in mathematics, albeit tangential, would continue through other public servants, including **James Abram Garfield** (1831–1881), for instance, who proved the Pythagorean theorem based on properties of a trapezoid, five years before becoming the US president in 1881. A recent article shows how the Cauchy–Schwarz inequality can be derived wordlessly using a similar approach.[51]

Surveying

By the middle of the eighteenth century, the need for mathematics in surveying, astronomy, and navigation created an increasing need for a working knowledge of algebra, geometry, and trigonometry. However, this material generally remained beyond the grasp of most educated colonists, causing a need to import two surveyors from England, Charles Mason and Jeremiah Dixon, to lay out their eponymous line between 1763 and 1767 in order to settle the long-standing dispute over the boundary between lands granted to William Penn (Pennsylvania) and Lord Baltimore (Maryland). When Mason and Dixon were unable to finish the job after three years, David Rittenhouse and Andrew Ellicott completed it. Both of these men would have preferred to spend more time with astronomical and mathematical pursuits, but instead they were kept busy with surveying, public office, and supporting their families.[52] Rittenhouse contributed by reducing data required for astronomical computations involved in its construction, another instance of his ability with numerical analysis. Ellicott will appear more prominently in the Chapter 3.

I cite here two other accomplished American surveyors who carried out surveys in the middle of the eighteenth century. The first president of the United States,

George Washington (1732–1799), can hardly be regarded as a mathematician, even by eighteenth-century standards. However, between the ages of 13 and 17 the teenager produced two cyphering books that provide evidence of latent mathematical ability. First, a word about terminology is in order. The word "ciphering," which was common in the eighteenth century, refers to arithmetical computation. I will adopt instead the convention due to the groundbreaking work of Ellerton and Clements of using "cyphering" in place of "ciphering."[53] Moreover, the Library of Congress employs the term "copy book," whereas for us it will be "cyphering book." Because textbooks were too expensive for most families in the colonies, even Washington's, those that could afford it, and his could, would purchase bound blank books that served as their own textbooks.

Jared Sparks (1789–1866), Harvard graduate, mathematics tutor, and Harvard president (1849–1853), compiled the first comprehensive edition of Washington's writings in 1837. Sparks recorded that Washington's early "exercises in arithmetic and geometry [indicate] the strong bent on his inclination to mathematical studies."[54] As a preteen, the future US president learned the arithmetic of whole numbers from the 1727 textbook *The Young Man's Companion* by William Mather. While his two cyphering books contain exercises on decimal numbers, there are none on integers because he already possessed the Mather book. Overall, Washington's two cyphering books recorded what he was learning about decimal arithmetic, geometry, trigonometry, logarithms, and surveying.

Without bothering with punctuation, Washington stated one exercise as follows:[55]

> If a Block of Marble be 9 foot long 19 Inches broad & 14 Inches Thick
> how many Solid feet doth it Contain

Calculations accompanying the exercise show that he initially worked in terms of inches by demonstrating that the block contains $(9 \times 12) \times 19 \times 14 = 28,728$ square inches. Because 1728 square inches compose a square foot, the answer is $\frac{28,728}{1,728}$ cubic feet. Washington expressed this volume as $16\frac{15}{24}$ cubic feet, having apparently reduced the fraction $\frac{1080}{1728}$ to $\frac{15}{24}$ on a slate but not including it in the cyphering book. It is not known why he did not reduce $\frac{15}{24}$ to lower terms, but we surmise that in practical surveying one could lay out a 15-inch part of a two-foot length. These computations show that Washington knew his 12-times multiplication table. In fact, this exercise shows that he could not be fooled by statements of problems that mix different units, a trap that some present calculus students fall into when exposed to optimization problems.

George Washington was granted a license as a surveyor in Virginia at age 17; most of his mathematical knowledge is associated with the needs of that profession. By its charter, the College of William and Mary was empowered to grant licenses to land surveyors in return for which the recipient was to pay one-sixth of the fees received. Such an appointment was equivalent to a degree in civil engineering. The prospective surveyor had to pass an examination, and Washington was probably examined in part by John Graeme, then professor of mathematics at the college.[56] The examination required knowledge of trigonometry and logarithms, two topics that were generally unknown in the American colonies, which leads to a description of the type of mathematics that was used in surveying at that time, as well as an examination of Washington's early education in mathematics.[57]

The first of Washington's two cyphering books begins with a 19-page section on geometry, with exercises like the one above, followed immediately by a section on surveying that appears to be written at a later time. Washington's approach here is to triangulate the region. He stated the appropriate rule more generally:[58]

> If your figure is not a Square or a Triangle it must be reduced to Squares or Triangles and the content found of every Square or Triangle Severally & added together for ye content of ye whole Plot.

In the early eighteenth century, most writers employed y^e for the word "the."

A painstaking analysis by the trio of Theodore Crackel (recently retired as editor-in-chief of the *Washington Papers*) and the historians of mathematics Fred Rickey and Joel Silverberg revealed that in the ten pages of Washington's cyphering books devoted to trigonometry, he stated the law of tangents and used it to determine the area of an oblique triangle when two sides and the included angle are given. This law, which has virtually disappeared from modern accounts, can be recast in modern notation as follows:

$$\frac{a + b}{a - b} = \frac{\tan\frac{1}{2}(\alpha + \beta)}{\tan\frac{1}{2}(\alpha - \beta)},$$

where a and b denote lengths of the sides of a triangle with opposite angles α and β, respectively. In another place, Washington evaluated $\log_{10}(\sin(35° \; 30'))$ entirely differently from what is done today because, up to about 1750, tables of trigonometric functions did not use the standard radius $r = 1$, but rather a radius of $r = 10^{10}$.

Washington never explained how he obtained his surveying answers. He often provided no clue as to how to obtain an answer. The three "detectives" Crackel, Rickey, and Silverberg believe that, in the practice of surveying, Washington eschewed trigonometry and logarithms in favor of physical measurements with dividers based on a scale of 100 perches to the inch. (There are 160 square perches to an acre, and four roods to an acre.) The trio concluded:[59]

> Washington followed the practice of surveyors of his time. He carefully made a scale drawing of the region, divided it into triangles, and then measured the sides and altitudes of the triangles so that he could approximately compute the sum of their areas. This was a practical and easy task compared to the cumbersome task of exact trigonometric computations. This contrast between knowing the theory but not using it in practice was common at the time among land surveyors in colonial Virginia.

To a modern reader, "This process seems tedious, but, in practice, it is quite easy and after one has done it a few times, it becomes routine."[60]

How did George Washington obtain such relatively advanced knowledge, having had a very limited education that did not produce one formal degree? From ages 3 to 8, he was tutored at home by John Hobby, a young convict who was transported to Virginia for stealing from a London silversmith. From about 1740, Washington attended a school taught by a local Anglican minister. It was expected that he would soon study at the Appleby School in England, like an older brother, but his father's death three years later prevented this. He then attended another nearby school until he almost reached the age of 16. During this time, he seems to have obtained a copy of the

Young Man's Companion; or, Arithmetic Made Easy, by William Mather. Washington's cypher books contain several direct quotations from this work. Like most textbooks from the eighteenth century, Mather's was aimed for the master, not the pupil, because books were expensive and scarce.

Washington treasured his cypher books and kept them for a lifetime, as they provided a ready reference for surveying. Indeed, 47 of the 179 pages (26%) in those books deal with surveying, suggesting he used them in preparation for the examination he took in July 1749 at age 17 that resulted in a commission as Culpepper County Surveyor. Over the next 3-1/2 years he conducted roughly 200 surveys until being appointed captain and adjutant for the southern district of Virginia. As is well known, this appointment changed the contour of Washington's career dramatically, resulting in his role as General of the Army in the American Revolution and later his election as the first president of the country for two terms (1789–1797). In spite of the demands for Washington's time and energy, he still managed to perform surveys throughout his lifetime, particularly on his own homestead at Mount Vernon. In 1785, four years before being elected to the highest office, he began surveying land on Four Mile Run but "after having run one course & part of another, My Servant William (one of the Chain Carriers) fell, and broke the pan of his knee wch. put a stop to my Surveying."[61] Finally, in November 1799, the month before he died at age 67, George Washington carried out a two-day "Survey [of] my own Land" on Difficult Run in Fairfax County.[62]

Benjamin Banneker. The other notable figure in surveying, **Benjamin Banneker** (1731–1806), is sometimes referenced for his surveying exploits, yet his work in astronomy was based on knowledge of mathematics. Banneker serves as another example of a "rugged individualist," a mostly self-educated colonist who basically worked without the support of a community. In many respects he resembles his contemporary, David Rittenhouse, with one critical difference: Banneker was black. He is shown on a US stamp honoring his contributions to American science. Banneker's heritage enables us to discuss conditions which faced not only some of the slaves who were brought to America but also some of the English lower-class of whites.[63] His maternal grandmother was the white woman Molly Welsh, who was found guilty of theft and shipped to the American colonies and sold as an indentured servant. She worked for seven years to pay for the cost of the voyage, gaining her freedom around 1690. She rented a small farm in Maryland and two years later purchased two slaves but, opposed to slavery, she freed them within three years.

In 1696, Molly married one of them, Robert Banneky, a dangerous action because of strict miscegenation laws prohibiting mixed marriages. The couple remained on their secluded farm and raised four daughters, with security always in jeopardy. Banneky died at a relatively young age, leaving Molly to raise their four daughters. The second, Mary, Banneker's mother, was born around 1700. Although Molly had a very fair complexion and blonde hair, all of her children were black. About 1730, Mary married Robert, an African slave who had gained his freedom from a wealthy but liberal master. Having no surname, he adopted hers and became Robert Banneky. I do not know when the name became Banneker. Their first son, Benjamin, was born there the next year (1731). While he was still a child, his parents purchased a small tract of farm land that had been part of a large plantation.

Figure 2.5. Benjamin Banneker

Benjamin Banneker's grandmother Molly taught him how to read and write at an early age. He was precocious and evinced a great interest in statistics and in all mathematical endeavors. He attended a one-room country school run by a Quaker schoolmaster during winter months, where several white children attended as well as two or three blacks. But as soon as he was old enough to work full-time on the farm, he was forced to stop attending classes. After that, he had to educate himself.

Like David Rittenhouse, Benjamin Banneker became equally interested and skilled in mechanical craftsmanship as much as in mathematics. Rittenhouse constructed a wooden clock at age 17; Banneker at 22. Both viewed their achievement as a mathematical puzzle requiring precise calculations for the relative size and number of teeth for each of the wheels and pinions. While Rittenhouse's achievement of untutored craftsmanship thrust him into a career, Banneker was restricted by farm duties that only increased when his father died six years later. Full responsibility for the management of the farm then fell entirely in his hands because he was the only male in the family.

A fortuitous turn of events occurred when two brothers, John and Andrew Ellicott, moved into the area in 1771 to purchase land to build mills. Occasionally, Banneker ventured to the project to observe how the milling operation was completely automated. In this way Banneker met with members of the extended Ellicott family at a makeshift country store they constructed nearby.

The Revolutionary War played no role in Banneker's life, but the Ellicott men took up arms despite their Quaker beliefs. While the older men were away, Banneker became friendly with George Ellicott, son of the founding father Andrew. By 1788 George became aware of his 57-year-old friend's ability and interests, and opened his library to him. Banneker read voraciously; until then the only book in his house was a Bible he had purchased. George Ellicott also introduced Banneker to instruments and telescopes to accompany the astronomy texts among his holdings. He also provided him with a massive table to hold the telescope for observations and a candle holder to supply light when recording times during these nightly sightings.

By this time, Banneker's grandmother and mother had died, leaving him in charge of his sisters, yet, like Rittenhouse, he conducted his own self-education whenever the opportunity presented itself. He set out to make a projection of a solar eclipse, which required mastery of logarithms needed for the calculations. When Ellicott saw how quickly Banneker had advanced, he encouraged him to calculate an ephemeris for an almanac, a project that required more complicated calculations. (An *ephemeris* is a table giving the coordinates of celestial bodies at given times over specific periods. The plural is *ephemerides.*)

Banneker, like all farmers, was familiar with almanacs, which played an important role in colonial family life. By late 1790 he completed an ephemeris for each month of 1791. He contacted three publishers about printing his calculations, but none consented. Serendipitously, his dealings with one publisher led to correspondence with Andrew Ellicott, who was back in his native Philadelphia. This, in turn, brought Banneker to the attention of the Quaker "Society of Free Negroes unlawfully held in Bondage." This instrumental group sought to publicize Banneker's unusual achievement for its value in demonstrating the mental capability of a black man when given sufficient opportunity. Unfortunately, by the time the Society was able to campaign for publishing the almanac, it was too late in the year, and the project had to be postponed.

By this time Banneker was almost 60 years old. It might have seemed that this setback would have marked a dead end on his meager academic road, but a fortuitous turn of events occurred when Major Andrew Ellicott was commissioned to survey a ten-mile square of land for the new capital of the US. In February 1791 Ellicott asked his cousin George Ellicott to be his assistant, but George pleaded he had no time because of business pressures and recommended Banneker instead. Furthermore, Secretary of State Thomas Jefferson also encouraged Andrew Ellicott to employ Banneker, though the future president had never met Banneker. In spite of having no familiarity with surveying, and no field experience, Banneker was hired as Ellicott's assistant because of his knowledge of astronomy and mathematical skill. It is noteworthy that neither Banneker's race nor age prevented his hiring. Benjamin Banneker thus became Andrew Ellicott's assistant for two months in what was unquestionably the most important surveying project of the time.

Banneker soon learned that he had to digest a great deal of astronomy and perform precise numerical calculations for determining the exact boundaries of the new city. By the time he departed the project in April, his laborious calculations had helped determine the first stone marker on the trapezoidal boundary. By the end of 1791, another 13 stones were placed on the Virginia line, and by the end of the next year all 26 stones rested on the Maryland side. A Georgetown newspaper reported:[64]

> Mr. Andrew Ellicot [*sic*]…is attended by Benjamin Banniker [*sic*], an
> Ethiopian, whose abilities, as a surveyor, and an astronomer, clearly
> prove that Mr. Jefferson's concluding that race of men were void of
> mental endowments, was without foundation.

Perhaps this notice accounted for Banneker's reputation as a surveyor. In reality he played a minor role in the project, which served as his only experience with surveying. When he returned home he turned full attention to astronomy, the field where he made a more important mark.

With renewed vigor Banneker set about performing calculations for producing an ephemeris for 1792. He purchased a new manuscript journal of 300 pages to record his observations; it remains the only source of his work besides the almanacs he published. Working feverishly and ignoring farm duties, Banneker completed the ephemeris by the beginning of June 1791. Then he began searching for a publisher. This time Banneker received support from high places. A prominent member of the Pennsylvania Society for the Abolition of Slavery, James Pemberton, who learned of the work from George Ellicott's brother Elias, supported publication of the almanac to champion the antislavery cause. Seeking advice from a competent scientist, Pemberton forwarded the copy Banneker had sent him to David Rittenhouse. As noted above, Rittenhouse was working on exactly the kind of mathematics as Banneker, mostly for astronomy. He was impressed. "I herewith return to you a very extraordinary performance, considering the Colour of the Author," Rittenhouse replied to Pemberton. He added, "Every Instance of Genius amongst the Negroes is worthy of attention, because their oppressors seem to lay great stress on their supposed inferior mental abilities."[65] William Waring, a second authority, agreed with Rittenhouse's assessment. Waring was a mathematics teacher and popular scientist in Philadelphia.

This support by Rittenhouse and Waring interested Philadelphia publishers in producing the almanac, but they demurred because in the meantime Banneker had arranged with a Baltimore publisher to print the work. Even to get to that point, however, Banneker needed the help of a very prominent citizen.

Despite 60 years of modesty, prudence, and dignity, Banneker took a proactive role in advancing his almanac by sending a copy to Thomas Jefferson, who knew about Banneker's role in surveying the District of Columbia. In his accompanying letter, Banneker likened slavery to the enslavement of the American colonies by the British Crown. Jefferson, who was then in the nation's capital of Philadelphia, replied in only four days' time. He was so impressed with the work that he forwarded his copy to the Marquis de Condorcet, one of France's foremost mathematicians, to have it reviewed by a committee within the Royal Academy of Sciences in Paris. Unfortunately, Condorcet was frantically involved with revolutionary politics at the time—he committed suicide in prison in 1794—and he probably did not have the opportunity to present Banneker's ephemeris to the Royal Academy.

Nonetheless, the support of Jefferson encouraged a Baltimore publisher to print Banneker's almanac, and the agreement to publish it had been settled before the Philadelphia publishers had become involved. Ultimately, two separate editions of the *Almanack and Ephemeris for the Year of our Lord, 1792* appeared. The 24-page edition published in Baltimore was a huge success; its first printing sold out, requiring a second. A Philadelphia publisher printed a shorter edition (18 pages), omitting an introduction by Maryland Senator James McHenry (of Fort McHenry fame) that described the author and his career.

Once Banneker's almanac appeared, and especially since it was a commercial success, several publishers contacted him about printing future almanacs. Ultimately, his ephemerides were published for the years 1792–1797. Banneker's work for 1793 was also published in separate editions in Baltimore and Philadelphia. This time his exchange of letters with Thomas Jefferson was included. As well, a plea by renowned

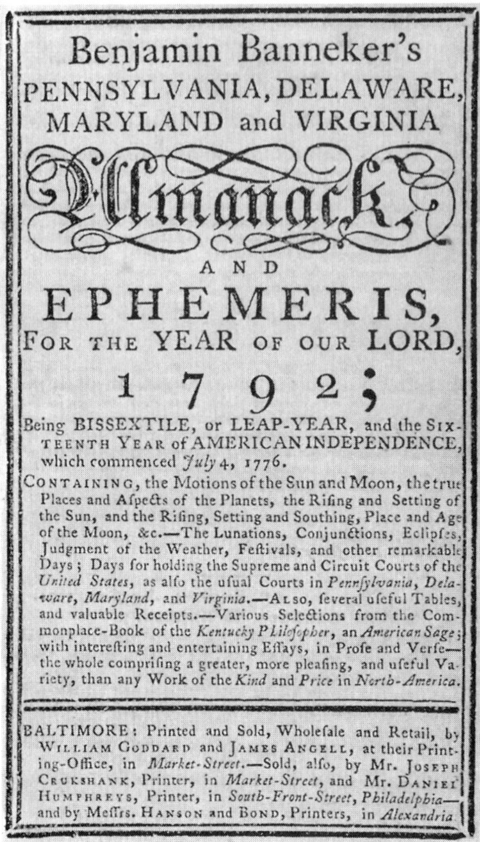

Figure 2.6. Title page of Baltimore edition of Banneker's 1792 almanac

physician Benjamin Rush for the new republic to establish a Peace Office was also included, adding prestige to the publication. This almanac seems to be the most important of the six that Banneker published, and it vastly outsold its competitors. Copies were sent to England with an ephemeris for that country as well. Demand for these works reached a peak in 1795, when nine separate editions appeared up and down the Atlantic coast. The Baltimore publisher for 1795 cast a purported portrait of Banneker, based on a woodcut bust, on the cover of that edition.

By the time Banneker's final almanac appeared in 1797, he was somewhat rich and famous. Yet he still lived alone on his farm amidst very modest surroundings. He was 66 years old and his many years of hard labor had taken a toll on his body and spirit. Although Banneker continued to record calculations for ephemerides in his manuscript journal, he did not publish any of them.

In light of limitations on the precision of his results due to the inexact nature of the astronomical instruments Banneker used, it is not surprising that errors crept in. Yet a 1987 study comparing his calculations with those compiled by his three main competitors—including Andrew Ellicott and William Waring—concluded, "Banneker's ... data compared very favorably with that published by his contemporaries."[66]

The most desirable and useful feature of Banneker's almanacs was the inclusion of tide tables he produced not only for the nearby Chesapeake Bay but for ports as far away as Boston and Halifax. These tables were particularly useful for river pilots, but their construction required virtually no mathematical expertise. Consequently, to examine Banneker's ability in mathematics, it is germane to examine his manuscript journal.

Banneker's house burned to the ground on the day of his funeral, thus destroying his laboratory, the clock he had made at age 22, and all records of his achievements except the one manuscript journal that was not in the house at the time. Banneker used this journal for all kinds of purposes, including details of the huge amount of calculations required to determine even a single eclipse—at least 68 for each eclipse. But, of more interest for us here, he also included mathematical problems he created as a challenge for others and those others had posed to him. Consider one particular solution of a problem posed to him by Andrew Ellicott. Banneker's journal states the problem and its solution in the following form:

> Divide 60 into four Such parts, that the first being increased by 4, the Second decreased by 4, the third multiplyed by 4, the fourth part divided by 4, that the Sum, the difference, the product and the Qutient [*sic*] shall be one and the Same Number—
>
> Ans. first part 5.6 increased by 4 ( ) 9.6
> Second part 13.6 decreased by 4 ( ) 9.6
> third part 2.4 Multiplyed by 4 ( ) 9.6
> fourth part <u>38.4</u> divided by 4 ( ) 9.6
> 60.0

This problem would seem to be an appropriate example for a high-school algebra course by searching for numbers w, x, y, and z satisfying

$$60 = w + x + y + z,$$

where

$$w + 4 = x - 4 = 4y = z/4 = N$$

for some number N.

Although Banneker had taught himself algebra (as well as geometry, trigonometry, and logarithms) needed for his astronomy calculations, insufficient information is given to be able to solve this problem using algebra alone. Moreover, his manuscript journal provided only the statement in the form presented above, with the solution given only implicitly in what appears to be a check of the answer. A recent analysis by Beatrice Lumpkin suggests that Banneker used the method of false position, because he was known to be familiar with this approach.[67]

A solution to the problem using this method can also be used as an application of the history of mathematics in the classroom, especially since the method of false position is no longer part of a typical curriculum. Roughly, the procedure is to guess a solution, calculate what students call "a fudge factor," and multiply this factor by the guess to produce the desired outcome. For instance, guess $N = 10$. Then from the second set of equations one easily gets $w = 6$, $x = 14$, $y = 2.5$, and $z = 40$. Then

$$w + x + y + z = 62.5.$$

This sum is larger than the desired one in the first equation, leading to a fudge factor of $\frac{60}{62.5}$. Therefore the desired solution is $\frac{60}{62.5} \times 10 = 9.6$, which can be seen in Banneker's statement of the answer in the journal.

Banneker, and several of his supporters, wanted to show by example that African Americans were as intellectually gifted as any other people. Solutions like this one were particularly influential since Banneker was basically self-taught. However, one other example of an African American with innate mathematical abilities preceded him by a few years.

Thomas Fuller. The Quaker Benjamin Rush printed his appeal for a Peace Office within the federal government in Banneker's almanac for 1793. Rush was a renowned physician, psychiatrist, professor of chemistry, and signer of the Declaration of Independence. But he was also an active member (and secretary) of the Pennsylvania Society for the Abolition of Slavery. It was this latter role that induced him to record an account of William Hartshorne and Samuel Coates, "two men of probity and respectable character,"[68] who traveled to Virginia about 1780 to test the arithmetical ability of a 70-year-old African American slave.

The calculating savant was **Thomas Fuller** (c. 1710–1790), who "was probably one of the 150,200 slaves shipped in the 1720s from somewhere between Liberia and Benin."[69] At age 14 he was brought to the Colony of Virginia, where he was bought and lived for the rest of his life. Although unable to read or write, Fuller gained a reputation for a remarkable ability to perform arithmetical operations. Hence there were numerous offers to buy him. Nevertheless, as Benjamin Rush wrote, "He spoke with great respect of his mistress, and mentioned in a particular manner his obligations to her for refusing to sell him, which she had been tempted to do by offers of large sums of money, from several curious persons."[70]

By 1780 Fuller was totally gray-haired and said that his memory had begun to fail him when Hartshorne and Coates visited him to ascertain whether the rumor about his remarkable ability was correct. They posed three questions. First, "How many seconds are there in a year?" Within two minutes Fuller answered 47,304,000. Second, "How many seconds has a man lived who is 70 years, 17 days, and 12 hours old?" It took Fuller only 1-1/2 minutes to reply 2,210,500,800. One of his questioners responded that this total was too large, whereupon Fuller corrected him—the questioner had neglected to account for leap years. The third question was, "If a farmer has six sows, and each sow has six female pigs, the first year, and they all increase in the same proportion, to the end of eight years, how many sows will the farmer then have?" Besides the complexity of the calculations required to answer this question, its solution requires the kind of innate knowledge of the multiplication principle that Benjamin Franklin had exhibited. That might explain why it took Fuller ten minutes to answer the question, but actually the longer time was due to the fact that he did not understand the wording of the question at first. His final answer of 34,588,806 was entirely correct.

Clearly, Thomas Fuller was a calculating savant. But what kind? Was he like the famous Karl Gauss, who demonstrated extraordinary arithmetical ability as a child and later developed into one of the most accomplished mathematicians in the history of mankind? Or was he like *idiot-savants*, who demonstrate no other mathematical insight or creativity? Or did his ability lie somewhere between these two extremes? As the historians John Fauvel and Paulus Gerdes concluded, Fuller's "exceptional abilities cannot be understood except through closer examination of the cultural context that stimulated their development."[71] To date no such examination has been carried out.

Benjamin Rush, however, had found the evidence he needed. Fuller's performance, as well as another one in front of two other questioners on a separate occasion, provided Rush with sufficient evidence to counter the charge that blacks were intrinsically, naturally, inferior to whites. Rush found an even better example in Benjamin Banneker, not only with regard to arithmetical calculations but in his mastery of mathematical concepts as well.

The questions posed to Fuller and the problem solved by Banneker provide two examples of historical material that can be used profitably in a classroom today. Taken together with earlier examples due to John Winthrop, Isaac Greenwood, David Rittenhouse, and Benjamin Franklin, it becomes clear that mathematics carried out by colonists remains relevant even after the passage of more than two centuries.

Canada. In the realm of surveying, the French Canadians fared better than the Americans. The Collège de Québec established a chair in hydrography to deal with problems arising from drawing accurate maps and fixing land boundaries. As well, in the first half of the eighteenth century, Jesuit priests who held the chair taught courses in geometry, trigonometry, physics, and naval theory. Unfortunately, the Collège was shuttered when the Jesuits were expelled in 1760, thus putting an inglorious end to what had been a promising start to higher education in Canada. Of related interest, it was not until 1807 that the US federal government got into the business of drawing accurate maps and fixing boundaries when it established its first federal institution, the US Coast Survey, in Washington, DC.

Cryptology

The term "cryptography" basically refers to what can be called "encoding," where a message (specifically, "plaintext") is transformed into a secret form that anyone can see but only the recipient knows how to decode. The contrasting term, "cryptanalysis," refers to the task of determining the inverse transformation by someone who does not know the key to the encoded message. "Cryptology" is the umbrella term for cryptography and cryptanalysis. The early cryptologists mentioned here were concerned with both aspects of the subject.

Cryptology began in America before there was a United States. It involved two traitors. In September 1775 General George Washington was handed an encrypted letter sent by Dr. Benjamin Church to a female acquaintance. The letter appeared to be suspicious, so Washington entrusted two patriots with trying to decrypt it. One was Elbridge Gerry, the fifth vice president of the US, known today for the word "gerrymander." The other is more important.

Samuel West (1730–1807) was born, raised, and educated in the Boston area, being prepared for the entrance examination to Harvard College by his minister. West's performance was so impressive that he was awarded a Hollis scholarship, graduated in 1754, and became a school teacher and preacher until being forced to retire from ministerial duties in 1803. It is not known how he became interested in cryptology, although he had studied natural science at Harvard, and later in life he was one of the charter members of the American Academy of Arts and Sciences. Nor is it known how Gerry became accomplished in cryptology. In 1788 West was elected a delegate to the Massachusetts Convention to ratify the new US Constitution, where his Harvard classmate John Hancock presided. It seems ironic that Church was another member of that class of 1754.

But in September 1775 General Washington chose West and Gerry to decode Church's message. They succeeded admirably. Their solution indicated that Church was sending information to Thomas Gage informing the British commander of colonial proceedings. Accordingly, he was tried for treason. In defense, Church argued that he supplied this information in order to gain the trust of the Tories. The damning piece of evidence, however, was the final sentence, which read, "Make use of every precaution or I perish." The final two words were prophetic because, based on the cryptanalysis, Church was found guilty of treason, imprisoned, and then exiled to the West Indies. However, his small schooner apparently capsized at sea, making Benjamin Church the first American whose demise was caused by cryptanalysis.

The other Tory traitor was the more well-known Benedict Arnold, who apparently handled his own encoding duties. Although no patriot was able to decode his clandestine correspondence about the betrayal of West Point to the British, his loyalties were revealed in other ways.

The tasks of coding and decoding were carried out by both the British and the colonists, as well as the French, throughout this period. Perhaps the most important codebreaker during the revolutionary era was **James Lovell** (1737–1814), whom historian David Kahn said "may be called the Father of American cryptanalysis."[72] Like Samuel West, Lovell was a Harvard graduate (1756) but he obtained excellent preparation beforehand at the Boston Latin School, where his father was headmaster. As often happens, war within a country pits family members on opposite sides of the battle; James Lovell was a Whig, his father a Tory. The son was, in fact, arrested as a dissident in 1775 and imprisoned in Boston for nine months before being transferred to Halifax, NS, where he spent another nine months in jail. Lovell was exchanged for a British colonel and returned to Boston in December 1776. He was then elected to Congress, so he spent the next six years (allegedly as a womanizer) in Philadelphia, a location that proved fortunate for the cause of the patriots. In the fall of 1781, James Lovell was able to decode messages exchanged between British commanders Cornwallis and Clinton in their campaign to conquer Virginia and the Carolinas. Unfortunately, by the time the decrypts reached the proper military authorities, the tactical situation had changed too much for the information to be of practical use.

However, Lovell's most important contribution occurred shortly thereafter, in October 1781, when one of Clinton's boats delivering a message to Cornwallis was captured near Little Egg Harbor in present-day New Jersey. The intercept was delivered to Philadelphia, some sixty miles away, where Lovell was asked to decode it. Fortunately, the key was the same as the earlier messages, enabling him to read it at once. The information gleaned from the decoded message allowed colonial militia to prevent Admiral Graves and General Clinton from relieving Cornwallis. In the packet that Lovell had sent with the decoded message, he wrote, "If they [the British forces] fail, it appears here that they are disposed to give up the contest for North America."[73] This statement was prophetic. Cornwallis surrendered five days later, an event that signaled the final victory in the US war for independence.

Some prominent early Americans were also interested in cryptology. Thomas Jefferson, for instance, devised a cipher wheel that was the most advanced in its day.

When he was in France (1785–1790), he and James Madison encoded part of their correspondence. Jefferson also used invisible ink extensively in his coded messages. Here a method due to Benjamin Franklin is cited, yet another indication of the mathematical ability of this founding father. In 1781 Franklin sent a message from France that began "I HAVE JUST RECEIVED A 14, 5, 3, 10, 28, 50, 76, 203, 66, 11, 12, 273, 50, 14." He included a long passage in French that began "voulez-voussentirladifferenc." The recipient of the message had to know the key, which assigned consecutive integers to the French text consisting of 682 letters and punctuation marks. The 2×28 matrix below is a modern way of understanding the key.

v	o	u	l	e	z	-	v	o	u	s	s	e	n	t	i	r	l	a	d	i	f	f	e	r	e	n	c
1	2	3	4	5	6	7	8	9	10	11	12	13	14	15	16	17	18	19	20	21	22	23	24	25	26	27	28

In the message, 6 corresponds to z, while 5, 13, 24, and 26 correspond to e (which effectively nullifies an easy statistical analysis of Franklin's method). Because the French alphabet contains no w, Franklin used consecutive u's to denote this letter. Thus 14, 5, 3, 10 represents "new." I leave it as an exercise for the interested (or tantalized) reader to decode 28, 50, 76, 203, 66, 11, 12, 273, 50, 14, knowing that

o	i	m	i	j
50	66	76	203	273

Confederation

In the post-war period the most important social/political document was the Constitution, which is relevant to these proceedings. The Constitutional Convention met at Independence Hall in Philadelphia in May 1787 with the aim of revising the Articles of Confederation because, as many patriots recognized, it did not provide a sufficiently durable model for the new country. After a month of discussion and debate, the delegates came to a mathematical conclusion: instead of merely amending the Articles of Confederation (the present set of axioms), it was incumbent to replace them with an entirely new document (a different set of axioms). The delegates spent the entire summer drafting a new frame of government, eventually reducing the document to four pages that were signed into law on September 17, 1787. The preamble reads:

> **We the People** of the United States, in Order to form a more perfect Union, establish Justice, insure domestic Tranquility, provide for the common defense, promote the general Welfare, and secure the Blessings of Liberty to ourselves and our Posterity, do ordain and establish this Constitution for the United States of America.

Seven articles followed the preamble, thus forming the axiomatic system for the government of the US. The list of axioms has grown since then.

There were only two notable events in the development of American mathematics from the end of the Revolutionary War until 1800. The first was the founding of the second scientific organization in America in 1780 and its official journal five years later—the American Academy of Arts and Sciences and its *Memoirs*. The other was a notable textbook published by Nicholas Pike in 1784. In general, the state of mathematics in the young republic was woeful. Before turning to the American Academy and Pike, however, in this section we consider changes in American colleges and introduce two important figures who deserve mention besides Pike—Walter Minto and Robert Patterson.

Colleges. At least 16 colleges were founded in the US during 1782–1800, joining the nine that had been established from 1636 to 1776. As the historian Brook Hindle commented, "The newest element in the collegiate picture was the interest demonstrated by the state governments."[74] Of particular note were the University of Georgia in 1785 and the University of North Carolina four years later. Yet none of these 25 institutions exerted much influence on American science generally, mainly because so few students attended college, resulting in very few professorships. Even the three oldest colleges in the US stagnated. As noted in Chapter 1, no significant mathematical activities took place at the second oldest college in the US, the Royal College of William and Mary. Similarly, mathematics at Harvard encountered hard times for some 30 years after the death of John Winthrop in 1779.

Yale too struggled in the post-war period. **Nehemiah Strong** (1730–1807), a 1755 graduate who stayed on campus for three years as tutor before serving as pastor of a church in New Haven, was appointed to the first chair of mathematics and natural philosophy in 1770; he held it until 1781. Strong then resigned his position to study law. The only contribution he seems to have made to science was the 1784 book *Astronomy Improved*.

Yale did not appoint a successor for 13 years. However, **Josiah Meigs** (1757–1822) accomplished little of consequence in mathematics during his tenure 1794–1801. Meigs came to the post with great promise, having graduated from Yale in 1778 and serving as tutor 1781–1784, but he then practiced law for the next ten years until accepting the Yale chair. In 1801, he moved from New Haven to Athens, GA, as president of the fledgling University of Georgia. That university had been incorporated in 1785 but was not established until 1801 when a land site was selected. As soon as Meigs was named president, work began on the first building, originally called Franklin College in honor of Benjamin Franklin, but it is now known as Old College. Georgia graduated its first class in 1804.

While the newer universities in the US also struggled to find firm financial footing, two older ones benefited from outstanding appointments—Princeton and Pennsylvania. However, mathematics at these institutions, located on opposite sides of the Delaware River, improved only marginally. The first notable appointment occurred in 1779 when Robert Patterson accepted a professorship in mathematics and natural philosophy at the University of Pennsylvania ("Penn" for short). Because Philadelphia occupied a strategic location during the Revolutionary War, the conflict took a heavy toll on Benjamin Franklin's college. Although Penn remained open during hostilities, the war dramatically reversed progress that had occurred beforehand. Nonetheless, Patterson stood out in the post-war period.

Born and raised in Ireland (to parents of Scottish descent), **Robert Patterson** (1743–1824) set sail for America at age 25 in 1768. His first job was teaching school near Philadelphia, where one of his prize pupils was Andrew Ellicott. Two years later, Patterson moved into Philadelphia to offer instruction to navigators anxious to learn how to calculate longitude from lunar observations. In 1772 he opened a country store in New Jersey; it failed, but he met his wife there. Next, Patterson taught at the Wilmington (DE) Academy, where he trained three companies of a militia that expected a war with England to break out. Ironically, he had served for a year with the military service in Britain before emigrating. During the Revolutionary War, he joined the army as an assistant surgeon and then volunteered for a militia, serving until 1778. Penn had closed when British troops occupied the city. It reopened the next year when those troops withdrew, but the previous faculty was dissolved and new professors were hired. Robert Patterson was one of them, appointed professor of mathematics and natural philosophy effective December 1779. He mainly taught elementary classes on arithmetic and geometry, but also an advanced course, "Oblique Spheric Trigonometry," that included such topics as stereographic projection of the sphere, conic sections, surveying, navigation, spherical geometry, and "fluxions for rectifying a curve." It is not surprising that his approach to calculus was based on Newton's fluxions and not the system favored by Leibniz and his continental adherents.

Robert Patterson published 11 papers in the *Transactions of the APS* from 1786 to 1818, mostly on astronomical observations and mechanical devices. He was also a frequent contributor of problems and solutions in emerging mathematical journals, one of which turned out to motivate an important discovery. In 1808 he proposed a surveying problem in Robert Adrain's the *Analyst* asking for a method of computing an area in which an error had been made in one of the dimensions. Ultimately, Adrain published a solution in which he derived the method of least squares, before either Gauss or Legendre had published their own derivations.

In 1790, concerned about the unfairness in transactions computed monthly, with January having about 10% more days than February, Patterson proposed, in a letter to Thomas Jefferson, a reworking of the calendar. Nothing ever came from that proposal. Nevertheless, two years later, "Patterson computed new insurance rates, and is credited with publishing the first set of premiums, based on actuarial science."[75]

Robert Patterson corresponded with Thomas Jefferson over several matters mathematical. For one, in 1798 the US vice president sought Patterson's opinion on the shape of a plough that would turn soil and cut it in lines. Patterson responded with a reference to that problem in a calculus text, impelling Jefferson to reply:[76]

> Tho I possess Emerson's fluxions at home, & it was the book I used in
> college [William & Mary], yet it had escaped me that he had treated
> the question of the best form of a body for removing an obstacle in a
> single direction.

In December 1801 Patterson proposed to Jefferson, now president, an intricate idea for encoding a message based on a key consisting of a word or name. An article from 2016 by Richard DeCesare explained the Patterson method in detail and cited a friendly challenge that he issued Jefferson: "In fact, it would be over two hundred years before someone would decipher it."[77] The most well-known aspect of the Patterson–Jefferson correspondence took place in March 1803, when the president asked the Penn professor to tutor Meriwether Lewis for two or three weeks before Lewis embarked on his famous journey of exploration and discovery with William Clark and 45 other men.

In 1805 Robert Patterson was appointed director of the US Mint. He continued to hold his professorship until becoming vice-provost at Penn five years later. He retired in 1813, and was succeeded in the professorship by his son. He served as president of the APS from 1819 until his death. His most recent biographer concluded that Patterson "rightfully deserves recognition as a mathematician who helped shape the future of the United States."[78]

Like Penn, Princeton had been founded in the middle of the eighteenth century and, after the Revolutionary War, had the good fortune to appoint a very competent mathematician almost simultaneously with the drafting of the Constitution. The chair of mathematics and natural philosophy had lain vacant for two years after the resignation of W.C. Houston in 1783. The next holder, **Ashbel Green** (1762–1848), had graduated in Houston's final year and then served as tutor for two years. However, Green resigned within two years to enter the Presbyterian ministry, leaving the chair vacant once again. Green returned in 1812 as president of Princeton, a post he held for a decade, but he played no further role in mathematics.

Green's successor was a notch above his predecessors. **Walter Minto** (1753–1796), born in Scotland of Spanish parents, typifies America's continued dependence on Great Britain. After attending the University of Edinburgh from 1768 to 1771, he left without taking a degree, most likely because at that time professors' personal certificates carried more clout than a degree.[79] In 1776 he moved to Italy to tutor the two sons of a wealthy businessman, and while there he furthered his own education by studying mathematics and astronomy at the University of Pisa. As a result, when he returned to Scotland in 1782, he was appointed professor of mathematics at the University of Edinburgh, the institution that awarded a doctoral degree to John Ewing seven years earlier. (Ewing will be highlighted in the next section, "Transition 1776.")

Just one year after returning to Scotland, Minto published a 72-page treatise on the recently discovered planet Uranus that provided formulas for determining astronomical magnitudes. He also collaborated on a biography of John Napier, the Scottish inventor of logarithms, but Minto had already immigrated to the New World a year before that book appeared in 1787. An ardent supporter of American independence, he traveled to Albany and Philadelphia before teaching at a boys' high school on Long Island for a year. In 1787 he accepted the professorship of mathematics and natural philosophy at Princeton at a salary of £200 a year plus room and board. He published just one work in the New World, based on an inaugural oration that was delivered on the eve of the 1788 commencement; it contains an account of the progress in mathematics up to that time and a persuasive statement of its importance as an intellectual discipline. However, Minto's biggest contribution might have been enabling juniors and seniors at Princeton to devote their time to mathematics, literature, and other subjects instead of Greek, Hebrew, and Latin. Since most American colleges were founded to train future ministers, it is not surprising that an emphasis on classical education was steeped in these three languages.

Walter Minto represented a local maximum in the chair of mathematics and natural philosophy at Princeton, as none of his predecessors or successors approached his (admittedly few) attainments. His immediate successor, **John Maclean, Sr.** (1771–1814), another Scot, came to the US in 1795 and established himself as a physician in Princeton. That summer he delivered a short course of chemistry lectures that impressed the trustees and led to his appointment as professor of chemistry in the fall. When Minto died the next year, Maclean assumed responsibility for instruction in mathematics and natural philosophy as well as in chemistry. Clearly, Princeton had little interest in producing mathematicians.

These developments at Penn and Princeton provide evidence of the ability and desire on the part of those colleges established in mid-eighteenth century to compete with the earlier established trio of Harvard, William and Mary, and Yale. All American colleges, however, experienced various levels of tension in three areas throughout the entire Revolutionary period 1776–1876:

1) classical languages *vs.* modern languages,
2) classical education *vs.* the sciences, and
3) a prescribed curriculum *vs.* a system with electives.

One indication of the low state of American mathematics at the end of the Period of Confederation is evident in Harvard's entrance requirements. When the College first instituted mathematics requirements in 1803, the only prerequisite was knowledge of arithmetic. Until then no knowledge of arithmetic was demanded at all. Algebra was required in 1820, but it was not until 1837 that arithmetic was eliminated from the list. Imagine—during this time Cauchy was placing analysis on a firm footing in terms of limits, Abel and Galois were revolutionizing the study of algebra, and non-Euclidean geometry was being discovered independently in three different countries by Gauss, Bolyai, and Lobachevski, yet America's foremost institution of higher learning required only the barest mathematical essentials for admission.

The American Academy of Arts and Sciences. US founding father John Adams became interested in the advance of science in the colonies while attending sessions of the Continental Congress in Philadelphia even before independence was declared. Consequently, he introduced a congressional resolution that each colony establish a "society for the encouragement of agriculture, arts, manufactures, and commerce."[80] While Adams was in France in 1778 and 1779, he found that "among the academicians and other men of science and letters I was frequently entertained with inquiries concerning the Philosophical Society of Philadelphia. ... These conversations suggested to me the idea of such an establishment at Boston."[81]

Upon his return to Massachusetts, Adams broached the subject of establishing a general learned society in Boston, where he gained considerable support from one of the Bay Colony's political leaders, James Bowdoin. As a result of their joint activities, in May 1780 the Massachusetts legislature established the second learned society in the US—the American Academy of Arts and Sciences (or the American Academy).[82] The American Academy charter included mathematics among its aims:[83]

> To promote and encourage the knowledge of the antiquities of America and of the natural history of the country, and to determine that uses to which the various natural productions of the country may be applied; to promote and encourage medical discoveries, mathematical dispositions, philosophical inquiries and experiments; astronomical, meteorological and geographical observations, and improvements in agriculture, arts, manufactures and commerce, and in fine, to cultivate every art and science which may tend to advance the interest, honor, dignity and happiness of a free, independent, and virtuous people.

The concluding sentence reflects the revolutionary zeal of the time when the document was written. The expression "mathematical dispositions" probably refers to demonstrations using mathematical logic.

Table 2.2. Some presidents of the American Academy of Arts and Sciences

President	Term
James Bowdoin	1780–1791
John Adams	1791–1814
Edward Augustus Holyoke	1814–1820
John Quincy Adams	1820–1829
Nathaniel Bowditch	1829–1838
—	—
Joseph Lovering	1880–1892
Edwin Bidwell Wilson	1927–1931

The name of the American Academy of Arts and Sciences was deliberately chosen to distinguish it from the American Philosophical Society (**APS**), although both societies were concerned with "useful knowledge." Although the titles began with "American," each organization was regional, the APS being mainly centered in Philadelphia and the American Academy in Boston. The American Academy began with a much larger membership, with 62 incorporating fellows of high standing in the republic, including clergymen, merchants, physicians, farmers, and public leaders signing their names to the charter. Of the 62 fellows, 60 had ties to Harvard. Incorporators included Hollis Professor of Mathematics and Natural Philosophy, Samuel Williams, as well as notables such as Governor James Bowdoin of Massachusetts, Samuel Adams, John Hancock, Charles Chauncy (the great-grandson of Harvard's second president introduced in Chapter 1), and James Winthrop (the Harvard librarian and son of eminent mathematician John Winthrop, who had died in 1779). The first class of foreign members included Euler and d'Alembert.

In spite of larger numbers than the APS, which met monthly in the summer and twice a month the rest of the year, the American Academy met four times a year—at Harvard in the summer and autumn, and at various places in Boston in the winter and spring. The American Academy sponsored an impressive initial project with Harvard to support an expedition to Maine to observe a solar eclipse. Even though dense fog prevented the observation from being made, the expedition discovered the astronomical phenomenon that became known as "Baily's beads" 50 years later; these are beads of light that appear along the edge of the Sun a few seconds before or after the totality of the eclipse.

James Bowdoin, who devoted himself chiefly to civil and natural history, served as the first president of the American Academy from 1780 until his death in 1790. He was succeeded by John Adams, who became its second president (from 1791 to 1814) while also serving as vice president of the US. Two other members of the Adams family have also held the American Academy presidency—his son John Quincy Adams and grandson Charles Francis Adams. Table 2.2 lists the first five presidents of the American Academy, ending with the death of Nathaniel Bowditch in 1838. The table then includes two notable figures to be introduced subsequently, the nineteenth-century mathematician Joseph Lovering and twentieth-century statistician E.B. Wilson; both were at Harvard when they held the office.

Like the APS, the American Academy felt the need to establish a journal to print articles *disseminating* knowledge of science, knowing that American authors could hardly *contribute* to knowledge. This led to the publication of Volume 1 of the American Academy's *Memoirs* in 1785. Its 622 pages were divided into three parts, the first comprising mathematics and astronomy. The Preface read, in part:[84]

> The astronomical and mathematical papers … will have the smallest number of readers. … Few, if any of [the papers], contain deep speculations and obstruse [*sic*] researches and calculations; but they are chiefly of a practical kind.

Overall, the aim of the *Memoirs* was to educate the American public in science. Most of the papers in Volume 1 were of a practical nature. Although some in the first part dealt with applications to geography and navigation, the majority concentrated on natural history (flora and fauna), astronomy, and medicine. As Greene wrote, "There was less experimental and theoretical science in the Academy's *Memoirs* than in the American Philosophical Society's *Transactions*."[85] Indeed, there were no articles in the *Memoirs* approaching the level of mathematics by David Rittenhouse in the *Transactions of the APS*. In 1848 the American Academy initiated a second journal, the *Proceedings of the American Academy of Arts and Sciences*. Overall, the two major learned societies catered to the scientific cultivator, practitioner, and researcher alike.[86]

Volume 2 of the *Memoirs* was divided into two parts. Part 1 appeared eight years later in 1793. Admittedly, no paper in either volume contained significant original research. The first paper in the *Memoirs* to contain advanced mathematics was written by Nathaniel Bowditch on lunar observations, and it was published in Part 2, which did not appear until 1804. Over half of the papers in these first two volumes were written by, or solicited by, Harvard professors. That changed dramatically with Volume 3. Nonetheless, along with the *Transactions of the APS*, the *Memoirs* of the American Academy represented an important first step toward the professionalization of science. The importance of these two general science periodicals is not so much about major contributions to science as much as the fact that they provide evidence of scientific matters being pursued by American practitioners.

Although the APS almost ceased to exist in the early 1780s, the formation of the American Academy and the publication of its *Memoirs* seems to have sparked a resurgence in the older society. These developments spurred the APS to begin construction of Philosophical Hall in 1785 and to publish the second volume of its *Transactions* the next year. It too lacked any articles on mathematics.

The APS and American Academy of Arts and Sciences have endured until today. However, other learned societies established during the Period of Confederation in Connecticut, New York, New Jersey, Virginia, and Kentucky were fleeting.

Pike's *Arithmetic*. Some critics have cited textbooks written in America during the period of confederation as further evidence of a woeful state of affairs. Indeed, one can even look beyond this period to 1818, when Robert Patterson published an arithmetic text for use at Penn, *A Treatise on Practical Mathematics*, about which the historian Florian Cajori opined, "Though lucid and ingenious, this arithmetic was rather difficult for beginners and never reached an extended circulation."[87] But a closer analysis of books on arithmetic penned after the Revolutionary War reveals some advances beyond those written beforehand. The publication of the best of the bunch constitutes the third notable event in this era, the 1788 textbook, *A New and Complete System of Arithmetic*, by **Nicholas Pike** (1743–1819), a Harvard graduate (1766) who spent most of his life as a teacher and magistrate in Newburyport, MA. The subtitle of this textbook reflects the patriotism that ran rampant throughout the new country at the time: *Composed for the Use of citizens of the US*. This theme continues in the Preface, where Pike distinguishes his book from those written by British authors, writing, "as the United States are now an independent Nation it was judged that a System might be calculated more suitable to our Meridian than those heretofore published." (A modern reader might be surprised at the use of the plural form "are" in that statement, but generally the US was viewed as a collection of states up to the Civil War.) "To shake off the shackles of colonialism,"[88] Pike added in the dedication, "The Federal Coin, being purely decimal, most naturally falls in after Decimal Fractions." As a result, the book's treatment of monetary systems, with rules for conversion between the new American system of dollars and cents and the older English system of pounds and shillings, is one feature that aided its popularity at once.

In his introduction, Pike wrote that "this work is the first of the kind composed in America." Clearly, he was unaware of Greenwood's *Arithmetic, Vulgar and Decimal* written 59 years earlier. In fact, Pike attended Harvard just 35 years after Greenwood was dismissed, yet he never heard mention of that *Arithmetic*. This is not a criticism of Pike as much as an indictment of a mathematical community that was unable to understand the material in Greenwood, and hence let it disappear from collective memory within one generation. Like the Greenwood work, the title of Pike's textbook is misleading, because it contains material on algebra as well as arithmetic. In fact, the Pike textbook was very extensive and complete for its time, containing, in addition to the usual account of operations on numbers, sections on various other topics. Some are no longer taught in US schools, such as taking square roots and cube roots of numbers. The book also includes a small table of powers of numbers. While topics in arithmetic do take up the majority of the pages—339 out of 512—the material is not always elementary. In fact, right after explaining how to perform division of integers, Pike defines prime numbers and lists the first ten perfect numbers, ending with

$$191, 561, 942, 608, 236, 107, 294, 793, 378, 084, 303, 638, 130, 997, 321, 548, 169, 216.$$

Pike's *Arithmetic* also contains a section on permutations and combinations, wherein a particular problem determines the probability of obtaining all ones when four dice are thrown. I do not know of any earlier work in the country that mentions probability.

Various topics from natural philosophy are followed by 15 pages of problems, one of which determines the age of the moon and another the dates of Easter for the years from 1753 to 4199. After that, the text provides overviews of plane geometry and rectilinear (as contrasted with spherical) trigonometry. The term "mensuration" has been used several times so far without definition, even though the term is hardly known today beyond historians. Generally, mensuration refers to measuring, and in Pike's textbook it refers specifically to formulas for the area of polygonal regions and the volume of regular figures.

Pike's textbook also remains significant for its inclusion of algebra, signaling the first advance beyond mere arithmetic in the country since Greenwood's work at Harvard. In 34 rather dense pages, the section begins with easy examples like $\frac{x}{2} + \frac{x}{3}$ and $\frac{x}{3} - \frac{2x}{11}$ but moves quickly to $\frac{x+a}{2x-2b}$ divided by $\frac{x+b}{5x+a}$. Just a few pages later Pike invokes

"Sir Isaac Newton's rule for raising a binomial to any power whatever" by raising $a + x$ to the fifth power and $x - a$ to the sixth.

The highlight of *A New and Complete System of Arithmetic*, however, is a topic that seems to be missing in most secondary accounts of the book—infinite series. The inclusion of series as the last topic in the book marks a great leap forward for an American audience. Although few of his readers could probably understand the treatment in the final 20 pages of the book—indeed, probably few teachers could—just the fact that Pike felt justified in including it shows marked improvement. The treatment is not rigorous by today's standards, which is not surprising considering that it is presented with the same kind of intuitive approach pioneered by Leonhard Euler, who had died just five years before Pike's textbook was published. As an example, the first problem, whose curious wording read, "Let $\frac{1}{1+x}$ be thrown into an infinite series," is solved by long division without any restrictions on x or any concern for convergence. To gain an appreciation of the extent to which Pike took his readers, on only the second page of the coverage of infinite series, he presented the example, "Extract the square root of $a^2 + x^2$ in an infinite series." Readers with knowledge of Maclaurin series might pause to perform the power-series representation of the function $f(x) = \sqrt{a^2 + x^2}$. Before moving beyond Pike's peak (so to speak), it is germane to mention his convention taken for granted today, one originated by Descartes some 150 years earlier: letters at the beginning of the alphabet stand for constants, and those at the end for variables.

Nicholas Pike's *Arithmetic* received a strong commendation from a famous American. After receiving his personal copy, George Washington wrote to the author, "I flatter myself that the idea of its being an American production and the *first of the kind* which has appeared, will induce every patriotic and liberal character to give it all the countenance and patronage in his power."[89] Yet, as often happens with innovators, Pike was ahead of his time. When Chester Dewey, the professor of mathematics and natural philosophy at Bowdoin College, brought out the fourth edition of Pike's (posthumous) *Arithmetic* in 1825, he omitted several sections that, as he wrote in the preface, "were so briefly treated by Mr. Pike, as to possess little value." Dewey cited sections on logarithms, trigonometry, algebra, and conic sections, but he neglected that he omitted infinite series as well. The online file "Web02-Pike" expands coverage of Pike's *Arithmetic*.

Up to 1800, Pike's work was the only arithmetic book written by an American that gained wide appeal, although several others were published by such authors as Thomas Sarjeant, John Gough, John Vinall, Gordon Johnson, and Erastus Root (later a member of Congress). More importantly, Pike's *Arithmetic* served as a model for numerous subsequent textbooks on arithmetic. Two merit mention. **Nathan Daboll** (1750–1818), a teacher in New London, CT, wrote the second popular arithmetic in 1800, *The School-Master's Assistant*. Although this book covered the same basic ground as Pike's, the arrangement of topics introduced federal money immediately after the addition and multiplication of integers, thus showing the reader how to find the value of goods without having to master operations on rational numbers. Since the adherence to the British system of currency was regarded as offensive by many people in the fledgling republic, most American textbook authors presented the American system instead. This applies to both *The School-Master's Assistant* and the *Scholar's Arithmetic* (1801) by **Daniel Adams** (1773–1864), a physician who had graduated from Dartmouth just four years before his *Arithmetic* was published. Adams taught school in Boston 1806–1813 and served as a state senator. He subsequently published several other arithmetic books, including one that adopted the Pestalozzi analytic or inductive method of teaching. Daniel Adams was not related to John Adams, who famously wrote in 1780, "I must study politicks [*sic*] and war that my sons may have liberty to study mathematicks [*sic*] and philosophy."[90]

Pike, Daboll, and Adams were regarded as the three "great arithmeticians" in America before 1800.[91] There is no need to pursue the dozen or so other arithmetics from this time any further, because all were soon superseded by more advanced books that began to appear in the new century. The historians Karen Parshall and David Rowe summarized the period of confederation, 1783–1800, as follows:[92]

> The United States achieved political independence as a result of the conflict [Revolutionary War], but the cultural dependence engendered by common ancestry and language proved much more difficult to shake.

Transition 1776:
The Patriot

The structure adopted in this work for presenting the history of mathematics in America is to partition the overall development into parts determined by outstanding events, beginning with the American Revolution. Each of the parts is followed by a transition section aimed at emphasizing the evolution that occurred from one part to the next. But this transition section is different from the others. Subsequent parts will be bridged by transition sections from one to the next.

While Chapter 1 described American history up to the Revolutionary War and Chapter 2 considered the quarter century after its outbreak, the line of demarcation is artificial. On the one hand, the Revolutionary War did play a role in impeding progress of higher education in the country, in general, and mathematics. Yet most of the major players after the war were the same as before, so the demarcation line is not abrupt. To emphasize the evolution that occurred, I consider the life and accomplishments of someone with one leg in each period.

In the fall of 1804 Meriwether Lewis was asked by President Thomas Jefferson to lead a journey of exploration and discovery that has become known as the Lewis and Clark expedition. The president suggested that Lewis study under specialists on navigation and astronomy before embarking on his journey the next spring. Thus, during late winter 1805, the dutiful army officer set off to learn navigation from Robert Adrain in Lancaster, PA, and geographical charting from Robert Patterson in Philadelphia.

Why charting? Because in the days before GPS it was not so easy to know your location. Latitude could be determined with little effort, by using the angle between the horizon and the North Star. But determining longitude—the measure of east-to-west direction—involved higher mathematics and the location of the moon to determine one's location. Therefore, Jefferson recommended that Lewis be tutored on lunar positioning by Robert Patterson, because Patterson was the new country's expert on these matters. Patterson was then the professor of mathematics at the University of Pennsylvania.

How did Patterson become an expert in mathematics and astronomy?

The answer to that question involves Patterson's mentor John Ewing, who is little known in the history of mathematics. I regard Ewing as a model of an intelligent, mainly self-educated colonist, the type of person encountered several times in Chapter 1, but whose accomplishments after the war morphed smoothly into the period covered in Chapter 2. Overall, Lewis's saga involves conditions in the US both before and after independence, containing lessons about higher education during this period and the role of mathematics in this education. The story of Ewing is told first before returning to Patterson and his role in it.

John Ewing (1732–1802) was born in Nottingham (Cecil County), Maryland. Few details are known about his ancestors, who came from Scotland, settled along the banks of the Susquehanna River, and engaged in farming. Such a life called for hard work and long hours for the children, so Ewing and his four brothers spent most of their time doing farm chores. There was little time for frivolity, like learning. (It is not known if Ewing had any sisters. He and his twin, James, were the youngest of the sons.)

Apparently, Ewing's ability and drive to learn became apparent early because during nonfarming seasons he was allowed to attend school. Unfortunately, the closest one was several miles away, so it took him over an hour to walk in each direction, but his desire to learn turned this apparent obstacle into an advantage; later in life he commented that vigorous exercise allowed him to live over 60 years before he had any need for a physician.

However, a one-room country school had little to offer this enterprising youngster, and within a short time he switched to a school in New London, located right over the Pennsylvania border from Maryland. The director of the school was the clergyman Francis Alison. I do not know how long Ewing remained at Dr. Alison's school, or whether diplomas were even granted, but Ewing's formal education came to an abrupt halt when his father died. In those days, the law of primogeniture prevailed, so the oldest son, William, was granted the entire farm. Each of the other sons received 20 pounds, which was not enough to sustain Ewing further, so he had to seek employment. Fortunately, the school director was so impressed with the ability and character of his prize pupil that he hired him as a tutor, a position that Ewing held for three years. Perhaps Dr. Alison had been schooled in mathematics, because Ewing continued private lessons with him while serving as a tutor.

Robert Patterson wrote, "Under the kind care of Dr. Alison he [Ewing] made considerable progress in his favourite pursuit, the study of mathematics."[1] Pursuing higher mathematics was not easy. Printed books were scarce in the colonies, though there were private libraries in cities, such as Benjamin Franklin's and James Logan's in Philadelphia. Such collections, however, were rare outside such cities, forcing Ewing to scramble once again to pursue his passion. How? He rode 30 to 40 miles to consult a book whenever he heard that someone within that range had purchased a copy. All such books were written in England by English authors; the only work on mathematics up to that time written by a colonist was *Arithmetic, Vulgar and Decimal*, and even that one was published anonymously in 1729. (It took impressive detective work from the early-twentieth century historian of mathematics David Smith to determine that the author of that textbook was Isaac Greenwood, the Hollis Professor of Mathematics at Harvard up to 1738.) Ewing could not have met up with Greenwood, and there is little chance he would have seen Greenwood's *Arithmetic*. Although it is not known which books Ewing was able to consult in this manner, reputedly he found errors in many of them. This ability suggests that Ewing read these works like a mathematician— independently and critically. He did not accept a writer's logic on authority alone, rather holding the subject to rigorous, independent scrutiny, causing Patterson to conclude, "In the science of mathematics, he was self-taught."[2]

In the summer of 1754 at age 22, Ewing ended his tutorship and headed east to Princeton to continue his studies at the fledgling College of New Jersey. The president of the college at the time was Aaron Burr, whose son tied Thomas Jefferson in the electoral vote for US president in 1800 (and lost in the vote in the House of Representatives, whence he became vice president—one way in which the country's executive branch of government has changed in the last 200 years). But the role of a college president has changed too, dramatically. At that time a college president was, in essence, *the* college. Presidents taught most of the classes, and hired tutors to teach the others. They also interviewed all prospective students and determined whether they were qualified for admission and, if so, where they would be placed in the four-year curriculum. The focus of the admission interview was knowledge in reading and writing Latin. Fortunately, Ewing was well schooled in this classical language, because he was admitted and placed in the senior class. Yet Ewing needed employment to support himself, so while conducting his studies he also taught classes at the grammar school connected with the college. Nonetheless, he graduated the following June with a bachelor's degree.

Figure 2.7. John Ewing

Ewing remained at Princeton as a tutor after graduation, but a short while later he returned to New London, PA, to resume lessons under Dr. Alison, this time to study for the clergy. Indeed, almost all higher education in the colonies was then aimed at producing clergymen. After the usual period of preparatory study, Ewing was licensed to preach the gospel by the presbytery of New Castle, DE, located about 35 miles southwest of Philadelphia.

This was in 1758 when Ewing was 26 years old. And his timing was perfect, because the provost at the Academy in Philadelphia (later the University of Pennsylvania), Dr. William Smith, was taken ill, causing the college to seek someone to teach his classes. Who better than Ewing for the job? He readily accepted and, before delivering even one sermon in New Castle, found himself in charge of fulfilling Smith's duties as professor of ethics.

Philadelphia was then the center of learning in the colonies. Just 15 years earlier, the printer Benjamin Franklin had founded the American Philosophical Society

(APS). Ewing began attending APS meetings at once, ultimately rising to become vice president and publishing three articles in the *Transactions of the APS.*

Word of Ewing's pastoral abilities quickly spread, because he was asked to become the pastor at his native town's Presbyterian congregation. Honored to be selected as the moral authority for the community, coming as it did from those with whom he had grown up, he nonetheless declined the offer, opting instead to provide the type of instruction he had missed in his youth. In rejecting the offer, Ewing expressed the view that a liberal education was to be praised over wealth. Nonetheless, his reputation quickly spread throughout the City of Brotherly Love, and he accepted a similar offer at the First Presbyterian Church the next year, 1759, a position he held until his death 43 years later. His weekday routine remained the same over that period, serving as professor by day and pastor by night, all the while engaging in constant study of natural philosophy (science) and teaching those topics to his students as he learned them.

A sudden change occurred in 1773, when he left the University of Pennsylvania to go to Great Britain to solicit subscriptions for a new academy in Newark, DE (which later became the University of Delaware). Accompanying him was Dr. Hugh Williamson (1735–1819), who would go on to serve in the Continental Congress, assist in the wording of the Constitution, and occupy a seat in the first US Congress. Ewing traveled about England, Scotland, and Ireland, where he gained respect as a scientist and a minister. As a result, he was awarded a Doctor of Divinity from the University of Edinburgh, and henceforth became known formally as Dr. Ewing. Later he was invited to write astronomy articles for the *British Encyclopedia*, an honor bestowed on only a few colonists.

Figure 2.8. Hugh Williamson

A fellow Philadelphian in London at that time was Benjamin Franklin. As you might imagine, there was a lot of talk in general society about the repulsive acts of some of the radicals in the colonies. Patterson later summarized Ewing's stay in Great Britain: "When he first visited England, the approaching contest with his native land was a topic of conversation in every society. ... He had frequent offers of reward from men, high in power, yet his knowledge of the causes of the revolution forbade him to listen to them."[3]

Ewing was clearly a patriot. And his patriotism would soon be challenged by the famous writer **Samuel Johnson** (1709–1784; also known as Dr. Johnson), a towering figure of eighteenth-century English literature. Johnson became a national sensation with the publication of *A Dictionary of the English Language* in 1755. Overall, he gained fame from his conversation and wit as much as from his writings, with much of his social life spent at dinner parties where he engaged in intellectual discussions. Ewing was invited to one of these gatherings at the home of a London book seller known only as Mr. Dilly, who warned him beforehand not to disagree with Johnson: "You must not contradict him." But Ewing could not resist responding when Dr. Johnson bluntly asserted, "Sir, what do you know in America? You have no books there." The very tone of the question, and the implied accusation, caused the usually retiring mathematician to answer in a firm but gentle way: "Pardon me, Sir, we have read [Johnson's] *The Rambler*." Apparently, the civility pacified Johnson, who generally referred to the colonists as rebels and scoundrels, because the two men ending up chatting amiably past midnight.

Ewing returned to Philadelphia in the fall of 1775 and resumed his former professorial and pastoral duties. He spent mornings and afternoons at Penn before tending to his flock all evening. The latter tasks included visiting the sick, meeting with parishioners, and preparing for the one or two sermons he delivered every Sunday. He credited his robust character with allowing him to rise at daybreak and to carry out all these missions without tiring.

Then came the cry for independence, and with it King George's troops. When the British army approached Philadelphia in 1777, Ewing moved his family to his native town in Maryland, returning only after those troops were repulsed back to New York. Two years later, the 47-year-old Ewing was appointed provost of the University of Pennsylvania, its highest academic office. Although he probably would have preferred to spend all his time pursuing investigations in natural philosophy, the fame he had attained led to his appointment on several commissions. For instance, he served as a surveyor to 1) define the boundaries of Delaware, 2) settle a dispute between Massachusetts and Connecticut, and 3) settle a dispute between Pennsylvania and Virginia. In addition, he and David Rittenhouse were appointed by the state of Pennsylvania to lay out the first turnpike in the country, linking Lancaster to Philadelphia (and paving the way, so to speak, for Meriwether Lewis's commute between these two sites).

Despite these incursions on his time, over the next 20 years Ewing continued his manner of reading topics in natural philosophy and then lecturing on them in his classes. By 1795 he decided to assemble his lectures on natural history into book form and publish them, but the following summer he was afflicted with a "violent disorder" that left him unable to walk, probably a stroke. Ewing never was able to complete his project. In the summer of 1802 he moved his family out of Philadelphia and into the house of a son who lived in nearby Montgomery County in order to avoid the yellow-fever scourge that had plagued the city. The violent disorder that had stuck in 1796 returned in more deadly fashion, and Ewing died on September 8, 1802.

In 1779 the Philadelphia College and the Philadelphia Academy had been reorganized into the University of Pennsylvania. John Ewing selected Robert Patterson (1743–1824) the as professor of mathematics, a position Patterson held 1779–1810. The

Figure 2.9. Robert Maskell Patterson

two became close and trusted friends, developing a symbiotic relationship regarding progress in science. Patterson knew of Ewing's intention to publish his lectures, so he worked on them after Ewing's death and finally arranged to have them published posthumously in 1809. The resulting work is entitled *A Plain Elementary and Practical System of Natural Experimental Philosophy*.

The seal on the inside cover of the book indicates how dates were recorded at the time, with the Clerk of the District of Pennsylvania asserting that it was deposited "on the twenty-second day of November, in the thirty-fourth year of independence of the United States of America, A.D. 1809." The book begins with a 19-page biography of Ewing by Patterson, followed by Ewing's lectures on natural philosophy, which were divided into two distinct parts, physics and astronomy. The first part contains material on magnetism, electricity (think Benjamin Franklin), gravitation, mechanics, motion of bodies, hydrostatics, optics, and light, among other topics. The second part consists of various aspects of astronomy. Altogether the lectures comprise more than 500 pages and serve as a living testament to the lifetime achievement of John Ewing, a patriot who straddled the line between pre- and post-Revolutionary America.

Part II

New Republic, 1800–1876

Introduction to Part II

Developments in mathematics that took place in America took a leap forward from 1800 to the watershed year 1876 although, once again, a war put a crimp on advances, this time the American Civil War. Two mathematicians towered over the American landscape—Nathaniel Bowditch and Benjamin Peirce. Bowditch had one foot in the previous period, dominated by British mathematics, but the other foot in newer advances from continental Europe. Two of his books, one from 1802 (based on a British work) and the other from 1829 (based on the French), reflect the patriotic fever that swept the US after the Revolutionary War and, later, the War of 1812. Peirce, on the other hand, had more in common with the revolutionary transformation that took place in American mathematics in the last quarter of the nineteenth century.

Overall, Part II provides a (mostly) chronological tour through the period from 1800 to 1876. Although no revolutionary changes occurred, the place of mathematics in America did experience incremental improvements that would set the stage for the country's relatively rapid ascension into the ranks of international status. Because of the dominant roles that Bowditch and Peirce played, I divide the 76-year span in half. Chapter 3 describes some major developments that occurred in the Age of Bowditch, when navigation and celestial mechanics took huge leaps forward with his work. The chapter includes accounts of the founding of West Point and notable changes at Yale, Virginia, and Toronto. Moreover, it discusses the first stage in the evolution of statistics that occurred with the founding of a successful statistical society.

Chapter 4 is centered on Benjamin Peirce, who can be regarded as America's first research mathematician. In what I have defined as the Age of Peirce, 1838–1876, he contributed to both mathematics and mathematics education in sundry important ways. Mostly, however, he worked in isolation, like four of his notable contemporaries—George Hill, Willard Gibbs, Simon Newcomb, and Christine Ladd. Toward the end of this timeframe, doctoral programs began to develop at a few select universities, and another important figure emerged, Hubert Newton.

3

The Age of Bowditch

Nathaniel Bowditch contributed to almost every development that occurred in American mathematics over the first third of the nineteenth century. Therefore, I label the period 1800–1838 "The Age of Bowditch." In addition to Bowditch, two notable individuals engaged in some mathematics on the international stage and sought to contribute to current progress taking place nationally—Robert Adrain (who founded four mathematics journals) and John Farrar (who brought French scholarship to these shores).

The export of French mathematics to America was important because England's isolation, still simmering from the bitter priority dispute between Isaac Newton and Gottfried Leibniz over the discovery of the calculus, had resulted in the scientific insularity of Great Britain from the Continent. After 1800 France had become the undisputed leader under the direction of the "three L's"—Lagrange, Laplace, and Legendre. However, even France soon faltered and was overtaken by Germany, with its establishment of research departments and special seminars for training future researchers. For instance, the Prefatory Notice in the 1830 issue of *Transactions of the APS* stated:[1]

> The contents of this volume partly belong to the physical and partly to the moral sciences. In this the society has followed the example of several learned societies in Europe, and particularly of the Royal Academy of Berlin.

Although a (roughly) 50-year lag endured between advances in Europe and America, by 1876 the New World possessed a community sufficiently sizable and prepared for a mathematical revolution, one that occurred exactly one century after the start of the Revolutionary War.

The Age of Bowditch also saw important changes at three American universities:

(1) the delineation of the professionalism of the term "mathematician" at Yale ,

(2) the establishment of the curriculum at Thomas Jefferson's University of Virginia, and

(3) the founding of the University of Toronto as the leading institution of higher learning in Canada.

Another notable development was the founding of the American Statistical Association (**ASA**) in 1839. Like the American Academy of Arts and Sciences, the ASA was initiated in Boston and still exists today. Several leading statisticians emerged, although none pursued mathematical statistics.

The great statesman Thomas Jefferson (1743–1826) was president of the US (1801–1809), but his influence, in both politics and science, extended for another two decades. Per historian Todd Timmons, "The presidency of Thomas Jefferson is symbolic of [the] new zest for scientific work."[2] There were two overarching developments during Jefferson's first term in the White House. One was the publication of specialized journals for the first time in the history of America. The first to appear in the sciences was *The Mathematical Correspondent* in 1804. The most important figure involved with mathematics journals was a leading *Correspondent* contributor, Robert Adrain. The other primary development was the founding of the United States Military Academy at West Point. The early years of the Academy, with its emphasis on mathematics, will be traced in some detail. Among the more important consequences for mathematics instruction in American colleges were the increasing influence of French authors and the efforts of American authors to write their own texts, not just to adapt British (and, in the latter part of the present period, French) texts to an American audience. The prime textbook authors and translators for university students across the US were John Farrar and Nathaniel Bowditch.

West Point

In 1802 an Act of Congress authorized the creation of a Corps of Engineers, but the desire to establish a military academy in the US was first expressed at the time of the American Revolution. In 1776 the Continental Congress debated the creation of a "Military Academy for the Army," and six years later a "Corps of Invalids" was established at a fort located at West Point, whereby lame veterans instructed younger officers. However, it took another 20 years before the academy was formed. President Thomas Jefferson signed legislation in March 1802 creating the United States Military Academy at West Point as a Corps of Engineers to separate its mission from the artillery. The Military Academy (shortened sometimes to "West Point" or the "Academy") dates its founding from the passage of this law. Its location was strategic; Britain had tried unsuccessfully to win this vital bend in the Hudson River to sever communication between New York City, located 50 miles to the south, and the interior of the country.

However, the first mathematics class actually preceded the founding of the Academy when, in September 1801, George Baron was appointed a civilian "teacher of mathematics."[3] What a distinction—mathematics had a professor at West Point before the Academy even started! That explains why the mathematics department at the Academy sponsored a seminar in 2001 to celebrate West Point's bicentennial one year in advance of the institution's official celebration. Jefferson appointed Jonathan Williams, a grandnephew of Benjamin Franklin, as the Academy's first superintendent. The Academy became the most important college for the study of mathematics in the US during the period 1815–1850.

In one sense, the appointment of Baron as a mathematics instructor at West Point was beneficial to the country because he introduced a new technology—teaching with chalk on a "standing slate." In addition, he adopted the two-volume textbook *Course of*

Mathematics (1798) written by the Englishman Charles Hutton. Yet Baron's behavior got him into hot water. Is this a surprise? After all, recall a couple of other renegade American mathematicians who were initial mathematics professors at their colleges—Lefevre at William and Mary and Greenwood at Harvard. Baron was dismissed the following February due to several unsavory incidents, one of which resulted in an almost violent confrontation with a cadet. Because George Baron was "canned" one month before the Academy was established, perhaps his initial appointment was not a distinction for mathematics after all.

However, matters improved quickly when President Jefferson appointed two mathematics instructors to the new Academy at once. **Jared Mansfield** (1759–1830) graduated from Yale in 1777 and then taught in New Haven and Philadelphia before entering the regular army as captain of engineers in May 1802. He mainly taught algebra at West Point, but that year he also published *Essays, Mathematical and Physical*, one of the first science textbooks in the US. However, after serving as the first head of the department for less than two years, he was appointed surveyor general of the US and sent to the Northwest Territory where he remained until 1812. (The "Northwest Territory" then comprised the five present-day states of Ohio, Indiana, Illinois, Michigan, and Wisconsin.[4]) At that time Mansfield returned to the Academy as professor of natural and experimental philosophy, a position he held until 1828. His 274-page book with an onerous title was printed in New Haven in 1802: *Essays, Mathematical and Physical: Containing New Theories and Illustrations of some Very Important and Difficult Subjects of the Sciences. Never Before Published.*

The other mathematics instructor appointed in 1802, **William Amherst Barron** (1769–1825), was quite a character. The son of a British army surgeon on the medical staff of Lord William Amherst (hence the first two names), Barron graduated from Harvard in 1787; John Quincy Adams was his classmate. Barron (not to be confused with Baron) received an AM degree in 1792 and then served as a tutor until 1800. He left Harvard in 1800 to accept an appointment in the Artillery and Engineers. Two years later he was transferred to the Corps of Engineers at West Point as Baron's replacement. Apparently, Barron had a real flair for teaching and was quite popular with cadets; he mainly taught geometry at the Academy. However, in 1807 improper behavior—he was charged with "suffering prostitutes to be the company of his quarters"[5]—led to a threat of court martial. Although Barron resigned before being brought to trial, he was yet another smudge on the face of mathematics. Did our mathematical forefathers *not* know how to behave outside the classroom?

In summary, Mansfield left the Academy after only two years and Barron after five. **Ferdinand Hassler** (1770–1843) succeeded Barron and brought temporary relief to instruction in mathematics. One of many immigrants who contributed vastly to American mathematics, Hassler achieved later fame at the US Coast Survey. He attended the University of Bern and worked in geodetics before emigrating from his native Switzerland in 1805 to take part in a Utopian community. When that idealistic mission failed, he made use of his personal contacts to lead a survey of the American coast, but the project fell through and freed him to succeed Barron. Hassler was unlike Barron in two ways, one bad, one good. The bad news was that Hassler was not rated highly as a teacher; the good news was that his behavior was exemplary. Nonetheless, in 1808 the US secretary of war declared that civilians would not be permitted to teach

Figure 3.1. Ferdinand Hassler

at the Academy, so Hassler was forced out after only two years. He became professor of mathematics and natural philosophy at Union College in Schenectady, NY.

Ferdinand Hassler proposed the formation of a federal installation for performing geodetic surveys of the coastline to be used for, *inter alia*, establishing clearly demarcated shipping lanes. In 1807 Congress authorized funds for the US Coast Survey but delayed its establishment until 1816. (Its name was expanded to the US Coast and Geodetic Survey in 1878.) Hassler was chosen to head the agency that played a major role in the development of mathematics in the country because of the way it "functioned more as a small scientific school than as a producer of coastline maps."[6] However, in a fashion typical of federal treatment of science in all periods of US history, the project was not funded properly for several decades. Hassler published two articles in the *Transactions of the APS*, one of which from 1825 was a 180-page report on his work with the Coast Survey. Indeed, he was responsible for annual reports of the Coast Survey beginning in 1837.

The dismissal of Ferdinand Hassler left the Academy with one mathematics instructor, **Alden Partridge** (1785–1854). An 1806 graduate of the Academy,[7] Partridge made such a positive impression that his first military assignment was to remain at West Point as the assistant professor under Barron. He then taught with Hassler after Barron's dismissal. At that early time in its history, the Academy had no admissions standards, little academic structure, and no consistent set of graduation requirements. Superintendent Williams petitioned the secretary of war to address these deficiencies, but his proposals were ignored. Instead, in 1810 the Secretary ruled that Academy graduates were no longer entitled to commissions. Is it any wonder the Academy foundered? The crisis reached a low point in 1812 when the last professor resigned, leaving Partridge as the sole instructor. How could he handle all the instruction? That was easy, because by then the corps consisted of a single cadet. But a conflict intervened—in this case one that paid dividends for mathematics—when the threat of war, coupled with the US's notorious lack of preparedness for hostilities, motivated Congress to expand the Army. The resulting Act of April 1812 authorized funds to reform the Military Academy by expanding the corps size to 250, setting academic standards, undertaking physical improvements, and establishing three professorships, including one in mathematics. Ostensibly, Partridge benefited greatly, being appointed the first formal "professor of mathematics" at a salary of $20 a month, which doubled

Figure 3.2. Alden Partridge

his previous salary. But he remained in this position only six months. Although he stayed at West Point, he was appointed the nation's first professor of engineering. During his tenure, Partridge emphasized two practices that became the hallmark of his teaching for the rest of his career: 1) an emphasis on practical mathematics, and 2) active student participation in learning mathematics. Today Partridge is remembered as an "educational innovator."[8]

It might appear that Partridge's reassignment would decimate instruction in mathematics at the Academy, but instead the appointment of a 60-year old surveyor, **Andrew Ellicott** (1754–1820), as professor in 1813 brought a sense of stability until his death seven years later. Ellicott's credentials were impeccable, as he had studied navigation and surveying under Robert Patterson (1768–1772), and then conducted numerous surveys, notably the newly established District of Columbia and the boundary between the southernmost states and Florida (then owned by Spain). He published 15 articles in the *Transactions of the APS* from 1793 to 1818, all on surveying or astronomical observations.

Ellicott remained at West Point until his death. He was known to be somewhat aloof, called "Old infinite series" by the cadets. Throughout his tenure, he taught from a version of Charles Hutton's monumental work, *A Course of Mathematics: Composed for the Use of the Gentlemen Cadets in the Royal Military Academy of Woolwich*, which shows the continuing influence of British texts in America. Different versions of this text were used at West Point from Baron's tenure in 1801 until 1823, when new texts by American authors supplanted it.

An historic event happened at West Point in 1815, surrounding **Sylvanus Thayer** (1785–1872). Born of Puritan stock that extended back to the seventeenth century and included several patriots who fought in the Revolutionary War, Thayer graduated first in his Dartmouth class of 1807. However, he was unable to deliver the valedictory address at commencement because he had already enrolled at West Point. He completed his requirements within a year and received his commission. From 1810 to 1812, Thayer served as assistant professor of mathematics but, with the outbreak of the War of 1812, he was sent to the Canadian frontier and then to Norfolk. When the inconclusive conflict between the US and Great Britain ended in December 1814, Thayer returned to West Point.

Figure 3.3. Sylvanus Thayer

A significant event for mathematics occurred when an academic building housing all classrooms and the library was completed in 1815. The next year Sylvanus Thayer and William McRae were dispatched to Europe and charged with examining European universities in order to equip the new building with books, maps, and instruments. However, timing is everything, and in this case Thayer's timing was poor, because he arrived just as Napoleon met his Waterloo and the École Polytechnique in Paris was closed. Fortunately, this outstanding institution reopened a year later; it served as a gateway to France's military, engineering, government, and technology professions. In the meantime, Thayer toured other establishments throughout Europe. When he returned to Paris, he met with French mathematical leaders Legendre, Laplace, and Monge, whose educational philosophy and structure left a deep and enduring impression. Consequently, when Thayer returned to West Point from his European junket in the spring of 1817, he brought with him not only seven trunks transporting about a thousand books, but, more importantly, an academic model for advanced scientific training. Moreover, the books were equally distributed among British and French authors, the first significant indicator of a shift in primary external influence from Great Britain to France. Thayer continued to build the West Point library throughout the 1820s. By 1830 one third of the Academy's mathematics holdings were works by French authors.

Thayer's trip to Europe paid further dividends for mathematics when he was appointed superintendent of the Academy in 1817, a position he held until 1833. During this time, he sought to adapt the pedagogy and curriculum of the École Polytechnique to American culture. His meetings with Laplace, Legendre, and Monge had imbued him with the expectation that professors should "advance science through their own research," an endeavor that was a far cry from the ability of American practitioners at the time. Moreover, he established a full, well-defined, four-year course of study with a

Figure 3.4. Claudius Crozet

specified calendar. The resulting "Thayer System" emphasized mathematics, science, and engineering, and by the end of his tenure, West Point was regarded as the leading engineering school in the country. Today, Thayer is known as the "father of the Academy," and a modern visitor to West Point is encouraged to stay at the luxurious Hotel Thayer, located right after the entrance onto campus. The Thayer System curriculum required every cadet to take the following courses in the mathematical sciences:

First year: algebra, geometry, trigonometry, mensuration, analytical geometry;
Second year: analytical geometry, descriptive geometry, shades and shadows, calculus, surveying;
Third year: mechanics, acoustics, optics, astronomy;
Fourth year: civil engineering, military engineering.

It is notable that arithmetic is absent from the curriculum. American mathematics education was moving forward.

While in Paris, Thayer had met **Claudius Crozet** (1789–1864), who was appointed professor of engineering in 1816, one year before Thayer became superintendent. In the sense that Walter Minto typifies British influence on America, Claude Crozet represents the shift to French influence on the development of mathematics in the US. Having studied under Gaspard Monge at the École Polytechnique, Crozet was familiar with the descriptive geometry developed by the French master for designing fortifications, which had been kept secret for military reasons. Crozet sought to introduce this nascent study into the Academy's curriculum, and since no English text on this subject was available, he wrote his own book, which was published in 1821: *A Treatise on Descriptive Geometry for the Use of the Cadets of the United States Military Academy. Part I. Containing the Elementary Principles of Descriptive Geometry, and its Application to Spherics and Conic Sections.* Crozet left the Military Academy in 1823 when appointed chief engineer for the Commonwealth of Virginia.

These advances at West Point represent one forward step in mathematics from colonial times in America. Another step involved the publication of a journal devoted to a specific subject.

The Mathematical Correspondent

A scientific journal serves two major roles. The obvious one is to advance science, but equally as important is the purpose of acquainting investigators of common interests with one another. The latter purpose was particularly significant in America in 1800 because of postal and transportation difficulties. George Baron deserves gratitude for forming a group of men in New York City to learn and disseminate mathematics. To achieve this goal, Baron published a mathematics periodical that would cater to a far narrower and more focused audience. The American Philosophical Society had initiated its *Transactions* in 1771 and the American Academy of Arts and Sciences its *Memoirs* in 1785, but these were general scientific periodicals. Baron instituted the revolutionary idea of restricting publication to one scientific field, mathematics, modeling it after a publication with the somewhat misleading title, *The Ladies Diary*, a British periodical devoted to problem solving that had been in existence since 1704.

George Baron (1769–1818) was born in England and spent much of his early life at sea. His navigational experiences required him to learn mathematics, leading to the position of master of a mathematical academy. At age 25, Baron set sail for the US, first settling in Hallowell, ME, and then New York State. In 1801 he was appointed a civilian "Teacher of the Arts and Sciences to the Artillerists and Engineers" at West Point, but his tenure lasted only five months, after which he started his own academy in New York City.

Over the next couple of years, George Baron apparently formed the New York Mathematical Club. Was there a critical mass of people interested in the study of mathematics at the beginning of the nineteenth century? In a word, No. But Baron located a community of enthusiasts and inspired one practitioner. Eventually, this community would evolve from enthusiasts to practitioners to mathematicians, according to my definitions of these terms:

- An *enthusiast* was a problemist who engaged in posing and/or solving problems.
- A *practitioner* published at least one article on mathematics or mathematics education in a periodical that was original with the author but not original to mathematics.
- A *mathematician* is someone who contributed an original piece of research mathematics.

The distinctions among these three terms will be used throughout this book.

Baron's New York Mathematical Club decided to publish the country's first periodical devoted strictly to mathematics, with Baron as editor-in-chief and a cast of associate editors—neither named nor numbered—assisting him. Together they established *The Mathematical Correspondent* as a quarterly publication with issues of 24 pages appearing in May, August, November, and February.[9] Sometimes, additional pages were added. The first issue appeared on May 1, 1804. Baron's role with the publication ended with a double issue numbered 9 and 10 that appeared in November 1806 and which completed Volume I. Volume II commenced the next year under the editorship of Robert Adrain, but only one issue was published, and the entire project folded. This short period indicates the lack of a critical mass of mathematical practitioners in the early part of the nineteenth century.

Baron wrote the verbosely titled initial article, "A New elucidation of the principles of the Rule of Proportion in Arithmetic; applied to the resolution of practical questions and to the invention of general rules for making in the neatest and shortest manner possible many of the most useful calculations that daily occur in the counting house," that began the first issue of *The Mathematical Correspondent* in 1804. However, the journal's emphasis was on problems, which accounted for all of its space during its brief existence, except for five papers, one note, and two reprints. Baron himself wrote three of the five papers, and Robert Adrain the other two. The *Correspondent*'s importance, however, lies not so much in its level of mathematics but with the community of individuals that contributed to it by submitting or solving problems. With the exception of Adrain, these contributors were not mathematicians. Frederik Nebeker described them as follows:[10]

> [Scientists who] were directly engaged in the practical tasks of surveying, cartography, time determination, almanac making, navigation, and hydrography. Others, such as those pursuing meteorology, climatology, seismology, or the study of terrestrial magnetism, believed that their work would one day be of great practical value.

Nonetheless, many of the journal's contributors were mathematical enthusiasts who devoted their preciously scarce leisure time and resources to posing and solving problems. Altogether 63 different mathematical enthusiasts proposed or solved problems. The most active group, besides the mathematical practitioner Baron and the mathematician Adrain, consisted of individuals who are virtually unknown today: John D. Craig, John Craggs, Neil Gray, Thomas P. Irving, William Lenhart, Thomas Maughan, John Smithis, Diarius Yankee, and Noah Young. I know that Craig was a teacher in Baltimore, but little else is known about any of them, except for their location. Some relevant details are known about six of the mathematical enthusiasts who contributed to the problem section; additional facts are available online at "Web03-MathCorr."

William Lenhart (1787–1840) was strongly influenced by the *Correspondent* as a teenager. That he achieved such success was pure serendipity, because when he began his education at age 13, his hometown York County (PA) Academy was run by Robert Adrain! Although Lenhart's father withdrew his son after 18 months, his talents were obvious to Adrain. Therefore, Lenhart was a 17-year-old farm boy when he first solved problems in the fledgling journal. In early 1805 he moved from York to Baltimore, where he was employed, in his words, "selling knob-locks and butt-hinges."[11] This is where he resided when he won a Baron Prize for solving the hardest problem in that issue, the most prestigious award that existed for mathematics at that time. Lenhart never held an academic position, and he did not engage in mathematics from 1812 to 1825 because a spill from a carriage left him disabled. However, his mentor, Robert Adrain, convinced him to contribute to his latest journal, the *Mathematical Diary*. Lenhart openly admitted that his taste for mathematics was restricted to "old fashioned pure Geometry and the Diophantine Analysis," yet the historian of mathematics Florian Cajori listed him among "the more prominent mathematicians of America."[12] One can only wonder what Lenhart might have accomplished if not for the tragedy that befell him.

Robert Maskell Patterson (1787–1854) was another young active problemist deeply influenced by the *Correspondent*. His educational opportunities contrasted sharply with Lenhart's, as his father Robert Patterson had been professor of mathematics at Penn since 1779. The younger Patterson graduated from Penn in 1804 shortly after Meriwether Lewis had studied global positioning with his father. Upon graduation the son enrolled in the country's first medical school at Penn. This means that he posed two problems and solved 11 others while pursuing the MD degree he obtained in 1808. After studying in Europe for three years, he held chairs at Penn and Virginia. The elder Patterson subscribed to the *Mathematical Correspondent* but did not otherwise participate, although he did contribute vigorously to successor journals.

Many *Correspondent* problem solvers attained considerable recognition in other fields. Four of these mathematical enthusiasts indicate some of the obstacles these early Americans faced. **Walter Folger, Jr.** (1765–1849) spent his entire life on the island of Nantucket, attending only elementary school but educating himself sufficiently in mathematics and science (as well as French) to stay in touch with major developments on the mainland. At age 23 he constructed "Folger's astronomic clock," which designated the time, day, and year, showed the paths of the sun and moon, and demonstrated various other planetary phenomena. Subsequently, he became a land surveyor, telescope maker, almanac author, and lawyer. It was during his time as a lawyer that he proposed one problem in the *Correspondent* and solved eleven.

Moreover, Walter Folger asked George Baron to republish a British account of differential calculus based on the American edition of the Rev. Samuel Vince's book, *The Principles of Fluxions*. Here are four results taken from pages 136–138 of the Vince article in the *Correspondent*:

> Prop. III: If the fluxion of x be denoted $\dot{x}$, the fluxion of ax will be $a\dot{x}$.
> Prop. IV: The fluxion of $x \pm a$ is $\dot{x}$
> Prop. V: Given ($\dot{x}$) the fluxion of x, to find the fluxion of x^n, n being a whole number.
> Prop. VI: To find the fluxion of $x^{\frac{n}{m}}$, m and n being a whole number.

A reader with knowledge of first-semester calculus should recognize these results, though probably not in this form, and should be able to complete Propositions V and VI. Vince supplied proofs using the definition of a fluxion as a limiting ratio. For Proposition V he concluded, "Therefore $nx^{n-1}\dot{x}$ represents the cotemporary fluxion of x^n." Using this as a guide, can that reader deduce the form of Vince's conclusion to Proposition VI?[13]

Another contributor to the problems section in the *Correspondent* was **Enoch Lewis** (1776–1856). Lewis, who ended his formal education by age 15, worked on a survey for a year in western Pennsylvania under Andrew Ellicott before returning to Philadelphia as a mathematics teacher at the Friends Academy. In 1799 he moved to the newly opened Westtown Boarding School, located about 20 miles west of the city. In addition to teaching duties, he found it necessary to engage in farming to support his 15 children. Nonetheless, he was able to submit correct solutions to ten of the 11 problems in one issue of the *Correspondent*.

The Westtown Boarding School bore ripe fruit in one of its students, **John Gummere** (1784–1845). At age 19 Gummere began a career as a teacher, but stopped the next year to study in Westtown under Enoch Lewis for six months. He then resumed teaching in Burlington, NJ, where he resided when he submitted solutions to problems in the *Correspondent*. Papers on astronomy in the *Transactions of the APS* and the

Memoirs led to his election to membership in the American Philosophical Society. He also wrote the books *A Treatise on Surveying* (1814), which went through 22 editions, and *Elementary Treatise on Astronomy* (1822), which was adopted at several colleges, including West Point. As a result of these publications, John Gummere should be regarded as a mathematical practitioner, not just an enthusiast. In 1833 he was appointed the first mathematics professor at the newly opened Haverford College.

Our final enthusiast, **John Eberle** (1787–1838), was schooled at home in Lancaster, where he spoke a dialect of Pennsylvania Dutch, not English, until age 12. Yet he studied medicine in Lancaster before enrolling in medical school at Penn in 1806. This means that Eberle was farming by day and studying by night when he proposed and solved problems in the *Correspondent*. In 1824, he and two other physicians founded Jefferson Medical College.

The *Correspondent* contained a subscription list of 345 individuals—including New York mayor DeWitt Clinton and well-known public servant Alexander Hamilton. Yet, like all other such attempts up to 1878, this journal had a short duration and little impact. In fact, it failed miserably, not just because its chief activity was problem solving, but because Baron and several contributors engaged in personal attacks on other individuals in its pages—including Nicholas Pike, Jared Mansfield, and Nathaniel Bowditch. For instance, the second issue contained an advertisement for a lecture by Baron with "a complete refutation of the false and spurious principles, ignorantly imposed on the public, in the 'New American Practical Navigator' of Nathaniel Bowditch." A scholar writing 70 years after the demise of the *Correspondent* commented, "These Editors, being of Hibernian descent, were prejudiced against American authors."[14] Seventy years after that, in 1940, *American Mathematical Monthly* founder B.F. Finkel found Baron's stinging words so malicious that he described him in one place as "an irascible and egotistic character"[15] and in another as an "egotistic, imperious, and sarcastic editor."[16] However, to Baron's credit, the *Correspondent* encouraged other individuals to establish their own journals, and not just in mathematics; for instance, the physician John Eberle was associated with five fledgling medical journals in his lifetime.[17]

The *Correspondent*'s publication community consisted of 63 different contributors, five of whom were authors, 38 proposers, and 58 solvers. If individual subscribers are included and intersections factored out, one is left with a surprisingly large group of 362 participants. However, the diverse set of subscribers—ranging from scientists to politicians—would not be repeated in later journals, mainly due to the proliferation of specialized scientific periodicals beginning about 1810. The professions of this extended group of *Correspondent* participants typify mathematical investigators in the country up to the founding of Johns Hopkins University in 1876.

In spite of a proliferation of colleges and universities founded in the first half of the nineteenth century, relatively few mathematics professorships arose, so most participants continued to find employment outside academia. But even in the nation's colleges, almost all mathematics professors—including *Correspondent* participants—taught other subjects. Only three of the six leading problemists taught in a college at one time or another R.M. Patterson also served as director of the US Mint while both Robert Adrain and John Gummere also taught in private schools. Moreover, Enoch Lewis taught in private schools and farmed, John Eberle practiced law and held elected office, and William Lenhart was a bookkeeper/accountant. Overall, *Correspondent* contributors were doctors, lawyers, surveyors, clergy, judges, clock-makers, farmers, teachers, and professors, while the ranks of subscribers included important scientists and political figures.

Misspelled names in the *Correspondent* indicate that some contributors were unknown before their initial participation. For instance, Eberly in the *Correspondent* was corrected to Eberle in Adrain's *Analyst*. Even Ferdinand Hassler was initially called John Hasler in the *Analyst*, an error corrected in the next issue, when he was listed as professor of mathematics at the US Military Academy at West Point.

Robert Adrain

Several professors of mathematics and natural philosophy subscribed to the *Correspondent*, including George Blackburn at William and Mary (and later South Carolina), Joseph Caldwell (North Carolina), Collin Ferguson (Washington College), John Maclean (Princeton), Robert Patterson (Penn), and Samuel Webber (Harvard). The most qualified, however, was **Robert Adrain** (1775–1843), arguably the first creative mathematician in the US. Born in a coastal town near Belfast, Ireland, Adrain's father was a schoolmaster and maker of mathematical instruments who set him on a classical education intended for the ministry. However, the youngster's formal education stopped at age 15 when both parents died, leaving him in charge of four siblings. He then assumed his father's position as a teacher and began a program of self-education, especially in mathematics.

However, Adrain's life was altered dramatically in 1798 when he joined the insurgency of the Society of United Irishmen, a coalition of Catholic and Protestant forces opposed to British rule. He was shot in the back during that ill-fated rebellion and left for dead. Friends nursed him back to health, and he narrowly escaped to America with his wife Anna Pollock and their infant daughter. The star-crossed Adrain jumped out of the pot and into the fire, because he landed in New York in the midst of a cholera epidemic. Some luck did intervene, however, when he found refuge with the widow of the founder of the United Irishmen in Princeton, where he served as the master of mathematics at Princeton Academy (not the university). It is reported that he was no stranger to the rod there;[18] however, "he was also described as a kind and patient teacher, who would gladly tutor students who sought him out for assistance."[19] Two or three years later, Adrain moved from Princeton to York (then called York-Town), PA, to become headmaster of York Academy. This is where he resided when he heard about the *Correspondent*. He continued his activity with the journal when he moved to Reading, PA, in 1805 to accept a similar position as principal.

The *Correspondent* lit a fire in Adrain so bright that he solved every one of the 82 problems, and he was the only person to win the Prize Problem more than once. (The person submitting the best solution to the last problem in each issue was awarded a vase valued at $6. Even given the rise in inflation, few people in the history of mankind have gotten rich by solving mathematics problems.) Moreover, Adrain wrote two very good papers in the journal, with an article on Diophantine algebra being the first of its kind published in America. Therefore, it seemed natural for the associate editors to choose Adrain to succeed Baron as editor. The editors promised a second volume, and Adrain produced the first issue after a one-year hiatus, but he was so unhappy with the quality of the printing that he did not publish another. Although Adrain assured

Figure 3.5. Robert Adrain

Correspondent readers that "nothing unbecoming a Christian and a gentleman shall be suffered," the journal ceased publication after that one issue in 1807. This put a formal end to the country's first mathematics journal.

The *Correspondent* did not produce original mathematics judged by European standards—its only substantial paper was Adrain's "A view of the Diophantine algebra," which appeared in the final issue. Similarly, in the 70 years from the demise of the *Correspondent* to the opening of Johns Hopkins, only a handful of Americans carried out true research in mathematics. However, Adrain had caught the journal bug from the *Mathematical Correspondent*. The following year he established another mathematics journal, the *Analyst; or Mathematical Museum*. Although he generally referred to it only as *The Analyst*, I use *Mathematical Museum* to distinguish it from subsequent periodicals also called *The Analyst*, one of which Adrain established in 1814 as a revival of the *Mathematical Museum*.

The preface to the *Mathematical Museum* shows Adrian's wide reading of the masters—he mentions Pascal, Leibniz, the Bernoullis, Huygens, Newton, Euler, and Lagrange as well as British stalwarts Maclaurin, Emerson, Simpson, Hutton, and Vince. Adrain also cites two continuing enthusiasts from the *Correspondent*, the Baltimore high-school teacher, John D. Craig, and Thomas Maughan of Quebec, about whom nothing seems to be known. The first issue of the *Mathematical Museum* was an exact copy of the final *Correspondent* issue Adrain had edited in 1807.[20] Apparently, he was so distraught with the printing at Reading that he moved the whole enterprise to Philadelphia, where it was set by the printers Fay and Kammerer. He then republished the entire previous issue at his own expense. It seems clear that Adrain aimed his journal at a higher level than its predecessor, beginning with a continuation of his paper "View of Diophantine algebra," which was followed by an essay titled, "Observations on the study of mathematics." The second issue of the *Mathematical Museum* was devoted mostly to problems, but they too offer evidence of a slightly higher level. For instance, one problem asked for solutions to the equations $x^x = 100$ and $x^{x^{x^x}} = 10$. The third and fourth issues followed the template established in the *Correspondent* of sandwiching an article with new results between problem sections, with Adrain writing both articles. The one in the third issue, "Researches concerning isotonous curves, inspired by Rittenhouse's hygrometer," reflects the enduring influence of David Rittenhouse.

The fourth, and final, issue of the fleeting *Mathematical Museum* is prominent both for the quality and originality of the paper that appeared and for its lack of recognition. Adrain's 1808 article "Research concerning the probabilities of the errors which happen in making observations," introduced the least squares method (LSM). His motivation was to solve a problem in surveying posed by Penn's Robert Patterson that has been described as follows:[21]

> The problem concerns a land surveyor who traces the boundary of a polygonal field, measuring each of its five sides by traversing a prescribed distance on a prescribed angular bearing. At the end, the survey should return to the starting point, forming a closed pentagon, but instead there is a small gap. ... The idea is to choose from among all possible adjustments those that put the vertices in their most probable positions.

Later in life Adrain used the LSM to calculate the ellipticity of the Earth, which he computed to be 1/319, an improvement on Laplace's 1/336, but one that was refined to the accepted value today of 1/297 when determined by Nathaniel Bowditch.

Who deserves priority for inventing the LSM? Today credit is generally given to Karl Gauss, the German superstar of the first half of the nineteenth century, who claimed priority in 1809, one year after Adrain's publication. However, the accomplished Frenchman Adrien-Marie Legendre had invented the method, named it, and published it in a book on comets three years before Adrain published his solution. Soon Legendre's senior colleague in the French Academy of Sciences, Pierre-Simon Laplace, also contributed to the development. Unfortunately, the *Mathematical Museum* had virtually no readership within the US, let alone outside its borders, so Adrain's discovery went unnoticed. According to the late-nineteenth century mathematician/engineer Mansfield Merriman, "Adrain's paper was unknown to mathematicians until 1871, when it was republished in *American Journal of Science*, Vol. **I**, pp. 412–414."[22]

Adrain, in fact, provided two demonstrations of what he called the law of errors. Although few copies of the *Mathematical Museum* are available today, his proofs were reproduced in an article by the *Monthly*'s founder Benjamin Finkel in a journal that, like the *Mathematical Museum*, is difficult to access in print today but, thanks to modern technology, is available on JSTOR, thus enabling a modern reader to access it easily.[23]

Nonetheless, the fact that an unknown person working in a virtual American mathematical wasteland can also be mentioned in this regard is quite impressive, though admittedly Adrain's accomplishments "do not put him in the first rank of nineteenth-century mathematicians."[24] A recent account of Adrain described the necessity of a *community* by asserting, "It takes more than a village to raise a scientist. It takes a village full of scientists." However, two contributors to Adrain's journal deserve mention—this chapter's namesake, Nathaniel Bowditch, and Ferdinand Hassler.

In order to satisfy mathematical enthusiasts who contributed problems and solutions to the *Correspondent*, Adrain continued the problems column in the *Mathematical Museum*. John Eberle continued to solve problems while completing his MD program at Penn. In addition, Adrain attracted a new bunch of enthusiasts, including

John Garnett (ca. 1748–1820), a farmer from New Brunswick, NJ, who gained fame as an astronomer of distinction. In 1809 the APS awarded him "an Extra-Magellanic Premium of a Gold medal" for a paper he published that year in the *Transactions* on a nautical chart for navigation. Garnett resided near Burlington, NJ, where John Gummere operated his school. Several other mathematical enthusiasts from that area participated actively with the *Mathematical Museum*, suggesting the existence of one of the country's first problem-solving groups. Interestingly, the seventh and final child in the Adrain family was named Garnett Bowditch Adrain.

The New Brunswick connection, along with his budding reputation, served Adrain well, as he was offered a professorship at Queen's College in 1809, where he remained for four years. In 1812, and despite rising nationalistic feelings due to the outbreak of another war with Great Britain, Adrain published an American edition of Hutton's *A Course of Mathematics for the Cadets of the Royal Military Academy of Woolwich*, whose original had appeared in 1801. Charles Hutton was a professor of mathematics at the Royal Military Academy for 34 years. Even his students had trouble with it, which led Adrain to add a large number of explanatory notes and several additions to make it more appropriate for American students.

After Queen's College, Adrain taught at Columbia College (King's College up to 1784) from 1813 to 1826. In 1821 Columbia became one of the first American colleges to require algebra for admission. His students at Columbia referred to him as "Old Bobbie," and the class of 1822 presented him with a self-portrait. While at Columbia, he contributed mathematical material to several periodicals, including the Charles Hutton–edited *Ladies' and Gentlemen's Diary*. That same year, a reprinting of his version of the Hutton compendium contained a 62-page essay on descriptive geometry.

Upon leaving New York City, Adrain accepted his old position at Queen's College, which had been closed due to financial distress but was renamed Rutgers College in 1825. However, after only one year, he returned to Philadelphia as professor of mathematics and natural philosophy at the University of Pennsylvania. He maintained residences in New York City and Philadelphia; therefore, it is no wonder that, "Adrain was always pressed for funds to support his large family (seven children) and his work."[25] While at Penn in 1832, when he also served as provost, he found time to edit a book on globes by Thomas Keith. The author claimed that the Andes were the highest mountains in the world, but Adrain suggested that "Peak 15," renamed Mount Everest in 1865, held that distinction. Adrain resigned his position at Penn two years later, taught at a private school in New York, retired in 1840, and died at the family home in New Brunswick three years later.

Yet Adrain was able to launch two more journals during these peregrinations. In 1814 he established *The Analyst*, which consisted of only one issue and can be regarded merely as a continuation of the *Mathematical Museum*.[26] Once again Adrain appealed to mathematical enthusiasts, writing in the preface, "It will be proper … to assign a suitable portion … to problems within the reach of the studious though less experienced contributors." But apparently this was not enough to attract a sufficient number of subscribers. There are two notable aspects of *The Analyst*, the first being Adrain's article, "Algebraic method of demonstrating the propositions in the fifth book of Euclid's *Elements*." The other is on a lighter note: J. Roosevelt of New York proposed a problem a century before two other presidential Roosevelts tried to solve problems of a different ilk.

William Marrat was another enthusiast whose effort to found a mathematical journal failed miserably. Marrat had edited *The Enquirer* in England 1811–1813 before immigrating to the US. He initiated the *Monthly Scientific Journal* at Perth Amboy, NJ, in 1818. However, only seven monthly issues appeared that year from February through August, and two in July and October 1819, before the initiative folded. Marrat returned to England a few years later.

Robert Adrain was undeterred by the collapse of two of his journals—the *Mathematical Museum* after one year, and the (continuation of the) *Analyst* after one issue—as well as the *Monthly Scientific Journal*. Nor did the three-issue duration of the *Ladies' and Gentlemen's Diary or United States Almanac* (edited by Melatiah Nash) in 1819 faze him. In 1825, Adrain founded yet another journal that adopted one word from the title of Nash's entry. His *Mathematical Diary* survived for seven years and became the first American mathematics journal to include reviews of mathematical publications, including some that appeared in Europe. The title page indicates that the *Diary* was published by James Ryan at his bookstore on Broadway in New York City. Perhaps the reason for the relatively long duration of the *Diary* is that it was mostly concerned with problem solving, thus lacking the high quality of the *Analyst*. At least two solutions to problems in the *Mathematical Diary* were signed "Mary Bond;" this was not a woman, rather it was a pseudonym used by William Lenhart to get more of his contributions published.[27]

It is noteworthy that Benjamin Peirce was credited with solving a problem in the July 1827 issue, the first time the name of this Harvard student appeared in print. Five years later, when Peirce was professor of mathematics and natural science at Harvard, he published what remains as the very most important contribution of the *Diary*, a paper on perfect numbers that B.F. Finkel republished in its entirety in two parts a century later.[28]

Robert Adrain served as the editor of the *Mathematical Diary* for only one year because his move from Columbia to Rutgers in 1826 prevented direct involvement with the endeavor. The journal's publisher, **James Ryan**, took over as editor the next year, although Adrain continued to contribute to the journal. Little is known about Ryan, including birth and death years, but apparently he wrote textbooks on algebra and calculus, as well as a book on astronomy.

The *Diary* was a quarterly for the first two years but an annual publication after that, resulting in a run of 13 issues altogether. The final issue of the *Diary* appeared in 1832. Its title page stated that it was "Conducted by James Ryan, AM," indicating he had earned, or had been awarded, a master's degree. That issue included a short biography of Lagrange written by Σ, a pseudonym of **Samuel Ward**, 3rd (1814–1884). Ward had attended the influential private Round Hill School from age nine to 15, when he entered Columbia. He graduated from Columbia in 1831 after only 2-1/2 years, but during this time he wrote articles and solved problems in the *Diary*. Then he moved to Boston to be tutored by Nathaniel Bowditch for several months. While there, he also met Benjamin Peirce. In early 1832, a large share of the editing the *Diary* fell to Ward. Moreover, Ward's wealthy father helped with the flagging finances of the journal. Nonetheless, Ward proved to be disastrous for The Diary. A writer from 1870 vibrantly described the scene of the disaster:[29]

> For six years he [Ryan] ably conducted The Diary, and he would have
> continued to do so for many years longer, but for an unfortunate quar-
> rel among the mathematicians. Mr. Samuel Ward 3rd ... had in part
> the management of the last number [issue]. In it he caused to be in-
> serted a Dialog, written by himself, wherein he exhibits in a ridicu-
> lous light, Dr. Henry J. Anderson, then Mathematical Professor in
> Columbia College. ... Both the Editor and Dr. Anderson were highly
> indignant at this performance. The parties met at the mathematical
> book-store of James Ryan, and high words passed between the par-
> ties and their friend. The result was the complete breaking up of the
> *Diary.*

Therefore, one reason for the dissolution of the *Diary* was the same as the *Correspon-dent*—personal attacks. That incident was regrettable because Samuel Ward was a capable mathematical practitioner. He edited an American edition of J.R. Young's *Algebra* while assisting Ryan with the *Diary*. Ward had so impressed the authorities at West Point that he was offered a position, but he felt the need for higher education in mathematics and military engineering before accepting, so he traveled to Europe, where he studied with Cauchy and Poisson in Paris before moving to Germany to study with Gauss. Relatively few Americans traveled abroad at the time, but Ward's family supported his sojourn, which culminated with a doctorate in mathematics from Tübingen in 1834. Samuel Ward thus became the first American to achieve that feat, though admittedly, criteria for obtaining the degree were much easier than later in the century. Upon his return to the US, he donated to a New York library the personal libraries of Legendre and Halley that he had purchased during his travels. Ward never did teach at West Point. Instead, he left mathematics altogether. Historian Edward Hogan wrote, "After his father's death, Ward lost his fortune by excessive speculation."[30]

In spite of the drama involving the end of the two journals *Correspondent* and *Diary*, the major lesson to be drawn from the experience of trying to establish a successful mathematics journal—a forewarning that would be repeated for the next fifty years— is that true progress cannot be made until a *community* of scholars exists. The critical mass for success is greater than one might imagine. As already noted, the subscription list for Baron's *Mathematical Correspondent* contained 345 individual names (and two institutions), which seems like a sufficiently large base. However, even taking into account the journal's emphasis on problem solving, only 63 individuals actively engaged in solving or proposing problems.[31]

Robert Adrain appears to have been a lone wolf crying in the wilderness. His legacy would seem to lie with his dual roles as a teacher and a mathematician. Harvard mathematician Julian Coolidge concluded his study of Adrain as follows: "There can be no question as to his outranking every American mathematician who was really his contemporary."[32] More recently, Frank Swetz concluded his investigation: "Ultimately, he should be recognized primarily as a mathematics educator, perhaps America's first mathematics educator."[33]

Only two different authors wrote articles for the *Correspondent*. This is not surprising since most scientifically oriented people in America at the time pursued observational studies such as botany, zoology, and geology, which do not require the background necessary for understanding mathematical sciences. In this sense, mathematics had not progressed much since the appearance of Harriot's *A Briefe and True Report of the New-found Land of Virginia* some 200 years earlier.

A recent study, which sheds light on the paucity of Americans engaged in writing articles on mathematics in the New Republic, substantiates these assertions. Todd Timmons studied all mathematics papers that appeared in three nonspecialized scientific journals during 1771–1834—the *Transactions of the APS*, the *Memoirs of the American Academy of Arts and Sciences*, and the *American Journal of Science and the Arts* (*AJSA*).[34] He was able to identify only 81 authors of mathematics papers in that 63-year period, amounting to roughly 1.3 per year. By contrast, the editor of the *American Mathematical Monthly* revealed that in the year 2014 he received roughly three submissions *per day*! Yet the individuals who wrote papers in the Timmons study were a step up from the mathematical enthusiasts who solved problems in the *Correspondent*. Timmons labeled them "mathematical practitioners," the term I borrowed to contrast with enthusiasts and mathematicians. Every American engaged in mathematical research during 1776–1876, except six giants—Robert Adrain, Nathaniel Bowditch, Benjamin Peirce, George Hill, Josiah Gibbs, and Simon Newcomb—can be regarded as a practitioner.

A prototypical example of a mathematical practitioner from this period is **Parker Cleaveland** (1780–1858), an honors graduate from Harvard (1799) who was appointed professor of mathematics and natural philosophy at Bowdoin College in 1805 (11 years after the college was founded) and who remained there for the rest of his life. The first building on that campus was constructed in 1799 and was called Massachusetts Hall. Maine ceased to be a part of the Commonwealth of Massachusetts in 1820 when it became a state, so Bowdoin College is located in Portland, ME, today.

A Bowdoin graduate from 1820 described the situation of the college and Cleaveland's role in it during its formative years:[35]

> Until near the close of our college life we had but one professor with the president and two tutors. Professor Cleaveland added to his duties as teacher of the natural sciences … those of instructor in mathematics; although … in mathematics, one of the tutors from part. … Of the books they used, the first, and that which went with us … from the freshman year to the senior, was Webber's *Mathematics*. The only other book of pure mathematics was Playfair's edition of Euclid.

The Webber series was soon replaced by a series of books from John Farrar of Harvard. That same student described the typical method of instruction at the time:[36]

> We were required to study the prescribed lesson in the book, then to repeat it, not of course word for word, but distinctly, to the professor or tutor at the recitation.

Parker Cleaveland was professor of mathematics 1805–1835 and published a few papers in mathematics early in his career. However, his major interest soon turned to geology and mineralogy; according to Cajori, Cleaveland "earned for himself the enviable reputation of 'Father of American Mineralogy'."[37] His 1816 book, *An Elementary Treatise on Mineralogy and Geology*, was not only the first American textbook on these two subjects but was uniformly lauded at once and resulted in a two-volume second edition just six years later.[38] Overall, he was one of the few people to write articles in more than one of the general science journals of the time, publishing four in the *Memoirs* of the American Academy and two in the *American Journal of Science and the Arts* (*AJSA*).

A mysterious person who also serves as a mathematical practitioner from this period is A.B. Quinby. Although he published 12 articles in the *AJSA* between 1824 and 1828, almost nothing is known about this New York resident whose "identity continues to be a mystery."[39] Quinby also published papers in Robert Adrain's journal, the *Mathematical Diary*.

One other mathematics journal allegedly appeared about midway during the *Diary*'s run and aimed to compete with that journal. Based on the evidence, the *Mathematical Companion* can be regarded as a pre-journal. In 1828, an eight-page pamphlet appeared with a title page resembling a journal. It stated that *Mathematical Companion* was "Conducted by John D. Williams in New York City." However, the inside of the front cover states that the first issue would be published on May 1, 1829. One source stated that Harvard was the only library with a copy of issues that ran from 1829 through 1831. But in 1940, B.F. Finkel reported that 31 years earlier, William Byerly, one of the first figures to earn a PhD in mathematics from an America university, had loaned him a copy of the pamphlet. Byerly wrote to Finkel:[40]

> I have had the Harvard library searched high and low for the *Mathematical Companion* with the result of unearthing among our stray pamphlets the enclosed copy apparently a prospectus of the proposed publication. It is the only number of the *Companion* that we possess.

The pamphlet was divided into two parts. The first five pages consisted of 36 problems, some of which were lifted from the *Diary*. The final problem was a Prize Question:[41]

> Suppose a perfectly elastic uniform circular hoop, of very small thickness and density is suspended by a given pivot on its plane vertical. It is required to determine its form supposing all its parts to be acted upon by a uniform gravity.

All eight pages consisted of problems, with the second part numbering 27 altogether.[42]

Little is known about John D. Williams. He was an occasional contributor to the *Diary* but not to subsequent mathematical journals. He revised and corrected at least four elementary mathematics textbooks, but his most important work was the 605-page book *An Elementary Treatise on Algebra, in Theory and Practice* (Boston, 1840).[43] The book's title belies the word "elementary," as a great deal of space was devoted to infinite series and Diophantine analysis. An example of a problem on Diophantine analysis taken from a list of 14 questions that Williams proposed in 1832 was the following:

> Make

$$\left(m^2 + n^2\right)^2 x^2 \pm \left(m^2 + n^2\right) x = \Box,$$
$$\left(m^2 - n^2\right) x^2 \pm \left(m^2 - n^2\right) x = \Box,$$
$$4m^2 n^2 x^2 \pm 2mnx = \Box.$$

Williams wrote, "A certain teacher in this city … has offered to solve the question for $5.00; I now offer him $20.00 to prove either its possibility or *impossibility*, and show all the conditions that can exist and those that cannot."[44] ($20 in 1832 was the equivalent of about $570 in 2014.)

Apparently, Williams challenged all mathematicians in America to solve the 14 problems. Curiously, he excepted three leading figures from the challenge: Nathaniel Bowditch, Eugene Nulty, and Theodore Strong. Moreover, in his challenge Williams specifically named Robert Adrain at Penn, Henry Anderson at Columbia, and Samuel Ward 3rd from New York City, among others. All 14 problems were solved by several of the listed mathematical practitioners, including the fact that the Diophantine problem stated above has no solutions.

Textbook series

While the appearance of a specialized journal and the founding of the Military Academy were important developments at the beginning of the age of Bowditch, a significant change concerned textbooks, a topic intertwined with the evolution of mathematics at Harvard and Yale. Even after the Revolutionary War, most books on arithmetic continued to reflect a decidedly English influence. However, books on algebra and geometry showed a pronounced French influence.

A notable development with textbooks took place at Harvard with the two immediate successors of the estimable John Winthrop, though neither Hollis Professor matched his accomplishments. Like Winthrop, **Samuel Williams** (1743–1817) came from a family of patriots; his grandfather William Williams had signed the Declaration of Independence as a representative of the Colony of Connecticut. Samuel Williams was a 1761 graduate of Harvard whose proficiency in mathematics induced Winthrop to select him as his companion to observe the famous transit of Venus in June of that year. After that, Williams studied theology while teaching school, was licensed to preach in 1763, and became pastor of a church two years later. He remained in this position until accepting the Hollis Professorship of Mathematics and Natural Philosophy in 1780. However, he resigned eight years later to return to pastoral duties without having produced any notable mathematics outside the many manuscripts he wrote for use in his classes in mathematics, astronomy, and natural philosophy.

While Williams's successor **Samuel Webber** (1759–1810) left a more impressive legacy at Harvard, most of it was due to his term as president of the College from 1806 until his death four years later. As Webber came from a farming community, his entrance into Harvard was delayed until age 20. He graduated four years later, in 1784, taking high rank in mathematics. After that he remained on campus to study theology for two years, and was ordained a Congregational minister in 1787. That same year he was appointed tutor, a position that had changed 20 years earlier, when the university replaced the previous policy in which tutors taught all branches of study, to one where they were responsible for only one subject. For Webber that meant mathematics. He remained tutor until Williams resigned the Hollis Professorship, whereupon Webber was appointed to the position he held until assuming the presidency of the College in 1806.

Webber had made his mark in mathematics before becoming president with the 1801 publication of the two-volume compendium, *Mathematics: Compiled from the best authors, and intended to be the text-book of the course of private lectures on these sciences in the university at Cambridge.* This work was apparently the second mathematics book written for colleges by an American author, preceded only by Greenwood's *Arithmetic, Vulgar and Decimal.* Pike's *New and Complete System of Arithmetic,* like all books on arithmetic up to that time, was written for academies. Webber's compendium marked quite a feat, as each volume numbered 460 pages. The topics covered were arithmetic, logarithms, algebra, geometry, plane trigonometry, mensuration of surfaces, mensuration of solids, gauging, heights and distances, surveying, navigation, conic sections, dialing, spherical geometry, and spherical trigonometry. (In navigation, the term *gauging* means determining the position of a vessel relative to the wind or to another vessel.) Notice that a significant portion of the compendium was devoted to applied topics, called "mixed mathematics" at the time. The influence of Webber's compendium was limited by the inability of American students to understand its contents, even at Harvard. Fifty years later, in 1853, a student recalled, "Pure mathematics was a very unwelcome study. Nine tenths of every class had broken down in quadratic equations; seven eights did not get so far."[45]

Although this work and its second edition (1808) were used as textbooks at Harvard for teaching algebra, it represents Webber's only contribution to mathematics. However, its popularity was short lived, exceeded by a Yale series in 1814. Webber's appointment as president left the Hollis chair vacant for a while but it was soon occupied by John Farrar, who turned out to be an important cog in the production of textbooks for college students. But before Farrar got involved with texts in Cambridge, several notable developments took place in New Haven.

American Journal of Mathematics. In 1802, Yale created a new professorship of "chymistry" and natural history (which in those days referred to geology, mineralogy, botany, and zoology) for **Benjamin Silliman** (1779–1864), who had received his BA degree from the college in 1796. In preparing to teach chemistry, Silliman realized that his meager knowledge prevented him from learning it sufficiently independently, so he spent two winters studying at the Medical College of the University of Pennsylvania. In his spare time he performed experiments in a laboratory set up in a cellar kitchen of his boarding house. Silliman also felt the need for further studies in natural history. Because America housed no experts, he appealed to Yale to support him for a year abroad. Thus he made a four-week voyage across the Atlantic to spend 1805–1806 in England and Scotland, notably at the University of Edinburgh. Silliman emerged "with a solid, up-to-date background in theoretical and experimental chemistry, and a practical knowledge of geology, mineralogy, zoology, and medical subjects."[46]

Benjamin Silliman left two enduring contributions to American mathematics. One was the journal he founded, the *American Journal of Science and the Arts* (*AJSA*), which he began publishing the next year with four annual issues. *AJSA* was like the *Transactions of the APS* and *Memoirs* of the American Academy in two ways—it was a general science periodical and it emphasized the utility of science (at least in the beginning). However, *AJSA* was unlike its predecessors in three ways—it appeared on a regular basis, was not regional, and was not associated with a learned society. *AJSA* soon became an important outlet for mathematical practitioners, including two notable ones. The recent book by Todd Timmons analyzed the nascent American publication community by examining mathematical papers that appeared in these three general-science journals at the time.[47]

The first four volumes of *AJSA* included only two papers on mathematics, both by Theodore Strong, then at Hamilton College, but both articles were judged to be "of a rather simplistic nature."[48] **Theodore Strong** (1790–1869) entered Yale in 1808

Figure 3.6. Benjamin Silliman

when Jeremiah Day held the chair of mathematics. Strong graduated four years later and then became a tutor in mathematics at Hamilton. He was appointed professor of mathematics and natural philosophy in 1816, and remained at Hamilton for 11 years. During that time, in language from 140 years ago,[49]

> Strong early formed the acquaintance, either personally or by corre-
> spondence, of the most eminent men addicted to pure science on this
> side of the Atlantic, including Dr. Nathaniel Bowditch and Dr. Robert
> Adrain, and others of like proclivities and pursuits.

Strong moved to Rutgers College in 1827 (Queen's College until 1825; Rutgers became the land-grant college in New Jersey in 1864) and remained there until retiring in 1861. His very best student was George W. Hill, who enrolled at Rutgers in 1855 and made profitable use of Strong's library of classic mathematics books. Strong was selected as a charter member of the National Academy of Sciences (**NAS**) when it was formed in 1863.

Theodore Strong wrote about 60 papers on mathematics, including 22 published in each of the *AJSA* and *Mathematical Miscellany* as well as seven in the *Cambridge Miscellany*. He also read five papers at the NAS, including "On the integration of differential equations of the first order and higher degrees" in 1864 and "A new theory of the first principles of the differential calculus" in 1865. In addition, Strong wrote two textbooks: *A Treatise on Elementary and Higher Algebra* (1859) and *A Treatise on the Differential and Integral Calculus* (1869). None of Strong's contributions was of a high quality, so he is regarded as a practitioner and not a mathematician in the modern sense. However, he was ahead of his time regarding the education of women, as one of his former female students stated:[50]

> He believed in the thorough education of woman [*sic*], and I think
> it was a source of deep regret to him that none of his daughters gave
> any promise of emulating the fame of Mrs. Mary Somerville, whom
> he often spoke of as an example of what woman [*sic*] may achieve in
> the domain of science.

Although most of Strong's papers dealt with mathematics education, one paper in an early issue of *AJSA* is typical of volumes published between 1822 and 1834 (when more papers on mathematics that reflected a more sophisticated level) appeared than in any other journal, including those specializing in mathematics. Many of the *AJSA* papers addressed the usage of calculus and its logical foundations, like those written by Theodore Strong. Todd Timmons devotes an entire section of his book to this topic.[51] Although some papers dealt with mathematics *per se*, not one found permanent importance in the field. Nonetheless, an impressive paper from an 1822 issue of *AJSA* shows how much mathematics had evolved in America over the first part of the nineteenth century. It also demonstrates the role *AJSA* played as an outlet for young, aspiring mathematicians.

It is quite possible that the author, **Alexander Metcalf Fisher** (1794–1822), might have become America's first research mathematician, a label usually reserved for Harvard's Benjamin Peirce. At the age of ten, the precocious Fisher found his school's arithmetic text unsatisfactory, so he wrote his own, *Practical Arithmetic*, now stored in the Fisher Collection at Yale. He entered the university four years later and graduated "first scholar" in the class of 1813 at age 18.[52] After attending Yale's Divinity School, he was appointed tutor in 1815. Two years later, Alexander Fisher was named professor of mathematics and natural philosophy, a position he held until his untimely death. During his tenure, he developed a full course of lectures in natural philosophy—both theoretical and experimental—and began publishing in Silliman's *AJSA*. However, Fisher drowned in a shipwreck off the Irish coast while sailing to France to meet with some of the world's finest scientists.[53] Thus America lost a promising mathematician shortly before the world lost the equally young and extraordinarily talented Niels Henrik Abel (1802–1829) and Evariste Galois (1811–1832).

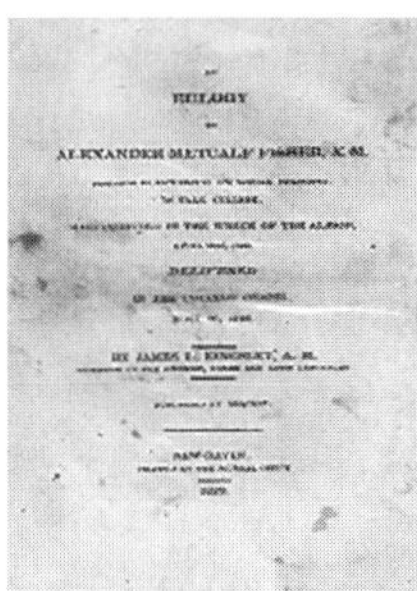

Figure 3.7. Cover page of memorial article on Alexander Fisher

Alexander Fisher should be better known in the history of mathematics, not for what he accomplished—his output was minuscule—but for what he might have achieved if not for the tragic accident that cut his life short. Figure 3.7 reproduces a copy of the cover page of a memorial article on Fisher.

Like Cleaveland, Fisher published papers in more than one of the general science journals of the time, four in *AJSA*, and one in the *Memoirs* of the American Academy. Todd Timmons summed up Fisher's role in American mathematics as follows:[54]

> His impact on American mathematics, had he lived, cannot be known. However, his abilities and interests certainly had Fisher pointed in the direction of leading this generation of American mathematicians into the new age of modern mathematics.

In August 1818, one year after having been promoted to professor at Yale, Alexander Fisher submitted a paper that was published four years later in *AJSA*. Titled "On maxima and minima of functions of two variable quantities,"[55] it presented a method for optimizing a function $u = u(x, y)$ subject to $v = v(x, y)$ being constant. Fisher stated his method as a theorem without proof:

> If u and v are functions of x and y, and u becomes any simple function of v (that is, vary as v^n, $\log v$, a^v, etc.), when $\dfrac{x}{y}$ is supposed constant,
>
> either of the equations $\dfrac{d\left(\frac{u}{\varphi v}\right)}{dx} = 0$ or $\dfrac{d\left(\frac{u}{\varphi v}\right)}{dy} = 0$ gives u a maximum or minimum to a given value of v.

Notably, this statement shows the preference for Leibniz's differentials over Newton's fluxions. However, the theorem is difficult to understand. Moreover, Fisher provided no proof, nor did he define φ. However, he illustrated his method with 15 "problems," one of which states: given the solidity [volume] of a cone, determine when the inscribed sphere is maximum. He concluded that the [volume of the] sphere is maximum when the [volume of the] cone is double the sphere.

This tragic figure's predecessor as professor lived a much longer life, and hence left a more voluminous legacy even though he was not a mathematical practitioner.

Jeremiah Day. Although not a mathematician, Benjamin Silliman played a second significant role in the development of the subject in America with an appointment that provided a forward step in delineating mathematics as a profession because it freed **Jeremiah Day** (1773–1867), the chair of mathematics and natural history at Yale, from having to teach these other subjects as well.[56] Day had entered Yale at age 16 in 1789, but withdrew due to health problems two years later, so he did not graduate until 1795. He returned to campus three years later as a tutor and was appointed professor of mathematics and natural philosophy in 1801.

A serious illness prevented Jeremiah Day from performing his duties for two years, but as soon as he began teaching, he became disenchanted with Webber's compendium. In this sense, Day preceded Harvard's John Farrar, who arrived at the same conclusion a few years later. While Farrar turned to continental Europe for appropriate texts for his students, Day wrote his own versions based on classroom experiences. The first was *An Introduction to Algebra*, published in 1814. Its subtitle carried a significant intention: "Being the first part of a course of mathematics, adapted to the method of

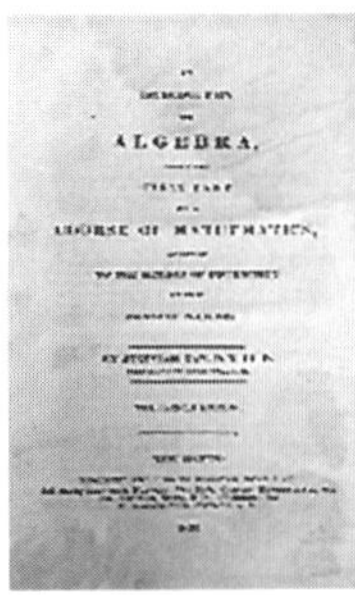

Figure 3.8. Cover page of Jeremiah Day's *Algebra*

Figure 3.9. Jeremiah Day

instruction in the American colleges." Day's book was the first *Algebra* by an American, although he "followed the British synthetic style."[57] Louis Karpinski called it "the foremost American algebra for fifty years."[58] (Figure 3.8 shows the title page of my copy of the 1834 edition, one of the 67 it ran through.)

Day's *Algebra* began with an advertisement that elaborated on the subtitle:

> This course of Mathematics, when completed, it is expected will consist of the following parts:
>
> *Algebra*, now published.
>
> I. Plane Trigonometry, including Analytical Trigonometry, and the nature and use of Logarithms.
> II. The Mensuration of Superficies and Solids.
> III. Navigation and Surveying.
> IV. Conic Sections.
> V. Spherical Geometry and Trigonometry.
> VI. Fluxions.
>
> Most of the succeeding parts will be considerably smaller than the *Algebra*; the whole amounting to two or three volumes.

Within a few years, Day's series resulted in four popular textbooks that essentially defined the undergraduate curriculum and were adopted at numerous colleges. As often happens, however, an initial plan had to be reshaped, and in this case *Mensuration of Superficies and Solids* followed *Algebra* that same year. Next came *Plane Trigonometry* in 1815 and *Navigation and Surveying* two years later. These texts became, after Webber's compendium, the second series of full-length mathematics texts for American students. But that only accounts for four of the proposed seven works. What happened to the last three? It turns out that Day's days of authoring texts came to a resounding halt when he was appointed the fifth president of Yale in 1817. So, like

Webber, his work in mathematics was cut short by an administrative promotion even though he continued to lecture on the subject. While the field of mathematics might have suffered, many mathematicians in the early years of the New Republic assumed high administrative positions in their universities. Nonetheless, Day's output during 1812–1817 was certainly impressive. When he accepted the presidency at Yale in 1817, he was succeeded in his professorship by Alexander Fisher.

Day's *Algebra* was not a "cookbook," and its coverage of algebra extended well beyond what appeared in Pike's *Arithmetic*, written 26 years earlier. *Algebra* begins with a five-page preface supplying the author's reasons for writing a book for an American audience. He divided British textbooks into two categories—extended treatises and texts for beginners—praising the former for their coverage but criticizing their voluminous size, while commending the latter for providing an outline of the subject, yet disparaging their conciseness. In a tone surely approved by twenty-first-century mathematicians, Day also intoned that his aim was to provide the principles of the subject, not merely to present the tools for practical parts. What were his sources? He wrote, "Free use has been made of the works of Newton, Maclaurin, Saunderson, Simpson, Euler, Emerson, Lacroix, and others," but quickly warned, "Original discoveries are not for the benefit of beginners." Day felt that the beginning student would be better served by qualified expositors. A brief overview of the contents of Day's *Algebra* follows. Further details are available online at "Web03-DayAlgebra."

Algebra begins with a nine-page "Introductory observations on the Mathematics in general." Here he lists 22 articles (themes) that explain the nature of mathematics, which he defines as "the science of quantity." The introduction states that mathematics is also valuable for applications, such as mercantile transactions, surveying, mechanics, architecture, fortification, gunnery, optics, astronomy, history, government, chemistry, mineralogy, music, painting, and sculpture. Generally, the contents of *Algebra* are what one would expect to find in high-school or college algebra courses today. Two topics deserve attention. For one, he provides a chapter on Mathematical Infinity that could serve as an excellent introduction to calculus, discussing the notions of the infinitely large and the infinitesimally small in an intuitive manner, without mentioning limits. This chapter precedes coverage of division by compound divisors, where the concept of infinity enters at a critical juncture. A later chapter in the 296-page book introduces infinite series. Here Day takes pains to distinguish convergent from divergent sums, using an approach based on algebra and not limits, which would have been a revolutionary step forward at that time anyway, as it would have preceded the historic discoveries of Bernhard Bolzano (1781–1848) and Augustin-Louis Cauchy (1789–1857) by several years.

John Farrar. Samuel Webber's resignation as Hollis Professor to assume the Harvard presidency in 1806 left that endowed position vacant. The Harvard Corporation initially offered it to Nathaniel Bowditch, but he turned it down in favor of continuing his work with the Nautical Office. Next, Harvard offered it to Bowdoin College president, Joseph McKeen (1757–1807), but he too declined. Finally, their third choice accepted: John Farrar was only 27 years old, had only minimal qualifications as a scientist, and had but one year's college teaching experience. Yet, Farrar turned out to be a sound choice—for Harvard and for the American mathematical community.

John Farrar (1779–1853) played a central role in the history of American mathematics in the first third of the nineteenth century. Born and raised in Lincoln, MA, Farrar received his BA from Harvard in 1803. He was awarded a master's degree three years later. Like most American college graduates at the time, Farrar had been trained to be a minister and "was well on his way to a career in the pulpit when he was unexpectedly offered the position of Tutor in Greek at Harvard in 1805."[59] This appointment dramatically changed the arc of his career and led to the offer of the Hollis Professorship two years later, a position he held from 1807 to 1836.

During the late 1810s, American interest in the analytical style of French mathematics, as opposed to the synthetic style of the British, increased markedly. Textbook selections at West Point as well as the Fisher paper in the *AJSA* reflect this. At about the same time that Fisher was writing his paper on optimizing functions of two variables, Farrar began translating texts with the aim of bringing modern methods to American students, thereby breaking the traditional British hold on American colleges. Whereas Fisher made his publication mark with papers, Farrar did it with books, publishing an incredible seven books in only six years, 1818–1824, three of which appeared in 1818 alone.

The first book that John Farrar translated was the *Elements of Algebra* by the towering, eighteenth-century figure Leonhard Euler, written in German in 1765 but not published for another five years. Farrar titled this popular book *An Introductions to the Elements of Algebra, Designed for the Use of Those Who Are Acquainted Only with the First Principles of Arithmetic*. This was the first foreign-language mathematics book translated and used as a text in America. It predates the more well-known translation by British mathematician John Hewlett in 1822.

The second translation to appear in 1818 was *An Elementary Treatise on Arithmetic by Sylvestre Lacroix*. Farrar made many adaptations of the original work, first published in 1797, including the use of American weights, measures, and currency. Although he added some sections to make the work more accessible for Harvard students, he also deleted material he thought would be too advanced for this audience. Just four years later, knowledge of the contents of the Euler and Lacroix translations was required for admission to Harvard.

The remaining Farrar translation from the busy year 1818 was Lacroix's *Eléments d'algèbre*, which was more advanced than Euler's *Algebra*, including a rudimentary account of the theory of equations. Farrar added numerous explanatory notes to aid his American audience. The full title of the translation was *Elements of Algebra / by S.F. Lacroix; Translated from the French for the Use of Students of the University at Cambridge, New England, by John Farrar*. Negative quantities still caused problems for mathematicians at the time, and the Lacroix text began with a subsection in the introduction that went into considerable detail about the meaning of negative numbers.[60]

The following year Farrar published a translation of a portion of the book *Elements of Geometry* by the famous French mathematician A.M. Legendre. It too had an expansive title indicating its intended audience at Harvard. Farrar began with a short introduction that provided necessary background for American students. The work showed that he had one foot in the older tradition of synthetic proofs of the Greeks but the other foot in the new analytic methods of the French. "[The] algebra textbook … rivalled Day's in immediate influence if not in enduring popularity."[61]

In 1820 Farrar combined two sources into one work: trigonometry from the *Course of Mathematics* by Lacroix and applications of algebra to geometry from the book *Algebra* by Étienne Bézout. It bore the expansive title *An elementary treatise on plane and spherical trigonometry, and on the application of algebra to geometry from the mathematics of Lacroix and Bézout. Translated from the French for the use of the students of the university at Cambridge, New England.* Once again he prepared the translation for an American audience by supplying extra explanations and references.

Two years later Farrar published a book that assumed knowledge of his 1820 work. Its title described the contents: *An Elementary Treatise on the Application of Trigonometry to Orthographic and Stereographic Projections, Dialing, Mensuration, Navigation, Nautical Astronomy, Surveying and Leveling; Together with Logarithmic and Other Tables.* The "Tables" in the title are taken from the 1802 book *American Practical Navigator* by Nathaniel Bowditch.

In 1824 Farrar published the book *First Principles of the Differential and Integral Calculus, or The Doctrine of Fluxions, Taken Chiefly from the Mathematics of Bézout.* This work was based on the first part of Volume 4 of Étienne Bézout's multivolume compendium from 1795. The inclusion of "fluxions" is curious because Farrar adhered to the notation of Leibniz throughout, and only added notation and concepts from the British approach to calculus in short notes at the end of the book.

The aforementioned seven books by Farrar—the three from 1818, geometry from 1819, trigonometry from 1820, applications of trigonometry from 1822, and calculus from 1824—formed a six-book series (the two on trigonometry being combined into one text) that became known as the Cambridge Course of Mathematics. This was the third series of full-length mathematics books written for American students (after those compiled by Samuel Webber and Jeremiah Day). Knowledge of the first two books in the Cambridge Course was required for admission to Harvard; the other four were used as textbooks there.

Following this, Farrar produced the Cambridge Natural Philosophy Series, used as primary textbooks at American colleges for some 20 years. Also, in 1831 he published a translation of Louis Bourdon's *Éléments d'Algèbre*, which was adopted at several colleges (including West Point), and became the most popular American textbook on algebra at the time. In assessing the overall age of Bowditch (1800–1838) Todd Timmons concluded, "Although his translations of classic French mathematical works were soon outdated and superseded, ... Farrar provided the groundwork for much of what was to come in American mathematics."[62]

Unfortunately, however, Farrar carried out much of this work under poor lighting conditions that caused severe eye problems, tribulations exacerbated by his tendency toward overwork. This in turn affected his emotional stability. Adding to his woes was the fact that his physicians administered opiates and morphine, which made him even sicker.

Farrar's savior through this excruciating and long ordeal was his second wife, the successful author of children's books, **Eliza Ware Rotch** (1791–1870), who devoted her life to her husband's health until his death in 1853. Daughter of Nantucket Quakers who had immigrated to France (where she was born) but subsequently fled to England during the Reign of Terror, she received an excellent education and grew up among eminent Europeans and Americans. The family moved to New Bedford, MA, when her father lost his fortune in 1819. She married John Farrar nine years later. Eliza

Table 3.1. Curricula at Harvard before and during Farrar's tenure

	1788	**1830**
Freshman	arithmetic	Euclidean and solid geometry, algebra
Sophomore	arithmetic, algebra	trigonometry, topography, calculus
Junior	Euclidean geometry, trigonometry, conic sections	natural philosophy
Senior	spherical geometry, trigonometry	natural philosophy

Farrar's most famous children's books were *The Children's Robinson Crusoe* (1830), *The Story of the Life of Lafayette as Told by a Father to His Children* (1831), and *The Young Lady's Friend* (1836). Her "writing … was very high and greatly appreciated by her public."[63] A dozen years after her husband's death, Eliza Farrar wrote *Recollections of Seventy Years* (1865), a collection of anecdotes that today provide glimpses into her life in England and France between 1783 and 1819. That work enlivened her husband's sickroom for the final 14 years of his life.

What was John Farrar's legacy? He certainly played a critical role in changing the landscape of American mathematics from Britain to continental Europe, yet his true interest lay in natural philosophy, not mathematics.[64] In fact, Farrar was actually a reluctant translator, having entered into the enterprise only because he was unhappy with the series he inherited—the Webber compendium—despite of the fact that Samuel Webber had become president of the university at the time of Farrar's appointment. From our vantage point today, the major drawback to Farrar's activity was that his translations were based on outdated works.

In addition, Farrar was an influential teacher throughout his 29 years at Harvard. Included among the 275 students who wrote theses on mathematical subjects during his tenure were such stalwarts as George Bancroft, George Emerson, Warren Colburn, Sears Cook Walker, and Wendell Phillips. The most outstanding, however, was undoubtedly Benjamin Peirce. In this sense, Farrar provided an element of continuity from the achievements of colonial leaders John Winthrop and David Rittenhouse to the community of researchers that emerged in the last quarter of the nineteenth century.

John Farrar also played a prominent part in modernizing the curriculum. Table 3.1 contrasts the courses in the mathematical sciences required of all Harvard students in 1788 and 1830. The 1788 curriculum was still in place when Farrar accepted the Hollis Professorship in 1807. By removing arithmetic, he was able to alter the curriculum dramatically.

As the historian Helena Pycior concluded, "Day and Farrar belonged to the first generation of Americans exposed to the rival algebraic styles,"[65] synthetic and analytic. In retrospect, the two authors seem like a much earlier generation than the one that followed and was headed by Charles Davies and Nathaniel Bowditch.

Charles Davies. It was noted above that American books on algebra and geometry began to display an emergent French influence in the first third of the nineteenth century. John Farrar and Charles Davies engaged in this pursuit, and their students at Harvard and the Military Academy were immediate beneficiaries.

Charles Davies (1798–1876) was born in Connecticut but was raised and educated in local schools in northern New York State. At age 14 his ability was discovered by General Joseph Swift, the first graduate of West Point and its superintendent 1812–1817. With Swift's support, Davies enrolled at West Point in December 1813, in the midst of the War of 1812, so he was rushed through to graduation and was commissioned two years later. Because there were no openings in the Corps of Engineers at the time, he took a less desirable position in artillery, serving a year on garrison duty before returning to West Point in December 1816 as assistant professor of mathematics. Initially, Davies taught only mathematics, but his assignment was soon enlarged to natural and experimental philosophy as well. Seven years later he succeeded Claude "Claudius" Crozet as professor of mathematics when Crozet left the Military Academy to become chief engineer for the Commonwealth of Virginia. With this promotion Davies followed in the footsteps of excellent teachers—George Baron, Jared Mansfield, Ferdinand Hassler, and Andrew Ellicott. Davies held the position 1823–1837.

Charles Davies became a prolific textbook writer, beginning with eight books published in the eleven-year period 1826–1837. In this sense, he was even more productive than John Farrar. As noted earlier, Claudius Crozet had come to West Point in 1816 and published a book on descriptive geometry five years later. This was the first outward sign of discontent with the Hutton compendium of texts then in use at the Academy. But three years after Crozet left West Point in 1823, Davies published his own book on descriptive geometry. It was followed in 1830 with a text on surveying. His translation of a geometry treatise by Legendre appeared in 1834; its 28th edition was published 56 years later, in 1890. Generally, texts by Davies enjoyed vast critical and commercial success.

Overall, Charles Davies wrote two exceptionally influential textbooks—the algebra text is discussed here, and the calculus text in Chapter 4. By the late 1820s the faculty at West Point had become disenchanted with Farrar's translation of the Lacroix algebra as a textbook. They examined an 1828 translation by Augustus De Morgan of about one-fourth of the tome *Eléments d'algèbre*, published by the École Polytechnique mathematician Louis Pierre Marie Bourdon in 1817. But the Academy sought a more appropriate translation (even though all cadets studied French) because of the desire to cover more material than De Morgan had prepared for students at the University of London. A former professor obliged.

Edward Coke Ross (1800–1851) was raised and educated in Milford, PA. He enrolled at West Point at age 17 and graduated four years later in 1821. Therefore, both Edward Ross and Charles Davies cut their mathematical teeth on Hutton's *Course of Mathematics*. In fact, Ross joined Davies as an assistant professor for two years before being transferred to Fort Monroe as an artillery school instructor. In spite of being distant from West Point, Ross took it upon himself to publish a more extensive translation of the Bourdon algebra in 1831.

However, in 1835 Charles Davies revised and adapted the Ross version for American students under his own name: *Elements of Algebra: Translated from the French of M. Bourdon. Revised and Adapted to the Course of Mathematical Instruction in the*

United States. The preface stated that it was "an abridgement of the Bourdon, from the translation of Lt. Ross." The Davies version was so suitable for its intended audience that it ran to 40 printings, the last in 1901. It was no mere translation: in the preface Davies stated his goal "to unite … the scientific discussions of the French, with the practical methods of the English school." In this sense, the Davies *Algebra* was a mix of the synthetic approach adopted by Jeremiah Day and the analytical style favored by John Farrar in their algebra books. The historian Helena Pycior analyzed the Davies algebra by describing his approach to negative quantities (such as "the product of two negative expressions is positive").[66]

The series of 13 textbooks by Davies, collectively called the Course of Mathematics, became the best-selling collection of college and high school mathematics texts during the middle of the nineteenth century. This compendium began with *First Lessons in Arithmetic*, included works on algebra, geometry, drawing, trigonometry, surveying, and linear perspective, and concluded with the popular text on calculus described in the next chapter. "The publisher, Alfred Smith Barnes, personally promoted [the Course] in a two-year tour of American towns and villages."[67]

Unfortunately, the exertion exacted a toll on his health, and Davies resigned in May 1837 due to a severe bronchial infection. All of those textbooks were used at West Point and adopted at many other institutions, partly due to the mass of West Point graduates who taught mathematics across the country, but mainly because they were vastly superior to competing works. Boasting a line passing from Thayer to Crozet to Davies, it is no wonder that "unlike all other colleges of the first half of the century, West Point specialized in mathematics."[68]

After leaving West Point, Charles Davies sailed to Europe to restore his health. Upon his return, he was appointed chair of mathematics at Trinity College in Hartford, CT, holding the post 1839–1841. Once again, illness forced him to resign. After recovering, he was appointed paymaster in the US Army, and then served as treasurer at West Point 1841–1846. In 1848 he became professor of mathematics and philosophy at New York University. However, he left after only one year to complete his series of textbooks. He then taught at a normal school in Albany, NY, before accepting a professorship at Columbia in 1857, where he remained until retiring in June 1865. His last work was the treatise *The Metric System* in 1870. Davies continued to work on the many editions of his commercially successful textbooks up to his death six years later.

Nathaniel Bowditch

This section introduces the dominant figure in the period 1800–1838. In the 1840s, students at the University of Virginia learned celestial mechanics not directly from the world's leading expert, Pierre-Simon Laplace, but from a masterful translation of his classic *Mécanique Céleste*. The translator, Nathaniel Bowditch, is the second of seven rugged individualists singled out in Part II; Robert Adrain was profiled earlier. The five others appear in Chapter 4—Simon Newcomb, Benjamin Peirce, George W. Hill, J. Willard Gibbs, and Christine Ladd.

Nathaniel Bowditch (1773–1838) is regarded as one of the top mathematicians in America before 1876. Bowditch was unable to receive a thorough formal education, his school days being curtailed when at age 12 he was apprenticed to a ship's chandler. But he read voraciously during his free time, devouring books from the Irish chemist

Richard Kirwan's library, which had been on a boat captured by a Yankee privateer during the Revolutionary War and which was then brought back to Bowditch's hometown of Salem, MA. Joe Albree described Salem as "a compact seafaring town along the picturesque north shore, sixteen miles north of Boston."[69] Bowditch also taught himself Latin in order to read Isaac Newton's masterpiece, *Principia Mathematica*. During five lengthy sea voyages, he devoted his free time to the daunting task of reading the masterful *Traité de Mécanique Céleste* by Simon Laplace, whose first two volumes were published in 1799. He instructed some of his shipmates in this material, yet he never held an academic teaching position. Therefore, he is viewed today as an "early American autodidact and polymath."[70]

Nathaniel Bowditch became the foremost American scientist from the time of the publication of the first of his two major works, *New American Practical Navigator* (***NAPN***), published under his name in 1802. This classic work was initially aimed to be the third edition of an earlier book on navigation by **John Hamilton Moore** (1738–1807), who was born in Edinburgh, educated in Ireland, and subsequently joined the Royal Navy at Plymouth.[71] In 1772 Moore published his popular book, *The New Practical Navigator and Daily Assistant*. Later he not only taught navigation but ran a business selling charts and nautical instruments. Despite this book's popularity, Bowditch found so many errors—reputedly about 8000—that, as he wrote in his preface, "It was concluded to take up the subject anew." Though the title Bowditch adopted might suggest otherwise, his resulting book was no mean maritime manual. *NAPN* begins with brief treatments of decimal arithmetic, geometry, trigonometry, and logarithms, and then turns to extensive applications in astronomy, geography, sailing, astronomy, surveying, and gauging. *NAPN* ends with a set of tables that accounts for over half of its pages.

NAPN became an indispensable guide for navigation almost instantly and ultimately developed into a timeless classic, yet in the annals of mathematics it contained neither new theories nor conceptual advances. However, Bowditch played an instrumental role in the American Academy, starting with a paper in the Academy's *Memoirs* in 1804 while continuing to write for the rest of his life. That first article, "New method of working a lunar observation," was the first of seven papers in the section on astronomy and mathematics in Part 2 of Volume II. Two others of note were "A method of finding the area of a field arithmetically" by Elizur Wright and "Remarks on Mr. Winthrop's paper on the duplication of the cube" by George Baron. All seven articles presented but minimal contributions to mathematics.

However, papers by Bowditch that appeared in the *Memoirs* from 1809 to 1818 cast him as a mathematician, not a mere practitioner. Moreover, his contributions to Volumes III (1809) and IV (1818) demonstrate the impressive extent of his scientific knowledge. Each of these volumes was published in two parts, and overall, Bowditch published more than 25% of the papers on mathematics in these four issues. Moreover, many of the remaining articles arose from his communications with the authors. His involvement thus provides documentation for naming this chapter after him. Indeed, Todd Timmons's book is subtitled "The Bowditch Generation."[72]

The first four articles in Volume III, Part 1, were by Bowditch. The title of the lead paper, "Observations of the comet from 1807," belies its content, as evidenced by the statement, "Early in November I began to calculate the elements of the orbit by the method given by La Place, in … the 'Mecanique Celeste'."[73] This shows that Bowditch was already becoming acquainted with the celestial mechanics of Laplace to calculate the orbit of the given comet at this early date. Eight years later he commenced his famous translation of the Laplace classic on celestial mechanics.

Bowditch wrote seven of the 23 articles on mathematics when the next issue of the *Memoirs* appeared in 1815 (Part 2 of Volume III). In them he displayed mastery of Newton's fluxions and even solved systems of differential equations. The best of the lot was the 23-page "On the motion of a pendulum suspended from two points," which contained his most original contribution to mathematics, a new class of sine curves he introduced to solve a system of second-order differential equations.[74] Around 1850 this class of curves was rediscovered and named after Jules Antoine Lissajous (1822–1880), who applied them to acoustics. The fact that Lissajous was born seven years after Bowditch published his paper warrants two comments. First, Bowditch deserves credit for (at least independent) discovery. Second, American journals were virtually invisible during the first part of the nineteenth century; clearly the *Memoirs* did not enjoy wide readership. But what our brief coverage of this paper does not convey here is the level of mathematical sophistication Bowditch possessed that far surpassed his American contemporaries.

It is also significant that many of the articles Bowditch communicated to the *Memoirs* were submitted by academics, including Parker Cleaveland (Bowdoin), James Dean (University of Vermont), John Farrar (Harvard), and Alexander Fisher (Yale). As Timmons commented:[75]

> Interestingly, many of these communications came from Professors of Mathematics and Natural Philosophy at the leading colleges [and were] addressed to an insurance businessman. [These] indicate the relative lack of the standing of science in academia in early nineteenth-century America.

Nathaniel Bowditch assumed the presidency of a marine insurance company in Salem from 1804 to 1823. Recall that he turned down Harvard's offer of the Hollis Professorship in 1806. The reasons for the rejection are unclear, but in general, positions at American colleges at the turn of the nineteenth century required far too much teaching and the compensation could not compete with business. Later he declined offers at the University of Virginia (1818) in spite of entreaties by Thomas Jefferson, and at West Point (1820) in spite of an appeal from Sylvanus Thayer. Unlike Europe, America did not possess royal courts or academies that offered alternate employment opportunities for scholarly activity. Figure 3.10 is part of an 1835 painting by Charles Osgood that hangs in the Peabody Museum in Salem today. A bust of Laplace is placed in the upper-left corner.

In 1823 Bowditch moved from Salem to Boston, where he worked as an actuary for another insurance company. He brought 2500 books with him. On his final voyage at sea (1802–1803) he purchased a copy of Laplace's *Traité de Mécanique Céleste* and began devoting all his free time to mastering it in November 1803. Apparently, Bowditch completed the first four volumes in 1806. (Laplace's fifth, and final, volume did not appear until 1825. The first had been published in 1799 and the fourth in 1805.) Laplace's aim in his *magnum opus* was "to reduce all the known phenomena of the system of the world to the law of gravity, by strict mathematical principles." The first two volumes dealt with general laws of equilibrium and the motion of solids and fluids, as well as motions of the planets. The main mathematical approach was to set up differential

Figure 3.10. Nathaniel Bowditch

equations and, by solving them, to describe the resulting motions. Bowditch's paper on the 1807 comet shows that he was in command of that material by that date.

By 1815 Bowditch got the idea to translate the work into English, and by 1818 he had completed most of it. When the first volume of the translation finally appeared in 1829, it was titled *Celestial Mechanics: Translated from the French, with a Commentary, by Nathaniel Bowditch* (**CM**). Over the ensuing eleven years, he elaborated the work with numerous explanatory notes. Therefore, *CM* was not merely a translation. Instead, Bowditch sought to provide a fuller explanation of Laplace's often sketchy proofs, writing "[I] insert the intermediate steps of the demonstrations, necessary to enable [the reader] easily to follow the author in his reasoning." This was no easy task. As Bowditch added, Laplace "has been found difficult to be understood by many persons, who have a strong and decided taste for mathematical studies, on account of the time and labour required." This too presented a formidable hurdle. The translator is famous for stating a notorious refrain among undergraduate students: "I never came across one of Laplace's '*Thus it plainly appears*' without feeling sure that I have hours of hard work before me to fill up the chasm and find out and show *how* it plainly appears." For instance, Laplace listed without justification, the set of partial derivatives,

$$\frac{dr}{dx} = \cos\theta\,, \qquad \frac{d\theta}{dx} = -\frac{\sin\theta}{r}, \qquad \frac{d\omega}{dx} = 0,$$

as a solution for the attraction of spheroids in terms of the standard spherical coordinates,

$$r = \sqrt{x^2 + y^2 + z^2}, \qquad \cos\theta = \frac{x}{r}, \qquad \tan\omega = \frac{x}{y}.$$

Bowditch added 15 lines of explanation to supply the intermediate steps.

Another way in which Bowditch converted *CM* into a textbook on celestial mechanics appropriate for American scientists and advanced students was by adding a list of trigonometric formulas that Laplace used without reference. The addition of explanatory notes and commentary to the Laplace version resulted in roughly twice the number of original pages. The first edition of *Mécanique Céleste* comprised 1508 pages.

CM served two other purposes. For one, Bowditch provided references to earlier works that Laplace had used, since the French mathematician was wont to credit others. Secondly, Bowditch was able to update *CM* with discoveries and mathematical techniques that were unknown when the original version appeared. For instance, Volume III of *CM* added a 150-page appendix giving "important improvements made by Gauss, Olbers, and others."

The updates were made possible by Bowditch's delay in publishing from 1818 to 1829. One reason for the delay is that he expected Laplace to produce a second edition with corrections and updates. That never happened. A second reason for the gap of 11 years was the cost. Although the American Academy, as well as several individuals, offered financial assistance, Bowditch insisted on paying for the entire work himself, and ultimately spent about one-third of his life savings to complete the four volumes that appeared in 1829, 1832, 1834, and 1839. Bowditch's extensions produced material that was similar to what Laplace included in his fifth volume, published in 1825, making it unnecessary for Bowditch to add this final volume to his collection.

The publication of *CM* cemented Bowditch's scientific reputation. However, he minimized his importance as compared to Laplace. He told his son, Henry, that "Laplace originates things which it would have been impossible for me to have originated."[76] In this sense, Bowditch claimed that he was to Laplace what Euclid was to Archimedes—an organizer, compiler, and educator but not an original investigator. Consequently, while his impact in America was significant, it was negligible in Europe. Nonetheless, *CM* was important for changing European perceptions of American science, as exemplified by positive comments it garnered by Baggage and Hershel in England, Lacroix and Legendre in France, and Bessel in Germany. In the US, Michigan's Louis Karpinski called *CM* "the most monumental American mathematical publication to 1850."[77]

As Bowditch would undoubtedly have admitted, he was no match for the contemporary titan Gauss, let alone Cauchy and Laplace. Indeed, as Bowditch learned Gauss's contributions to mathematics and astronomy, he "became convinced that America should emulate the German system in science and scholarship."[78] In this sense, Bowditch was some 40 years ahead of his time.

Nathaniel Bowditch did not work entirely alone on the Laplace translation. Rather, he enlisted the services of the Harvard undergraduate student, Benjamin Peirce, to assist him with proofreading. Illness prevented Bowditch from completing the fourth, and final, volume, but the faithful Peirce concluded it for him; the finished product appeared just one year after Bowditch's death in 1838.

Perhaps Bowditch's most impressive accomplishment was that he was the first native-born American able to understand mathematics at a research level and to elucidate it to his fellow countrymen. At a time when American colleges were just beginning to introduce elementary algebra into the curriculum, he was solving differential equations. Along with John Farrar, Bowditch played a crucial role in alerting American mathematicians to British works (as seen in his *NAPN*) and continental European advances (as seen in his *CM*). Timmons summarized Bowditch's role as follows: "The scientific community in America, in its embryonic stage in the early nineteenth century, owed a significant part of its future development to Nathaniel Bowditch."[79]

Colleges

This section traces the founding of three colleges that played important roles in the development of mathematics starting in the first third of the nineteenth century—the Naval Academy, University of Virginia, and University of Toronto. It also discusses an important change for the teaching profession at Yale.

Army–Navy. The Naval Academy opened formally almost a half-century later than the Military Academy but, like its counterpart at West Point, its roots lie in the late eighteenth century when the Continental Navy was formed during the American Revolution to match Great Britain's Royal Navy. However, the Continental Navy was demobilized by Congress in 1785. Nine years later, President George Washington persuaded Congress to authorize a new naval force to combat the growing menace of piracy on the high seas. Over the next 50 years, small naval schools were established in Philadelphia, New York, Norfolk, and Boston, but President John Quincy Adams took a bolder step in his First Annual Message of December 1825, urging Congress to establish a "naval school of instruction, corresponding with the Military Academy at West Point, for the formation of scientific and accomplished officers." His proposal, however, was not acted upon for another 20 years. By that time, the Philadelphia Naval School offered the premier training. However, Secretary of the Navy George Bancroft decided to move it to "the healthy and secluded" location of Annapolis, MD, to rescue midshipmen from "the temptations and distractions that necessarily connect with a large and populous city."[80] The relocated naval school was established without Congressional funding on a 10-acre Army post in Annapolis in 1845, with seven professors and a class of 50 midshipmen. As a sign of the respect accorded the Philadelphia facility, four of the original seven faculty members came from the Philadelphia school. The curriculum included mathematics and navigation, gunnery and steam, chemistry, English, natural philosophy, and French. The Naval School became the United States Naval Academy in 1850, when a new curriculum went into effect requiring midshipmen to study at the Naval Academy for four years and to train aboard ships each summer, a format that forms the basis for a far more advanced and sophisticated curriculum today.

Once again, a mathematician was instrumental in establishing a college, this time the Naval Academy. **William Chauvenet** (1820–1870) was born in Milford, PA, the only child of a French father who met his wife in Boston after immigrating to the US. The family moved to Philadelphia, where Chauvenet received an excellent education at a private school run by Dr. Samuel Jones. As a result, Chauvenet entered Yale at age 16 and graduated four years later with high honors in both mathematics and classics. Over the next year he began his academic career by assisting a professor at Girard College (a private, secondary school for orphans in Philadelphia) in a series of magnetic observations. In 1841 he was appointed a professor of mathematics in the US Navy and for a few months served on the *USS Mississippi* (a steamer), where he taught midshipmen. The following year he was put in charge of the Philadelphia

Figure 3.11. William Chauvenet

Naval School in his hometown. When the Naval School was established in Annapolis in 1845, he was one of the professors who moved onto that campus, whereupon he and five other naval officers drafted a plan to form the Naval Academy to correspond to West Point, whose offerings in mathematics and engineering were among the best in America up to the American Civil War. That plan for a four-year program was put into effect in 1851, by which time the school was called the Naval Academy and located in Annapolis. Chauvenet was appointed head of a new department of astronomy and navigation in 1853. He was especially successful in establishing this unit over the next six years, when he was recruited elsewhere.

While associated with the Naval Academy 1845–1859, Chauvenet did more than anyone else to establish the scientific foundation of the fledgling institution. As a result, in 1916 the Naval Academy placed a bronze tablet in the library to commemorate his founding contributions. Nine years later, the Mathematical Association of America (MAA) established the Chauvenet Prize for Mathematical Exposition. Today this prestigious prize is awarded for a noteworthy expository or survey article published in a North American journal or in an anthology published by the MAA, with preference given to works that can be read profitably by its members. Several of Chauvenet's books served as paragons of exposition, notably *A Treatise on Plane and Spherical Trigonometry* (1850), *A Manual of Spherical and Practical Astronomy* (1863), and *A Treatise on Elementary Geometry* (1870).

However, by 1859 William Chauvenet was being actively pursued by two institutions, his *alma mater* Yale and Washington University in St. Louis. Ironically it was a former Yale classmate who persuaded Chauvenet to accept the offer of the professorship in mathematics at Washington University.

Yale. Mention of Chauvenet and the Naval Academy takes us out of chronological order. This section updates his *alma mater*, Yale, where the chair of mathematics and natural philosophy regressed after the drowning of Alexander Fisher in 1822 at age 28. Two of his successors also died young. The first, Reverend **Matthew Rice Dutton** (1783–1825), shared two important traits with Fisher, studying divinity after graduation and serving as tutor before being named professor of mathematics and natural philosophy. But Dutton died in 1825 after having held the position for only three years. In the year before Dutton's death, he published *An Elementary Treatise* *on Conic Sections, Spherical Geometry and Spherical Trigonometry*, which was aimed to form the fifth and sixth parts of the compendium compiled by Jeremiah Day. Like Fisher, Dutton might have become Yale's first professional mathematician if not for his untimely death at age 42; a history of the university up to 1831 asserted that as "logician and mathematician [he] would have left valuable legacies for the literary benefit of his successors."[81]

Dutton was in turn succeeded by **Denison Olmsted** (1791–1859), who carried out original investigations, but mainly in astronomy and physics (then called natural philosophy). Olmsted graduated in the same class as Alexander Fisher (1813) and named his third son Alexander Fisher Olmsted. Like his classmate, Denison Olmsted returned two years later as a tutor. But in 1817 he was appointed professor of chemistry and mineralogy at the University of North Carolina, where he performed the first state geological study in the US. Olmsted remained at UNC until returning to Yale yet again in 1825, this time to replace Dutton as professor of mathematics and natural philosophy. A notable event for mathematics occurred in 1836 when, at Olmsted's request, the chair was divided into two distinct professorships. Olmsted retained the chair of natural philosophy. He was regarded as a renowned teacher but not an original investigator.[82] Nonetheless, he was a very successful textbook author; his works in astronomy and natural philosophy sold more than 200,000 copies in his lifetime.

The title "professor of mathematics" appeared for the first time in the Yale catalog for 1841–1842. **Anthony Dumond Stanley** (1810–1853), yet another Yale graduate (1830), was a teacher at a school in Hartford before returning to his *alma mater* as tutor two years later. He accepted the notable professorship in 1836 and held it until his untimely death. While engaged in the care of fruit trees on his homestead in the fall of 1849, he came down with a severe illness that developed into bronchitis. He recuperated sufficiently enough after spending the ensuing winter and spring in Southern Europe, Egypt, and Palestine (present-day Israel) to resume duties for the fall terms in 1850 and 1851, but he never fully recovered and died at age 43.

Although Stanley published one book on spherical geometry and trigonometry, and another on tables of logarithms (of numbers and of trigonometric functions), all the while editing an edition of Jeremiah Day's *Algebra*, he did not otherwise contribute to mathematics. In spite of being a minor figure in American mathematics, Stanley nonetheless holds the distinction of being America's first professor of mathematics having no duties in any other field, a designation that marked a significant step in the professional status of "mathematician."

Although the contributions of Fisher, Dutton, Olmsted, and Stanley were minimal, Yale's next professor of mathematics played a vital role in the emerging community of American research mathematics in the last quarter of the nineteenth century. I call Hubert Newton the "grandfather of American mathematics" because of the exploits of his student, E.H. Moore, who is called the "father of American mathematics."

In contrast to Stanley's professorship of mathematics, during the first part of the nineteenth century about half of the community of American mathematical practitioners held positions as professors of one or more areas in addition to mathematics at some point during their careers. And most of their tenures were short-lived. **Elizur**

Wright (1804–1885) serves as a good example. His father, Elizur Wright senior (1762–1845), was known for mathematical erudition but also for his devotion to the Presbyterian faith. The son graduated under Olmsted at Yale in 1826 and became a professor of mathematics and natural philosophy at Western Reserve College three years later. (A federation of Western Reserve College, founded in 1826, with Case Institute of Technology, founded in 1881, resulted in today's Case Western Reserve University in Cleveland, OH.) However, Wright held the post only four years. In 1833 he attended a convention in Philadelphia that led to the formation of the American Anti-Slavery Society, whereupon he left academics to become the corresponding secretary of the fledgling organization. Nonetheless, Wright's mathematical inclinations re-emerged about the middle of the century when he became interested in life insurance reform. His campaigns in this field, as well as his development of actuarial tables, earned him the title of "father of life insurance."[83]

What about student (meaning undergraduate) life at Yale in 1830? Admissions requirements were stringent for Latin and Greek but much lighter for mathematics, with knowledge of the contents of arithmetic books by Barnard or Adams the only stated prerequisite. The minimum age for enrollment was 14, which might explain why the prodigy Alexander Fisher did not matriculate earlier. All instruction was in the hands of the president, five professors, and eight tutors; in addition to the professor of mathematics, natural philosophy, and astronomy, there were professors of chemistry, mineralogy, and geology; Hebrew, Greek, and Latin; rhetoric and oratory; and divinity.

Like today, Yale had a four-year curriculum with each year divided into three terms. Unlike the present, classes were conducted six days a week with students attending three recitations or lectures each day, except for Wednesday and Saturday when only two were required. The first two years were entirely prescribed but the final years offered limited options. The first two terms in the freshman year were devoted to Day's *Algebra*. The third term of that year and the first of the next covered the first five books of Euclid from the text by John Playfair. The remaining two terms in the sophomore year involved various topics in Day's series, *Course of Mathematics*:

Term II: plane trigonometry, nature and use of logarithms, mensuration of superficies and solids, and isoperimetry; mensuration of heights and distances, and navigation; and

Term III: surveying; conic sections; spherical geometry and trigonometry.

No mathematics *per se* was offered in the junior year, but Olmsted's book on natural philosophy and mechanics was covered during the first two terms. In addition, students took courses in astronomy and logic in the third term, but they also had three choices for another course, one of which was a Newtonian approach to calculus in terms of fluxions. No mathematics was offered in the senior year, but students could select among chemistry, mineralogy, geology, and astronomy.

University of Virginia. Another notable development in mathematics in the first part of the nineteenth century was the opening of the University of Virginia in 1825. Thomas Jefferson's university has had a checkered history in mathematics.[84] Matters got off to a good start with the initial appointment of **Thomas Hewitt Key** (1799–1875), an 1821 graduate of Cambridge University, where he was the nineteenth wrangler on the *Mathematical Tripos*. (At Cambridge, a wrangler is a student who achieves first-class honors in the third year of the undergraduate mathematics program. The highest-scoring student is the "senior wrangler," the second highest the "second wrangler," etc. Cambridge ended those rankings in 1910.) In the summer of 1824 an agent from the State of Virginia went abroad in search of four professors for the fledgling University of Virginia, and Key was offered the professorship of mathematics, probably due to his relatively high score on the Tripos. He came to the US and was a huge success as a teacher. But Key remained in the US only two years, most likely departing due to harsh, humid summer weather, much to the dismay of the Charlottesville school. He returned to London and one year later was appointed professor of Latin at the newly founded University of London.

Key was succeeded at Virginia by another Englishman, **Charles Bonnycastle** (1796–1840), who was born in Woolwich, England, where his father John was professor of mathematics at the Royal Military Academy. Bonnycastle came to the University of Virginia in 1825 along with Key but as a professor of natural philosophy. When Key resigned, Bonnycastle moved into his position of professor of mathematics, holding it until his untimely death in 1840 at age 43. During this time he shifted the mathematics program away from the antiquated synthetic method to the modern analytic approach. His 1834 textbook *Inductive Geometry* united the two approaches to geometry for a college-level audience. Although his contributions to mathematics were modest, with papers appearing in the *Transactions of the APS*, he established a program unsurpassed by any other university in the country at the time, exposing Virginia students to an enriched version of calculus, the study of curves and surfaces, mechanics, and celestial mechanics (based on the Bowditch translation). Unfortunately, few students were able to complete the program. Once again a key ingredient was missing—a *community*—this time a critical mass of advanced students.

Thomas Jefferson emphasized life-long learning and, as a result, students were not called freshmen, sophomores, etc.; rather, they were (and still are) called first-year, second-year, etc. Undoubtedly, Jefferson would be proud of the mathematics program at the university he had founded and the individuals who have provided instruction in the subject in Charlottesville.

University of Toronto. While these institutions of higher education were developing in the US, stirrings were taking place in Canada as well, notably at what was called King's College (today, the University of Toronto). The history of this institution can be traced in the context of the history of Canada up to 1863. The online file "Web03-Canada" provides an outline of these developments.

It is notable that conferences held in 1864 at Quebec City and Charlottetown (on what is now Prince Edward Island) united the British colonies in North America into a federation that formed the basis for the London Conference held two years later. Delegates to the London Conference drafted the British North America Act and presented it to Queen Victoria in 1867. She granted Royal Assent to this act that March, setting July 1 as the date for the formation of the Dominion of Canada. Canada was thus formed in 1867 as a federal dominion consisting of four provinces: Ontario, Quebec, New Brunswick, and Nova Scotia. The term "dominion" was chosen to signify Canada's status as a self-governing colony of the British Empire. The British Parliament then enacted a law defining the Province of Canada, New Brunswick, and Nova Scotia as a federated kingdom. Thus began an accretion of provinces and territories and a process of increasing autonomy from the British Empire, which became official in 1931 and completed in 1982. The Canada Act of 1982 severed the vestiges of legal dependence on the British Parliament.

The Toronto area was home to such Native American nations as the Seneca, Mohawk, and Cayuga before the first permanent European presence, a French trading fort established in 1750. The first large influx of English occurred 26 years later when Loyalists fled the American Revolution; Britain formed New Brunswick for these Loyalists in 1784. Seven years later the Province of Quebec was divided into Upper Canada (which was mainly English) and Lower Canada (mainly French). In 1794 York was named capital of the new colony of Upper Canada; it was burned by US troops during the War of 1812. Six years after that a treaty established the 49th parallel as the border between the US and Canada. York was incorporated and renamed Toronto in 1834. It became a major destination for immigrants to Canada during the nineteenth century, its population surpassing Montreal in the second half of the twentieth century. To complete the idea of a transcontinental union, British Columbia, including Vancouver Island, was formed from remnants of Hudson's Bay Company following the Oregon Treaty, and it joined Canada in 1871.

The University of Toronto dates it founding to 1827 but its roots go back to 1792, when the first lieutenant governor of Upper Canada, **John Graves Simcoe** (1752–1806), echoed thoughts from Loyalists about establishing "a college of the higher class."[85] In his plan for the governance of Canada, Simcoe added that such an institution "would be eminently useful and would give a tone of principles and manners, that would be of infinite support to government." Land for this purpose was set aside in 1798, but there was no effort to take advantage of it for three decades. At this time Canada's population was growing rapidly and included many well-educated people in government. Simcoe, for instance, had attended the prestigious Exeter Grammar School, Eton College, and Oxford University.

The first step in establishing Simcoe's long deferred plan for a college took place in 1826, seven years after **Sir Peregrine Maitland** (1777–1854) had been appointed lieutenant governor. That year he chose the Anglican clergyman **Rev. John Strachan** (1778–1867) to be his emissary to London. Some 13 years earlier, Strachan had played a pivotal role in negotiating with the government of the US to save York after the War of 1812. When Strachan returned to Canada in March 1827, he bore the royal charter approved by King George IV establishing the University of King's College, so this stands as the founding date for what would become the University of Toronto in 1849.

King's College was unable to find a permanent structure until 1842, when it was ceded a temporary home in the old Parliament Building in Kingston, ON, while funds were being sought for a new building. Classes began in October 1843 with 27 students. Admission standards were rigorous. Applicants had to be at least 16 years old and were required to know the first two books of Euclid as well as linear and quadratic equations. The curriculum was entirely prescribed and lasted three years (a fourth was added later). It was modeled after Trinity College in Dublin.

By 1849 funding still had not been secured, and King's College was ejected from its quarters when the government was forced to move its legislature to Kingston after the parliament buildings in Montreal were burned. Other space was soon acquired, but in the 22 years since the royal charter had been granted a great deal of animosity had been engendered because the University of Toronto, as it was now called, was in the hands of the Church of England. In the meantime, Methodists established Victoria

College in Coburg, ON, while Presbyterians and Catholics founded Queen's College and Regiopolis College, respectively, in Kingston, ON. (Regiopolis–Notre Dame is a secondary school today.)

In an attempt to heal these wounds, the name of the institution was changed to University College in 1851, but that did little to appease differences for another 38 years. Finally, in 1889, with the proclamation of the Act of Federation, St. Michael's and Victoria joined University College. Trinity was added in 1905 to yield the University of Toronto. Queen's College remained a separate university in Kingston.

Mathematics was unaffected by these religious controversies, but it did not get off to a propitious start. **Richard Potter** (1799–1886) was selected as one of six charter professors of the university. When classes opened in 1843, he became the first professor of mathematics and natural philosophy. A sixth wrangler at Cambridge in 1838, he taught at University College (London) for two years before coming to Canada. However, he resigned his position in April 1844 because he had not been compensated for "expenses and crossing and re-crossing the Atlantic, [and] for the purchasing [of] the Apparatus for the National History Museum."[86]

Little is known about the activities of Potter's successor during his tenure in this professorship, but a few facts are known about his earlier career. The **Rev. Robert Murray** (ca. 1795–1853) was a graduate of the University of Glasgow, and was ordained a minister of the Church of Scotland. During 1824–1834, he was master of the Edinburgh Commercial and Mathematical Academy, publishing a very successful textbook on commercial arithmetic that was later reprinted in Upper Canada. He immigrated to Canada in 1836 to become a minister of the Church of Scotland in Oakville, ON. Six years later he was appointed Canada's first assistant superintendent of education, but for political reasons he was removed from this post in 1844. Potter's convenient resignation that year enabled the Upper Canada governor to appoint Murray as professor of mathematics and natural philosophy. Unfortunately, little is known of Murray's life from this point except that he died in 1853 after a prolonged illness.

Nonetheless, Murray made a very good choice for his assistant in 1850 when he chose **John Bradford Cherriman** (1823–1908), who succeeded him as professor of mathematics and natural philosophy, holding the position until 1875. Cherry, as he was known to his students, was born in England, entered St. John's College, Cambridge, in 1841, and graduated four years later as sixth wrangler on the Mathematical Tripos. For the next two years, he was assistant master at a secondary school, and in 1848 he received a master's degree at Cambridge.

J.B. Cherriman was appointed assistant professor of mathematics at University College (Toronto) in October 1850. He was promoted to chair three years later upon the death of Robert Murray. Cherriman published many papers during his 25 years in Toronto, mostly in mathematical analysis and applications to the social sciences. Most his papers appeared in the *Canadian Journal of Industry, Science, and Art*, but one was published in Silliman's *American Journal of Science and Arts*. Although mathematics was his main interest, the professorship required him to teach the physical sciences (heat, optics, electricity, magnetism, mechanics, hydrostatics, pneumatics, acoustics, and astronomy) even though there were no laboratories for demonstrations at the time. Cherriman left the university in 1875 to assume a federal position as an actuary with the Canadian government in Ottawa. He maintained this post until 1885, when he returned to England.

During Cherriman's tenure at Toronto, graduating classes were generally small, ranging from 20 to 30. Only men were permitted to matriculate until 1885, when women were permitted to attend lectures. All classes continued to be held in the one building that also served as a residence hall. The faculty generally consisted of six professors and three tutors. In 1878 the university established a School of Engineering that offered courses in the natural sciences with laboratories.

Genesis of statistics

The discipline of statistics was only beginning to develop worldwide in the first half of the nineteenth century, so it is not surprising that the American effort lagged behind Europe. Nonetheless, there were stirrings, mainly in Boston, about the collection and interpretation of data in the social sciences. This section describes the nascent movement, introduces some of the leaders, and examines several statistical societies established in the New Republic. We begin with a second scientific society that has lasted until today and was also local, but, importantly, it was more restrictive in its aims.

American Statistical Association (ASA). In addition to the APS and the American Academy, a substantial number of regional learned societies were formed in the major cultural centers of the US over the next 50 years. It was seen above that there were some attempts to form groups of scientists interested in mathematics, but not one survived very long. In a certain sense, that changed in 1839 with the founding of the American Statistical Association (ASA), although the development of *mathematical* statistics did not occur for almost another century. An important monograph on the history of American mathematics stated, "While societies of a more regional character worked well in their limited venues, national organizations would better serve to set and maintain scientific standards, to encourage and promote research and publication, and to provide a network for communication."[87]

Of interest here, the *Memoirs* of the American Academy did publish articles on life expectancy. These works attest to a rising interest in statistics in the New Republic but they were not written from a mathematical standpoint, as such a development did not occur until later in the nineteenth century, and not in America until the twentieth. However, a related development also took place in Boston, with the founding of yet another local society that, importantly, was devoted to a specific subject.

In late November 1839 a group of five physicians, lawyers, and ministers met at an American Education Society meeting to organize a society devoted strictly to statistics. All five were graduates of three New England colleges: Brown, Dartmouth, and Harvard. The person most responsible for founding the group was Lemuel Shattuck. A noted public health official with strong interests in statistics and genealogy, Shattuck will be introduced in more detail below. **Richard Fletcher** (1788–1869), a former member of the US House of Representatives and prominent leader of the Boston bar, chaired the meeting. Shattuck served as secretary, but he was the driving force behind the founding and early development of the society. The other three founders were the Rev. **William Cogswell** (1787–1850; a Dartmouth graduate who, like Shattuck, was a charter member of the New England Historic Genealogical Society, and the General Agent of the American Education Society, which raised funds for a better educated ministry), **Oliver Peabody** (1799–1848; a lawyer and clergyman who was the Register of Probate in nearby Suffolk County), and **John Dix Fisher** (1797–1850; a graduate of Brown with medical work at Harvard, and an outstanding physician who played a prominent part in establishing the Perkins Institution and the Massachusetts School for the Blind).

Shattuck was the only one of these five founders who had shown any sustained and compelling interest in statistics. He alone attended all early meetings. According to Cornell statistician **Walter Francis Willcox** (1861–1964), "We may surmise that the other four men were hand-picked by Shattuck ... in his profound devotion to the public welfare and to social reforms."[88]

At the organizational meeting it was moved that the five founders prepare a constitution to be presented at a second meeting to be held two weeks later. At that subsequent meeting, a draft of a constitution was reported and then approved. The constitution stated that the organization would be known as the American Statistical Society and that its aim was "to collect, preserve and diffuse Statistical information in the different departments of human knowledge."[89] Note the capital letter in Statistical. It is also noteworthy that the word "American" was used in the official name, indicating that Shattuck envisioned a national organization even though no evidence of interest in statistics in America existed outside the northeast.

The constitution delineated four types of members: fellows, corresponding members, honorary members, and foreign members. Although members were called "Fellows," that term is now restricted to individuals who have an established reputation in some aspect of statistical work and who are elected by a select committee. Initially, fellows and foreign members were chosen by ballot and had to be approved by 4/5 of the fellows present at a meeting, as long as there was a quorum of ten members. The constitution also enumerated the officers: president, vice president, recording secretary, home secretary, foreign secretary, treasurer, and librarian. In addition, there were nine counselors who formed a board of directors along with the officers.

The society met for the third time one week later, December 18, to select officers. Shattuck was selected home secretary but continued to be the driving force. The fourth gathering occurred in early January 1840, when it was voted to make application to the legislature of Massachusetts for an act of incorporation. All three meetings after the organizational gathering in November were held in the office of the founder, William Cogswell.

One of the articles in its constitution stated that the society would meet annually in Boston on the first Wednesday in February, so the first annual meeting was held on February 5, 1840. At that gathering the membership "Voted, by recommendation of the Directors, that the name of the society be altered" to the present American Statistical Association (ASA),[90] perhaps because that abbreviation did not carry the possible offensive connotation of the original acronym. An initial group of 54 fellows was selected as well, including Nathaniel Bowditch, Benjamin Peirce, and Daniel Treadwell (1791–1872), an American inventor who held the Rumford professorship at Harvard. The 36 honorary members included US president Martin Van Buren and two American scientists introduced above, Parker Cleaveland and Yale professor Benjamin Silliman.[91]

At the first annual meeting the membership adopted 31 bylaws. One bylaw stated, "It shall be deemed the duty of every fellow to prepare at least one article a year on some statistical subject."[92] That one requirement set the American Statistical Association

Table 3.2. The first six ASA presidents

President	Term	President	Term
Richard Fletcher	1839–1845	Francis Walker	1883–1897
George Shattuck	1846–1851	Carroll Wright	1897–1909
Edward Jarvis	1852–1882	Simon North	1910

apart from the APS and the American Academy. Like those predecessors, membership in the ASA was strictly by invitation. Yet the membership the second year, 1841, reached 109. Nonetheless, the by-invitation-only restriction limited total membership at about 100 until 1890.

The first quarterly meeting of the ASA was held in April 1840. Numerous documents were presented, with Lemuel Shattuck reporting on the most. For example, he demonstrated forms that were used in France for the registration of marriages, births, and deaths that he would later adopt for the US. He also read an article on the statistics of Saxony (Germany). The second group of foreign members elected at this meeting included the famous Belgian statistician Adolphe Quetelet. Another famous European statistician, Florence Nightingale, was elected at a subsequent meeting and donated two of her statistical reports on sanitation in India to the ASA library.

While APS meetings were always held in Philadelphia, ASA meetings, like those of the American Academy, were conducted in Boston. The first ASA meeting held outside that city did not occur until 1908, when the annual gathering was held in Atlantic City, NJ. Walter Willcox asserted that the previous year the ASA had "voted unanimously at my suggestion to strike out the provision requiring [annual] meetings to be held in Boston. ... The society then became peripatetic, affiliated with the cognate societies."[93]

Table 3.2 lists the first six ASA presidents.[94] Notice that the first five served multiple-year terms. Presidents have held one-year terms since 1910. Richard Fletcher was apparently only a figurehead, as Lemuel Shattuck was responsible for most ASA activity during the first six years. (I am unable to find any relationship between Lemuel Shattuck and the second president.) Edward Jarvis was ASA president for 31 years and managed to keep the ASA afloat during a turbulent period in US history.

As the historian of mathematical statistics Patti Hunter wrote, "Jarvis's interests reflected the mid-nineteenth-century view that medicine should function to prevent disease 'by specifying those social and environmental conditions that promoted the maintenance of health'."[95] With this focus in mind, Edward Jarvis became an authority on vital statistics, and this led to his becoming a central figure in the first connection between the ASA and the US Census Bureau. The two organizations maintained close ties over the first 70 years of ASA's existence; those ties remain strong today. **Simon Newton Dexter North** (1848–1924), the chief statistician of the 1900 census, became the first director of the permanent Census Bureau, 1903–1909. One year later he was the first ASA president to serve a one-year term.

The third ASA president, Edward Jarvis, who specialized in mental diseases and was deeply interested in the treatment of insane patients, made a close study of the sixth federal census of 1840, in which he found many serious errors. Three years later, he was one of the three members of an ASA committee that examined those census figures, and reported that several northern towns had more black insane people than the total black population. In 1844 Jarvis published a pamphlet entitled *Insanity among the Coloured*

Population of the Free State, yet nothing was done to address this issue because of the slavery question. The ASA did become closely involved with the decennial censuses after that, however. One commentator wrote, "Jarvis was appointed by the Secretary of the Interior to prepare and edit the vital statistics of the 1860 federal Census ... thus placing vital statistics on a more comprehensive and scientific basis."[96]

Overall, the ASA, like both the APS and the American Academy, was not formed as a group of specialists, as the term "statistical" in its title might indicate, but instead its members came from various walks of life. Therefore, the articles that members presented at meetings dealt with the enumeration, and sometimes the description, of demographic or social data. In fact, as early as April 1840 Lemuel Shattuck published a circular seeking donations for the library of any book, document, report, or statistical table.[97]

A major achievement of ASA was the publication in 1847 of articles presented at meetings up to that time, bearing the title, *Collections of the American Statistical Association*. This massive volume consisted of a short preface describing the historical development of statistics followed by three separate parts that had appeared earlier. The first part, "Statistics of towns in Massachusetts" (120 pages) had appeared in 1843. The second part, "Statistics of population in Massachusetts" (216 pages) had appeared in 1845. The third part, "Statistics of taxation in Massachusetts including valuation and population" (375 pages) had appeared earlier in 1847. Material for a second volume was apparently prepared 25 years later, in 1872, but the Great Boston Fire destroyed the manuscript.

In 1888 the ASA initiated the journal *Publications of the American Statistical Association* almost 50 years after its founding, under the direction of president Francis Walker. In 1922 the title was changed to the *Journal of the American Statistical Association* (***JASA***), which today is one of the most important periodicals of statistical science. The journal was published in two-year volumes under both titles until 1924, when it became an annual.

Early Statisticians. In 1900, ASA president Carroll Wright wrote a short history of US censuses that singled out three "eminent statisticians" for their work on the 1850 census: Lemuel Shattuck, Jesse Chickering, and Edward Jarvis.[98] **Jesse Chickering** (1797–1855) does not rank on the same level as the other two. Chickering was among the initial group of ASA fellows and was elected to the first set of counselors. He studied theology at Harvard and became a Unitarian minister after graduating in 1818. A decade or so later he pursued medical studies at Harvard and received his diploma in 1833, when he began practicing medicine. All along, Chickering was interested in political economics, as evidenced by three works he wrote: the books *Statistical View of the Population of Massachusetts from 1765 to 1840* (published in 1846) and *Emigration into the United States* (1848), as well as the "Report on the census of Boston" (1851).

The remainder of this section introduces Shattuck and Jarvis, another influential ASA member (Francis Walker), and someone who played no role in the ASA but represented an entirely different part of the US regarding the development of statistics in America (James DeBow). Additional details on the lives and careers of these four early statisticians are available online at "Web03-Statists."

Undoubtedly, **Lemuel Shattuck** (1793–1859) was the most influential member of ASA during its early period, even though he never served as president. Shattuck was born in Massachusetts but a few months later his family moved to New Hampshire. His schooling was limited to no more than six weeks a year, though he engaged in a program of self-education that enabled him to become a teacher at the end of the War of 1812.

Shattuck taught in schools in Detroit and in the state of New York. Due to poor health, he returned to the family home in Concord, MA, in 1822 and remained there until 1833. Concord was one of the rapidly growing seaboard cities in the US. During this time, he wrote a book on the history of Concord from its founding in 1635 to 1832, highlighting the fact that the town served as a temporary refuge for both the legislature and for Harvard during the opening days of the Revolutionary War.[99] Of interest here is a special chapter devoted to his statistical analyses based on church and municipal records. This investigation showed his interest in exact numerical statements when he discovered that imperfect records of births, marriages, and deaths had been recorded by various localities.

Shattuck lived in Cambridge (1834–1835) and then in Boston for the rest of his life. He made a living as a bookseller and publisher on a modest scale, yet within five years was able to retire and spend the rest of his days devoted to vital statistics. Because of errors he had discovered while researching his book on the history of Concord, he used his membership on the Boston Common Council to correct errors in censuses, first on a local scale and then on a national stage.

Shattuck enlisted the support of three bodies for a more effective system of registration: the state legislature, the Massachusetts Medical Society, and the American Academy. He achieved success in 1842 when the legislature passed a law requiring more exact registration of births, marriages, and deaths. Accuracy was achieved by following the British practice of making each *person* the census unit instead of each family, thus recording the name, age, birthplace, marital condition, and occupation of each individual.

In 1846, Lemuel Shattuck published the results of a statistical study he had directed in the 274-page book, *Report to the Committee of the City Council Appointed to Obtain the Census of Boston for the Year 1845*. A distinguishing feature of this report was the inclusion of an interpretative introduction, a standard practice today. In 1900 ASA president Carroll Wright commented:[100]

> Shattuck … recommended that a central board of three persons … should be organized at Washington, to be selected "not for their political opinions, but for their scientific attainments and knowledge of the matters they are to investigate."

Shattuck clearly favored knowledge of statistics over political connections.

All the while, Shattuck maintained correspondence with leading statisticians in Europe, notably Adolphe Quetelet but also the president of the London Statistical Society (Sir Rawson William Rawson) and the first Registrar General of England (Thomas H. Lister). Statistical practices in Europe, particularly Britain, were ahead of those in America. These contacts made Shattuck aware of the spate of local statistical societies that had sprouted up in France and Britain. In particular, the constitution of the Statistical Society of London served as the model for the ASA when he spearheaded its founding in late 1839. Nearly all of the European societies had been named for cities, but Shattuck chose the national name American, although he insisted on the parochial requirement of holding annual meetings in Boston.

In 1849 Shattuck was elected to the Massachusetts legislature, where he was selected as chair of a special committee that thoroughly revised the state registration laws. His bill was then enacted and soon became the model for every other state in the union. Walter Willcox wrote, "For our present country-wide registration system Shattuck deserves more credit than any other man. ... The United States census of 1850 was the Boston census of 1845 writ large."[101]

As chairman of a commission in 1849, Shattuck made a sanitary survey of Massachusetts, arguably his crowning achievement in social service. This important statistical undertaking resulted in the publication of a report[102] that he planned, implemented, and wrote. This work proved to be a milestone in the improvement of American public health, and it serves as the culmination of Shattuck's lifetime interest in public health.

In 1959, the centennial year of his death, Lemuel Shattuck was called the "prophet of American public health"[103] and more recently, in 2008, "the architect of American public health."[104] Overall, though not a mathematical statistician, he should be remembered for four accomplishments: (1) founding the ASA, (2) establishing an effective system of registration for population studies, (3) using statistics in census practice, and (4) improving American public health and preventive medicine. Because of these achievements, he stands out as the most influential American statistical investigator of his time.

Edward Jarvis (1803–1884) was an early American statistician who served the longest presidency of the ASA, 31 years, and managed to keep it afloat during the American Civil War. Born and raised in Concord, MA, Jarvis graduated from Harvard in 1826 and then attended the University of Vermont as a medical student before earning his MD degree in 1830 at Harvard Medical School, then called Massachusetts Medical College. He was interested in mental illnesses and it was his study of the 1840 census, ably assisted by his wife, which led to his work on the black population in the US. As a member of a commission charged with exploring and studying the problem of handling the insane, Jarvis prepared a 600-page report in 1855 that attracted wide attention. This also led to subsequent dealings with the census bureau, as he edited the vital statistics section in the 1860 federal census and also served as a consultant on the census ten years later.

The fourth ASA president, **Francis Amasa Walker** (1840–1897), had a varied career. Born in Boston, Walker graduated from Amherst College in 1860 and then spent a year studying law at a private practice before entering the Civil War. During 1865–1868, he taught mathematics, Greek, and Latin at a seminary. In 1869, at age 29, he was appointed chief of the Bureau of Statistics and Deputy Special Commissioner of Revenue. Walker reorganized the Bureau of Statistics on a scientific basis, leading to his appointment as superintendent of both the 1870 and 1880 federal censuses. In 1874 he edited a government publication entitled *The Statistical Atlas of United States*. His biographer stated, "This pioneering Atlas, making skillful use of colors and graphic presentations, took advantage of recent developments in color printing and photography, and was the forerunner of wider employment of charts in the future."[105]

Walker entered academia in December 1872 when he left the census bureau to teach economics and statistics at Yale, where he wrote three books on economics. In 1877, he became the first person to lecture on economics at Johns Hopkins, one year after that historic university was founded, and he repeated those lectures the following

Figure 3.12. Francis Amasa Walker

year. He was president of MIT (1881–1897) when he published five books applying statistics to various domains. Walker foreshadowed the need for statistics courses at American universities in "The study of statistics in colleges and technical schools,"[106] an article from 1890:

> It cannot be long, however, before the growing interest in economics and history will compel the recognition of statistics as a distinct and an important part of the curriculum of every progressive institution. The main difficulty will be to find the men who have had the training at once severe and liberal, which will qualify them to inspire and direct the studies.

Walker's views on the need for training in statistics were emphasized in his address at the 1897 annual meeting of the ASA, just five days before his sudden death: "All those who have had anything to do with American statistics came into the service comparatively late in life, without any elementary training."[107]

Francis Walker was very influential in turning the ASA into a national organization during his term as president, 1883–1897. In 1888, the ASA initiated today's *Journal of the American Statistical Association* during his presidency. Notably, it was during Walker's term that the exclusive, by-invitation-only membership requirement was waived, resulting in dramatically increased national interest and a growth in membership from 160 at the beginning of his term to 500 at the end.

The final mid-nineteenth century American statistician introduced here is **James Dunwoody Brownson DeBow** (1820–1867), who was born in Charleston, SC, never lived in New England, and played no role in the ASA. DeBow attended the College of Charleston 1839–1843, graduating as valedictorian of his class. He became associated with a very important periodical in its day, the *Southern Quarterly Review*. Even though he became editor, he left that position in 1845, after just one year, and moved to New Orleans to found his own magazine, *The Commercial Review of the South and West*, whose first issue appeared in January 1846. In it, DeBow expressed his intention "to collect, combine, and digest in permanent form, for reference, important statistics" relating to the South and West. In an article he published two years later, "Universities in America," he urged the teaching of commerce and statistics. The periodical became

Figure 3.13. James DeBow

known as *DeBow's Review* after his death in 1867, and it continued publication until 1880.

James DeBow was appointed to an endowed professorship in 1848 at Tulane University. This institution traces its origin to the Medical College of Louisiana in 1834, which became part of the public University of Louisiana 13 years later. When that institution was reorganized in 1884, Tulane emerged as a private university. Back in 1848 the university had difficulty attracting students. As professor of commerce and statistics, he offered the course "Statistics of population and wealth in their application to commerce, agriculture, and manufacture," but he attracted no students.

Later that year, DeBow accepted the position of head of the newly created Bureau of Statistics in the State of Louisiana. He was one of a handful of statisticians who were consulted for the 1850 census. When difficulties arose with the politician charged with conducting the census, President Franklin Pierce appointed DeBow to the position. DeBow reorganized the census work and carried it through to its completion in 1855, including a very important, 1200-page compendium, *Statistical View of the United States*. In it, he recommended the establishment of a permanent office of the census in place of the policy of dismantling that office after it completed its work every ten years. However, it was not until 1902 that a permanent Bureau of the Census was finally established in the US.

Short-lived societies. In his 1957 paper describing some leading American statisticians in the nineteenth century, the economist and historian of political/social thought Paul FitzPatrick wrote:[108]

> The winning of our independence in 1783 brought forth many problems, political, social and economic. To solve these problems we had at that time very little statistical data and records available. Hence it was a difficult situation to solve these problems objectively. It became necessary to take an inventory of our resources, both human and material.

That informative source described several statistical almanacs or yearbooks that addressed these growing problems at both the national and local level from 1794 to 1828.

Table 3.3. The five statistical societies in America
formed up to 1888

Founded	Name	Success
1836	New York Statistical Society	Died in infancy
1839	American Statistical Association	Continues today
1847	General Statistical Society for the United States	Not one meeting
1848	Goliad Statistical Society	Died in infancy
1851	American Geographical and Statistical Society	Changed to Geography after 1871

In a separate paper from that same year, FitzPatrick described several attempts by various individuals to form statistical societies in America shortly after that.[109] He began by citing some such societies that arose earlier in nineteenth-century England, like the British Association for the Advancement of Science (BAAS), which was founded in 1831 due to dissatisfaction with the rather aristocratic Royal Society. Two years later the BAAS established a statistical section. Shortly thereafter, in September 1833, the Manchester Statistical Society was founded. Six months later the Statistical Society of London was organized; its name was changed to the Royal Statistical Society in 1885 to reflect its national character. Subsequent statistical societies were established from 1836 through 1847 in Bristol, Ulster, Liverpool, and Leeds in England; Glasgow, Edinburgh, and Aberdeen in Scotland; and Dublin in Ireland.

The idea of forming such a society seems to have crossed the Atlantic. Table 3.3 lists five societies that were formed in America during 1839–1851, the only ones up to 1888. Only the American Statistical Association, founded in late 1839, can be considered a success. However, the ASA was not the first such society to be formed in the US. That honor, so to speak, goes to the New York Statistical Society (**NYSS**), which was formed in May 1836. Its constitution stated, "The object of the said society shall be the collection, arrangement and diffusion of statistical knowledge."[110] Apparently Charles Sanderson, an American member of the Universal Statistical Society of France, instigated the attempt to form NYSS, writing to Yale professor Benjamin Silliman in 1835, "I respectfully desire to propose for consideration the establishment of a general statistical society in the United States."[111] In spite of Sanderson's worthy intentions, for unknown reasons the NYSS never functioned, perhaps because the legislature required a $50,000 capital stock requirement.

The third column in Table 3.3 indicates that the NYSS had but a brief existence. The same is true of the Goliad Statistical Society, born in Goliad, TX, in 1848 but it died in its formative stage. There is no evidence that the General Statistical Society for the United States ever held one meeting. **George Tucker** (1775–1861) was the person most responsible for proposing the establishment of this society. A statistician and versatile polymath, Tucker was born in Bermuda but came to the US at age 12 to live with an uncle of the same name who was professor of law at William and Mary. Tucker graduated from William and Mary in 1797, spent a year studying law under his uncle, and practiced law until being elected a congressman for three successive terms (1819–1825). Instead of running for a fourth term, Tucker accepted Thomas Jefferson's offer to join the faculty at the newly opened University of Virginia as professor of moral philosophy. He was chosen the first head of the faculty and remained until retiring in 1845 at age 70.

During his 20-year tenure, Tucker published two works that indicated a growing interest in statistical matters. His 1839 book *The Theory of Money and Banks Investigated* makes considerable use of statistics. However, his 1843 work *Progress of the United States in Population and Wealth in Fifty Years, as Exhibited by the Decennial Census* earned him a reputation as "a pioneer in this country in advocating the use of statistics to place economics on a sound scientific basis."[112] The preface to the book states that studying federal censuses from 1790 through 1840 taught him important lessons:[113]

> They have conducted [me] to important inferences on the subjects of the probabilities of life, the proportion between the sexes, emigration, the diversities between the two races which compose our population, the progress of Slavery, the progress of productive industry.

Although in 1940 Walter Willcox described this book as "the most important American work on statistics to appear in the first half of the nineteenth century,"[114] 16 years later Paul FitzPatrick put its importance in a more balanced perspective:

> Some are inclined to regard this work of Tucker as probably our first American statistics text. While it obviously did not deal with statistical tools such as averages, dispersion, correlation, and the like, it did undertake to explain various aspects of population statistics which at the time was one of our most important types of statistical data.

George Tucker retired from teaching in 1845 and moved to Philadelphia, where he remained for the rest of his life. Here, he wrote one of the first comprehensive histories of the United States, the four-volume *History of the United States, from Their Colonization to the End of the Twenty-Sixth Congress, in 1841*. In December 1847 he proposed forming a statistical society but his idea never materialized. At the first meeting of the American Academy in Philadelphia the next year, he proposed a section of statistics and political economy within the Association but the statistics part was deleted. It is somewhat surprising that a statistics subsection was not put within the Mathematics and Physics section of the American Academy.

As the bottom row in Table 3.3 indicates, the American Geographical and Statistical Society (**AGSS**) endured for 20 years, 1851–1871. It was formed in October 1851 in New York City with a constitution that read, "The American Geographical and Statistical Society of New York is instituted for the collection and diffusion of geographical and statistical information."[115] A prominent early statistician with AGSS was the Scottish-born **Archibald Russell** (1811–1871), who graduated from the University of Edinburgh before immigrating to the United States in 1836. Three years later he published the influential book, *Principles of Statistical Inquiry*. Russell assisted a special committee of the US Senate in preparing the agricultural schedule for the 1850 census.

The AGSS published four issues of a *Bulletin of the AGSS* from its inception in 1851 through 1857. Two years later the AGSS initiated a monthly *Journal*. However, the American Civil War took its toll on publications—a single issue appeared in 1860 and then not another until 1870. The crowning blow to AGSS occurred in 1871 when the legislature of New York State amended its charter to remove the words "and Statistical" from if the title. From that time forward, the Society emphasized geography at the expense of statistics.

Statistics courses. As noted above, in 1848 James DeBow offered a course on statistics at Tulane but enrollment was insufficient to actually give the course. That same year, DeBow published an article urging the teaching of commerce and statistics. This advocacy was the forerunner of our schools of business today, the first of which was the Wharton School at the University of Pennsylvania. There had been some movement to offer courses in statistics at institutions of higher learning just a couple years earlier. The earliest mention of the word "statistics" appeared in the catalog of Transylvania University (Lexington, KY) for 1843–1844, but there is no record of that course ever having been offered.

The University of Virginia was actually the first institution in America known to give a course on statistics. In fact, its department of moral philosophy offered three during 1845–1846. That year's catalog stated, "The Intermediate Class studies Political Economy, Statistics, and the Philosophy of Social Relations, or 'Ethics of Society'," so apparently third-year students were required to take the course "Political economy, statistics, and philosophy of social relations." The course was taught by the Rev. **William Holmes McGuffey** (1800–1873), a professor of moral philosophy, best known for writing *McGuffey's Readers*. The main textbook was George Tucker's *Progress of the United States in Population and Wealth in Fifty Years, as Exhibited by the Decennial Census*, published two years earlier.

It seems to have taken another 25 years before a successful course on statistics was conducted. From 1873 to 1881 Francis Walker taught the course "Public Finance: Statistics of Industry" at Yale. Unfortunately, no further information is known about the course, including any text that might have been used.

4

The Age of Peirce

I define the "Age of Peirce" to extend roughly from the death of Nathaniel Bowditch (1838) up to the pivotal year for research in mathematics in America—1876. This chapter divides the Age of Peirce into three periods—Antebellum, Civil War, and Reconstruction.

The Antebellum period, 1846–1860, was dominated by the most accomplished American scientist up to 1876, Benjamin Peirce, whose outstanding contributions were somewhat ahead of their time. A student of Nathaniel Bowditch during his formative years, Benjamin Peirce dominated the American scientific arena for almost four decades, from the time he burst onto the scene about 1830 until 1876. The Antebellum period of the Age of Peirce includes such important developments as the formation of the American Academy of Arts and Sciences, the establishment of high-level schools of science within Harvard and Yale, and the founding of a journal with tenuous ties to the popular one of today, the *American Mathematical Monthly*.

Sections then follow on the American Civil War and the period of Reconstruction; together, they span 1860–1876. Cryptology links the Civil War to the Revolutionary War. In parallel with the earlier war, the Civil War exacted a devastating toll on mathematics before and after its outbreak, so the US needed a decade to regain strides made by the late 1850s. However, during the period of Reconstruction (1866–1876) a community of sufficient size and interest set the stage for the mathematical revolution that followed. During this time, George W. Hill and J. Willard Gibbs emerged as two leading mathematicians who made original investigations; *The Analyst*, another mathematics journal, launched the careers of several notable figures; and Artemas Martin engaged in forming several journals. The final figure to be introduced, Hubert A. Newton, "the grandfather of American mathematics," is cited for his role in the production of the first PhDs in mathematics in America.

Where were such devotees in this community employed in the Age of Peirce? With only a few exceptions, colleges did not provide a supportive environment for mathematicians to carry out original investigations. Instead, many individuals with a strong interest in mathematics worked at the US Coast and Geodetic Survey in Washington, DC, or at the Nautical Almanac Office, initially in Cambridge, MA, and later also in

Washington, DC. Even there, however, the positions were designed mainly for consultation and practical research. The notable scientists who dabbled in mathematics at the time were supported neither by private sponsors (think of the Microsoft Research Group today) nor the federal government (think of the National Security Agency or the National Science Foundation).

Recall Thomas Brattle's exasperation at working "here alone by myself, without a meet help in respect to my studies." This chapter reveals that the leading contributors to mathematics over the next 100 years worked in a similar kind of isolation, with few compatriots able to appreciate, let alone understand, their work. I refer to these leading figures as "rugged individualists," thereby borrowing the phrase popularized by the American novelist Wallace Earle Stegner (1909–1993), who described the American West as a desirable landscape for rugged frontiersmen standing tall and self-reliant—particularly, stereotypical loners sporting Stetson hats. The rugged individualists in mathematics possessed creative minds beneath those wide brims, but there were few outlets for publishing their original investigations. Throughout this period several individuals attempted to establish journals geared specifically to mathematics, all of which suffered short durations due, primarily, to the lack of a cohesive community. Similarly, there were a few failed attempts at founding learned organizations of mathematical practitioners.

Calculus textbooks

The publication of calculus textbooks by American authors provides a natural transition from the Age of Bowditch to the Age of Peirce.[1] Overall the most successful calculus books have been known by their authors' names rather than their titles, like Davies and Loomis (today it might be Granville, Thomas, and Stewart). Calculus texts existed in the US even before Charles Davies published his in 1836. The first calculus book penned by an American author was published in New York in 1828, *The differential and integral calculus*. Few copies of this 328-page text are extant, and I know nothing about the author, James Ryan.

A large demand for calculus books did not materialize until the early twentieth century. For instance, up to the Civil War, two of the leading academic institutions in the US, Harvard and West Point, devoted only about one-third of the sophomore year to differentiation and integration. All curricula were entirely prescribed at the time; hence, all students were required to successfully pass this course for graduation. Also, Yale offered a course on fluxions as an option in one of the three sophomore-year terms, but that too confirms that calculus could not have been developed in much depth.

At the beginning of the nineteenth century, American editions of British works were adopted for use in the classroom. Generally, only the professor owned a copy; students were required to transcribe into their notebooks the material that the professor recited or wrote on the board. When *The Principles of Fluxions* (by the Rev. Samuel Vince) was first printed in Philadelphia in 1812, the subtitle read, "The first American edition, corrected and enlarged." A more important author for the American audience was Charles Hutton, whose expansive *A Course of Mathematics in Two Volumes for the Use of Academics as well as Private Tuition* appeared in 1812. In 1831 the title page of the ninth edition of this popular compendium revealed "the additions of Robert Adrain … the whole corrected and improved." In fact:[2]

> Adrain edited, revised and condensed the British text and in 1812
> published an American version consisting of two volumes. ... [It]
> remained in use at West Point until 1823.

Although Adrain was fluent in French and generally adopted Continental notation, the calculus in this book was based on fluxions; indeed, calculus did not appear until midway through the second volume.

By 1820 America began to move away from total reliance upon Great Britain toward Continental advances, particularly France. Over the next two decades, Americans debated the relative merits of approaching calculus via fluxions or differentials. The historian Todd Timmons devotes an entire section of his recent book[3] to discussing many articles attacking the foundations of analysis in the pages of the *AJSA* during that time. However, few French texts were available in the New World, including French Canada. Therefore, John Farrar set about translating appropriate texts. In 1824, he adapted a text by Étienne Bézout for domestic consumption, with its verbose title indicating the overlap of British and French approaches: *First principles of the Differential and Integral Calculus, or The Doctrine of Fluxions, intended as an introduction to the physico-mathematical sciences; taken chiefly from the mathematics of Bézout, and translated from the French for the use of the students of the university at Cambridge, New England*. Cauchy's *Cours d'Analyse*, published three years earlier, made a revolutionary leap by basing calculus on limits, but apparently Farrar was either unaware of this critical advance or had not had sufficient time to master it, because his text is based on infinitesimals.

Charles Davies had been teaching the methods of the French and other continental mathematics at the Military Academy since his appointment in 1816. One year before leaving West Point in 1837, he published *Elements of the Differential and Integral Calculus*, which became such an instant success that it made James Ryan's earlier textbook superfluous. Davies published so many successful texts in algebra, geometry, and trigonometry as well as calculus that his "name [was] known to nearly every schoolboy in our land."[4] His was the first commercially successful calculus text in America, with a new edition appearing every year or two between 1836 and 1860. Overall, Davies wrote at least four different versions of his calculus books, with the last edition appearing in 1901, 25 years after his death.

Davies revealed that his initial source was an 1828 book by the French mathematician Jean-Louis Boucharlat, which had been used as a textbook at West Point.[5] Davies seems to base his definition of the derivative on the limit concept, defining it as $\frac{u'-u}{h}$, where u' is the incremented value of u. However, in practice, he used infinitesimal arguments to evaluate the difference quotient, as in his specific examples for $u = ax^2$ and $u = x^3$. Davies did not introduce differential notation; instead he used the phrase "limiting ratio of the increment of the variable to that of the function." Using this definition, Davies proceeded to today's standard fashion of developing the usual rules for differentiation, including the chain rule. He included Taylor's theorem and applied this material to curve sketching.

The Davies textbook also avoided limits in integration. He defined the integral as "the method of finding the function which corresponds to a given differential," that is, as an antiderivative. Here he adopted the standard integral sign and noted that it initially represented a sum in earlier times. However, Davies never dealt with the integral as the limit of a sum. He then covered techniques of integration and applied them to arc length, area, and volume, with double integrals introduced for the latter.

In his study of calculus texts by American authors published before 1910, George Rosenstein identified seven that appeared after Davies in 1836 and before the onset of the Civil War.[6] The only author on that list who was a professional mathematician was Benjamin Peirce. As sometimes happens, however, the best mathematicians do not write the most successful textbooks. Rosenstein refers to Peirce's two-volume *An Elementary Treatise on Curves, Functions, and Forces* as "pedagogically painful ... [it] would delight the mathematician but horrify the sophomore."[7] Nonetheless, the work ran through three editions, the first in 1841 and the last 21 years later. Peirce also based the derivative on infinitesimals.

The only other commercially successful American author of calculus textbooks besides Davies in the period 1836–1860 was **Elias Loomis** (1811–1889),[8] the only one having direct contact with first-class mathematicians abroad. Loomis enrolled at Yale 1826–1830 when Denison Olmsted was professor of mathematics and natural philosophy. Upon graduation in 1830, he taught at an academy for a year and then studied divinity. Yale president Jeremiah Day hired him as a tutor in 1833, a position he held for three years. During this time, Loomis conducted recitations on mathematics, natural philosophy, and Latin, carried out research on measuring the Earth's magnetic field, and joined Olmsted on an expedition to observe Halley's Comet.

However, in 1836 Elias Loomis left Yale for a professorship of mathematics and natural philosophy at Western Reserve College. Like Sylvanus Thayer at West Point 21 years earlier, Loomis's first duty in this new position was to travel to London and Paris to purchase instruments for his laboratories. While in Paris, he attended courses of lectures by François Arago and Jean-Baptiste Biot. Once back in Cleveland, Loomis oversaw the construction of an observatory; over the next seven years he published ten papers in the *Proceedings of the APS*, a quarterly established in 1838. But the college ran into financial difficulties and his pay fell into significant arrears. Consequently, in the spring of 1844, Loomis left Western Reserve for the City College of New York, where he began writing textbooks, two of which became enormously successful, *Treatise on Algebra* (1846) and *Elements of Plane and Spherical Trigonometry* (1848). The 76th edition of the latter appeared in 1881.

Elias Loomis remained at City College (1844–1860), except for one year at Princeton when, ironically, he was replaced by Charles Davies. During that year, Loomis published weather maps using a new method of representing meteorological data that turned out to be useful in predicting storms. But he did not feel welcome in Princeton society, so he returned to New York City. It was there, in 1851, when the first edition of his textbook *Elements of Analytical Geometry and of the Differential and Integral Calculus* appeared. This book sold 25,000 copies by 1874. More importantly from our perspective, it turned out to have international ramifications for American mathematics. Even though Loomis is a minor figure in the history of mathematics, a detailed article on the history of mathematics in China declared, "In the history of mathematical developments internationally, especially in Asia, Loomis was a figure of considerable influence."[9]

China has a long and impressive history of mathematics that was carried out in virtual isolation from the rest of the world. However, after a particularly strong period that saw such advances as the solution of systems of linear equations, the vast country suffered a period of stagnation and decline after the fourteenth century. Western mathematics did not reach China until 1607, when the Italian Jesuit priest Matteo Ricci joined Guang-qi Xu in 1609 to translate the first six books of Euclid's *Elements* into a Chinese language. Interest in Western mathematics continued for another 150 years, but then waned again for another century. It was the aftermath of the Opium War of 1840 when China opened commerce and trade with Great Britain, and this in turn brought mathematics to the Qing government.

The catalyst was someone otherwise unknown in the history of mathematics. The Englishman **Alexander Wylie** (1815–1887) came to Shanghai in 1847 as a representative of the British and Foreign Bible Society. Shortly thereafter, he engaged **Shan-lan Li** (1811–1882) to help him translate mathematics works from English into Chinese. Li began a systematic study of Western mathematics with Wylie that ultimately led him to become the most important Chinese mathematician in the nineteenth century. Their first joint venture was to complete the Ricci–Xu project by translating the remaining seven books in the *Elements*, plus two spurious books. Next, they translated the book *Algebra* by the well-known Augustus De Morgan.

After that, Li and Wiley tackled the Loomis *Calculus.* Why this work? In the preface to their translation, published in 1859, Li wrote:[10]

> Loomis, a famous American mathematician, combined algebra, differential and integral calculus into a whole book, its title reflecting all of these, and set up an orderly system. This book is of benefit to the students. Mr. Wylie knew and appreciated this very much. He made great efforts to obtain this book, and asked me to translate it into Chinese with him.

In the preface to the Loomis text written eight years earlier, the author asserted:

> It was written not for the mathematician, not for those who have a peculiar talent or fondness for the mathematics, but rather for the mass of college students of average abilities … it appears to me that I have here developed it in a more elementary manner than I have before seen it presented.

This translation was a significant undertaking—the first book on calculus ever to appear in China. An important addition was an appendix listing 330 English–Chinese terms that became standard and are still in use today. It remains surprising that despite the earlier and more extensive impact of Europe on China, it was the US, not yet a century old, that exerted the major influence on modern Chinese mathematics. Beyond that, the Li–Wylie book was soon translated into Japanese, and for many Japanese mathematicians, this was their only source of knowledge of calculus at the time. This explains why Japanese books use the same Chinese characters for calculus today, leading two modern researchers to conclude, "It is no exaggeration, therefore, to say that Loomis has exerted an extraordinary influence throughout East Asia on the development of modern mathematics … thanks to the Li–Wylie translation of his calculus."[11]

Curiously, it is unknown if Elias Loomis was ever aware of this translation. He did not mention it in 1872, the year when the last edition of his calculus text appeared. Just two years later he published *Elements of the Differential and Integral Calculus*, which used limits as the foundation for the derivative; it was republished as late as

1902. Loomis was at Yale (having succeeded Denison Olmsted in 1860) when the latter calculus text first appeared in 1874. Olmsted, in fact, was another successful textbook author who sold over 200,000 textbooks in his lifetime. Loomis's array of books tripled those sales! In his will he left a bequest to Yale of $300,000 from royalties.

Journals

The preceding section mentioned the debate among American mathematical practitioners before 1840 over calculus in the *AJSA*. That general science publication prospered, but at the same time, attempts to establish periodicals devoted solely to mathematics by George Baron and Robert Adrain ended in abject failure. This section describes two closely linked publications that attempted—unsuccessfully—to achieve the same feat early in the period 1836–1876. The first was established in the mid-1830s by a third native Englishman, while the other was a kind of successor in the early 1840s initiated by the Harvard professor who dominated this entire period, Benjamin Peirce. Later sections will describe a pattern of subsequent sputtering attempts from 1843 right on up to, and even beyond, the first legitimate research journal that specifically targeted a mathematical audience in 1878.

Charles Gill (1805–1855) became the third immigrant to initiate a mathematical journal. Born in England, Gill was a country school teacher when he became a frequent contributor to the problem section in the popular British periodical, *The Ladies Diary*. English mathematical practitioners, after all, were accustomed to having outlets for their activities. Three years after emigrating to the US in 1830, Gill was appointed professor of mathematics at St. Paul's Collegiate Institute in Flushing on Long Island, NY. An article from 1875 called Gill "one of the best Diophantists in America. His speculations on problems relating to Polygonal numbers were profound and interesting."[12]

In 1836, just a few years after Adrain's *Mathematical Diary* had fizzled, Charles Gill launched the *Mathematical Miscellany* as a semi-annual publication whose first issue appeared that February.[13] He felt confident that the American populace was ready to support a periodical devoted solely to mathematics, but he miscalculated. One of his friends, a high-school teacher named Evans Hollis, warned presciently:[14]

> I wonder at your obtaining even sixty subscribers for in this country
> there are but few who can appreciate such a work and people have
> not yet acquired wealth or spirit enough to patronize works in which
> they are not personally interested.

Hollis was right. Gill began facing financial challenges just months after the first issue appeared. (The subscription rate was $5 per year—equivalent to approximately $150 today). To remedy the situation, after the first issue the editor divided the *Mathematical Miscellany* into two parts. He also wrote a section called "Hints to young students" in the Junior Department (one of four sections of the journal), which also included problems and the solutions to earlier problems. Benjamin Peirce wrote, "[I] cordially recommend it to our young friends as one of the best means of drawing out the mathematical talent in the country."[15]

Like the *Correspondent*, the "Senior Department" of the *Miscellany* was a mixture of articles and reprints of European works. The level of mathematics in those articles in the *Miscellany*, however, show that American mathematics had made significant gains over the intervening three decades. Most of the problems in the *Miscellany* also reflect this advance, as indicated by solutions in which Benjamin Peirce made use of concepts from differential geometry (which had been discovered by the "prince of mathematics," Karl Gauss) and in which Theodore Strong cited Abel's theorem on integrals of algebraic functions.

It is also noteworthy that **Nancy Buttrick** may have been the first woman to publish a mathematical work in an American journal when she submitted and solved problems in the *Miscellany*. It is difficult to pinpoint the exact "first" woman because many contributors used pseudonyms at the time, so work that appeared in Adrain's *Mathematical Diary* under the name Hypatia might have been a woman. Buttrick used "A Lady" in signing her submissions. Little is known about her, but she married Hamilton College's professor of mathematics and natural sciences **Oren Root**; their son Elihu Root served as Secretary of State under President Theodore Roosevelt and was awarded the 1912 Nobel Peace Prize.[16]

Between 1804 and 1806, the *Mathematical Correspondent* enjoyed a subscription list of 345 individuals, yet some 30 years later, Charles Gill could not count on even 60 subscribers when he began his run of eight issues of the *Miscellany* from February 1836 through November 1839.[17] Had the mathematical community shrunk? No. The difference in those figures is due to the evolving specialization of the sciences. The *Correspondent* subscriber list included numerous general scientists, then called natural philosophers, whereas by the time Gill published the *Miscellany*, the only individuals interested enough to subscribe were mathematical practitioners, a community exceedingly small, not then exceeding 100.[18] Besides, few Americans had been trained formally in mathematics. Even the talented practitioner William Lenhart lamented to Gill:[19]

> I must confess, and I feel ashamed in doing so, that I have but few
> Mathematical books; and not having an opportunity of seeing any,
> either periodical or otherwise, I know but little of what is doing or
> what has been done in the mathematical world.

Moreover, few professors of mathematics and natural philosophy participated in the journal in any way. Most of the contributors lived in New York, with a particularly large contingent in the Clinton area upstate.

The historian of mathematics Deborah Kent commented on another challenge Gill faced:[20]

> Gill worked, as any editor of a mathematical journal might, to recruit
> colleagues to help him generate problems and evaluate submissions
> for publication.

Few volunteered. Thus, pecuniary concerns, the pending closure of the Flushing Institute, the arrival of his wife and daughter from England, and the expected birth of a second child compelled him to scramble to find an editor to take the reins of the *Miscellany*. Consequently, Charles Gill stands in line with those individuals who published short-lived journals devoted to mathematics alone.

Despite his full plate, 33-year-old Benjamin Peirce launched the most ambitious periodical at this time, the first founded by an American-born editor. Kent added:[21]

> Peirce did as much as he could, but he was fully occupied by writing
> textbooks, teaching mathematics, introducing the elective system at

Figure 4.1. Joseph Lovering

Harvard, caring for his recently widowed mother, and experiencing
poor health.

Peirce intended the quarterly *Cambridge Miscellany of Mathematics, Physics, and Astronomy* to be, in some way, a continuation of Gill's journal. The first issue appeared in April 1842 with Peirce as the sole editor. His Harvard colleague Joseph Lovering became a co-editor with the second issue. Peirce was such an outstanding nineteenth-century mathematician that I named the period 1836–1876 in his honor. His co-editor is not so well known.

Joseph Lovering (1813–1892) was born and raised in Charlestown, now part of Boston. James Walker, the pastor of his church and later the president of Harvard, was so impressed with the young Lovering's talent that he advanced him money to attend Harvard. Lovering entered in 1830 in the sophomore class, and graduated three years later in a class that produced five other Harvard professors, and four professors at other institutions. Lovering taught in a private school for a year after graduation, and then studied divinity for two more years. During the latter period, 1834–1836, he served as Benjamin Peirce's assistant the first year and as proctor the second. In 1836, he was appointed tutor in mathematics and lecturer in natural philosophy. Two years later, Lovering succeeded John Farrar as Hollis Professor of Mathematics and Natural Philosophy, a chair he ended up holding for 50 years. However, most of his publications were in astronomy and physics. In addition:[22]

> Between 1867 and 1876 Professor Lovering was connected with the
> United States Coast Survey, and had charge of the computations for
> determining differences of longitude in the United States, and across
> the Atlantic Ocean, by means of the land and cable lines of telegraph.

Furthermore, Lovering served as permanent secretary of the American Academy (1854–1873) and then as president for 1873. He edited 15 volumes of the American Academy *Proceedings* as well as six volumes of its *Memoirs* and three of the *Proceedings* of the National Academy of Sciences.

Despite the established reputations of the two editors, the *Cambridge Miscellany* "failed to appeal either to the secondary school teachers on the one hand or to the college instructors on the other, and so the publication had only a brief existence."[23]

Indeed, the *Cambridge Miscellany* lasted only four issues, numbering but 192 pages. The release date of the last issue was January 1843. The price seems reasonable as the "Terms of Publication [were] $2 per annum—payable in all cases in advance," approximately equivalent to $60 today.[24] Each issue was divided into four sections: Junior Department of Mathematics, Senior Department of Mathematics, Astronomy and Physics, and Meteors and Meteorology.[25]

The Junior Department began each issue of the journal. It consisted of five problems and solutions aimed at high school or college students. However, two articles appeared, "Rules of false or double position" and "Solution of numerical equations." Problems were posed in one issue and solutions supplied in the next; they had up to six different solutions. The problems and solutions were mostly submitted by readers, although Peirce himself posed a few, such as "Find the spherical triangle, which is exactly equal to its polar triangle." A list of the names of all solvers followed. Other problems in this section involved finding the roots of an equation, using the derivative to find maxima or minima values, solving systems of equations, locating singular points on a curve, manipulating complex numbers, and using integration to find areas. Most of the questions are within reach of an undergraduate mathematics major today with knowledge of physics.

The Senior Department of Mathematics followed the Junior Department. This section consisted of six problems and solutions at a higher level, the final one being an open problem. There were no articles, but these two departments included editorial commentary and didactic notes aimed at educating the readership. Problems in this section were concerned with summing infinite series, solving systems of partial differential equations, probability, physics, and geometry. Here are three examples:

- Professor ___ was unable to spin a top upon a hard steel floor, the point being sharp like that of a needle; while he found no difficulty in spinning one whose point was blunted so as to be nearly hemispherical. Can this difference be explained by analysis? [*Cambridge Miscellany* **1**, p. 22]
- A given hemispherical shell, of uniform density, and thickness, is placed with its convexity in contact with a horizontal plane. It is required to find the equation of the motion of a given sphere, of uniform density, placed within the shell; and the small oscillations of the ball and shell. The friction is supposed to be just sufficient to make the shell role on the plane and the ball role in the shell, without sliding. [*Cambridge Miscellany* **3**, p. 109]
- Solve the spherical triangle, of which the surface is equal to that of its polar triangle, one of the angles is the supplement of its opposite side, and one of the two right triangles, formed by letting fall a perpendicular from the vertex of the angle which differs the least from 90 degrees upon its opposite side, is isosceles. [*Cambridge Miscellany* **4**, p. 159]

The first problem was submitted and solved by Benjamin Peirce, whose list of 23 variables in the solution required almost an entire page; the solution itself ran to six pages. He was the only solver.

Some of today's mathematics journals that have problems departments feature solutions by high-school students. It was the same with the *Cambridge Miscellany*. Of the 75 who submitted solutions, the highest concentration came from Central High School in Philadelphia, where the principal was Alexander Dallas Bache. Another active group from Philadelphia was located at the US Naval Asylum, where William

Chauvenet had perhaps assigned classroom work to the three midshipmen who submitted solutions.[26] Students at both Harvard and St. Paul's College (where Gill was on the faculty) also participated in the problem sections. Of interest, a mathematics professor at the Clinton (NY) Liberal Institute, identified only as Birdsall, solved 13 senior problems, and inspired seven others at the school to participate. One of the Clinton problemists, the only identifiably female contributor, signed ten submissions, "Melissa."

The third and fourth sections of the *Cambridge Miscellany* consisted of papers written mostly by the editors. For instance, the titles of the six papers by Benjamin Peirce are

- "American astronomical and magnetic observations,"
- "Distances of the fixed stars,"
- "Meteors,"
- "Varieties of climate,"
- "The barometer,"
- "On Espy's theory of storms."

Note the applied nature of these articles, then categorized as "mixed mathematics." The same is true for the four papers by Joseph Lovering,

- "On the internal equilibrium and motions of bodies,"
- "On the applications of mathematical analysis to researches in the physical sciences" (in two parts),
- "Encke's comet,"
- "The divisibility of matter."

The remaining eight articles were a combination of papers by American and European contributors, as well as reprints from continental sources,

- "The southern continent" (Charles Davies),
- "Observations on the diurnal variation of the magnetic needle" (Dutrochet, France),
- "Extract of a letter from Prof. Encke of Berlin to Mr. Airy, Astronomer Royal at Greenwich,"
- "On the law of storms" (H.W. Dove, Berlin),
- "History of the present magnetic crusade" (Rev. Humphrey Lloyd, Dublin),
- "Theory of colors" (Goethe),
- "Memoir on the general principle of natural philosophy" (M.G. Lamé, France),
- "Atmospheric electricity" (Mr. Spencer, England).

Mathematics at Harvard benefited from Peirce's role in editing the *Cambridge Miscellany*. In early 1842, two months before its debut, he ensured that the library would subscribe to the top four scientific European publications: *Crelle's Journal* (formally, *Journal für die reine und angewandte Mathematik*) and Liouville's journal for mathematics (formally, *Journal de Mathématiques Pures et Appliqeées*), as well as *Comptes rendus* and *Astronomische Nachrichten* for general science and astronomy. "Peirce was, in fact, 'so desirous that they should all be taken in the college library,' that he volunteered personally to contribute half the combined subscription cost."[27]

The *Cambridge Miscellany* failed for basically the same three reasons as the *Mathematical Miscellany*. One was the lack of time the editors could devote to other projects, including family demands. Two other reasons—finance (securing enough subscriptions) and scholarship (securing a suitable number of contributors)—continued to be sources of frustration. In short, there was still a glaring lack of a critical mass of mathematical investigators, practitioners, and supporters in America necessary to form a viable community.

The one notable figure in the US at the time who might have benefited from the *Cambridge Miscellany*, which, in turn, might have profited from his activity, was the Englishman J.J. Sylvester. One year after Charles Bonnycastle died in 1840, the 27-year old Sylvester succeeded him as professor of mathematics at the University of Virginia. Unfortunately, Sylvester's initial foray to the United States was not a happy one. Just one year earlier, a Virginia student was charged with murdering a professor. Apparently, the authorities at Virginia had not gotten a full grip on the situation by the time Sylvester arrived, because he too encountered severe problems with the threatening antics of American undergraduate students, being accosted by one.

Therefore, Sylvester made a hasty retreat from Charlottesville after less than one year. He then moved to New York City to live with his brother, all the while seeking employment in the New World. However, he was unable to secure an academic position at Columbia, the University of South Carolina, any college in the Washington, DC, area, and even Harvard, where he was in touch with Benjamin Peirce. To compound matters, his marriage proposal was rejected. This was the tipping point, so in the fall of 1843, Sylvester returned to his native London. Fortunately for the US, this experience did not prevent him from returning 33 years later, though under considerably more propitious circumstances.

Benjamin Peirce

Benjamin Peirce has appeared as an author of a calculus textbook and an editor of a mathematics journal. Those activities constitute but a minor part of his career. An earlier connection was his relationship with Nathaniel Bowditch, initially as a student and later as an aide in the translation of Simon Laplace's *Mécanique Céleste*. This section supplies details on the life and accomplishments of this towering figure, who achieved fame during an important transitory period in the history of mathematics in America.

Benjamin Peirce (1809–1880; pronounced "purse") was the first great professional mathematician in the United States. In fact, with regard to mathematical creativity, he can be considered the only native-born researcher until at least 1880. Care is necessary when encountering the name Peirce, because the sons James Mills Peirce and Charles Sanders Peirce, as well as distant relative Benjamin Osgood Peirce, all became accomplished mathematicians. Although no other family member was as adept as the scientific patriarch, in 1901 the eminent mathematical astronomer Simon Newcomb called the son, C.S. Peirce, "the most original and the most versatile intellect that the Americas have produced so far."[28] So, while world history trumpets the Brothers Bernoulli, the US can boast of the Peirce Pride.

Despite Benjamin Peirce's lofty achievements, he is not as well known in the history of mathematics as he should be. How good a mathematician was he? Well, he certainly was no Gauss. But who was? Karl Friedrich Gauss (1776–1855) was not only the most dominating contemporary mathematician (and astronomer) in the world,

Figure 4.2. Benjamin Peirce

but the Göttingen giant towered over all his contemporaries and is often regarded, along with Archimedes and Isaac Newton, as one of the three greatest mathematicians of all time. Another outstanding contemporary was Gauss's student, G.F. Bernhard Riemann (1826–1866). Peirce cannot be regarded on this level either. Nor that of Augustin-Louis Cauchy (1789–1857). But I feel strongly that a case can be made to rank Peirce on a par with most other European contemporaries whose accomplishments are much more well known—such as Joseph Liouville (1809–1882) in France, Carl Gustav Jacob Jacobi (1804–1851) in Germany, and Arthur Cayley (1821–1895) in England.

Like several American mathematicians mentioned so far, Benjamin Peirce's roots run deep in the colonies, reaching back to Watertown, MA, in 1637. He was descended from John Pers, a weaver who emigrated from Norwich, England, to the colonies one year after Harvard was established. The genealogy from Pers suggests why the name Peirce is pronounced "purse" and is at variance with the usual name Pierce (which sounds like "fierce"). It is unknown when, or why, the spelling changed, nor why it varies from the normal spelling. One of his grandfathers made a substantial fortune with the East India Company.

The Peirce Pride became a family of American intellectuals whose ties to Harvard lie as deep as their native roots. Peirce's parents were cousins. Our Benjamin Peirce's father (of the same name) graduated from Harvard in 1801 as valedictorian of the class. Moreover, the father's senior thesis was titled "Solutions of algebraic problems." The elder Benjamin Peirce served as Harvard's librarian from 1826 to 1831, and his highly regarded history of the university was published posthumously. Later, he served in the Massachusetts Senate and House of Representatives.

Benjamin Peirce was born in Salem, MA, then one of the busiest seaports in the New Republic, and it was actively engaged in worldwide commerce. The modern realtors' slogan, "location, location, location," surely applied to Peirce, whose hometown happened to be the residence of Nathaniel Bowditch. Moreover, Peirce attended the private Salem Grammar School, where a classmate was Bowditch's son, Ingersoll. One day, Ingersoll showed Peirce his father's solution to a mathematics problem. Upon scrutiny, the young Peirce found a mistake in the solution, sending his friend sprinting to his father. The elder Bowditch replied that he must meet the young man who had corrected his mathematics. From that time henceforth, Bowditch served as a mentor for Peirce, educating him in the theory and practice of mathematics, astronomy, and navigation, and imbuing him with the spirit of conducting independent research. Therefore, overall, not only did Peirce receive an outstanding education at Salem, he came into contact with America's most accomplished mathematical practitioner on a regular basis.

The moniker "rugged individualist" surely fits Benjamin Peirce, who did not proceed directly from Salem to college. First, he enrolled for a year at the prestigious Phillips Academy in North Andover, a thoroughly conservative boarding school founded to educate the sons of wealthy citizens in the New Republic. After one year at Phillips, Peirce entered Harvard in the fall of 1825 as a 16 year old. He was not an unknown quantity. Andrew Peabody, a senior that year and later a professor of ethics at the university, provided an insider's view of the child prodigy, gushing:[29]

> Even in our senior year we listened, not without wonder, to the reports that came up to our elevated platform of this wonderful freshman, who was going to carry off the highest mathematical honors of the University.

During Peirce's time at Harvard (1825–1829) he assisted Bowditch by reading the proof sheets of the first volume of his Laplace translation. This constituted almost all of Peirce's instruction in mathematics during his student days because Harvard's prescribed curriculum permitted little time for the sciences. Despite the Hollis Professor being John Farrar, Peirce said very little about his instruction at the college, though Farrar was one of the most effective teachers of his time. As a result, Peirce not only learned calculus and differential equations from Bowditch but advanced topics and applications as well. In addition to the mathematical expertise Peirce had gained from the Boston insurance executive, "Bowditch had a mathematical library of 3,000 volumes. This was probably the best library in the exact sciences in the country."[30]

Consequently, Peirce displayed an impressive mathematical prowess to the outside world while still an undergraduate, solving three problems in the first issue of Robert Adrain's *Mathematical Diary* during his freshman year. This journal had an important influence on his early mathematical development. Over the next few years, Peirce was credited with solving several problems and submitting one to the *Diary*. His senior thesis, dedicated to Bowditch, consisted of solutions to every one of the problems posed in the November 1828 issue, one of which required knowledge of Lagrange's mechanics. It seems that almost all successful research mathematicians from that time to the present got their start by contributing to the problems sections of journals, and for Peirce the *Diary* sparked his genius and industry.

Peirce had just turned 20 when he graduated in 1829 and was elected to Phi Beta Kappa. One of his classmates was future Supreme Court justice Oliver Wendell Holmes. What could Peirce do for a living at this age? He accepted a teaching position at the Round Hill School in nearby Northampton, MA, founded in 1823 by George Bancroft and Joseph Cogswell. Bancroft was an 1817 Harvard graduate who obtained a doctorate in classics from Göttingen in 1820. Later he became Secretary of the Navy, and was instrumental with establishing the Naval Academy in 1845. Peirce remained at Round Hill for two years before returning to Harvard in the fall of 1831 as a tutor in full charge of all mathematics work because Farrar was in Europe that academic year.

Moreover, Peirce continued these duties his second year, as Farrar's failing health prohibited him from returning to his post. Peirce spent the rest of his life at Harvard.

The year 1833 was critical for Peirce. Professionally, he was appointed University Professor of Mathematics and Natural Philosophy. This position was created by Harvard president Josiah Quincy, probably at the urging of Nathaniel Bowditch, who was then a Fellow of the Corporation that governed the university. I do not know why Joseph Lovering was awarded the Hollis Professorship (when John Farrar relinquished it in 1838) instead of Peirce, but as soon as the endowed Perkins Professorship of Astronomy and Mathematics was established in 1842, Peirce accepted it with alacrity and held it until his death in 1880. It is noteworthy that Peirce was never a professor of mathematics only, and that the title of the Perkins position begins with astronomy. Indeed, although Peirce's contributions to mathematics were exceptional, only about one quarter of his publications were in mathematics; the rest dealt with various topics in what was then called "natural philosophy" (astronomy, physics, chemistry, and geology).

The other major event in Peirce's life in 1833 was his marriage to Sarah Hunt Mills, who became his lifetime companion and the mother of four sons and a daughter. The two eldest sons, James and Charles, became respected mathematicians, the third, Benjamin was a mining engineer, and Herbert was a career diplomat. The daughter, Sara, stayed at home and took care of her parents into their old age.

The first paper in mathematics that Peirce wrote was published in the *Mathematical Diary* in 1832, when he was a tutor. Titled briefly "On perfect numbers," it proved that all odd perfect numbers must have at least four prime divisors. This result, though somewhat minor, shows nonetheless the extent to which the American mathematical enterprise was developing. Yet because of the journal's inaccessibility, when the towering American algebraist L.E. Dickson wrote his three-volume *History of the Theory of Numbers* during 1919–1923, he was unaware of this paper. Instead he attributed the result to V.A. Lebesgue from a paper that appeared in 1844, 12 years after Pierce.[31] Nonetheless, the publication may have played some role in validating Peirce's qualifications for appointment as professor the next year, though probably Bowditch's influence alone was sufficient.

Peirce took to his initial professorship at once, producing a series of textbooks requested by Harvard's president, Josiah Quincy, who felt that textbook reform was an important part of mathematical education. At first these efforts seem to be unenthusiastic, as his *Plane Trigonometry* (1835) and *Spherical Trigonometry* (1836) ran to only 90 pages and 71 pages, respectively. Part of this brevity may have been a reflection of the way Peirce was reputed to teach—with spartan terseness. However, he also published a 220-page volume, *Sound*, in 1836, so the effort expended on this project might also account for the slim feature of his first two books on mathematics. *Sound* is based on the work of the Englishman John Herschel (1792–1871); its list of titles connected with musical matters is valuable. Textbook publication continued for the next decade, beginning with two in 1837, *Algebra* and *Plane and Solid Geometry*. These two were more expansive than the earlier ones, running to 276 pages and 159 pages, respectively. Three years later he published two more, on *Plane Geometry for Use of the Blind* and *Plane and Spherical Trigonometry with Applications to Bowditch's Navigation*. The first of these is notable due to the intended audience, while the latter reflects Bowditch's enduring influence. Overall, these books were considered abstract and abstruse, so they were not popular, even at Harvard.

The 1841 book *Curves, Functions, and Forces* took Peirce's texts to a new level even though it was pedagogically challenging as a calculus textbook. Until then, his topics were from the lower part of the undergraduate curriculum—algebra, geometry, and trigonometry. But this 304-page work covered analytic geometry and differential calculus, basing them on functions, which were only then beginning to be regarded as a concept much broader than equations. The second volume in the series, which appeared five years later, contained the subtitle, "Imaginary quantities and integral calculus," thus informing the reader that the primary objects are functions of a complex variable. This too was a topic of current interest at the time, an indication that Peirce kept abreast of modern developments abroad.

Although Peirce was quite active writing textbooks during 1835–1846, there was little production of research papers during this period, outside the *Cambridge Miscellany*. He did publish "Researches in the integral calculus" in the *Philosophical Transactions* but that was his only journal article over those 12 years. Where was he to publish stateside? The only mathematics journal was Charles Gill's *Mathematical Miscellany*, where he was quite active with the problem section during its run 1836–1839. But that journal's demise left Peirce with no outlet for the fruits of his ongoing investigations. As a result, at the end of 1842, he enlisted his colleague, the Hollis Professor Joseph Lovering, to help him edit a new quarterly journal, *Cambridge Miscellany*. To keep the enterprise afloat, Peirce wrote one long article in every issue that year, but when it became clear that the subscription list was not going to support its continuation, the editors halted its publication, thereby tossing it into the trash heap along with every other mathematics journal that had been founded since the *Mathematical Correspondent* in 1804.

Another activity that kept Peirce from conducting research through the mid-1840s was his effort to introduce elective courses into the curriculum. Only in 1824 were Harvard students allowed to make decisions—a substitute for Hebrew for juniors and a choice between calculus and chemistry for seniors. In 1839 Peirce proposed having electives for mathematics courses on a two-year trial basis. That recommendation was apparently so successful that, four years later, all courses beyond the freshman year were electives. However, in 1846 some of those gains were curtailed, the sophomore year returning to a prescribed set of courses in addition to the freshman year, a practice that remained in place until 1883.

American Academy for the Advancement of Science (AAAS). By the late 1840s Benjamin Peirce had established a national reputation that catapulted him into leadership positions in a new organization of American scientists. The American Association for the Advancement of Science (AAAS) was established in 1848 from a pre-existing group of scientists.[32] (Here we remind the reader that we refer to the American Academy of Arts and Sciences as the "American Academy" and reserve AAAS for the American Association for the Advancement of Science.) Since the mid-1830s, various scientists in the US corresponded with each other about the possibility of forming a broad-based organization like the APS and the American Academy, but national in scope. Local, successful societies already existed in the three leading cities in the US: the Academy of Natural Sciences in Philadelphia, the Society of Natural

History in Boston, and the Lyceum in New York City. Yet, ironically, the AAAS grew out of a narrow organization whose membership was national.

The Association of American Geologists and Naturalists had been founded in 1840 and had met annually in various cities through 1847. It had been successful as a specialized society yet was unable to maintain an ongoing publication. Consequently, a few dedicated members, including Benjamin Silliman, decided that year to transform the society into the AAAS. The model was the British Association for the Advancement of Science (**BAAS**), which had been formed in 1831. The BAAS held peripatetic annual meetings aimed at informing scientists, across loosely defined disciplinary boundaries throughout Great Britain, of each other's discoveries. The first annual AAAS meeting was held in Philadelphia in September 1848. The constitution stated three aims:[33]

> The objects of the Association are, by periodical and migratory meetings, to promote intercourse between those who are cultivating science in different parts of the United States; to give a stronger and more general impulse, and a more systematic direction to scientific research in our country; and to procure for the labours of scientific men, increased facilities and a wider usefulness.

In short, the goals of the AAAS were (1) to sponsor meetings to bring leading scientists together, (2) to encourage research in science, and (3) to obtain funding for that research. The Association was divided into two sections, one of which consisted of general physics, mathematics, chemistry, civil engineering, and the applied sciences generally.

Benjamin Peirce was a particularly active member of the AAAS. At the initial meeting, he read two papers, one dealing with mathematics proper, "Upon certain methods of determining the number of real roots of equations, applicable to transcendental as well as to algebraic equations," and the other of related interest, "Communication on the general principles of analytical mechanics."

One other paper of relevance at that first gathering, "On the fundamental principles of mathematics," was presented by **Stephen Alexander** (1806–1883), a Princeton mathematical astronomer who had graduated from Union College in upstate New York at age 18. In 1832 he accompanied his brother-in-law, Joseph Henry, to Princeton, where Henry was professor of natural philosophy. Alexander was appointed tutor in mathematics the next year and professor of astronomy in 1840. He remained at Princeton for the rest of his life. He served as president of the AAAS in 1859 and was chosen as one of the original 50 members of the National Academy of Sciences in 1862. His paper from the 1848 meeting dealt with the meaning of truth in science. The account of his presentation by AAAS secretary Benjamin Silliman included Alexander's claim that:[34]

> mathematics had not to do with things, but the relations of things, and it was sufficient that those relations should be supposable; and that the certainty of mathematical reasoning rested upon the fact that those relations could be more readily understood and completely defined, than the properties of the things themselves.

The second annual AAAS meeting was held at Cambridge, MA, in August 1849. Benjamin Peirce was a major organizer of this gathering as a member of the local committee that also included internationally known Harvard scientists Louis Agassiz and Asa Gray as well as Lieutenant Charles H. Davis. Peirce was involved with the meeting in various ways. For instance, a paper read by Rev. Thomas Hill discussed the method of circular coordinates that Peirce had introduced first in Gill's *Mathematical Miscellany* and then in the first volume of his book, *Curves and Functions*. At the conclusion of the presentation, Peirce observed that this system of coordinates had enabled him to solve problems which were impossible to do by former systems. Then he added that he found the system of coordinates to be "especially useful in solving all problems relating to flexible surfaces."[35]

More directly, Benjamin Peirce presented two papers of scientific interest, "On the connection of comets with the solar system" and "On the relation between the elastic curve and the motion of a pendulum." Apparently, he had planned on presenting the paper, "A new demonstration of the parallelogram of forces," but he had not completed it in time for the meeting. Another relevant paper was "Mathematical investigation of the fractions which occur in phyllotaxis," which dealt with Fibonacci numbers. Its introduction may have been delivered with tongue in cheek because the opening paragraph read:[36]

> The Association may wonder what a mathematician can have to do with Botany, and what right he has to discuss such a subject as vegetable morphology. But let me assure you that the geometer is somewhat omnivorous in intellect, and although he has lived and thriven for centuries upon the sun and moon, the planets and comets, and other such inorganic food, he is already aspiring to a vegetable diet, and may ere long be whetting his teeth for flesh and blood. But, in the present case, the botanists have provoked the invasion by undertaking to demonstrate that plants grow according to exact mathematical laws. They have presumed to measure with minute accuracy, and exact measurement must open the path to geometry. They have dared to use our numbers and fractions, and we must reclaim them with interest.

Table 4.1 lists the locations of the week-long AAAS meetings from 1848 through 1860. Due to the Civil War, no meetings were held from 1861 through 1865. Table 4.1 indicates that annual AAAS meetings met throughout New England and the mid-Atlantic states (with one in the South, one in the Midwest, and another in Canada). There was a proposal to hold the 1860 meeting in Nashville, TN, then considered the western part of the US, but it was defeated in favor of Newport, RI. An external review of the first four meetings stated that the "Association meets every year and sometimes oftener, not in one fixed place. … Its meetings are public and are always well attended, by persons of both sexes."[37]

AAAS membership was open, unlike the APS and the American Academy. This was part of the dual goals of supporting research while simultaneously popularizing new findings among the increasing population of graduates of the growing number of American colleges. Historian of science Sally Kohlstedt commented, "The Association's early membership was a heterogeneous mix of practicing scientists, interested participants, and onlookers."[38] Nonetheless, over the 13-year formative period, Kohlstedt singled out 52 AAAS leaders who held at least two offices in the Association and presented at least two papers at annual meetings.[39] Among this group were seven with ties to mathematics: Alexander Dallas Bache, William Chauvenet, Elias Loomis,

Table 4.1. Locations of AAAS meetings, 1848–1860

Date	Location	Date	Location
September 1848	Philadelphia, PA	April 1854	Washington, DC
August 1849	Cambridge, MA	August 1855	Providence, RI
March 1850	Charleston, SC	August 1856	Albany, NY
August 1850	New Haven, CT	August 1857	Montreal, Canada
May 1851	Cincinnati, OH	July 1858	Baltimore, MD
August 1851	Albany, NY	August 1859	Springfield, MA
July 1853	Cleveland, OH	August 1860	Newport, RI

Joseph Lovering, Denison Olmsted, Benjamin Peirce, and Benjamin Silliman. Peirce was the only one in this group to be elected president of the AAAS, and Lovering the only secretary.

Based on the scientific activity of individuals during this period, the largest group came from mathematics and physics (which were lumped together), with 21 of the 59 having one of these subjects as their major field and another 15 as their secondary field. In 1900 Charles Peirce recalled that, for his father and his father's friends, "The word science ... did not mean ... 'systematized knowledge' ... but the devoted, well-considered life pursuit of knowledge; devotion to truth."[40]

As is customary today, a local committee was in charge of all arrangements for each meeting. In addition to organizing the program of lectures, these committees also sponsored social gatherings that caused gripes from some members who wanted only to hear research findings and viewed such events as time-consuming distractions. Most of the assembled, however, recognized the importance of fellowship with colleagues. On the other hand, scientists were generally critical of the way the meetings were covered in the public press, as reporters then—as now—were not always accurate in their portrayal of scientific presentations and, moreover, tended to concentrate on idiosyncratic personalities.

The AAAS published an annual issue of the journal *Proceedings*, which printed papers delivered at its meetings. While in general this was a positive development, the selection of which papers to include led to considerable controversy within the scientific community and to animosity between amateurs and researchers. Both situations caused much friction during the 1850s and were part of the evolution of the professionalization of science that gained traction after the Civil War.

Astronomy. Beginning in 1843, Benjamin Peirce turned his attention to astronomy. His public lectures on astronomy in Boston that year stimulated financial support for the installation of the Harvard Observatory four years later. That year, he published a paper in the *Transactions of the APS* on the perturbation of meteors. It was followed by a dramatic pronouncement at a meeting of the American Academy in March 1847. Up to then, Peirce's findings on perturbations had earned him a considerable reputation in the US, although he remained virtually unknown outside the country. But at the meeting, he publicly criticized the findings of the French astronomer Urbain Jean Joseph Leverrier. The resulting international conflict was important in the history of science in the US.

Leverrier had used his own mathematical computations of the perturbations of Uranus to determine the orbit and, hence, predict the location of an unknown planet, even though that planet had never been observed beforehand. This was a seminal event in the history of science—the first time the trajectory of an unknown planet was calculated using only mathematical tools based on the perturbations of a known planet. In September 1846 the German astronomer J.G. Galle observed the new planet, which came to be known as Neptune, thus confirming Leverrier's prediction and validating his methods. Word of this event spread quickly across the ocean (if you can call a month "quick"), exciting American astronomers to search for Neptune. The first American to make a study of the new planet was **Sears Cook Walker** (1805–1853), a Harvard graduate and director of the observatory at Central High School in Philadelphia.[41] This observatory was the best in the country at the time despite being located at a secondary school. Walker was unable to account for discrepancies he found between some of Leverrier's predictions and Neptune's actual location, so he asked his friend, Benjamin Peirce, who thereupon initiated his own research. Shortly, Peirce concluded that "THE PLANET NEPTUNE IS NOT THE PLANET TO WHICH [mathematical] ANALYSIS HAD DIRECTED THE TELESCOPE ... its discovery by Galle must be regarded as a happy accident."[42]

It would be one thing if Peirce had written this in a private letter, but he stated it publicly at that March 1847 meeting of the American Academy. Moreover, he repeated his charge in an article on perturbations of Uranus that he submitted to the *Proceedings of the American Academy of Arts and Sciences.* Now, that article might not have garnered attention in Europe when it appeared in 1848, since American journals were sparse there, but that same year, Peirce allowed it to be reprinted in Europe's leading astronomy journal, *Astronomische Nachrichten.* Recall that when only a schoolboy, Peirce had challenged Bowditch. Apparently, he could not be intimidated by authority—including Leverrier. But he did not seek to infuriate the French astronomer, so he praised Leverrier's work while still discrediting the discovery. The tactic did not work. Leverrier was incensed and did not mellow with time, remaining annoyed at the "happy accident" charge. But back in August 1847, while on a carriage from Cambridge to Boston, Peirce held up a letter to a friend and Harvard colleague, and announced triumphantly, "Gauss says I am right."[43]

Not only did Peirce challenge Leverrier on this matter, he did not back down from Harvard president Edward Everett, the first American to earn a PhD (1817, in Greek literature) from a German university. At the American Academy meeting where Peirce presented his "happy accident" charge, Everett expressed the hope that the announcement would not be made public, declaring that such a coincidence was improbable. Unbowed, Peirce retorted, "It may be utterly improbable, but one thing is more improbable still, that ... the truth of mathematical formulas should fail."[44]

Was Peirce being arrogant? Perhaps. Was he confident? Definitely. As such, Peirce was the paragon of the emergent American scientist. Since colonial times, American scientists had bemoaned their inferiority. American mathematics textbooks had become a matter of national pride after the Revolutionary War and again after the War of 1812. However, by the mid-nineteenth century, Benjamin Peirce had shown that American scientists could no longer be ignored. His remarkably accurate work on the perturbations of Uranus and, later that year, also Neptune won him international recognition. In particular, the French physicist Jacques Babinet and the British astronomer George Biddell Airy even sided with Peirce in the controversy.

Would a dispute like this sap Peirce's energy, leaving no time for other matters? On the contrary, he had an enormous appetite for work, an impressive ability to focus, and a quickness of mind to master new material. While speed of thinking can be advantageous, it can have drawbacks too and, as a result, he was not known as an effective classroom teacher, though some of that might have resulted from his habit of seldom preparing lectures. His classroom *modus operandi* was to fill the slates with chalk, taking no time to ask for questions from his students, who were expected to take notes in class and digest them afterward. Harvard was all-male at the time, yet Peirce was not averse to teaching women. In a letter from 1879, he wrote:[45]

> I will do the same for the young women that I do for the young men.
> I shall take pleasure in giving gratuitous instruction to any person
> whom I find competent to receive it.

This viewpoint reflected a family tradition of feminism that harkened back to his parents.

Lawrence Scientific School. Benjamin Peirce was able to take advantage of an opportunity to raise the bar for mathematics instruction in the country when, in 1847, Harvard established the Lawrence Scientific School to provide instruction in engineering and the sciences at essentially a graduate level. From that time forward, Peirce ceased teaching introductory courses and offered instead courses elected by juniors and seniors that met three times a week. Charles Eliot wrote that, "In spite of the defects of his method of teaching, Benjamin Peirce was a very inspiring and stimulating teacher."[46] In addition to Eliot, other students inspired and stimulated by Peirce were Simon Newcomb, William Byerly, A. Lawrence Lowell, and Arnold Chace.

The Lawrence School enabled Peirce to offer a rigorous program in mathematics that enabled advanced students to attain research level in the field. Its three advanced offerings were

- *Curves and Functions*, using his own text of the same title, as well as books by Cauchy, Lacroix, Monge, Biot, and Hamilton;
- *Analytical and Celestial Mechanics*, using Bowditch's translation in addition to works by Lagrange, Hamilton, Gauss, Bessel, Airy, and—yes—Leverrier;
- *Mechanical Theory of Light*, based on texts by Cauchy, Neumann, Airy, and MacCullagh.

Most present-day mathematicians and many students will recognize several of these names. Clearly, American mathematics had come a long way in a short time. However, while the chemistry community was sizable enough to take advantage of the Lawrence School, it was rare for a mathematics class to enroll more than two students, so Peirce's numerical influence was minimized. Nonetheless, several of his students exerted profound impacts on mathematics in the country, including the four cited above. In this sense, Peirce's academic progeny was like his contemporary Gauss's publications: few but ripe. Besides, Peirce's course of study in the Lawrence School would not be replicated in the country for another 30 years.

Linear Associative Algebra. Benjamin Peirce wrote three noteworthy works. One was his two-volume *Curves, Functions, and Forces*. A third volume dealing with applications of analytic mechanics was planned but, in one sense, never appeared. Yet in another, it found light of day in the 1855 book *Physical and Celestial Mechanics ... Developed in Four Systems of Analytic Mechanics, Celestial Mechanics, Potential Physics, and Analytic Morphology.* Peirce dedicated this work to his mentor, Nathaniel Bowditch.

However, for mathematics, an 1870 publication marks the high point of Peirce's career and stands today as a testament to his originality. He was one of the 50 incorporators when the National Academy of Sciences (**NAS**) was established in 1863. Moreover, he was one of the first scientists in the country to read a paper the next year at the first NAS meeting, when he announced a paper on "Linear associative algebras." In 1866 he read a paper on part of his work on abstract algebra that landed on deaf ears. For Peirce, "Mathematics is the science which draws necessary conclusions,"[47] the statement that opened this famous work. One of the conclusions stated that $i^{-i} = e^{\pi/2}$. Nobody in the audience understood it. It was bad enough that he was using complex numbers, but raising a complex number to a complex power and getting a real number, was inconceivable, even if that real number involved the famous transcendental numbers e and π. Peirce was moved by this formula. In one of his advanced classes a few years later, after proving the theorem, he dropped his omnipresent chalk and eraser and stated, slowly and impressively:[48]

> Gentlemen, that is surely true, it is absolutely paradoxical, we can't understand it, and we don't know what it means, but we have proved it, and therefore we know it must be the truth.

Even though his colleagues and students struggled with these notions, Peirce persevered, presenting various aspects of his research on these abstract algebras to the NAS over the next four years. Just traveling from Boston to Washington to attend those meetings was challenge enough, though the train ride itself was rather comfortable. By 1870, at age 61, when mathematicians are reputedly past their prime, Peirce gathered his papers into a monograph. Unable to find a publisher willing to print it, he produced at his own expense 100 lithograph copies of *Linear Associative Algebra* (*LAA*) that he distributed to friends and colleagues. But the subject matter was too abstract, and the presentation too brief, even for them. So the work received scant attention until 11 years later in 1881 when his son Charles Sanders Peirce reprinted *LAA* in the *American Journal of Mathematics*. C.S. Peirce added nine pages of explanatory notes, and published another paper his father had written in 1875, "The uses and transformations of linear algebra," but he did not include his father's opening words in the dedication of the original, lithographed edition:[49]

> This work has been the pleasantest mathematical effort of my life. In no other have I seemed to myself to have received so full a reward for my mental labor in the novelty and breadth of the results.

An example of the novelty Peirce mentioned was the concept of quaternions, a revolutionary discovery the Irishman William Rowan Hamilton had made 25 years earlier. One of Peirce's students from the late 1860s wrote that Peirce believed "Hamilton's new calculus of quaternions ... was going to be developed into a most powerful instrument of research."[50] As late as 1938, the eminent American mathematician G.D. Birkhoff assessed *LAA* as follows:[51]

Peirce saw more deeply into the essence of quaternions than his contemporaries, and so was able to take a higher, more abstract point of view, which was algebraic rather than geometric.

Initially, Peirce described the quaternions as a cyclic algebra with the following:[52]

$$i^2 = -1, \qquad ij = -ji = k, \qquad ijk = -1.$$

Later in the paper, Peirce used multiplication tables to characterize all associative algebras of order ≤ 5, and certain ones of order 6. Among those of order 4, he listed an unconventional form of quaternions.[53]

	i	j	k	l
i	i	j	0	0
j	0	0	i	j
k	k	l	0	0
l	0	0	k	l

LAA made use of terms that are familiar today, like the *associative* law, which Peirce expressed as

$$ABC = (AB)C = A(BC),$$

and the *distributive* law[54]

$$(A \pm B)C = AB \pm BC,$$
$$(A \pm B)C = AB \pm AC.$$

However, he also introduced terms that confounded his contemporaries, but have become standard fare in abstract algebra since then. For instance:[55]

> When an expression raised to the square or any higher power vanishes, it may be called *nilpotent*; but when, raised to a square or higher power, it gives itself as the result, it may be called *idempotent*.

To define the new terms symbolically, Peirce added:

> The defining equation of nilpotent and idempotent expressions are respectively $A^n = 0$, and $A^n = A$; but with reference to idempotent expressions, it will always be assumed that they are of the form $A^2 = A$, unless it be otherwise distinctly stated.

Benjamin Peirce sent two of his lithographed copies to a fellow teacher from the Round Hill School, George Bancroft, who was the American ambassador to Germany in 1870. Peirce had pinned his fondest hopes for recognition on the one he asked Bancroft to donate to the New York Public Library. Peirce wrote Bancroft, "I also send you a copy for the Academy of Berlin. ... If it would be referred to a committee of geometers for report, I should be greatly gratified."[56] The location of that copy is unknown, but no committee of the Berlin Academy ever reported on it. The work was listed by title only in the authoritative *Jahrbuch für die Fortschritte der Mathematik*, and a note promised a report would appear in the next volume (vol. 14), but the promise was never kept. The historian Helen Pycior discussed the reception of *LAA* and concluded, "In America its potential audience was very small. ... Foreign indifference, however, is less readily explained."[57] English mathematicians William Spottiswoode and Arthur Cayley were aware of *LAA*, yet up to 1902 there seems to have been no knowledge of *LAA* on the continent.

That year, Yale mathematician **Herbert Edwin Hawkes** (1872–1943) wrote:[58]

> If Peirce's work is to receive its due recognition, three questions must be discussed. ... Second. What relation does this problem bear to that treated by Study and Scheffers?

Hawkes proceeded to address concerns raised by Eduard Study and to claim priority for Peirce over acknowledged work by G.W. Scheffers. Also in 1902, Hawkes wrote a second paper asserting the superiority of Peirce's account.[59] H.E. Hawkes studied in Göttingen before earning his PhD at Yale in 1900 with the dissertation "Examination and extension of Peirce's linear associative algebra," supervised by James Pierpont. Herbert Hawkes was dean of Columbia University from 1918 until his death 25 years later.

Unfortunately, Benjamin Peirce died in 1880, a year before his most important work in mathematics appeared in America's first mathematical research journal, and two years before it was published in book form. Even then it met with controversy. Ultimately, however, mathematicians were able to understand it more and more, and the paper came to be regarded as one of the classics in abstract algebra. Clearly, mathematics in America had developed dramatically in Peirce's lifetime—it was a virtual wilderness when he was born in 1809 but a brave new world at his death in 1880.

The Antebellum Period

This section begins with a recapitulation of the history of higher educational institutions in the country from a mathematical viewpoint. Harvard went through three distinct periods from the time of its founding in 1636. Mathematics was part of the first curriculum but, up to 1727, the level of instruction remained with arithmetic based on readings by the College's first president, Henry Dunster. The second period in Harvard's evolution began with the appointment of Isaac Greenwood in 1727 and continued through the term as Hollis Professor of his successor, John Winthrop. Next came a relative decline with Samuel Williams and Samuel Webber that was reversed with the appointment of John Farrar in 1806. Farrar resigned his position in 1836 and was followed by Joseph Lovering, who was also promoted from the ranks of tutors. However, no successor was hired to fill Lovering's tutorial shoes, and from this time forward, all mathematics classes at Harvard were taught by professors, marking another significant step in the evolution of mathematical instruction in the country's colleges.

About a dozen colleges were founded before the outbreak of the American Revolution, and two stood out in the late eighteenth century—Penn under Robert Patterson and Princeton under Walter Minto. The first half of the nineteenth century saw the founding of numerous small colleges, most of which were sectarian and had a pronounced mission of providing advanced biblical training. This is reflected in the duties of the faculty, where ministers and priests taught numerous mathematics courses. Even at the larger, established colleges there was no such thing as a professor of mathematics, as it is defined today, until A.D. Stanley at Yale in 1835. Until then, faculty members who taught mathematics were, at best, professors of mathematics plus another field, usually natural philosophy, physics, or astronomy. At worst, they were polymaths who happened to have some interest and expertise in the subject. The designation of "mathematician" as a profession did not occur until the latter part of the nineteenth century. Nonetheless, the curriculum at a few institutions reflected

Table 4.2. Courses in mathematics at Dartmouth
College around 1850

	First semester	**Second semester**
Freshman	plane geometry	algebra
Sophomore	trigonometry, analytics	calculus
Junior	natural philosophy	astronomy

an increasing level of sophistication, as shown in the offerings at Dartmouth College
toward the middle of the century (see Table 4.2). Keep in mind that only a few of
the established colleges were able to offer even this modest curriculum. Notice that
no mathematics courses were available for the fourth year, though some institutions
offered electives or the opportunity for individual instruction. It is notable that astronomy continued to represent the culmination of a program in mathematics. This is not
surprising; mathematicians and astronomers continued to meet professionally up to
the founding of the American Astronomical Society in 1900.

Table 4.3 lists the full curriculum at the University of Toronto around 1870, including courses in the natural sciences. The curriculum at this Canadian institution differs
from its US counterpart by having two levels of offerings, normal and honors. The
number of courses offered in the honors program in the third year is impressive; the
author of the text for the calculus course that year was the well-known mathematician
Augustus DeMorgan. The rigor of the curriculum shown in Table 4.3 indicates the
extent to which J.B. Cherriman revolutionized the teaching of mathematics in Canada
by the time he left his professorship in 1875.

Table 4.3. Courses in mathematics at University of
Toronto around 1870

	Normal	**Honors**
First year	algebra, Euclid, plane trigonometry	
Second year	statics, dynamics	conic sections; *Principia* (I, II, III); rudiments of calculus
Third year	hydrostatics, optics dynamics	calculus; analytic geometry, algebraic equations, analytical statics, geometrical optics, hydrostatics
Fourth year	astronomy, acoustics	spherical trigonometry, *Principia* (IX, XI), plane astronomy, lunar theory

Another noteworthy appointment in the US in the 1840s, besides Sylvester at Virginia, was to a nonacademic post—superintendent of the United States Coast and Geodetic Survey, which later evolved into the Bureau of Standards and is now called the National Institute of Standards and Technology. **Alexander Dallas Bache** (1806–1867), a great-grandson of Benjamin Franklin, was born in Philadelphia and received a classical secondary education there before enrolling at West Point as the youngest cadet in his class. After graduating from the Military Academy with highest honors at age 19, he remained as an assistant professor for one year, and then was assigned to engineering duty at Newport, RI, for another two years. During that time, he continued his studies of physics and chemistry, leading to his appointment as professor of natural philosophy and chemistry at Penn in 1928. During the period 1829–1836, he published his own research results mainly from astronomy (particularly terrestrial magnetism) but also from meteorology, chemistry, and physics, though, alas, not from mathematics.

In 1836 Bache was appointed president of the newly founded Girard College for Orphans, also located in Philadelphia. Before the school opened its doors, the trustees sent him to Europe for two years to study their most successful systems of education and methods of instruction. Upon his return, he wrote an extensive report, about which one of his biographers stated in 1869, "It has done more, perhaps, to improve the theory and art of education in this country than any other work ever published."[60] Ironically, that improvement took place in the public schools of Philadelphia and not at Girard College, because the trustees took so long to open the school that Bache became frustrated and resigned all connection in 1842 to return to Penn.

Although A.D. Bache's research was outside mathematics, he is relevant to this account because of his work with the US Coast Survey and the professionalization of science in America. After only one year at Penn, in 1843 he succeeded Ferdinand Hassler as superintendent of the Coast Survey and remained in that position until his death 26 years later. The Coast Survey had been recommended to the US Congress by Thomas Jefferson as early as 1807, but it did not begin for another ten years, Ferdinand Hassler served as superintendent of the Coast Survey from then until his death in 1843.

It is notable that Bache's managerial style was quite different from his predecessor in a way that was important for mathematics and its funding by the government.[61] Whereas Hassler was involved in repeated conflicts with Congress, from the time of Bache's appointment in 1843, he steered a path of stressing the utilitarian capabilities of mathematics in the interest of the country, thus winning huge increases for his agency's budget. Not only did the Coast Survey have the largest budget of any scientific institution in America, but it also employed more scientists, 42, whereas the Naval Observatory employed 12 and the Nautical Almanac Office 11. It took another 100 years before the federal government supported science at this level, with the founding of the National Science Foundation in 1945. All the while, Bache did not let his selling of basic science to Congress interfere with his major aim of advancing pure science, for he exacted the same meticulous standards for the Bureau of Weights and Measures as Hassler. The Coast Survey became known as "the best school in the world,"[62] thus lending credence to the contention that much of the progress in American mathematics made before 1876 was conducted outside the colleges.

This fact is borne out by the construction of the Harvard Observatory in 1847. The first observatory in America had been built at Williams College in 1836. Several others were erected within the decade 1835–1845, including the Naval Observatory in

1844. John Farrar had attempted unsuccessfully to rally institutional support for an observatory at Harvard for 20 years before Benjamin Peirce raised sufficient funds to begin construction of a new and more powerful telescope. In itself, the erection of a building for astronomical sightings might not be considered a significant step in the annals of mathematics, but its location in Cambridge attracted many scientists with an interest in the subject, marking the first time in the country's history that a sizable mathematical contingent coalesced in one place.

Of more immediate importance, the federal government located a new department called the Nautical Almanac Office in Cambridge. In 1849 Congress established the Nautical Almanac Office to generate an ephemeris of the moon and planets for the use of American mariners and astronomers. The first director, Charles H. Davis, was Benjamin Peirce's brother-in-law, which undoubtedly played a role in locating the Nautical Almanac Office adjacent to Harvard. It remained in Cambridge until moving to Washington, DC, in 1866. Davis often appealed to Benjamin Peirce as well as accomplished astronomers Maria Mitchell and Benjamin A. Gould. Another important astronomer soon brought on board was Joseph Winlock. The first volume of the *American Ephemeris and Nautical Almanac* appeared in 1852. It was an immediate success.

A majority of the human computers in the early years of the Nautical Almanac Office were Harvard graduates. That was not the case with **Simon Newcomb** (1835–1909). His father had met Peirce in early 1855, when he told him about his son and a problem he had just solved. But it was not until December 1856 that Newcomb received word of a possible vacancy at the Nautical Almanac Office. So, he journeyed to Cambridge armed with letters of introduction, including one from Joseph Henry, was offered a position, and accepted it the next month.

By that time, Joseph Winlock had succeeded Davis as director of the Nautical Almanac Office. At first, he assigned Newcomb to perform routine calculations, but shortly reassigned him to the important task of ephemeris production. By that summer, Winlock wrote:[63]

> Simon Newcomb ... has given evidence of Mathematical talent and
> knowledge very unusual for his age and limited opportunities. I have
> recommended in his case a salary of $500, with a view of ... his love
> for mathematics and his industry.

That recommended salary was a raise of $200, but it did not take effect until the following February.

Newcomb found that his work only required about five hours a day, so he matriculated in at the Lawrence Scientific School at Harvard to study under Benjamin Peirce. Newcomb received his BS just 18 months later in July 1858. One of the studies he carried out during his student days was to develop the series expansion for the perturbative function of a planet that was of theoretical interest only until confirmed by observations. This inspired a research program in celestial mechanics and positioned him as a legitimate mathematical astronomer.

After graduation, Newcomb's studies were generally self-directed, aided by scientists he met at American Academy meetings. He had received permission to attend such meetings through Peirce, who also arranged for him to borrow books from the Academy's library. This allowed Newcomb to read works by Karl Jacobi and François

Arago, among others. Newcomb also attended meetings of the AAAS, being elected a member in August 1859. The next year he presented three papers at meetings.

In retrospect, the Nautical Almanac Office served as the breeding ground for two illustrious mathematical astronomers, Simon Newcomb and George Hill. As a result of the location of both the Coast Survey and the Nautical Almanac Office, Cambridge became *the* recognized center of mathematical and astronomical studies in the United States from 1847 until 1876, when the center of gravity shifted south to Baltimore.[64]

Despite the availability of a community of scientists in the Boston area, Simon Newcomb soon began to search for other positions. Joseph Winlock had left the Nautical Office in August 1859 to succeed William Chauvenet as head of the mathematics department at the Naval Academy. Winlock recommended Newcomb for an assistant professorship, but Newcomb declined the offer. Instead, he learned of a vacant physics professorship at Washington University, where Chauvenet had moved, but he was not offered the position. Charles Davis was superintendent of the Nautical Almanac Office in September 1861 when asked about candidates for a vacant professorship at the Naval Observatory in Washington, DC. He supported Newcomb's application. Newcomb accepted, and moved there the next month as professor of mathematics at the Naval Observatory.

There were five professors on the Naval Observatory staff that formed a friendly group. Newcomb remained there for the next eight years. Even though the Civil War was raging all around the nation's capital, Newcomb was very productive with observations of the sun, moon, and planets, extending work he had begun at the Nautical Office. His guiding principle was the accuracy of observations, especially of Earth, because its precise position relative to other celestial objects was crucial. Consequently, Newcomb published numerous papers during the 1860s. By this time, his access to works by Laplace and Lagrange at the Smithsonian Institution had enabled him to build upon their investigations.

Yet despite these successes, and the fact that this position also afforded him an abundance of free time, Newcomb sought other positions as early as 1865, mainly to have even more freedom for his theoretical investigations. However, perhaps due to a somewhat abrasive personality that resulted in several personal disputes as well as his many large demands on administrators, he was not offered three positions that he sought: superintendent at the Nautical Almanac Office, a position as special theorist there, nor as the head of the Naval Observatory. Nevertheless, the head of the Naval Observatory valued his contributions and wanted to keep him on board, so he arranged a comfortable position for Newcomb that amounted to a special duty.

A substantial change in Newcomb's career occurred in 1870, when he was selected by the Navy to observe a solar eclipse. This took him to Gibraltar for the eclipse, but afterward he was able to visit the leading European observatories, where he met worldwide leaders in the field. Over the next few years, Newcomb's publications earned him increasing respect, resulting in his appointment as head of the Nautical Almanac Office in 1877. The following year he published important corrected tables of the motions of the moon and the planets Neptune and Uranus, whose accuracy exceeded all studies done theretofore.

At heart, Simon Newcomb was an astronomer but, importantly, a mathematical astronomer. The major reason for the emphasis on his astronomy here is that, when appointed head of the mathematics department at Johns Hopkins in 1883, he caused a dramatic change in emphasis away from pure mathematics, which had a dampening effect on the leading department in America at that time. The following year, an editorial in *Science* praising Newcomb's first seven years as superintendent of the Nautical Almanac Office stated:[65]

> Not only has the efficiency of the office been greatly promoted … but the policy of the office with regard to the conduct of original investigation has been greatly modified.

Due to this emphasis on research, Newcomb was chosen to succeed J.J. Sylvester at Johns Hopkins, where original investigations reigned supreme.

The coverage of Newcomb has taken us out of the late 1840s, when Cambridge became the leading center for science in America. Not to be outdone by Harvard, Yale too moved toward specialized graduate education in the Antebellum period. In 1847, the college formed a department of philosophy and the arts to provide graduate studies in liberal arts, but within a decade it began to offer advanced training in science and engineering. Until then, Yale consisted of an undergraduate college and three professional schools (law, medicine, and divinity). In 1861 the scientific component of the department of philosophy and the arts was named the Sheffield Scientific School, thus becoming Yale's analogue to Harvard's Lawrence Scientific School. This start was not as auspicious for mathematics, because advanced courses in the field were not offered until 1873.

Yale, however, had no one on its staff to match Benjamin Peirce. Three Peirce books show that he was acquainted with current developments throughout his career: *Curves, Functions, and Forces* (1841 and 1846), *Analytic Mechanics* (1855), and *Linear Associative Algebras* (1870). In addition, Peirce made many contributions to mathematical astronomy. But Peirce went beyond these bounds, being commissioned by the US Coast Survey in 1852 to work on the determination of longitude, and being appointed superintendent when A.D. Bache died. Peirce served in this capacity 1867–1874, maintaining his ties with Harvard all the while.

Peirce's *Cambridge Miscellany* was typical of all mathematics journals in the country up to 1878, lasting only a short period of time, in this case one year (four issues). Although the mathematical community was too small to sustain a journal devoted specifically to mathematics, three national organizations established periodicals that included some mathematical papers among articles on the physical sciences, and therefore helped to fill the void. In addition to the *Transactions of the APS* and the American Academy's *Memoirs*, the American Academy established its *Proceedings* in 1846. The American Association for the Advancement of Science, founded two years later, organized mathematics into Section A and published some mathematical papers in its journal *Science*. Also, the influential body of scientists called the NAS initiated its *Memoirs* in 1866.

Mathematical Monthly. One of the students that Benjamin Peirce inspired and stimulated at Harvard was **John Daniel Runkle** (1822–1902).[66] Born and raised on a farm in Root, NY, farm duties prevented him from entering college until age 25, when he entered the Lawrence Scientific School at Harvard. He was a member of the first graduating class four years later, in 1851, earning both a bachelor's degree and an honorary master's degree. During the year 1848–1849, he was listed as the only student in mathematics at Lawrence, although James Oliver was also there at the time. John

Runkle joined the staff of the Nautical Almanac Office in 1849 as a computer while a student at Harvard. He remained at the Nautical Almanac Office 1849–1884, when the local mathematical community included Simon Newcomb, George Hill, Truman Safford, and John Van Vleck. All the while, Runkle was active with Massachusetts Institute of Technology (MIT), starting with his membership on the "Objects and Plan of an Institute of Technology" in 1860 that led to the university's establishment five years later. He served as professor of mathematics at MIT from 1865 until his death in 1902, interrupting part of that tenure as president pro-tem 1868–1870 and as the second president, 1870–1878.

Runkle's legacy with mathematics mainly rests with the journal he founded in 1858, the *Mathematical Monthly*, which bears no relation to the popular journal of the same title today. The first issue appeared that October, but it was discontinued after September 1861. In that final issue, Runkle wrote, "On account of the present disturbed state of public affairs, the publication of the *Mathematical Monthly* will be discontinued until further notice."[67] Despite its short run, the *Monthly* published several serious articles, thus foreshadowing an emergent publication community that might have co-alesced into a legitimate mathematical community if not for the devastating war. Well-known contributors included Arthur Cayley, William Chauvenet, Benjamin Peirce, Simon Newcomb, and Truman Safford. An article by the physician David Hart in 1875 isolated several lesser-known authors: Rev. A.D. Wheeler (who wrote on indeterminate analysis and Diophantine analysis), Chauncey Wright (symmetry in honeybees' cells), and John B. Henck (the Theorem of Pappus).[68] Despite these articles:[69]

> Its most marked feature ... was the prize problem for students. ...
> one of the problems ... brought out a short paper from Cayley.

There were also prizes for essays. George Hill won the first such prize for his work on the theory of the figure of the Earth, about which Simon Newcomb commented, "It was almost his first paper published, but showed unmistakably the hand of the master."[70] Benjamin Finkel published a detailed analysis of the *Mathematical Monthly*, calling it "a first class mathematical journal, and one well suited to satisfy the diversified interests of all classes of Mathematicians."[71]

War and Reconstruction

The "Transition" sections in this book serve to emphasize that there are no absolute dividing points in the study of history. The topic of cryptology underlines this theme, not just in transitioning from one time period to the next, but falling within a time period. Thus, the present section begins with an account of some aspects of cryptology that occurred in the Antebellum, Civil War, and Reconstruction in the US. None of the leading mathematicians in this period played a direct role in cryptology, but two minor figures who did were Pliny Chase and Edward Holden.

Wars have generally interrupted the flow of mathematics, and in this case, the Civil War put a crimp in the arc of cryptology, which had taken a leap forward with an important paper published before the outbreak of hostilities in 1861. Nonetheless, two singular developments occurred over the four-year period of combat. During the war and the Reconstruction, graduate education began to emerge at Yale and Harvard. In this respect, Harvard president C.W. Elliot and two notable mathematicians who garnered international acclaim, George Hill and W. Josiah Gibbs, played notable roles.

Although the towering figures, Hill and Gibbs, continued to contribute to mathematics into the twentieth century, their lives belong more properly to the present period, so they are included here.

Cryptology. A monumental moment from the first half of the nineteenth century was Samuel Morse's invention of the telegraph in 1844. Although quite useful for personal and business communication, it soon became clear that securing transmissions was paramount because of the ease with which telegraph lines could be tapped. Due to increasing interest in this topic, the first paper on cryptology in an American mathematics journal appeared in the *Mathematical Monthly* 15 years later.

That paper was written in 1859 by **Pliny Earle Chase** (1820–1886), who spent the first part of his life in the Boston area and the second part in Philadelphia. Chase attended common schools in his hometown of Worcester, MA, before being sent to the Friends' Boarding School in Providence, RI. He entered Harvard College shortly after turning 15 and graduated four years later. Two former classmates recalled later that Chase was "distinguished while in college for general scholarship, and particularly for a remarkable proficiency in mathematics."[72] Chase taught in two Massachusetts schools until 1845, a year after being awarded an AM degree. He then moved to Philadelphia and resumed his teaching career, but his life took a dramatic turn three years later due to severe hemorrhages from his lungs. After partnering in a foundry business 1848–1861, he returned to the classroom at the start of the Civil War. Ten years later, in 1871, he was appointed professor of natural science at Haverford College near Philadelphia, where he remained until his death 15 years later. He also lectured at nearby Bryn Mawr College when it was founded in 1884.

Although Pliny Chase spent most of his professional career in academia, he was a businessman when he published his notable paper on cryptology in the very first volume of the *Mathematical Monthly*. After dismissing purely substitution schemes, even clever systems like the one adopted by Benjamin Franklin (illustrated in Chapter 2, p. 77), Chase described four encoding systems that "should not only conceal the message that it is intended to convey, but it should also effectually hide the key to the message."[73] For instance, even knowing that BKXVDI represents someone's name, it is not apparent who that person is. The online file "Web04-ChaseCrypt" illustrates the Chase system and ends by challenging the reader to decode the name BKXVDI.

The Chase system was based on the 3×10 tableau:

$$M = \begin{bmatrix} X & U & A & C & O & N & Z & L & P & \varphi \\ B & Y & F & M & \& & E & G & J & Q & \omega \\ D & K & S & V & H & R & W & T & I & \Lambda \end{bmatrix}.$$

However, even knowing this bit of information would not help the cryptanalyst decode ORHφTUVI.

Chase used the name PHILIP to illustrate one of his four encoding systems. Each letter was represented by a 2×1 vector $\begin{bmatrix} R \\ C \end{bmatrix}$ whereby R is the row and C the column of that letter in tableau M. Thus, P is $\begin{bmatrix} 1 \\ 9 \end{bmatrix}$ and, more generally, PHILIP is formed by placing each successive vector to the right of the preceding one, resulting in

1	3	3	1	3	1
9	5	9	8	9	9

Chase's ingenious approach was to multiply row 2 by 9 (to get 8639091) and form a new tableau by retaining row 1 and substituting this product as the new row 2 as follows.

	1	3	3	1	3	1
8	6	3	9	0	9	1

Now working from left to right, the initial vector $\begin{bmatrix} \\ 8 \end{bmatrix}$ can be any letter in the eighth column of the defining tableau M, so that letter can be L, J, or T. All other 2×1 vectors define unique letters, so that $\begin{bmatrix} 1 \\ 6 \end{bmatrix}$ is N, $\begin{bmatrix} 3 \\ 3 \end{bmatrix}$ is S, etc., producing *NSIφIX, where * can be L, J, or T. Notice that the numeral 0 in the vector $\begin{bmatrix} 1 \\ 0 \end{bmatrix}$ refers to column 10 in M.

In conclusion, LNSIφIX was one of three possible coded forms for the name PHILIP. It is notable that the encrypted form in the Chase system generally contains one more symbol than the plaintext, an aspect that minimizes (even educated) guessing. Chase's three-page paper concluded:[74]

> These illustrations, I think, are sufficient to show that a very simple arithmetical process may effectually conceal the meaning of a message from every one [*sic*] but the persons who hold the key to the cipher.

Pliny Chase was a very productive scholar, publishing some 250 articles in various outlets and across several fields, but none of the others dealt with cryptology. He published three other papers and one note in Runkle's *Mathematical Monthly*, as well as one in the *Proceedings of the Royal Society of London*. He also solved several problems in *The Analyst* after it was founded in 1874. Moreover, Chase regularly attended meetings of the APS after moving to Philadelphia, and he published many papers in its *Proceedings*, including one titled "On the mathematical probability of accidental linguistic resemblances."[75] However, his biographer lamented, "But there were as few who comprehended them as who could read Pierce's mathematics."[76]

Although the Chase systems are easy to operate, are fairly secure, and were published before the outbreak of the Civil War, it appears that neither side adopted them in the conflict. This is somewhat surprising since both the North and South employed coded messages, and most of the generals had been educated at mathematically oriented West Point. Chapter 1 described the use of cryptology during and after the Revolutionary War, so it is not surprising that it was refined by the time of the Civil War. However, the telegraph changed the nature of warfare by enabling military commanders to communicate with sizable armies spread over large areas. Hence, the security of transmissions became even more critical. Errors in transmission were a problem for both sides; error-correcting codes would not appear until Richard Hamming almost a century later.

Both the Union and the Confederacy employed telegraph transmissions on a large scale for military leaders to communicate with their troops and with governmental leaders. However, whereas the Union employed one encoding system and was centrally located in the War Department, housed adjacent to the White House, the Confederacy let each commander adopt his own code. Some of those choices were laughable. For instance, before the Battle of Shiloh in 1862, southern General Albert S. Johnston employed a Caesar substitution when communicating for military use. The easiest of all encodings, the Caesar system replaces each letter in the alphabet with the letter three places to the right (modulo 26), so decoding merely moves three letters to the left. Thus the inscrutable looking "wkhzrugmxqfwlrq" in one communication was decoded as "the word junction."

The following year, during the siege of Vicksburg, MS, the Union general Ulysses S. Grant's troops captured a rebel carrying a message to the Confederate lieutenant general Pemberton that ended, "How and where is the jsqmlgugsftve. Hbfy is your roeel." Grant sent the message to Washington, DC, "hoping that someone there may be able to make it out."[77] President Lincoln had employed three young telegrapher-cipher-operators in the second-floor quarters of the War Department, and this self-anointed "Sacred Three" soon found the proper version of the Vigenère system and deciphered the message at once to read, "How and where is the enemy encamped? What is your force?"

These three Lincoln aides were perhaps the youngest wartime cryptanalysts in history. In addition to military decrypts, they solved numerous political messages, one of which might have been their most important contribution. In December 1863 the New York City postmaster spotted an envelope addressed to someone in Halifax, Nova Scotia, known to be allied with the Confederacy. After clerks in the War Department pondered over the letter inside the envelope for two days without success, it was delivered to the Sacred Three, who unlocked the key in four hours. That letter made it clear that plates for printing Confederacy currency were being made in lower Manhattan. US Marshalls were called in to break up the plot and destroy the printing presses (as well as several million dollars of bonds), thus depriving the South of badly needed funds and plates for printing paper bills.

Clearly, in the Civil War the Union's mastery of cryptology was just as decisive as its success on the battlefield.

Cryptology was central to two notable events during the Reconstruction. The first occurred shortly after the war, when eight sympathizers of John Wilkes Booth went on trial for aiding the assassination of President Lincoln. Decoded messages were put into evidence, one of which read, "I am happy to inform you that Pet has done his work well. He is safe, and Old Abe is in hell."[78] As renowned historian David Kahn commented, "What connection all these displays had with the accused was never made clear, but they were hanged anyway."

Cryptology played one other important role in America during this period. In the presidential election of 1876, Democratic candidate Samuel J. Tilden held a clear lead over his Republican rival Rutherford B. Hayes in the popular vote but, like the election in the year 2000, contested electoral votes decided the outcome. This time there were conflicting returns from Louisiana, South Carolina, and Oregon, as well as the more recent Florida. An electoral commission awarded the votes in all four states to Hayes, who thus became the nineteenth president of the United States. Democrats were naturally incensed, and a Congressional committee was charged with investigating rumors of fraud. The committee subpoenaed 641 political telegrams that had not been burned by the Western Union Telegraph Company, which customarily destroyed all telegrams to protect privacy. By 1878 the committee had made little headway in decoding the messages, so when 27 were leaked to the *New York Tribune*, the Republican-leaning newspaper printed some of them and challenged readers to decode them.

One of the readers who took up the challenge was a mathematician whose solution matched the conclusion of two *Tribune* editors. The coded message showed that, ironically, it was the Democrats, not the Republicans, who had attempted to steal electoral votes in Florida, South Carolina, and Louisiana. The *Tribune* published this sensational discovery on October 16, 1878, just in time for midterm elections for Congress; consequently the GOP made emphatic gains in that election. Moreover, Samuel Tilden, who was seen as the leading candidate in the 1880 election for president, was implicated in the messages and, disgraced, left politics. Consequently, James A. Garfield won that election, though only narrowly. Sadly, he held the office for only eight months before being assassinated. Garfield might be better remembered as the only president to prove a theorem in mathematics.[79]

The mathematician who broke the code worked at the US Naval Observatory. **Edward Singleton Holden** (1846–1914) was born in St. Louis, MO, but attended a secondary school in Cambridge, MA, before attending the academy associated with Washington University in St. Louis 1860–1862. He then attended that university, graduating with a BS in 1866. While an undergraduate, he was inspired to pursue astronomy by the famous mathematician William Chauvenet, who employed him as an assistant in the observatory. Upon graduation, Edward Holden became a cadet at West Point, where he enrolled (1866–1870) and received a second degree (finishing third in his class) as well as a commission. He then spent a year as an artilleryman.

The year 1871 was critical for Holden, as he accepted an instructorship back at the US Military Academy and got married. He and his new bride remained at West Point for two years before he was appointed professor of mathematics in the US Navy at the Naval Observatory in Washington, DC, where he was an assistant to Simon Newcomb for four years and then Asaph Hall for five. It was during this time that Holden decoded the political messages. He moved to the Washburn Observatory at the University of Wisconsin in 1881 but left after four years to become president of the fledgling University of California at Berkeley. Although he held this post only from 1886 to 1888, he was simultaneously the director of the Lick Observatory on that campus, a position he continued (1888–1897). During that time, he became a trailblazer in the use of photography for locating astronomical bodies. Holden left California for New York City in 1897 and four years later was appointed librarian of the US Military Academy, the post he held until his death 13 years later.

Elected a member of the National Academy of Sciences, Edward Holden is mainly known today for his advances in astronomy. However, his NAS biographer did state that included among his wide range of accomplishments were "the celebrated cipher dispatches of 1876 relating to the election of a President of the United States in that year."[80] Over half of that 26-page biography is devoted to Holden's remarkable bibliography of writings from 1872 to 1905. The online file "Web04-Holden" provides more detail on Holden's life and accomplishments.

American Civil War. Several students of Benjamin Peirce played key roles in mathematics during the Civil War and its aftermath. One was **Charles W. Eliot** (1834–1926), a Harvard undergraduate (1849–1853) who played a central role in the development of mathematics at Harvard, whence the rest of the country. As a tutor and then an assistant professor at Harvard from 1854 to 1861, Eliot and his classmate James Mills Peirce (Benjamin's son and yet another student of his father) introduced written final exams over the protestations of many colleagues, one of whom wrote, "more than half of [the students] can barely write; of course they can't pass written examinations."[81] When the Civil War ended in 1865, Eliot accepted a professorship in chemistry at MIT, where Runkle was professor of mathematics. Eliot remained in this position only until 1869, when the Harvard Corporation appointed him president of the university.[82] It was in this role that he exerted his second major influence by instituting a system of free electives, a policy for which John Farrar had been an outspoken—albeit unsuccessful—advocate. This policy paid dividends within two years, with the Harvard class of 1871 including two students who would ultimately play pivotal roles in shaping the future of mathematics in the country, William E. Byerly and William E. Story. Both Byerly and Story elected to take all the mathematics courses that Benjamin Peirce offered. Byerly remained at Harvard after graduation, obtaining the first Harvard PhD in mathematics in 1873 for a dissertation on the heat of the Sun. Story received a bachelor's degree from Harvard and then studied abroad, earning a doctorate at Leipzig in 1875.

Eliot was the president of Harvard for 40 years, 1869–1909. He was also the editor of the 50-volume collection, the *Harvard Classics*, which were published 1910–1911. But he was not the only Peirce mathematics student to ascend to Harvard's presidency. **Abbott Lawrence Lowell** (1856–1943) was a student under Peirce from 1873 to 1877, when he received his bachelor's degree. One of his papers was read before the AAAS and published in its *Proceedings* in 1878. He succeeded Eliot as president of Harvard and steered the university during a time of significant academic growth, 1909–1933. This means that Peirce mathematics students held the presidency for 64 consecutive years, 1869–1933, an excellent return on the university's investment in him, during which time the university developed the main lines of the present undergraduate curriculum.

Another outstanding student of Benjamin Peirce was **Percival Lowell** (1855–1916), the older brother of A.L. Lowell who graduated from Harvard in 1876 with distinction in mathematics. Percival Lowell decided on a career in astronomy after traveling extensively in the Orient, and in 1894 he founded an observatory atop Mars Hill at Flagstaff, AZ, that still bears his name. Research at the observatory led to the discovery of Pluto and the first evidence of the expanding universe. The prominent Lowell family was particularly accomplished—the men's sister, **Amy Lowell** (1874–1925), was a famous poet and critic who won a Pulitzer Prize for poetry (for *What's O'Clock*) posthumously in 1926. Amy's alternative lifestyle was regarded as outrageous a century ago but would be accepted more readily today.

Another notable person who studied with Peirce was **Arnold Buffum Chace**, who became the Chancellor of Brown University from 1907 to 1931. Peirce's student W.E. Byerly wrote, "In his [Peirce's] personal relations with his students he was always courteous, kind, and helpful, if rather prone to overrate their ability and promise." It is unimaginable how Peirce could ever have overrated the likes of Story, Eliot, Lowell, Chace, or Byerly.

Because of the Civil War, no academic subject developed materially during the 1860s. However, two major events occurred during the war, one of which was the establishment of the National Academy of Science. The NAS was founded in 1863 by a congressional act of incorporation—signed by Abraham Lincoln—calling on the Academy to act as an official advisor to the federal government, upon request, in any matter of science or technology. It was established, and remains today, as a private organization of scientists and engineers dedicated to the furtherance of science and its use for the general welfare (sometimes called the "public weal"). The congressional bill included the names of four mathematicians among the initial group of 50 who were selected. Of course Benjamin Peirce topped the list. His groundbreaking work *Linear Associative Algebras* was published (posthumously) in 1870 in the *American Journal of Mathematics*. The other mathematicians selected with Peirce were William Chauvenet, Hubert Newton, and Theodore Strong.

The other major event at that time was a political action, the Morrill Land Grant Act of 1862, which has paid huge dividends for the American populace at large. This bill was signed into law by President Abraham Lincoln five years after it had first been proposed by the US Representative of Vermont, Justin Smith Morrill. It was ratified by Congress in 1859 but then vetoed by President James Buchanan. Its purpose was stated as follows:

> Without excluding other scientific and classical studies and includ-
> ing military tactic, to teach such branches of learning as are related
> to agriculture and the mechanic arts, in such manner as the legisla-
> tures of the States may respectively prescribe, in order to promote the
> liberal and practical education of the industrial classes in the several
> pursuits and professions in life.

The Morrill Land Grant Act of 1862 provided funding to create land-grant universities that have exerted a dramatic effect on academia generally, and on mathematics in particular. Cornell University, the University of Michigan, and the University of Wisconsin played central roles in the development of mathematics in the country within 50 years of the passage of the Morrill Act. Cornell, Yale, and MIT are the only private institutions partially supported by the Morrill Act.

Reconstruction, 1865–1876. A small step forward in the mid-1870s was the establishment of *The Analyst*, a journal devoted to mathematics that enjoyed a 10-year run before being succeeded by today's *Annals of Mathematics*. This section introduces two noteworthy contributors to *The Analyst*, Erastus De Forest and Christine Ladd, as well as Artemas Martin and two of the journals he founded. The chapter ends with an account of the first PhDs in mathematics awarded in America, a topic that introduces Hubert Newton as the grandfather of American mathematics.

The 1870s would become the most crucial period in the development of mathematics in America with the founding of Johns Hopkins University in 1876. In the earlier part of the decade the works of three individuals with no direct links to Johns Hopkins portray an emergent ascension to research levels in the country. The first individual was Benjamin Peirce, who classified all 150 algebras of dimensions 1 through 6 in his 1870 publication of *Linear Associative Algebras*. Peirce had presented his 153-page study to the NAS in installments, beginning in 1866, but when it garnered no attention, he had 100 copies of the work lithographed and distributed to his friends.

Later in the 1870s, publications by G.W. Hill and J.W. Gibbs appeared. Although the work of both men overlaps the critical period 1876–1900, their personalities caused them to play somewhat remote roles in the emergence of the American mathematical community. Hill and Gibbs followed in the footsteps of Adrain, Bowditch, Peirce, and Newcomb. Neither was a "rugged individualist" in the standard sense of the word, only the way it is used here—someone working almost entirely alone, without the support of a community of like-minded and able individuals.

George William Hill (1838–1914), got his start while an undergraduate student, publishing his first paper in Runkle's *Mathematical Monthly* in 1859, right before his graduation from Rutgers College. Hill had received an unusually good education at Rutgers under Theodore Strong, including knowledge of the classical works of Lagrange on analytical mechanics and of Laplace on celestial mechanics. In 1861 Hill moved to Cambridge, MA, to join the staff at the Nautical Almanac Office. However, this solitary figure soon gained permission to carry out his work alone at his home near New York City, where he remained until moving to the Almanac's Washington, DC, office in 1882. He remained in the nation's capital for ten years before returning to New York for the rest of his life. He lectured on celestial mechanics at Columbia University during the 1890s.

Figure 4.3. G.W. Hill

Two memoirs of epoch-making importance gained Hill an instant international reputation. The first was a 28-page manuscript, published privately in 200 copies in 1887. It contains the differential equation now known as Hill's equation:

$$\frac{d^2 p}{dt^2} + (\theta_0 + \theta_1 \cos 2t + \theta_2 \cos 4t + \theta_3 \cos 8t + \cdots)\, p = 0.$$

By an elegant method Hill solved the equation by calculating the determinant of the corresponding infinite system of linear equations correct to 15 decimal places. Hill's second paper, "Researches in the lunar theory," the lead article in the first issue of the *American Journal of Mathematics* in 1878, also dealt with an astronomical theme. Hill continued to publish after this, especially on theories of Jupiter and Saturn. In particular, his work calculating the influence of these two planets on the motion of the moon is a special case of the four-body problem. He was elected as the third president of the American Mathematical Society for 1895 and 1896, having served as vice president in 1894. In a poll taken in 1903, Hill was voted by mathematicians as second in the field (behind E.H. Moore) and by astronomers in a tie for first place (with Simon Newcomb).

Like Hill, **Josiah Willard Gibbs** (1839–1903) was a leading American scientist who garnered an international reputation for his work in more than one field, yet whose accomplishments in mathematics tie him closer to the period before the watershed year of 1876. Also like Hill, the influence of J. Willard Gibbs on the community that emerged in the last quarter of the nineteenth century was minimized, in this case by his reclusive nature.

Gibbs graduated from Yale University in 1858 with distinction in mathematics and Latin. He remained in New Haven at the Sheffield Scientific School, earning the first PhD in engineering in the country in the midst of the Civil War, 1863. The title of his dissertation was "The form of the teeth of wheels in spar gearing." Three years later he embarked on a three-year post-graduate study in Europe. First he attended lectures in Paris by M. Chasles on geometry, J.M.C. Duhamel on infinite series, J. Liouville on number theory, J. Serret on elliptic function theory and celestial mechanics, and G. Darboux on mathematical physics, but this overload of information exacted a toll that required a period of recuperation in a warm climate. He chose the same country, Italy, where the illustrious G.B. Riemann had fled in a fruitless search for good health. Feeling recovered, Gibbs traveled to Germany to attend lectures by L. Kronecker on number theory and quadratic forms, K. Weierstrass on determinants and complex analysis, and E. Kummer on probability theory.

Figure 4.4. Josiah Willard Gibbs

Imagine—American students were studying arithmetic in 1800 yet attending lectures given by a celebrated, international cast of figures on topics of current research just 70 years later. Consequently, when Gibbs returned to New Haven in 1869, he brought with him a comprehensive overview and deep understanding of current research in mathematics and physics. But what was even more important than the German example of professor-as-researcher was the exposure to an element of graduate education taken for granted today—seminars—which were nonexistent in America at the time. These small gatherings served the purpose of guiding advanced students from the role of passive listeners in a lecture hall to active participants in the development of a subject. Scientific seminars began in 1834 with a gathering run by the mathematician C.G J. Jacobi and the physicist F. Neumann at Königsberg (Germany). The first seminar devoted exclusively to mathematics was directed by Alfred Clebsch at Giessen (Germany at the time, now Kaliningrad, Russia) in 1864. By the 1880s such seminars were fixtures at all the major German universities.

In 1871 J. Willard Gibbs accepted a new (but unsalaried!) professorship in mathematical physics at Yale even though he had not published one paper in the field up to that point. It didn't take long. Just two years later he published two works that took the mathematical analysis of thermodynamics to new heights. Then his interests changed to chemistry, and in 1878 he completed a two-part paper "On the equilibrium of heterogeneous substances" that revolutionized the study of physical chemistry. In 1902 his book *Elementary Principles of Statistical Mechanics* revealed his creativity in another area of mathematical physics, but for mathematicians Gibbs will always be known for his work in vector analysis. He introduced his approach to vector analysis in an 1879 course on electricity and magnetism, but by 1884 he completed an 83-page manuscript that has earned him a permanent niche in the history of mathematics. Like Hill, Gibbs published his work privately, but unlike Hill he made sure to send copies to leading mathematicians and physicists in Europe. In 1902 a textbook written by Harvard's Edwin Bidwell Wilson (with Gibbs's permission) played a critical role in popularizing his former professor's contributions: *Vector Analysis: A Text-book for the Use of Students of Mathematics and Physics Founded upon the Lectures of J. Willard Gibbs.* The text runs to over 400 pages! The major source on the history of vector analysis concludes:[83]

> If Gibbs cannot be given great credit for originality of methods, yet he deserves praise for the sensitivity of his judgment as to what deletions and alterations should be made in the quaternionic system in order to make a viable system. His symbolism, now in the main accepted, also constitutes a significant contribution.... Though his booklet was difficult to read because of its compactness, it could and did form a basis for later writers.

The Analyst. The *Mathematical Monthly* ceased publication in 1861, and no periodical devoted to mathematics appeared until Joel Hendricks began producing *The Analyst* in Des Moines, IA, in January 1874. Hendricks was a relatively unknown, self-taught mathematician and astronomer whose family undertook some extraordinary efforts to publish this journal, which lasted from 1874 to 1883. It began as a monthly with 16 pages in each issue (though some individuals paid for additional pages beyond that limit), but in the second year switched to a bimonthly format of 32 pages, thus retaining the annual number of pages while reducing mailing costs. *The Analyst* might be better known today if not for some extraordinarily bad timing—the establishment in 1878 of the *American Journal of Mathematics*, a mathematics research publication with an internationally renowned editor that thrives to this day. However, *The Analyst* did spawn the founding in 1884 of *The Annals of Mathematics*, which ultimately became one of the leading journals in the world and has also prospered into the twenty-first century.

Moreover, *The Analyst* helped to boost the careers of several mathematicians while also publishing one of the first important series of papers on statistics (by E.L. De Forest) and launching the career the remaining rugged individualist, Christine Ladd. The remainder of this section introduces the founder of the journal, describes some of its features and major authors, examines contributions from De Forest and Ladd, and briefly assesses its place in the history of mathematics in America.

Joel E. Hendricks (1818–1893) was of farming stock, a lot in life generally unfavorable for someone born with a talent for mathematics. But his dogged pursuit of education enabled him to rise above this status and ultimately play a pivotal role in mathematics.[84] When he was six years old, his family moved from the Philadelphia area to a small town in Ohio and, as was the custom, all children were expected to perform farm duties. As a result, Hendricks could attend school only in winter, yet he excelled, and at age 18 he actually became a teacher at a rural school. The next year he bound himself as an apprentice but, unlike Newcomb's period of servitude, this one was with a respected mill-wright, lasted only two years, and included an agreement that he could continue teaching school in winter. During this time, Hendricks put his time to good use with an independent study program that included books on navigation and astronomy, resulting in knowledge of trigonometry so complete that he was able to calculate solar and lunar eclipses with facility. At age 22 he obtained a book on algebra that he digested in only five weeks, devoting two hours every night for solving every example. That year the county surveyor, who had an abiding interest in mathematics himself but, of more importance, an impressive library, gave Hendricks copies of Hutton's *Mathematics*, Newton's *Principia*, and Bowditch's translation of *Mécanique Céleste*.

In spite of what seemed like a natural progression into higher mathematics, Hendricks wanted to attend medical school. Luckily for mathematics, those plans were waylaid when he married in 1843 and began raising a family that grew to include six daughters—constraints that prevented him from ever attending medical school. From that time until the outbreak of the Civil War, he worked at a variety of jobs at the county level (surveyor, treasurer, and auditor) and served as US deputy surveyor for Colorado.

In 1864 Hendricks moved to Des Moines, IA, a time and a place that historian Deborah Kent called "the fringe of civilization" in a MathFest 2015 lecture. Yet it was there a decade later that he established *The Analyst*. Curiously, Simon Newcomb visited that city in 1869, but there is no record of the two men crossing paths. During this time Hendricks mainly engaged in surveying, a position sufficiently lucrative to allow him to devote his full time to the journal from 1874 through 1883.

Hendricks did not engage in vetting his idea the way that Runkle did with the *Mathematical Monthly*. In his first issue he stated that his goal was similar to Runkle's:[85]

> As the scientific character of the *Analyst* has not been fully explained
> by circular, we embrace this opportunity to state that … it is intended
> to afford a medium for the presentation and analysis of any and all
> questions of interest or importance in pure and applied Mathematics.

Yet Hendricks was no amateur, adding:[86]

> We are fully aware of the difficulty of publishing such a periodical
> … and of the apparent presumption of attempting it at this place,
> where we have no prominent institution of learning, nor the facilities
> for printing that might be obtained further east. Nevertheless, there
> seems to be an obvious want of a suitable medium of communication
> between a large class of investigators and students in science."

Roughly seventy percent of *The Analyst* pages were devoted to articles on mathematics, the remainder to queries and problems. Generally, those who submitted papers to *The Analyst* represented a new generation of mathematical practitioners from those who participated before the Civil War. In fact, many of them can be regarded as mathematicians in the modern sense that they carried out independent investigations on topics of current interest. Some of the contributors were second rate, however, and sometimes retractions had to be made but, for the most part, the papers in the journal were of a fairly high quality, notably those written by George Hill, Simon Newcomb, Robert Woodward, and William W. Johnson. Hill published several papers on mathematical astronomy on such topics as the motion of the center of gravity of the Earth and Moon, the secular acceleration of the Moon, and on Jupiter and Saturn. In addition, Newcomb wrote an article on limits,[87] future AMS president Woodward's chief paper concerned errors of interpolated values derived from numerical tables, and Johnson submitted several papers on such geometrical topics as the classification of plane curves with respect to inversion, pedal curves, and circular coordinates. Besides these established authors, younger figures included Wooster Woodruff, [88] Thomas Craig, [89] and Henry Eddy.[90]

The Analyst published two papers that have been useful for historians, David Hart's "Historical sketch of American mathematical periodicals" and Mansfield Merriman's "History of the method of least squares." The latter, in fact, reproduced some of the work done on the subject by Robert Adrain back in 1808. Not all authors in *The Analyst* were American; for example, Thomas Muir, who was born in Scotland and died in South Africa, is known for his work on determinants, including two papers that appeared in *The Analyst*.[91]

Mansfield Merriman (1848–1925) was a civil engineer born in Southington, CT, and was an 1871 graduate of the Sheffield Scientific School at Yale. He served as an assistant in the US Corps of Engineers (1872–1873) before becoming an instructor in civil engineering at Sheffield (1875–1878). After that, he moved to Lehigh University, where he held a professorship in civil engineering until 1907. While at Lehigh, he was also an assistant at the US Coast and Geodetic Survey (1880–1885). All the while, he conducted research on strength of materials and the design of bridges, consulted on several civil- and hydraulic-engineering projects, and pursued mathematics. Merriman published several books in these engineering fields; within mathematics, his *Method of Least Squares* was published in 1884 and ran to eight editions.

But the mathematician who seems to have gained the most from *The Analyst* was E.L. De Forest, who published more articles in the journal than any other author. The focus here is on his role in the development of statistics. He is an interesting character too from a personal standpoint, as his life's story might resonate with undergraduate students today even if separated from his time by some 150 years.

Statistics. *The Analyst* helped spawn the careers of two little known members of its publication community who wrote notable papers on mathematical statistics. Neither E.L. De Forest nor Charles Kummell devoted much time to problems. De Forest, for instance, solved but one problem and proposed another. Yet both of their careers were aided by publishing in the journal. The online file "Web04-Analyst" supplies further detail on their lives and careers.

Erastus Lyman De Forest (1834–1888) was a Connecticut Yankee whose family roots in the New World can be traced back to 1623. By the nineteenth century the family had acquired considerable wealth, and his father was a physician who had graduated from Yale. De Forest enrolled at Yale at age 16, and when he received a BA four years later, his father endowed the De Forest Mathematical Prizes at the university. De Forest then studied engineering at Yale's Sheffield Scientific School for the next two years, earning his PhB in 1856. During this time, J. Willard Gibbs was a student in engineering, and Hubert Newton was a young faculty member at Yale.

Like many students today, De Forest was quite unsure of his direction in life at graduation. But graduates from wealthy families do not have to join the work force at once; they have sufficient resources for taking time to travel to "find themselves." So the next February De Forest headed to Havana with an aunt. Shortly before sailing from New York, however, he disappeared, leaving his luggage and no clue of his whereabouts. The family panicked, their frantic state assuaged by neither speculation in the *New York Times* that their only child had met with foul play nor rumors that he had drowned in the East River. Yet the body was never found. Two years later his father received a letter from him postmarked Australia. Erastus De Forest explained that he had been depressed and therefore headed to California, where he worked in the mines and taught public school for a year before continuing on to Melbourne. From 1858 to 1860 he had been an assistant master at a Church of England grammar school, where he taught mathematics and surveying.

De Forest's letter informed the family that he would be returning to Connecticut by way of India and England, and indeed the relieved parents greeted him warmly upon his return in 1861. Even though De Forest remained unsettled for the next two years while the Civil War raged about him, it was a two-year trip to Europe (1863–1865) that seems to have settled him down and set him on a career path in mathematical statistics and, later, caring for his father; he never married.

From 1865 to 1867 De Forest published three papers on interpolation. While they were mainly concerned with meteorology, his interest changed to actuarial science as a result of the year (1867–1868) he spent working with his uncle Erastus Lyman, who was president of Knickerbocker Life Insurance Company of New York. De Forest was assigned the task of determining policy liabilities, which led to the problem of smoothing mortality tables, essentially death rates classified by age. Because the resulting tables were rough, he sought to smooth them in order to adjust death rates so that those for nearby ages would be more nearly equal than the raw figures. A leading authority on the history of statistics, Stephen Stigler, wrote:[92]

> Between 1870 and 1885 he wrote more than twenty papers on the graduation of series of numbers (smoothing by weighted averages), which were among the best and most perceptive on this topic to appear in the nineteenth century. In a *tour de force* that anticipated much of the work that would appear over the following half-century, he introduced formal optimality criteria for smoothness, and he borrowed statistical ideas from astronomy in developing and fully investigating the use of least squares methods in this area.

Initially, De Forest submitted the results of his investigation to the Smithsonian Institution, whose secretary, Joseph Henry, chose the English mathematician J.J. Sylvester as referee. Sylvester wrote a rather lukewarm review of the two papers, perhaps because he was sent an incomplete version of the work. This resulted in a delay in publication; the papers reached print in the *Annual Reports of the Board of Regents of the Smithsonian Institution* for 1871 and 1873. De Forest seems to have been so upset at this turn of events that his next research was printed privately as a pamphlet in 1876. Not surprisingly, it received little attention, so two of his advances remained unknown until the 1940s. One was his introduction of what is called the Monte Carlo method today; not having a table of random numbers handy, he constructed 100 identically sized, numbered pieces of cardboard, shuffled them in a bag, selected a sample, recorded the numbers, and repeated the experiment numerous times. De Forest's other chief idea was to use residuals for goodness-of-fit tests; this novel initiative was rediscovered by a group working on ballistics under John von Neumann in 1941.

De Forest must have come upon *The Analyst* the year after publishing his pamphlet, because over the next seven years, fifteen of his papers appeared in the journal, some in multiple parts that amounted to 25 articles altogether. His first contribution continued his study of optimality criteria for interpolation and smoothing problems.[93] However, his major advance in this field remained unknown until being rediscovered in the 1920s. Moreover, his work on actuarial mathematics seems to have eluded insurance mathematicians until 1924. After all, *The Analyst* had minimal readership outside the US and Canada, and few Americans were then capable of understanding his advances.

Another serious paper on actuarial mathematics from *The Analyst* was written by Robert Woodward, who would become an early president of the American Mathematical Society.[94] It has been called "the first serious paper written in this country," and it was on errors of interpolated values derived from numerical tables by means of first differences.[95]

In the 1880s De Forest turned his attention to symmetric and asymmetric error distributions in two and three dimensions.[96] In this, he partially anticipated correlation analysis for the study of associations in multivariate data that was developed after De Forest's death by Francis Galton, Francis Edgeworth, and Karl Pearson in England. However, credit for some of De Forest's advances from *The Analyst* went in the other direction; for instance, his results on testing goodness-in-fit by grouping signs of residuals had already been introduced by the Belgian statistician Adolphe Quetelet 25 years earlier.

When *The Analyst* ceased publication in 1883, De Forest turned to the *Transactions of the Connecticut Academy of Arts and Sciences* as the primary outlet for his research. However, his health began to fail in 1885, bringing an end to his career. Upon his death three years later, he bequeathed a fund for establishing a chair at Yale now known as the Erastus L. De Forest Professorship in Mathematics.

The other mathematical statistician in *The Analyst* publication community was **Gottfried Wilhelm Hugo Karl Kummell** (1836–1897), whose initial foray into the mathematics publishing sphere occurred in 1876 at age 40. Printed in two parts, his article showed that observational errors in an experiment are normally distributed.[97] He published an improved version of the proof of one result three years later.[98] A recent paper (2013) analyzed this important article and generally placed, in historical perspective, Kummell's contributions to the law of errors and to the least-squares method.[99] That paper singled out two other important aspects of this Kummell work: 1) the use of the quantity $h = \frac{1}{\varepsilon \sqrt{2}}$ as the measure of precision of a system of observations, and

2) the precisions of different systems of observations can be compared by means of probable error $r = 0.6745\epsilon$. (In modern terms, for normally distributed errors, the median absolute deviation is 0.6745 times the standard deviation).

Over his lifetime, Kummell published 30 papers, including several others in *The Analyst*. A three-part article dealt with Cauchy's theory of residues.[100] Two footnotes in it indicated problems that the editor Joel Hendricks encountered in typesetting. At the end of the second part of the article (p. 46), he added:

> Mr. Kummell has contributed another § to this paper but for want of suitable type we are not able to insert it at present, but hope to be able to do so before the close of the present volume.

When that third part did appear, he added (p. 175, footnote *), "For want of sorts, k is here … written instead of the Agate Greek π.—Compositor." Other topics that Kummell dealt with in the pages of *The Analyst* were least squares,[101] differential geometry,[102] elliptic functions,[103] and geometry.[104]

Kummell graduated from a polytechnic school in his native Germany at age 16 and then entered the University of Marburg, but left without taking a degree in early 1854. After teaching in Prussian schools, he left for Norfolk, VA, in 1866, whereupon he changed his name to Charles Hugo Kummell. He taught in Norfolk schools for the next five years before being appointed assistant engineer with the US Lake Survey in Detroit in 1871. This agency had been established by Congress 30 years earlier to conduct surveys of the northern and northwestern lakes, as well as to prepare nautical charts and other aides for navigation. Kummell left the US Lake Survey in 1880 and moved to Washington, DC, as a statistician and (human) computer with the US Coast and Geodetic Survey. He remained in these positions until his death 17 years later.

After moving to the nation's capital, Kummell became active in an organization that was short lived but which anticipated a national organization of mathematicians. The Philosophical Society of Washington (**PSW**) was founded in 1871 by Joseph Henry, the head of the Smithsonian Institution. Simon Newcomb served as one of its early presidents, and J.J. Sylvester lectured on the theory of quaternions during his first semester at Johns Hopkins five years later. In 1883 the PSW formed a Mathematics Section with 35 members that included C.S. Peirce, George Hill, and Simon Newcomb, as well as Kummell. The aim of the Mathematics Section was to discuss papers in pure and applied mathematics. At its first meeting the chair, Asaph Hall, advocated founding a new mathematical journal, but his proposal came to naught. A short while later, Artemas Martin proposed forming a national organization of mathematicians. It was moved that a committee be formed to report on the advisability of establishing such an organization, but the measure lacked a second, so the matter was postponed indefinitely. The American Mathematical Society (AMS) was founded in 1888, just four years after Martin's proposal. The Mathematics Section of PSW was disbanded in November 1892, probably because it no longer served any need due to the early success of the AMS.

Woman pioneer. Continuing in the early tradition of American mathematics journals, about 30% of the pages in *The Analyst* were devoted to problems, which were popular with many subscribers who supported the journal during its ten-year run. The

Figure 4.5. Christine Ladd Franklin

journal shows that the capability of problem solvers had improved decidedly, as indicated by a typical problem from Volume 5 (p. 160):

$$\text{Evaluate } \int_{-\infty}^{\infty} \frac{\cos mx}{a+x} dx.$$

The list of problem solvers included such prominent figures as G.B. Halsted, E.H. Moore, Alexander Ziwet, and Florian Cajori. Although Erastus De Forest devoted little time to the problem section of *The Analyst*, that was not the case with Christine Ladd, our last rugged, nineteenth-century individualist. Biographical coverage here is restricted to the first part of her life; the latter, far more important, part is detailed in the next chapter.

Christine Ladd (1847–1930) was born December 1, 1847, in Windsor, CT, to a merchant father and women's-rights activist mother who famously wrote that "women belonged not only in the pulpit, a place for which they were peculiarly suited, but also every place where a man should be." Ladd's mother often took her impressionable daughter with her to lectures on women's rights; the young Christine certainly picked up on these lessons. Unfortunately, Ladd's mother died when she was only 12, whereupon she moved to Portsmouth, NH, to live with her fraternal grandmother.

Ladd attended Wesleyan Academy in Wilbraham, MA, taking the same courses as her male counterparts who were being prepped for Harvard. She began keeping a diary that year, recording, "It is my great desire to have a college education & I shall use every means to bring my plans to a consummation."[105] Such a quest was out of the question for Harvard, even for such a talented prep-school student, because, like most colleges at the time, Harvard was all-male. However, Vassar College presented a viable alternative, having been established in 1861 to provide a similar experience for women. (It became co-educational in 1969.) Having had excellent preparation, Ladd entered the sophomore class at Vassar with tuition paid by her mother's sister, but financial difficulties forced her to drop out after only one year. She secured a teaching position, and during that year saved money and also "practiced the piano, read in three or four languages, worked problems in trigonometry, collected 150 botanical specimens, and published an English translation of Schiller's 'Des Mädchens Klage' in the *Hartford Courant*."[106] She then returned to campus for the final two years of the four-year curriculum, graduating in 1869.

Vassar greatly influenced Ladd's developing interest in science, as well as her passionate involvement in women's-rights activism. An astronomy professor, **Maria** (Ma-RYE-ah) **Mitchell** (1818–1889), served as an important role model and mentor for her during this period, encouraging her to pursue mathematics and science. Yet:[107]

> [Mitchell] also made decidedly unmathematical assertions. Once, she startled her students by stating, "We especially need imagination in science. It is not all mathematics, nor all logic, but is somewhat beauty and poetry."

For nine years after graduation, Christine Ladd taught high-school science and mathematics in three different states. She was interested in pursuing physics, but was deterred because women were not allowed in laboratories. Consequently, she conducted an independent study of mathematics that resulted in numerous contributions to *The Analyst* and the *Educational Times,* the latter published in London. Her first entry in *The Analyst* occurred in 1875 when she penned a report on a recent issue of *Crelle's Journal*. Ladd implicitly criticized the state of research in mathematics in the country, especially in the area with the greatest concentration of active workers at the time:[108]

> It is not greatly to the credit of the mathematicians of the vicinity that *Crelle's Journal* lies on the shelves of the Boston Public Library with uncut leaves.

During 1875–1880, Ladd proposed or solved 20 problems in *The Analyst*. As the historian of mathematics Florian Cajori stated:[109]

> Though the solution of problems is the lowest form of mathematical research, it is nevertheless, important, not for its scientific, but for its educational value. It induced teachers to look beyond the text-book and to attempt work on their own.

These words seem to fit Ladd to a "T." At one point the editor commented, "We have an elegant solution of this question by Miss Ladd."[110] To confirm Cajori's opinion, problem-solving activity seemed to induce independent activity on Ladd's part. Indeed, one can see her transition from a problem solver to a mature mathematician in the pages of *The Analyst*. In early 1877, while teaching in Boston and taking courses at Harvard, she wrote two one-paragraph notes, the first on the relation between sides and diagonals of the contra parallelogram, and the second on an unrelated result in geometry.[111]

Two papers written during 1877–1878 show that Ladd crossed the Rubicon from problem solver to independent investigator. An entry pointing out conflicting statements in works by algebraists Salmon (in England) and Serret (in France) stands as a testament to the progress she had made while taking courses with James Mills Peirce and William Byerly at Harvard.[112] More importantly, her incursion into research occurred with a three-page paper giving a résumé of quaternion equivalents of certain transformations from trigonometry.[113] Before leaving Boston, she submitted two more papers to *The Analyst*, one on geometry and the other on algebra.[114] After that, Ladd moved to Baltimore to enroll in the fledgling graduate program at Johns Hopkins. She published one more paper, on the nine-line conic, in *The Analyst*, which reflects her Hopkins training.[115]

Publication community. The number and quality of *Analyst* authors reflects the growth of a sizable mathematical publication community that expanded from practitioners who participated over its first few years of its existence, to a legitimate community of researchers over the final years of its ten-year run. Table 4.4 lists 21 leading contributors, and the number of entries taken from the "Index of Proper Names" that appeared with the final issue of *The Analyst*.

Six of the authors listed in Table 4.4 have already been introduced—Pliny Chase, George Hill, Simon Newcomb, Erastus De Forest, Charles Kummell, and Christine Ladd. Ten of the remaining 15 figures will be introduced in Chapters 5–7 as part of the initial American mathematical research community. Four of the ten were connected with Johns Hopkins, which was founded in 1876, when the third volume of *The Analyst* appeared. The entry for second faculty member in the Hopkins mathematics department, William Story, was a special notice in 1877 announcing his university's plan to publish America's first mathematics research publication, the *American Journal of Mathematics*, the next year.[116] The first graduate student at Hopkins, Thomas Craig, had received his PhD two years before publishing a paper in *The Analyst* in 1880, when he was employed by the US Coast and Geodetic Survey in Washington, DC.[117] That same year, Arthur Hathaway, then an undergraduate at Cornell, published two short notes.[118] The remaining *Analyst* contributor with ties to Johns Hopkins was Charles Van Velzer, who published solutions to problems using properties of determinants for which he became famous at the University of Wisconsin.[119]

Two of the six characters in Table 4.4 (who will be introduced in Chapters 5–7) had entries in Volume 1 of *The Analyst* (1874). One, William Woolsey Johnson of St. John's College in Annapolis, was the most active contributor. The other, Cornell chair James Oliver, sent a short communication. Calvin Woodward (Washington University) published a paper on hyperboloids of revolution.[120] Two authors were also active problem solvers: Henry Eddy[121] (University of Cincinnati) and Wooster Beman (Michigan), whose major work appeared in two parts.[122] Finally, it is notable that the astronomer Ormond Stone, director of the Cincinnati Observatory, published a mathematics paper,[123] replied to a query, and solved problems in *The Analyst*, because he continued the journal as the *Annals of Mathematics* in 1884.

Table 4.4. Publication community from *The Analyst*

Name	Entries	Name	Entries	Name	Entries
Beman	5	Hill	29	Newcomb	1
Chase	5	Hyde	17	Oliver	1
Craig	1	Johnson	40	Pratt	4
De Forest	26	Kirkwood	5	Stone	3
Eddy	6	Kummell	18	Story	1
Hall	23	Ladd	6	Van Velzer	2
Hathaway	2	Merriman	4	Woodward	3

That leaves five characters from Table 4.4 unaccounted for: Hall, Hyde, Kirkwood, Merriman, and Pratt. While all five participated in *The Analyst* in similar ways, their backgrounds and professionalization were quite distinct. The online file "Web04-Analyst" provides additional information on their careers and work, as well as those of De Forest and Kummell.

Asaph Hall III (1829–1907) was the most celebrated of the five, as indicated by a long memorial article on his life and accomplishments by one of the leaders of nineteenth-century American mathematics, G.W. Hill. Most of Hall's legacy lies in astronomy, but he was also proficient at mathematics.

Born in Connecticut, Asaph Hall left school at age 16, three years after his father died. He was then apprenticed to a carpenter for three years, after which he became a journeyman carpenter. Because such work was rare in winter, he set about studying algebra and Euclidean geometry. All the while, by the fall of 1854, he had saved enough money to attend Central College in McGraw, NY. He wanted to study mathematics, but no mentor was available, so he left after only 18 months. But during his first semester, he met Angeline Stickney, who graduated in 1855, joined the faculty, and became his geometry instructor. They married a year later. Central College was famous for educating blacks and whites together, including Charles L. Reason, but the school closed in 1860 due to financial difficulties and a smallpox epidemic.

The Halls moved to Ann Arbor, where he entered the University of Michigan in the sophomore class and excelled in astronomy, but they left after only three months for Ohio to run a school 1856–1857. During that time, he studied astronomy and mathematics independently. The next year, Angeline moved back home to teach school while Asaph moved east with the intention of enrolling at Harvard, where he attended Benjamin Peirce's lectures, all the while finding employment at the Harvard Observatory under William Cranch Bond. However, tension between the mathematics department and the observatory forced him to forego attending those lectures. Besides, Hall found Peirce's lectures too theoretical.

Over the next six years, Asaph Hall became expert at computing orbits of various celestial bodies, leading to his appointment at the Naval Observatory in 1862 in Washington, DC. He remained at the observatory until retiring in 1891. In 1877, he discovered the two moons of Mars, Phobos and Deimos, for which he is mainly known today. Five years after retiring, Hall accepted an invitation to lecture on celestial mechanics at Harvard. He delivered those lectures (1896–1901) before retiring for the second, and final, time.

The Hill memoir on Hall contains a complete list of Hall's publications, displaying a range of outlets that ran over several mathematics journals in the second half of the nineteenth century. His earliest paper on a mathematical topic appeared in Runkle's *Mathematical Monthly* in 1861, and dealt with the transformation of an infinite series into a continued fraction.[124] Earlier in the journal, he had won two prizes for solutions to posed problems. During 1867–1871, he published four mathematics papers in Artemas Martin's *Messenger of Mathematics,* including one on a method for approximating the value of π.[125] This paper described an experiment in random sampling that Hall had persuaded a friend to perform while recuperating from a wound. It involved repetitively throwing at random a fine steel wire onto a plane (a wooden surface) ruled with equidistant parallel lines. This precisely matched the experiment that Le Clerc de Buffon conducted in 1733, and it is called Buffon's needle problem today. Both were

early uses of the type of random sampling known as the Monte Carlo method since
World War II. As noted above, E.L. De Forest described a different approach to the
Monte Carlo method at about this same time.

Hall's first paper in *The Analyst*, the initial article in the journal's second issue of
1874, was on comets and meteors. That volume also contained his paper on the Bessel
function.[126] Two years later, he presented clever methods of numerical integration that
are appropriate for enrichment projects in a Calculus II class.[127] But the paper of his
I found most useful—for advising students about the kind of discipline necessary for
making progress in mathematics—appeared in 1881. Hall recalled that in 1858, two
years after reaching the Harvard Observatory, he began reading a translation of the
Theoria Motus by Karl Friedrich Gauss:[128]

> I well remember that at first the style of the work fairly took me off my
> feet, and seemed to leave me dangling in the air for a month or two
> before … the beauty and power of Gauss's methods were seen and
> felt. Having no teacher nor any one to assist me, I made it a rule to
> work out every equation and all the numerical examples before going
> on. … The whole reading occupied me nearly a year.

Hall's discipline paid dividends that year, resulting in two corrections to *Theoria Motus* that he published in the *Mathematical Monthly*. The aim of *The Analyst* paper,
Hall wrote, was to provide ten "of the points and reductions that gave me the most
trouble." He concluded the paper with advice for prospective readers of the classical
memoir: "It was by keeping the problem steadily before his mind for several years,
and carefully working out all its parts, that Gauss brought his solution at last to a form
almost perfect."[129]

In addition, Asaph Hall posed, and solved, numerous problems in *The Analyst*.
Two of them indicate how far American mathematicians had advanced since problems
posed in the *Mathematical Correspondent* 70 years earlier. One, for instance, called for
evaluating the determinant of a general 4×4 matrix. Another required a proof that

$$\int_0^a dx \int_0^x \varphi(x,y)\, dy = \int_0^a dy \int_y^a \varphi(x,y)\, dx.$$

The next member of *The Analyst* publication community, the virtually unknown
Edward Wyllys Hyde (1843–1930), deserves more prominent attention. His lifetime
extended from the Peirce generation to the onset of the founding of the Institute for
Advanced Study at Princeton. A native of Saginaw, MI, Hyde received a CE degree at
Cornell in 1872, shortly after the university was founded. He served as an instructor
on the faculty while a senior and continued over the next year. He then joined the
faculty at Pennsylvania Military College,[130] but left PMC in 1875 to become the first
(assistant) professor of mathematics (simultaneously instructor in civil engineering)
at the University of Cincinnati. The president of Cincinnati at the time was the mathematician Henry Turner Eddy. Hyde later served as president for three different terms
in the 1890s.

Edward Hyde was quite active in the emergent American mathematical community. He served as an associate editor of the *Annals of Mathematics*, 1896–1899. In addition, he was one of 28 signers of the 1896 circular "A call to a conference in Chicago"
that led to the formation of the Chicago Section of the American Mathematical Society.

The next year, he was elected to the AMS Council for the term 1897–1899, along with Woolsey Johnson and B.O. Peirce.

Hyde wrote three books, including two influential texts on mathematics. *The Directional Calculus: Based upon the Methods of H. Grassmann* (Ginn and Co., 1890) was the first textbook in English devoted to Grassmann's calculus. Called one "of Grassmann's [two] most important followers,"[131] Hyde published *Grassmann's Space Analysis* (Wiley, 1906) 16 years later. These two books on vector analysis "contain Grass-man's ideas in a simplified form, i.e., limited to three dimensions and with the stress on applications." The 1906 book was first published as a chapter in the *Higher Mathematics* by Mansfield Merriman and R.S. Woodward, which explains why its title page asserts "fourth edition." The roots of Hyde's 1890 text appeared a decade earlier with the four-part paper "Mechanics by quaternions," which appeared in *The Analyst* in 1880 and 1881. The long and detailed series began:[132]

> Owing to the fact that certain of the quantities treated in Mechanics possess *direction* as well as *magnitude*, and are thus in their very nature *vector* quantities, it appears that the Quaternion methods should be peculiarly fitted for dealing with mechanical problems. Such is indeed the case, and it is proposed in these papers to give an elementary quaternion treatment of the subject. [Emphasis due to Hyde.]

Hyde also published five papers on analytic geometry, two on synthetic geometry,[133] and one on integration. Curiously, his first article on analytic geometry lacked figures, but all four after that contained them.[134] The second paper proved a proposition that would make an appropriate project for an undergraduate course:

> Given four points [no one point lying within the triangle formed by the other three] to construct geometrically the axis and focus of the parabola passing through them.

In the paper on integration techniques,[135] Hyde presented a method for evaluating the triple integral,

$$V = \int_{x=0}^{x=a} \int_{y=0}^{y=f(x)} \int_{z=0}^{z=\theta(x,y)} dz\,dy\,dx,$$

by reducing it to a double integral.

Mansfield A. Merriman (1848–1925) graduated from the Sheffield Scientific School at Yale in 1871, and remained another year to earn a master's degree in civil engineering. He served as an assistant in the US Corps of Engineers 1872–1873 and then spent six months studying in Berlin, Dresden, and Hanover. Upon returning to New Haven, he became an instructor in civil engineering at Sheffield 1874–1878. During that time, he earned a PhD from Yale in 1876 for the dissertation "Elements of least squares." In 1878 he was appointed professor of civil engineering at Lehigh University, where he remained until 1907, when he joined a consulting firm in New York City. While at Lehigh, Merriman was also an assistant at the US Coast and Geodetic Survey 1881–1885. All the while, he conducted research on strength of materials and the design of bridges, consulted on several civil- and hydraulic-engineering projects, and pursued mathematics. Merriman published several books in these engineering fields; within mathematics, his *Method of Least Squares* was published in 1884 and ran to eight editions.

In 1877 Mansfield Merriman published a long bibliography of 408 articles on the theory of errors and the method of least squares.[136] One of the cited articles was Charles Kummell's *Analyst* paper from the previous year, and Merriman's comments caused the kind of bad blood that frequently happens when a reviewer criticizes an author's work. Initially, Merriman's statement seemed harmless. Regarding the Kummell article, he wrote, "Hagen's proof of 1837 is given abbreviated and improved, and the usual rules for normal equations and probable error are deduced."[137] However, later in 1877 another paper by Merriman[138] asserted that the Kummell article:[139]

> ... although very abbreviated, and requiring in its readers a previous knowledge of the subject, is very welcome to mathematicians, and it contains one or two modifications of the German method of presentation, which considerably shortens the algebraic work.

Quick to take offense, Kummell counterattacked:

> My paper is very abbreviated, as stated by Mr. Merriman, but is, nevertheless, clear and logical to any careful reader, and gives not a mere glimpse of the theory, but almost everything essential. Mr. Merriman's article contains a number of logical and theoretical blunders, which should not go uncorrected.

The matter got particularly testy when Kummell added, "Mr. Merriman writes that I have given Hagen's proof. Now who would like to be accused of such a thing?" Kummell claimed that his proof was totally original, and the controversy seems to have ended there.

Mansfield Merriman was essentially a civil engineer, as his first paper in *The Analyst* indicates. Its introductory statement exhibited a typical attitude that separates engineers from mathematicians today:[140]

> As a matter of purely mathematical interest I wish to give here, *without demonstration*, the relations between the reactions of continuous girders of equal spans resting on level supports" [my emphasis].

The other three papers that Merriman published in *The Analyst* dealt with the method of least squares (MLS). The first provided a history of the MLS, and was cited in Chapter 3 regarding the priority of Robert Adrain. Merriman noted, "The honor of the first publication of the method belongs to Legendre," from 1805.[141] He then cited 13 different proofs of the MLS, from the first by Adrain in 1808 to one by Crofton in 1870, including others by world renowned mathematicians Gauss (1809 and 1823), Laplace (1810), and Bessel (1838). It is curious that there was no mention of Kummell's paper published in *The Analyst* one year before Merriman's paper appeared in 1877.

Merriman did not have a copy of Adrain's 1808 paper from the earlier journal called the *Analyst; or Mathematical Museum* when he conducted his study of the MLS. However, later that year he lamented:[142]

> At the time the first paper appeared, in March, "I had not seen Adrain's original paper ... having lately been able to consult a copy ... I found that on pages 96 and 97, there is given a second deduction of the law of facility or error of an entirely different nature from that presented on pages 93–95. As this is a matter of considerable historical interest and as The Analyst for 1808 is quite rare I give the proof in Adrain's own words."

The author appended a page with data on the papers on the MLS he had catalogued. Regarding American mathematicians' limited access to journals abroad, Merriman wrote:[143]

> With better library facilities the number of titles in the Italian, Dutch, and Scandinavian languages would be much increased; and one who can consult the Russian and Hungarian literature might undoubtedly find a few titles to add.

The next member of *The Analyst* publication community, **Daniel Kirkwood** (1814–1895), authored the first paper in the journal, on astronomy.[144] Born in Maryland but educated at the York County Academy in Pennsylvania, Kirkwood was a principal at two academies (1843–1851) when he was appointed professor of mathematics at the University of Delaware. He had earned a master's degree at Washington College, PA, in 1850, and then an LLD from the University of Pennsylvania in 1852. He served as president of the University of Delaware 1854–1856, and then was appointed professor of mathematics at Indiana University. Daniel Kirkwood remained in Bloomington for ten years before accepting the same post at Washington and Jefferson College, PA, for one year, whereupon he was recalled to Indiana and remained there until 1886. He then moved to Stanford University as a lecturer, and lived in Palo Alto on an orange ranch for the rest of his life. An article in the first volume of the *American Mathematical Monthly* provides more details, as well as a list of his publications.[145]

Daniel Kirkwood published three other articles in *The Analyst*, one of which was also on astronomy.[146] His biography of the mathematician William Lenhart was cited in Chapter 3.[147] The remaining article was a short note on determining the length of a day.[148]

The remaining individual in Table 4.4 essentially published only one paper in *The Analyst*, but it led to three additional entries caused by yet more friction, this time between the editor and the author. **Orson Pratt** (1811–1881), then of Salt Lake City but born in central New York, is mostly known today for his role with the Church of the Latter Day Saints from the time he was ordained by Joseph Smith in New York at age 20. While serving on numerous missions, he conducted an independent study of mathematics (1836–1844) leading to his appointment as instructor at the University of Nauvoo, IL, when it was formed in 1841. Reputedly, he taught calculus there. Six years later, Pratt was the scientific observer for the Vanguard Company, led by Brigham Young, when it entered Salt Lake Valley as part of the cross-country campaign for Mormon colonization. Along the way, he invented the odometer, which he called a "roadometer." It seems that Pratt wrote a calculus book in the 1850s, but no copies are extant.

Orson Pratt was only involved with *The Analyst* during 1876–1877. Initially he proposed a set of six problems regarding velocities and forces of orbiting bodies that was published in November 1876.[149] He supplied solutions in the next issue, January 1877, but at the end of those solutions, journal editor Joel Hendricks criticized Pratt's propositions as a basis for a theory of gravity for its lack of a conceivable cause.[150] Pratt rebutted that assertion in the next issue, stating that his propositions had no bearing on, or reference to, the cause of gravity. Nonetheless, Hendricks dissented once again,[151] thus putting an end to the 66-year-old Pratt's involvement with the journal.

Earlier, Orson Pratt must have sent Joel Hendricks a copy of his 1866 book *New and Easy Method of Solution of the Cubic and Biquadratic Equations*. Hendricks listed it in *The Analyst* but wrote only, "We are not at present prepared to speak of the merits of this book, but insert a single paragraph from the author's preface which will indicate its character."[152]

Another notable figure from *The Analyst* publication community, though not listed in Table 4.4, was the (human) computer Artemas Martin, who published a brief note exhibiting a method of using natural logarithms to evaluate roots for computing purposes.[153] As an example, Martin proved that:

$$\sqrt[100]{2} = \sum_{n=0}^{\infty} \frac{(\log 2)^n}{n!\, 100^n}.$$

Not all papers in *The Analyst* maintained such a high degree of rigor. Regarding a purported proof in one of the papers that the basic Calculus I result $D\left(ax^2\right) = 2ax$, the editor politely wrote, "No logical inference can be drawn from the equation [in the proof]. ... The quantities themselves in such state elude our comprehension."[154]

Assessment. In the penultimate issue of *The Analyst*, Hendricks announced:[155]

> We regret to have to inform our readers that we have concluded to discontinue the publication of the Analyst on the completion of Vol. X. This determination has not been induced by any lack of interest in the publication manifested by our subscribers and contributors ... but wholly on account of declining health.

The Analyst, then, was the first mathematics journal in the country *not* to cease publication due to an insufficient critical mass of subscribers. In the very last issue, the rueful editor informed readers that although he had been unable to consummate a deal with a suitable successor, he harbored the hope that ongoing negotiations with interested parties would result in the journal's continuation.[156] Those hopes were soon realized, and *The Analyst* was continued as the *Annals of Mathematics* the next March, 1884.

What is the place of *The Analyst* in the history of American mathematics? In 1932, the historian of mathematics D.E. Smith wrote, "The best-known and most successful of the early mathematical magazines was *The Analyst*."[157] Nine years later, Benjamin Finkel published a detailed account of its contents.[158] Overall, *The Analyst* seems to

Table 4.5. Mathematical journals founded in America before 1878

Year	Journal	Editor
1804	*Mathematical Correspondent*	George Baron
1808	*Analyst; or Mathematical Museum*	Robert Adrain
1814	*The Analyst*	Robert Adrain
1818	*Monthly Scientific Journal*	William Marrat
1820	*Ladies' and Gentlemen's Diary*	Melatiah Nash
1827	*Mathematical Diary*	Robert Adrain
1836	*Mathematical Miscellany*	Charles Gill
1842	*Cambridge Miscellany*	Benjamin Peirce Joseph Lovering
1858	*Mathematical Monthly*	John Daniel Runkle
1874	*The Analyst*	Joel E. Hendricks

have played an important role as an outlet for mathematicians who would become vital cogs in the emergent mathematical research community from 1876 to 1900.

In addition to spawning today's prestigious *Annals of Mathematics*, *The Analyst* also played a leading role in permanently establishing the factorial notation. The first volume used both $n!$ and another notation for factorial, but the editor, Joel Hendricks, who set his own type, revealed that for simplicity of printing, he would henceforth use only $n!$, even if an author submitted an alternate notation. The factorial symbol has been standard ever since.

Table 4.5 lists the most prominent mathematics journals published in the US up to 1878, along with the name of the founder and the year its first issue appeared. The online file "Web04-19cJournals" contains my personal experience in consulting these publications.

Artemas Martin

The lifetime of **Artemas Martin** (1835–1918) spans three pivotal periods in the history of mathematics in the America. His example serves to highlight some of the critical traits that characterize each of these periods. For our purposes, his role in establishing journals links the period covered by this chapter and the next. Today, the Artemas Martin Collection at American University is a valuable repository of original mathematical works from the fifteenth through the twentieth centuries. He also proposed an organization similar to the AMS four years before its founding in 1888.

Martin was born in New York State in 1835. Two years later his family moved to Erie, PA, where he lived for the next 48 years. Martin's education was typical for the mid-1800s, meaning it was minimal, sporadic, and taken at local schools. A farmer's child could go to school only during nonfarming seasons, thus allowing him to attend classes in only three winters, after having been taught how to read and write at home. Consequently, he never earned a degree. Instead, he became a professional farmer and market gardener.

While in school, Martin became interested in problems in algebra and geometry. At age 18 he began to contribute to problem-solving columns by posing problems and solving some submitted by others. Within a few years, this interest enabled him to teach in the local public school, but only during nonfarming periods, of course. And even this foray into higher education lasted only four years. The life of a farmer was demanding, indeed. Yet during this time, his active involvement with the "higher mathematics" department of the *Normal Monthly* led to his selection as editor of the journal. Apparently, the Martin farm was growing fruit of another kind.

Artemas Martin felt strongly that communication was necessary for fostering enthusiasm for mathematics and for diffusing knowledge of it, so he carried out continuous correspondence with numerous mathematicians. By 1878 he felt the need to establish a journal devoted entirely to mathematics, even though Joel Hendricks had initiated *The Analyst* four years earlier. Curiously, this was precisely the year that the *American Journal of Mathematics* was founded at Johns Hopkins. Martin's *Mathematical Visitor* could not compete with this high-level publication, and research was not his aim. Rather there was still sufficient interest in such material for "run of the mill" mathematical amateurs to contribute articles and to solve problems that the *Mathematical Visitor* lasted 16 years. Not content with just this publication, in 1882 Martin

Figure 4.6. Artemas Martin

founded *Mathematical Magazine*, which was intended for an even more elementary audience.

Altogether, ten journals were founded in 1875–1900, five of which are still flourishing. Neither of Martin's journals survived into the twentieth century. *Mathematical Magazine* lasted only three years, but there is a good reason for its short duration. All of Martin's activities up to his fiftieth birthday took place in isolation, marking him as yet another rugged individualist. But in 1885 his life changed dramatically when he left his farm in Erie to become the librarian for the US Coast and Geodetic Survey in Washington, DC.

Besides mathematics, Martin possessed such an unbridled passion for books that he bought two neighboring homes in the nation's capital, one to live in, the other to store his books.[159] His collection is now housed in the rare-book room at American University. Among the standard calculus books of the day, it includes the first calculus text published in the US, Ryan's *Differential and Integral Calculus*, from 1828. The Artemas Martin Collection also contains such stalwarts as a 1488 copy of Boethius, a 1565 copy of Tartaglia's version of Euclid, and Descartes's *Opera Philosophica* from 1649.

It wasn't long before the US Coast and Geodetic Survey added another aspect to Martin's job description—computer—a title for someone who performed computations. Such calculations were not restricted to the numerical sort. Martin had become an expert in all sorts of averages, ranging from those involving functions of one and two variables to vector-valued functions.

Martin's position at the US Coast and Geodetic Survey brought him into daily contact with two major mathematicians, Simon Newcomb and Charles S. Peirce, thus effectively ending his period of isolation. Now Martin began to publish articles at a regular rate, on topics ranging from number theory to geometry to probability. He contributed a paper at the Chicago Congress of Mathematics in 1893, and subsequently presented papers at two International Congresses of Mathematicians.

Artemas Martin was not a research mathematician in the modern sense; he belongs with the amateur mathematicians of the first three quarters of the nineteenth century.[160] His time and place in the US up to 1876 all but prevented such a position. But during the last quarter of that century, he wrote basic mathematical work that added new procedures and provided fresh insight into the work of others. Moreover, after 1900 he presented his findings in front of international audiences. As a result, Martin's role of "mathematical ambassador" has earned him a place in the pantheon of tireless enthusiasts who set the table for those who followed in their footsteps. In this sense, Artemas Martin serves as a case study for the American mathematical community that worked in isolation before 1876, emerged as a community from then until 1900, whereupon it expanded its borders across international lines.

First PhDs

It was noted that William Byerly received the first Harvard PhD in mathematics in 1873 for a dissertation on the heat of the sun. But Harvard was not the first American university to offer a doctorate. Yale was. In 1860 Yale established a doctor of philosophy degree for its own graduates or for graduates of other institutions who met additional entrance requirements. To earn that degree, students had to spend two years in residence, study at least two branches of learning, pass a final examination, and produce "a thesis giving evidence of high attainment in the branches they have pursued."[161]

Initially PhD dissertations did not list an advisor. Furthermore, because universities were not divided into departments, these dissertations were not associated with a particular branch of learning. In fact, not all names of recipients of these degrees are identified and, for those that are, sometimes the title of the dissertation is lacking. Only in 1919, when Yale established departments, did it retrospectively classify dissertations and, as was stated at the time, "the guiding principle has been the subject of the dissertation as presented for degree."[162] Due to this classification, until quite recently it was accepted that **John Hunter Worrall** earned the first American doctorate in mathematics upon completion of his PhD dissertation in 1862. Even though its title was unknown, Worrall taught at various levels in West Chester, PA, for the rest of his life, and this accounted for the classification. Four years later **Charles Greene Rockwood** completed a dissertation entitled "The daily motion of a brick tower caused by solar heat," so it was therefore assumed that he was the second recipient of a PhD.

But those assumptions were overturned in a study conducted by Steve Batterson in 2008.[163] Batterson revealed that three dissertations were submitted in 1861, one year before Worrall, and he argues persuasively that one of them should have been assigned to mathematics. **Arthur W. Wright** wrote the dissertation titled "Having given the velocity and direction of motion of a meteor on entering the atmosphere of the Earth, to determine its orbit about the Sun." Like Gibbs's thesis two years later, Wright's dissertation was assigned to physics in 1919. However, the subject of the thesis matches the research of Yale's mathematics professor at the time, Hubert Newton, so it was probably written under his supervision. Why then was Wright's degree assigned to physics? Because after a year of study in Germany and a professorship at Williams College, he accepted the professorship of molecular physics and chemistry

at his *alma mater* in 1872, and five years later his appointment was changed to experimental physics. (The online file "Web04-Wright" supplies additional details on
Wright.) Unfortunately, Wright's dissertation is not extant, yet 1) its title fits into what
was regarded as mathematics at the time, and 2) his advisor was probably a mathematics professor, so I agree with Batterson that "with further hindsight, there is a strong
argument that Wright's degree was actually in mathematics."[164]

In any event, the first doctorate in the field was awarded at Yale in the early 1860s
with work probably supervised by Hubert Newton. Even though the first definite PhDs
in mathematics—who satisfied requirements virtually the same as today—date from
1878 at Johns Hopkins, the Yale model of not listing an advisor persisted through the
first quarter of the twentieth century.

The first doctorate awarded to an African American student also occurred in this
period, though not in mathematics. **Edward Alexander Bouchet** (1852–1918) grew
up in New Haven, CT, where he graduated from a prep school in 1870. He was then
admitted to his hometown university, Yale, and graduated four years later ranked sixth
out of 124 undergraduate students. He thus became the first black to be graduated from
Yale. The Philadelphia philanthropist Alfred Cope then paid for his tuition to continue
his studies at Yale, and Bouchet earned his doctorate in physics two years later in 1876,
thus becoming the first black to earn a PhD in any field. Upon receiving his degree,
Bouchet, like Charles L. Reason, was unable to obtain a faculty position at any college,
so he taught at Cheyney State University (then the Institute for Colored Youth located
in Philadelphia) 1875–1902. More than likely, Bouchet taught mathematics as well as
physics.

Bouchet was fired in 1902 at the height of a dispute between W.E.B. Du Bois and
Booker T. Washington over the need for an industrial vs. a collegiate education for
blacks. The all-white board voted for industrial education and thus replaced the other
faculty members. Bouchet then taught at a high school in St. Louis before becoming
director of academics at St. Paul's Normal and Industrial School (now St. Paul's College) in Virginia 1905–1908. Five years later he joined the faculty at Bishop College
in Texas. When illness forced him to retire three years later, he moved back to his
childhood home in New Haven. No African American earned a PhD in mathematics
at an American institution for almost another 50 years.

The grandfather of American mathematics

By 1853, Benjamin Peirce was firmly established as America's premier research scientist, mainly in astronomy and mathematics. Because there was not yet a formal organization devoted exclusively to those two specialties at either a national or local level,
any committed scientist active in these fields and seeking communion with similarly
interested souls had to be content with meetings sponsored by more general scientific
societies. One such organization was the AAAS, and it was at their fourth semiannual
meeting in Cleveland in 1853 when a fortuitous event for the history of mathematics
took place.

At the meeting, Peirce met a 23-year old tutor at Yale, **Hubert Anson Newton**
(1830–1896), who was virtually unknown then but who directed the first PhD dissertation in mathematics in America within a decade. Peirce had an eye for young talent and
he recognized Newton's potential at once. It is symptomatic of the low state of science

Figure 4.7. Hubert Newton

in the country in the 1850s that the meeting between the Harvardian Peirce and the Yalie Newton would take place in a city located so far from either Cambridge or New Haven. But their conversations illustrate the value of personal contacts that continue to take place on a regular basis today at the AMS, MAA, and SIAM. The influence that Peirce exerted on Newton turned out to be pivotal for Newton's career and thus salutatory for American mathematics, because it inspired Newton to return to Yale to write up the results of an astronomical investigation he had carried out. This resulted in his first publication, only one page long, but a harbinger of greater writings to come. And it appeared in an astronomy journal. But the thrill of independent discovery and having the conclusion published in a professional venue would thrust Newton into a lifetime of academic pursuit.

This was not the first auspicious event in Hubert Newton's career, but it was surely a deciding factor in his ascension into the ranks of researchers. The ninth of eleven children, Newton had entered Yale College in January 1847 at age 16. His young age and midyear enrollment are two conspicuous differences between the undergraduate experience in the mid-nineteenth century and early part of the twenty-first century. In fact, undergraduates today would hardly recognize the program that existed in Newton's time. For one thing, his entering class numbered about 100; few American colleges are that small today. And Yale's faculty numbered only eight—the president and seven professors.

More importantly, the undergraduate program in 1850 differed from the year 2018 in three vital ways: the curriculum, the manner of instruction, and the academic emphasis. First, all 100 students took the same classes through the first semester of their junior year. For mathematics, this meant that all students, regardless of major interest, studied algebra, Euclidean geometry, trigonometry, conic sections, and spherical geometry. It is impressive that every student had to be proficient in these areas. (Spherical geometry has not been part of the curriculum for almost a century.) Yet, aspiring mathematics scholars interested in pursuing science would be handicapped by a program that did not go beyond the rudiments of analytic geometry. It was still possible for science-oriented students to elect a calculus course in the second semester of their junior year, but generally such study had to be conducted independently, as formal courses in the subject were rarely offered.

A second major difference for today's college students was the nineteenth-century emphasis on rote learning. Students stood in class and recited material they had memorized; in mathematics, they went to the chalk board to solve problems using techniques that had been drilled in previous class meetings. Few questions were posed, and independence was hardly nurtured, although critical thinking was developed in debating societies. There was no such a thing as collaborative learning.

Thirdly, the emphasis was on the classical languages Latin and Greek, where long passages had to be memorized and regurgitated. Can today's students imagine a college experience without athletics? No sports teams yet existed. Instead, debating societies flourished, and Hubert Newton was active in one, although he was not particularly gifted in oratory.

It is not known whether Newton elected any mathematics courses during his final three semesters at Yale before graduation in August 1850. The academic calendar began in late September or early October, and ran through June, with summer vacation occurring in July and August. Like many college graduates today, Newton was unsure of his career path after graduation. He wrote with exasperation:[165]

> What I shall do after graduation I do not know. I may prepare to teach,
> I may study theology—perhaps to study engineering.

Notice the absence of mathematics in future plans. And note the absence of the pursuit of higher degrees. Why? Because the only advanced degree was the master's which, from the time of Yale's founding in 1701, was awarded three years after the bachelor's for a fee of $5 as long as one maintained a good moral character. There was no doctoral program in America in any academic field at any college or university throughout the 1850s.

Thus, Newton did what many college graduates do—he returned home, in this case to the family farm in Sherburne, a small town in central New York about half way between Syracuse and Schenectady. Over the long, cold winter he initiated a program of studying mathematics when not otherwise occupied with daily chores. I do not know the impetus for this initiative, but Newton did not expect his proposed program to be scintillating, admitting:

> I am now reading Analytical Geometry ... Such reading would of
> itself be too dry and for a change I shall read books upon the Natural
> Sciences. Afterwards I expect to avail myself of the direction of a
> teacher.

During the latter part of this nine-month period of independent study, Newton corresponded with **James Hadley** (1821–1872), a young scholar who had just been promoted to a professorship of Greek. Hadley's mathematical work, both before and after graduation from Yale in 1844, had attracted the attention of Benjamin Peirce, who expressed disappointment that Hadley chose philology over mathematics. Hadley initially demurred on Newton's request to study with him on academic grounds (a full schedule of teaching and studying philology) and for social reasons (an impending marriage). It might seem curious that Newton would write Hadley instead of Anthony Stanley, the sole mathematics professor at Yale since 1836. However, Stanley was in Egypt in a fruitless search for a cure from tuberculosis. The plight of Newton indicates the low state of mathematics in America in the mid-nineteenth century—Yale was one of the premier colleges in the country, yet nobody on campus was qualified to direct his study beyond basic mathematics other than a Greek professor.

Nonetheless, Hadley did agree to direct Newton's reading that summer, and apparently the arrangement worked out satisfactorily to both participants, as Newton repeated this schedule the following year, spending August to May at home in Sherburne and then returning to New Haven for additional study. By this time in 1852, however, Yale was in dire straits regarding mathematics. Stanley was not a factor; he would die in 1853. The general policy at leading institutions was to employ one professor in each subject area who would carry out teaching duties with the aid of a tutor. Tutors were generally recruited from the ranks of the best recent graduates, holding these positions for a few years in the hopes that an opening would occur at some college. Mostly, however, tutors were forced to seek other forms of employment.

Thus the timing was particularly serendipitous for Hubert Newton when he was offered the tutorship for mathematics in June 1852, beginning his duties the following January. In the meantime, Yale conducted a search for a successor to Stanley, with two Yale graduates on the short list of prospective candidates. Rutgers professor Theodore Strong was the only other American committed to research in mathematics besides Benjamin Peirce, but Strong was 66 years old. The other candidate, William Chauvenet, was not a researcher like Strong, but had established a reputation as a highly successful textbook author. In addition, Chauvenet was already being hailed for his decisive role in establishing the Naval Academy. Chauvenet was content with his position at Annapolis, yet he wavered when offered the prestigious Yale post. So Yale decided to forego seeking other candidates in the hope that the naval officer would change his mind and return to his *alma mater*.

It was Newton's great fortune that Chauvenet vacillated, and the American mathematical community benefited greatly, because Newton's academic genealogy includes more American PhDs than any other figure. Because of his mentorship of E.H. Moore, the acclaimed "father of American mathematics," Newton was baptized "the grandfather of American mathematics" by Steve Batterson in his 2008 article "The father of the father of American mathematics."

What this meant for Hubert Newton was that in 1852, at age 22, he was thrust into the role of the sole mathematics instructor for the entire program at Yale. Can you imagine? And then, just three years later, he was appointed professor of mathematics, a promotion apparently based on the one-page paper he had published to that point, the one resulting from his serendipitous meeting with Benjamin Peirce at the start of his career. Might there be a department head at a US college today who obtained that position on the basis of such a minimal record?

Few mathematics teachers conducted original research in the 1850s, and those who remained active in spite of heavy teaching loads generally wrote textbooks. Most practitioners were appreciative rather than original, with Chauvenet serving as a model of such professors. But Peirce had passed on his passion for original investigations to Newton, who, upon his appointment as professor in 1855, appealed to the president of Yale for a leave of absence for a year of additional training in Europe. Granted his wish, he became the first Yale mathematician to study abroad, but Newton accepted a stipulation that few of us faculty members would even consider today—the cost of hiring a replacement would come from his salary.

Where did Newton get the idea to go abroad for higher studies? Actually, there was a tradition going back several decades for aspiring American scientists to sail to Europe to pursue their dreams. Now, many Yale grads engaged in such pursuits, but I am unaware of any who studied mathematics or any who went to France instead of Germany. However, the Yale archives contain numerous letters from graduates reporting on their experiences in these distant lands in 1854, so Newton's sojourn the next year fell into this tradition. Such a journey was not for the faint of heart; the two- to three-week voyage lacked most of the amenities of modern sailing, let alone trans-Atlantic flights.

It is not clear why Newton chose France over Germany, surely the leading country in mathematics at the time. Gauss had just died but his successor at Göttingen was Dirichlet; moreover, Gauss's student Riemann offered a course during 1855–1856 on Abelian functions to just three students, one of whom was Dedekind. Therefore, Newton might have profited from acquaintances made there, although it is unlikely he would have comprehended the mathematics taking place in that rarefied air. A contributing factor was undoubtedly language fluency, as Newton had studied French during his two winters at home after graduation, while carrying out his program in analytical geometry. He had studied Latin and Greek as an undergraduate, but modern languages were not offered in the curriculum. Matters did not go so well when he arrived in Paris. For one thing, lectures commenced two months later than he had been told. Thus, although he spent September and October profitably by immersing himself in spoken French, little of that period was devoted to scientific pursuits.

Hubert Newton's correspondence back home to Yale presents an interesting contrast. On the one hand, letters to high administrators resonate with tributes to the leading French mathematicians Cauchy, Chasles, and Lamé. On the other, they contain stinging criticisms of certain other professors. He described one, perhaps Sturm, in rather unflattering terms:[166]

> He has a raw beefy looking countenance. ... With one hand in his pocket he chalks out diagrams and formulas with the other. He never looks up at the class.

Nor was Newton impressed with French classrooms:[167]

> The lecture room does not compare with ours for comfort, there being no backs for the seats and no alley so that to reach the front seats we walk down stepping on the seats. A clumsy arrangement that.

In the spring, Newton traveled from France to England, where he met the famous astronomer John Couch Adams. It is indicative of most of Newton's future endeavors that he would spend his time in observatories, as he had in Paris and would yet do so in Italy and Germany before returning to New Haven in August 1856. I doubt whether any inhabitants of Cambridge (England) mentioned the last name of this American in the same breath as the world-renowned Isaac Newton. To the best of my knowledge, there is no relation between them, even though the American's ancestors came from England.

Once Hubert Newton returned to the US, he would never travel abroad again. But his experience with cutting-edge science, and the culture that emphasizes research on the same plane as teaching, was a primary influence on the research program he would carry out over the next 40 years. Moreover, the lectures of Chasles in Paris on a synthetic approach to projective geometry inspired his first two papers on mathematics, both published in J.D. Runkle's *Mathematical Monthly*, the only American journal concentrating on mathematics at the time. Both papers were concerned with the problem of constructing a circle tangent to three given circles, the second of which (from 1860) was more substantial than the first (from Volume 1 in 1858). However, Newton's main focus after that was on astronomy, with 40 papers alone on the paths of meteors and planets. His colleague Denison Olmsted, who died in 1859, had also been concerned with meteors; in fact, Newton had sent Olmsted his observations from his home in Sherburne back in November 1850. Hubert Newton ultimately gained fame for his work on the November Leonid meteor showers, which were being investigated independently in England by the astronomer John Couch Adams.

Hubert Newton's assimilation of the research culture he had been exposed to in Europe in 1855 1856 set him on the publication path for the rest of his career. Ultimately, he thrust two of his best students along this same path, so radical at the time, J. Willard Gibbs and E.H. Moore, the latter is the shining star of a subsequent chapter.

Transition 1876:
Story vs. Klein

The main theme throughout much of the period 1776–1876 was rugged individualism, as the leading figures worked in relative isolation, receiving little support for their work, being deprived of a professional society, and having no journal devoted to mathematics. The work of Robert Adrain, John Farrar, Charles Davies, and Nathaniel Bowditch serves more to illustrate limitations than advancements, while contributions by Benjamin Peirce, George Hill, and Willard Gibbs were akin to voices crying in the wilderness. This somewhat grim picture of the state of American mathematics in 1876 awaited a solitary English mathematician who would organize the first *community* of American mathematical scholars whose coalescence ignited a virtual revolution on these shores. First, the education systems in America and Germany are examined through the lens of two participants who will ultimately play fundamental roles in our story.

By 1870 college education in the United States was evolving rapidly. In addition to older, established institutions like Harvard, Yale, and Princeton, a new breed of state schools emerged as a result of the Morrill Act, with several attaining international distinction a century later. How did the mathematics education at these public and private institutions compare to their counterparts abroad? A fascinating parallel between the formal training of two contemporary leading figures reflects the academic disparities between the US and Germany, the country that had become the acknowledged leader in the field by then.

Felix Klein (1849–1925) was a child prodigy who enrolled at the University of Bonn in 1865 at age 16. After completing two years of classical studies in Latin and Greek, he concentrated on mathematics, taking a number of courses from Rudolf Lipschitz in analytic geometry, number theory, differential equations, mechanics, and potential theory. At the time Klein aspired to become a physicist. **William Edward Story** (1850–1930) entered Harvard University at age 17. Up to that time the university's curriculum was completely prescribed, but in 1867 Harvard's president Charles W. Eliot instituted a system of free electives. This policy permitted Story to take, as he wrote, "all the courses then given,"[1] including one on elliptic functions and another devoted to Gauss's *Theoria Motus*, both taught by the estimable Benjamin Peirce.

Upon first glance, it appears that the mathematical education available to William Story at Harvard was comparable to its counterpart that Felix Klein pursued at Bonn two years earlier. However, Peirce was working in isolation; no other faculty members at Harvard would produce research in mathematics until the 1890s. Indeed, not one other person in the US or Canada was at Peirce's level. Besides, Harvard was by far the leading institution for mathematics in America; only Yale approached its offerings.

The mathematics faculty at the University of Bonn was different. Besides Lipschitz, the faculty boasted the famous algebraic geometer Julius Plücker. Today the University of Bonn might arguably be considered the leader in terms of mathematical research in Germany. A recent memorial article on **Friedrich Hirzebruch** (1927–2012) credited him with moving "the center of gravity of German mathematics from the traditional centers of Göttingen and Berlin to Bonn."[2] However, Bonn was not so highly regarded when Klein studied there, especially with the triumvirate of Weierstrass, Kronecker, and Kummer heading the faculty at the University of Berlin. In addition, the department at Göttingen, perhaps the most storied university in the history of mathematics in Germany, if not the world, included Clebsch, Brill, and Gordan. There were excellent faculties at Heidelberg and Leipzig as well.

No, the educational landscape ranged much broader and deeper for Klein than for Story, initial impressions notwithstanding.

What did Story do with the excellent, if narrow, education he had received when he graduated from Harvard in 1871? He did what any bright, young, aggressive, American student would do—he traveled abroad to continue graduate studies in Germany. He attended lectures in Berlin by such notables as Karl Weierstrass and Ernst Kummer, but preferred the mathematics faculty at Leipzig, which included Karl Neumann and Adolf Mayer. Story earned his doctorate in 1875, probably under Neumann. It is telling that Story traveled abroad more than ten years before such sojourns became fashionable for aspiring American mathematicians.

Meanwhile, Felix Klein began working under Julius Plücker at Bonn. Even before Plücker's death in 1868, Klein had come under the influence of Alfred Clebsch, yet he wrote his doctoral dissertation that year under Rudolf Lipschitz. Consequently, Klein not only earned his doctorate before he was 20 years old, he obtained it before Story was permitted to concentrate on mathematics. (Even today Harvard students "concentrate" in a subject area; they do not "major" like the rest of American students.) By the time William Story obtained his bachelor's degree and sailed for Germany in 1871, Felix Klein had already spent an intense period with his new-found friend, Sophus Lie, in Paris studying under Camille Jordan, among others, and had been appointed to his first position as a *privatdozent* at Göttingen. In 1875, just four years later, Story completed his doctoral studies at Leipzig and headed back to the US, whereas Klein was in the process of leaving his first professorial position at Erlangen for a second one in Munich.

One thin thread connects Story to Klein after their formal educations were completed. Story returned to Harvard as a tutor during the year 1875–1876 before becoming an associate at the newly formed Johns Hopkins University. One of the first graduate students at Hopkins was **Washington Irving Stringham** (1847–1909), who attended Story's two-semester course on higher plane curves and elliptic functions in 1878–1879 and 1879–1880. Stringham received his PhD in 1880 at Hopkins, ostensibly under department head J.J. Sylvester, but in reality under the supervision of Story. Stringham then won a traveling fellowship (under the aegis of Harvard University) for post-doctoral study abroad. With Story having matriculated in Leipzig, is it any wonder that Stringham would follow the same path?

Serendipitously, Klein had just moved to Leipzig, so Stringham attended Klein's very first course of lectures there, a one-year class on function theory from a geometric point of view. Stringham was one of 89 students who took the first semester of the course, and one of only 45 who survived into the second, thus affirming the rigorous education he had received at Johns Hopkins. As further evidence of this background, he received his second doctorate with a dissertation written under Klein's direction in 1882.

It is notable that both William Story and Irving Stringham traveled to Germany to pursue doctoral studies even though domestic PhD programs were available at Yale and Harvard at the time. But the level of scholarship at these two colleges was far inferior to what they would encounter abroad. This practice of enduring a multiweek voyage to Europe would become fashionable by about 1890, but it was only just beginning at the time Story and Stringham made their sojourns. In fact, both Americans traveled abroad with the aid of Parker fellowships. This section ends by examining the genesis and nature of the custom of traveling abroad for advanced studies, and in particular the central role played by Parker fellowships in providing needed financial support.

Throughout the nineteenth century American scholarship was inferior to that available in Europe in all subjects. Yale and Harvard led the way in encouraging individuals associated with them, either as professors or students, to enhance their studies by traveling to Europe. It seems that Yale recognized, and then supported, this need first. The initial American to make the arduous trans-Atlantic trip was Benjamin Silliman, known for founding the influential *American Journal of Science and the Arts.* Back in 1805, he became the first American to pursue post-graduate studies in Europe when Yale supported him for a year; he spent most of that time in Edinburgh.

By the 1840s it had become fairly common for bright, aggressive Yale students to continue their education in Europe upon graduation with one crucial proviso— their families had to have sufficient means to pay the substantial costs involved. I am not aware of the New Haven college sending any of its professors abroad for another 50 years after Silliman's journey in 1805–1806. As noted in Chapter 2, that person was Hubert Anson Newton, whose future exploits led to his being regarded as the "grandfather of American mathematics."[3] Newton had been the mathematics tutor at Yale from 1852 until 1855, when he was promoted to professor and awarded a one-year leave of absence with salary (minus the cost of a replacement) to study in Europe. Within ten years Newton had influenced two of Yale's first PhD students to continue their studies abroad—Arthur Wright and Willard Gibbs.

Although Yale seems to have been the first American university to support the study of higher mathematics abroad for its faculty members and doctoral recipients, it was its (usually) friendly rival Harvard that took this custom to a much higher level. In 1815, ten years after Silliman travelled abroad, Harvard named the 21-year-old **Edward Everett** (1794–1865) to its first chair in Greek literature with a provision for two years of additional study in Europe at full salary. Everett was joined by his friend George Ticknor, a Dartmouth graduate whose family possessed the requisite funds to pay for his educational experience out of pocket. After a prolonged stay that kept them in London until Napoleon met his Waterloo, the two young Americans finally reached Germany. They studied philology at Göttingen, where they might have come in contact with the great Carl Friedrich Gauss, who had a strong interest in the subject. When Everett received his doctorate two years later, Harvard awarded him a two-year extension.

Everett was overwhelmed with the high level of scholarship in Göttingen as well as the ethos of advancing knowledge. His traveling companion Ticknor summarized the gulf separating scholarship in the New World and the Old when he wrote to his father, "Every day I am filled with new astonishment ... what a mortifying distance there is between a European and an American scholar!"[4] As a result, Everett appealed to President John Kirkland to send a recent Harvard graduate to be imbued with this culture and thus return to campus to raise the bar on erudition. All along, Kirkland had been profoundly influenced by reports from Everett so he, in turn, created an *ad hoc* scholarship on the spot for a recent Harvard graduate.

George Bancroft (1800–1891), the recipient of Kirkwood's *ad hoc* scholarship, had entered Harvard at age 13 and upon graduation four years later sailed to Germany, where he studied in Göttingen in 1820. He then traveled throughout Europe for two years before returning to Harvard, first as a tutor and then as an instructor of Greek. However, Bancroft found it nearly impossible to reconcile his German educational experience with the limited interest and ability of the normal Harvard College student, so he soon left. His subsequent career was quite distinguished, however. Among other important governmental posts, he served as US secretary of the navy where, in 1845, he was instrumental in establishing the Naval Academy and in recruiting William Chauvenet to be its mathematics professor.

Meanwhile both Edward Everett and George Ticknor returned from their European jaunts, also having met such literary giants as Goethe and Lord Byron on their travels. Everett resumed his chair at Harvard, but remained for only five years before entering politics. Sandwiched around his return to campus as president of Harvard (1846–1849), he was elected to the US House of Representatives, as governor of Massachusetts, and to the US Senate. He also served as secretary of state and as US minister to Great Britain. Known as a great but longwinded orator, Everett spoke for two hours at the dedication ceremony at the National Cemetery in Gettysburg in 1863, right before Abraham Lincoln's famous, two-minute Gettysburg Address.

Ironically, of the three travelers abroad, it was the one without financial support from Harvard, **George Ticknor** (1791–1871), who purchased books for Thomas Jefferson while in Europe. Ticknor exerted the greatest impact on the College's scholarship during the 16 years he served as chair of French and Spanish. Yet, like the other two, he had little education in mathematics, and no direct influence on the subject. However, circumstantial evidence suggests that he played an important role in the development of the subject. In 1818 he was one of the original investors in the Massachusetts Hospital Life Insurance Company, where he subsequently became one of the vice presidents. Another vice president was Nathaniel Bowditch, a member of Harvard's governing fellows who was in the process of translating Laplace's classical work on celestial mechanics.

Yet a third vice president, and a large investor in the insurance company, was the successful wholesale merchant John Parker. His son, **John Parker Jr.** (1783–1844), had entered the family business at age 22 and also became quite wealthy. Although he had no ties to Harvard, his home near Boston Common situated him within easy walking distance of a small, closely connected group of a few dozen Bostonians of privileged economic means that included Bowditch, Everett, and Ticknor. Perhaps this accounts for a provision in the will that John Parker Jr. drew up, shortly after the death of his father in 1840 had left him with a sizable portion of the $2.2 million estate. From our perspective the important part of John Parker Jr.'s will reads:[5]

> Also at my wife's decease it is my will that the sum of fifty thousand dollars ... shall be paid to the President and Fellows of Harvard College in Cambridge to perform this my will.... To the instruction, education, and maintenance of one or more individuals as they may successively arise, of eminent natural talent or genius for some one or more of the sciences taught in said College ... at home or in foreign countries, for his or their most perfect education ... whose possessors, whether strictly poor or not, are not blessed with pecuniary means adequate to effecting the high state of improvement and advance in science for which they seem to be destined by nature.

Several parts of the will turned out to be important for mathematics. For one, it provided for one or more students of mathematics (as one of the sciences) to benefit from the program. Also, students had to possess potential for advancing their science materially.

Although John Parker Jr. died in 1844, Harvard did not receive his bequest until 1873, when his wife Anna died. The timing was propitious for mathematics, however, because Harvard had just initiated its graduate program the year before at the insistence of President Eliot, over the objections of a large percentage of his professors. Three donations to the college provided support for Harvard graduates in post-undergraduate education: the Harris program (initiated in 1868), Rogers program (in 1869), and Kirkland program (in 1871, based on a donation by George Bancroft). Kirkland fellows were permitted to "repair to a foreign country," a policy based on Bancroft's earlier experiences. This idea resonated with President Eliot in his aim to elevate the faculty's scholarship for the new graduate programs he had initiated.

Eliot appointed a faculty committee to help him draw up conditions for the Parker fellowships. Together they determined that three fellowships could be held at any one time, with $1,000 a year awarded to Harvard graduates of nonprofessional schools for up to three years. Although Parker's will stated that the study could take place at home or abroad, the committee favored students seeking to travel to a foreign country.

William Story graduated in 1871 and then had to use his own funds to travel to Germany for additional study in mathematics and physics, but he returned home in January 1874 after 2-1/2 years abroad without a degree. His timing was serendipitous, because at that point he was able to apply for a Parker fellowship. When awarded the second of these highly competitive fellowships, he used it to return to Leipzig that October, obtaining his PhD the next July for the dissertation "On the algebraic relations existing between the polars of binary quantic." Story thus became the first of numerous success stories in mathematics from recipients of Parker fellowships. B.O. Peirce was next, graduating from Harvard in 1876, winning a Parker fellowship the next year, and earning his Leipzig doctorate in 1879; he spent the rest of his career as a professor at Harvard.

Table 4.6 lists the five Harvard graduates who were awarded Parker fellowships in mathematics in the 1880s in addition to the three mentioned above. Part III details how all eight played prominent roles in the American mathematical research community from that time through the first four decades of the twentieth century. The column

Table 4.6. Harvard graduates awarded Parker fellowships in mathematics up to 1888

Recipient	Year Awarded	Doctorate	Universities
W.E. Story	1874	Leipzig, 1875	Johns Hopkins, Clark
B.O. Peirce	1877	Leipzig, 1879	Harvard
W.I. Stringham	1880	Leipzig, 1882	Berkeley
F.N. Cole	1882	Harvard, 1886	Michigan, Columbia
M.W. Haskell	1885	Göttingen, 1889	Berkeley
A.G. Webster	1886	Berlin, 1890	Clark
W.F. Osgood	1887	Erlangen, 1890	Harvard
M. Bôcher	1888	Göttingen, 1891	Harvard

"Universities" suggests how widely these fellowship holders made major contributions at American institutions located across the United States.

An irony here is that no donations were given to Yale at that time for similar purposes. This meant that E.H. Moore, the "father of American mathematics"[6] and Hubert Newton's best doctoral student at Yale, needed financial help from his mentor to bankroll his postdoctoral study in Germany after he received his PhD in 1885.

There was a stark difference in the sizes of mathematics classes in the United States and Germany in the early 1880s. Klein's class contained almost 100 graduate students, whereas Story was forced to leave the US to study for his doctorate because there was no legitimate graduate department then in existence in America. That situation would change radically by 1900.

Part III

Research Community, 1876–1900

Introduction to Part III

The Thirteen Colonies initiated a well known rebellion in 1776. A lesser known revolution broke out a century later that had nothing to do with government for the people, by the people. This one took place in higher education and was pulled off without one shot being fired. Up until 1876, higher education had pursued an evolutionary trajectory, with each generation building slightly upon the preceding one. The changes were not continuous—there were definite gaps, such as, for instance, the period of ten years following the shot heard round the world fired in 1776. Outside of such externally caused gaps, there was more or less steady, but very slow, progress in growing a mathematical enterprise from colonial times through the first third of the nineteenth century. The rate of progress increased, as we have seen, significantly in the middle of that century, not least because of the efforts of Benjamin Peirce. The year 1876 marked a jump discontinuity and a vastly increased rate of progress. True research communities emerged, starting at Johns Hopkins University and spreading relatively quickly through the US and Canada. By 1900 a national infrastructure and community existed. The next three chapters tell the story of its inception and early construction.

Three leaders of the 1776 revolt were George Washington, Thomas Jefferson, and Benjamin Franklin. Their legacy continues to this day. The trio who led the academic revolution in mathematics a century later—J.J. Sylvester, Felix Klein, and E.H. Moore—are hardly household names today. Yet they were at the forefront of a dramatic change in the educational landscape that began in mathematics and soon extended to all other fields of study. While Washington, Jefferson, and Franklin had their differences, one could not have imagined three more distinct characters than Sylvester, Klein, and Moore. We have already encountered Sylvester (in Chapter 3) when he crossed the Atlantic from his native England in 1841, only to be sent packing in about a year. Thirty-five years later, he was regarded by many as over the hill, his creative powers lost in the distant past, yet Sylvester thrived in the New World, and mathematics in America benefited enormously. Klein was highlighted in the "Transition 1876" section, which described his superior education in his native Germany. How could someone safely ensconced in the remote college town of Göttingen possibly affect mathematics in America? Chapter 5 discusses his enduring legacy on the American mathematical research community into the 1950s. Finally, the relatively obscure American E.H. Moore emerged from the best education the country could offer at Yale to head the best graduate program in America at the University of Chicago.

Part III concentrates mainly on these three luminaries—Sylvester, Klein, and Moore—and the lights they shone on communities that grew up around them, as mathematicians in America advanced from backwater observers to active participants with international aspirations. There was no way to predict such a development beforehand, at least not based on progress up to 1876, no way to project that the time was ripe for such a peaceful revolution. For instance, as late as 1870 Benjamin Peirce had to privately produce lithograph copies of his *Linear Associative Algebras*, which ultimately came to be regarded as one of the most impressive advances in research mathematics before the 1876 revolution. In 1870 J.J. Sylvester was in forced retirement in London, Felix Klein was in the midst of a postdoctoral study tour in Paris, and E.H. Moore was an eight-year-old schoolboy.

Yet these three became the star players in American higher education in mathematics during the last quarter of the nineteenth century. This 25-year period became *the* critical era in the history of mathematics in America. To structure the three distinct developments that formed the revolution, I partition the quarter century into three periods. The first took place 1876–1884, when J.J. Sylvester established a graduate *program* in mathematics at Johns Hopkins University, with a strong undergraduate program underpinning that extraordinary graduate program that produced the first true community of research scholars in America. From 1884 until 1900, the primary influence on American mathematics shifted overseas again, this time to Germany, and especially to Felix Klein, under whom a whole generation of influential twentieth-century American mathematicians received advanced training. The final part of this crucial era began in 1892 with the opening of the University of Chicago, whose mathematics department was led by the estimable E.H. Moore, whom I have called "the father of American mathematics."[7]

Yet, these are not the only developments that occurred during this period. Equally important was the founding of the first successful professional organization of mathematicians, the American Mathematical Society, in 1888. Chapter 7 describes its activities up to the initiation of a mathematical research journal in 1899. That chapter also charts developments in graduate education in mathematics that took place at such American universities as Stanford, Clark, Cornell, and Penn. Since some American students continued to go abroad for advanced studies, the chapter discusses some of their destinations outside Göttingen, notably Sophus Lie's program in Leipzig. While all this was happening, American women began to assert themselves to achieve access to programs for which they were qualified. I describe the accomplishments of ten who were awarded PhDs before 1900 and the institutions that supported them—as well as one who earned that right but was prevented because of her gender. Part III ends with an account of the emergence of statistics courses introduced at some leading American universities over the last two decades of the nineteenth century.

By 1900 universities provided the major employment for research mathematicians. Although few colleges provided private offices for faculty members, even for *the* professor in the department, these institutions featured expanded libraries and attractive seminar rooms. Leading contributors no longer worked in the total isolation endured by the former rugged individualists; rather, they tended to cluster in certain areas of the country. In addition, five publications geared entirely to mathematics provided outlets for research, and are still thriving today. In addition, the American Mathematical Society conducted regular meetings throughout the US, published the transactions of these gatherings, and affected the bifurcation of astronomers and mathematicians, a development that was part of the specialization that all the natural sciences experienced during this critical period in the nation's academic history.

But all was not perfect at the end of this quarter century. Though philanthropists provided generous support for institutions, no such funding would be available for individuals until the twentieth century. Similarly, support for individuals from the federal government was practically nonexistent and would remain that way until the Roaring Twenties.

5

Sylvester, Klein, AMS

Progress in mathematics in America changed dramatically with the opening of Johns Hopkins University in Baltimore in 1876. It featured two striking shifts: the first was a change in the mission of faculty at leading universities from teaching and administering to teaching plus 1) conducting research, 2) writing scholarly papers, and 3) training future researchers. The second striking shift was that the sciences attained equal status with the liberal arts (with modern languages replacing Greek and Latin).

This chapter is divided into three parts. The first is concerned with the Sylvester School at Johns Hopkins, the second with the Klein Klub in Germany, and the third with the founding of the first successful organization of professional mathematicians in America, the American Mathematical Society.

Sylvester School

The 1876 revolution in American mathematics began, innocently enough, in 1867 shortly after the end of the Civil War, when the septuagenarian and financier **Johns Hopkins** (1795–1873) floated the idea of establishing a new kind of university. Though he possessed little formal schooling, the Quaker benefactor developed a passion for higher education that translated into a will that allocated $8 million for education. When this industrial magnate died in 1873, his will bequeathed $7 million (the equivalent of about $143M[1]) to be split between a medical school/hospital and a university. Another $1 million was set aside for public schools in Baltimore. A Hopkins biographer wrote, "An abolitionist and a warm friend of negroes, he included attention to their needs in the hospital and an orphanage."[2] His bequest was the largest to higher education up to that time.

Unhappy with the largely classical education offered by colleges in the US and Canada, the trustees of the will set about establishing a radically different American institution in two ways—its curriculum and faculty. The trustees formed a group of three advisors, each the president of a prestigious university known for recent educational innovations. The trustees asked the advisors pointed questions about the type of school they envisioned—primarily a graduate school with as much emphasis on the sciences and modern languages as the classical liberal arts. This was a radical break

Figure 5.1. Daniel Coit Gilman

with the past—no longer would the primary mission be the molding of the sons of wealthy inhabitants into gentlemen. Sciences were put on an equal footing with liberal arts.

Of equal importance, the expectations for faculty members would change radically. They would teach fewer hours than the traditional load because hiring and subsequent rewards would be based on two interrelated activities: (1) carrying out original research, and (2) training future scholars. An American revolution in education ensued; scholarship reigned supreme. Curiously, only one of the advisors, Cornell's Andrew White, was receptive to the idea; the other two, Charles Eliot of Harvard and James Angell of Michigan, felt that a strong undergraduate program had to be developed as a feeder system first. Notably, Cornell and Michigan had reaped the benefits of the Morrill Act.

Johns Hopkins' trustees proceeded with their plan anyway, hiring as president **Daniel Coit Gilman** (1831–1908), who should be remembered for pioneering contributions to higher education in America. Gilman traced his ancestry back to 1638, when the first family members emigrated from Wales. Educated at Norwich Academy, he attended Yale 1848–1852 when the mathematics professor was A.D. Stanley. In 1853 Gilman traveled to St. Petersburg as an attaché along with a fellow 1852 Yale graduate, Andrew D. White, one of the two founders of Cornell University at the close of the Civil War in 1865. Gilman spent the next two years traveling throughout Europe and studying its educational systems, much like Sylvanus Thayer had done almost 40 years earlier for West Point. When Gilman returned to Yale in 1855, the 24-year old drew up plans for what became the Sheffield Scientific School six years later. Being one of the first educators in the country to fully perceive the significance of the Morrill Act, he made sure that Yale hosted US Representative Justin Morrill, and the college subsequently qualified for funds. In 1872 Gilman took the bold step of moving from the East

Coast to Berkeley, CA, where he accepted the presidency of the nascent University of California even though he had turned down Berkeley's offer two years earlier. Next, this institution's origins as a small college of modest means are described.

University of California, Berkeley. In 1866 the California legislature took advantage of the Morrill Act to charter the Agricultural, Mining, and Mechanical Arts College (**AMMAC**), whose aim was to teach agriculture, mechanical arts, and military tactics "to promote the liberal and practical education of the industrial classes in the several pursuits and professions in life."[3] Scientific and classical studies were not to be excluded but were of secondary importance. The college was confronted with one big problem, however—it had no land. Essentially it was no more than a paper institution with a board of regents.

Serendipitously, the College of California (**CoC**), a private school that had been founded in nearby Oakland in 1855, had land but insufficient funds. It was a perfect match. The founders and early faculty of CoC were mostly Yale graduates who enacted rather formidable requisites from the start. For instance, entrance requirements included knowledge of higher arithmetic and algebra up to quadratic equations, as well as expertise in reading and writing in English, Latin, and Greek. The curriculum was entirely prescribed, with algebra required during the first year (from quadratics to properties of equations) and trigonometry the second (including spherical geometry and spherical trigonometry, as well as the rudiments of surveying). The only mathematically related topic offered beyond that was astronomy. No wonder the first graduating class consisted of but four students. With such minimal numbers, it is also no surprise that CoC found itself seriously in debt.

Therefore, in 1867, after a difficult courtship, the CoC trustees voted to donate their assets to California to create a state university that would blend AMMAC's proposed practical training with CoC's classical course of study. The merged school was modeled after Yale and Harvard "with courses of instruction equal to those of Eastern Colleges."[4] Incidentally, Yale too was a land-grant institution because of its college of forestry.

The regents of the new university asked the CoC trustees to conduct classes during 1868–1869, to allow time to choose a new president and faculty members. Political considerations entered the fray from the very beginning. Civil War stalwart George McClelland was nominated as the first president, but voting adhered strictly along party lines: the twelve Democrats voted "yay" and the five Republicans "nay," but the general declined the offer anyway. The regents then set about conducting a short course in micromanagement by selecting all ten initial faculty members without a presiding officer.

By late summer 1869 faculty members were hired for all disciplines except one— mathematics. Initially, three candidates for the position were considered, but by the time of the regents' August meeting, physics professor John LeConte had added two more names to the mix. LeConte being a friend of Benjamin Peirce, it is not surprising that his two candidates were well known figures. One of them, Simon Newcomb, had graduated from Harvard's Lawrence Scientific School under Peirce and was well established by 1869.

The other, **William Woolsey Johnson** (1841–1927), was a Yale graduate who had taught one year at the US Naval Academy at Newport, RI, before accepting a position at the US Naval Academy in Annapolis, MD, in 1865. He left the Naval Academy five years later, moving to Kenyon College, OH, for two years and then St. John's College in Annapolis for nine. But in 1881 he returned to the US Naval Academy, where he remained for the next 40 years. Over his career, he published several papers and wrote numerous textbooks on mathematics.

Clearly, Newcomb and Johnson were worthy candidates. Yet the tally for the chair of mathematics at Berkeley showed one vote each for Bates and Parker, three for Johnson, and nine for Welcker. Apparently, Newcomb garnered no support. George Bates and William H. Parker are virtually unknown. What about the new chair of mathematics?

William Thomas Welcker (1830–1900) was appointed the first professor in the department of mathematics even though he had never taught a class at any school up to that time. An 1851 graduate of West Point (fourth in his class), the Tennessee native resigned his commission at the outset of the Civil War to engage as a captain in the Confederate Army. At Berkeley, Welcker provided all military instruction, an important consideration because the Morrill Act required land-grant institutions to offer "military tactic." I am unaware of any publication by Welcker, not even problem solving, except the textbook *Advanced Algebra* (1880), which, despite its title, was elementary.

Moreover, the regents also hired an assistant professor of mathematics to accompany Welcker in 1869, **Frank Soulé, Jr.** (1845–1913). A recent West Point graduate (1866) and California native who could trace his American roots back to the Mayflower, he was an acting professor at West Point before coming to Berkeley. Even though the Military Academy was still respected as an engineering school, its position as a leader in mathematics education had diminished significantly by then, so the two appointments are not as stellar as they might otherwise appear. Nonetheless, mathematics was the only department to garner two of the university's ten initial faculty positions. However, just three years later Soulé was appointed to a professorship in astronomy and civil engineering. He remained at Berkeley for the rest of his career, having been selected as the first dean of engineering when that school was inaugurated in 1898.

The University of California opened its doors in September 1869 with 38 students enrolled on the Oakland campus. The CoC's board of trustees had purchased land nearby on a site they named Berkeley in 1866, and this became the permanent location of the new university. Classes began in Berkeley in September 1873, when the university moved into two new buildings, one of which, South Hall, still stands. Enrollment had increased to 191 by then.

Like many colleges in the 1870s, the mathematics curriculum was similar to the earlier one at CoC for the first two years except that algebra, geometry, and trigonometry were squeezed into the first year, while analytic geometry and descriptive geometry were added for the second. A one-year sequence in differential and integral calculus was offered for third-year students, using the textbook *Elements of Differential and Integral Calculus* by Albert E. Church. When this book was first published in 1842, the author wrote in the preface that it was based on his "experience of several years in teaching large Classes in the US Military Academy,"[5] a statement providing yet another connection between the curricula at Berkeley and West Point. Church's text emphasized the techniques of calculus rather than concepts, with the integral being defined as an antiderivative and not as a limit of sums. In contrast with West Point, however, very few applications were included in the presentation.

This book was indicative of the low level of mathematics at Berkeley throughout the 1870s. Several regents were unhappy with this state of affairs and in 1880 formed a Committee on Instruction and Visitation to investigate the status of the entire university, including President John LeConte and the faculty. Included amongst the ten recommendations that the Committee issued was the dismissal of William Welcker, an action that harkened back to his interview for the professorship in 1869. At that time one of the regents, Horatio Stebbins, asked him if he could explain Taylor's theorem. Welcker demurred, saying that he could if given time to study it. Although he ended up being appointed, Stebbins was never satisfied, so when he became chair of the committee, he used his position to have Welcker removed the following year, 1881.

Welcker, nonetheless, exacted a measure of revenge, though not directly on Stebbins. The next year he was elected as California's superintendent of public instruction, thereby defeating the incumbent, who just happened to be one of the regents who supported Welcker's dismissal. Moreover, this position made him an ex-officio member of the Committee on Instruction and Visitation, which enabled him to lead a rearguard action that disbanded the committee in 1883.

It is refreshing to think that a regent, or a member of a board of trustees, would ever question a candidate about a specific result. The way that Stebbins posed the question suggests that he was aware of Church's calculus text or an equivalent because he asked Welcker to "explain" Taylor's theorem. The version in the Church book reads:[6]

> The object of Taylor's theorem is, to explain the manner of developing a function of the algebraic sum of two variables, into a series arranged according to the ascending powers of one of the variables, with coefficients which are functions of the other and dependent also upon the constants which enter the given function.

The online file "Web05-Taylor" examines the "explanation" of this theorem both to understand what this seemingly inscrutable description means and to illustrate how it relates to Taylor's theorem as taught to first-year calculus students today.

One other individual from the early days of University of California, Berkeley, was **George Cunningham Edwards** (1852–1930), who entered the university with the first class in 1869 and was one of the 12 students to graduate four years later. He was then appointed instructor and remained at Berkeley for the rest of his life, retiring in 1918 after 45 years on the faculty. The football stadium on campus is named for him.

However, it is for an entirely different Berkeley tradition that Edwards is singled out. From the time of his appointment he was charged with teaching the freshman college algebra course based on Davies's *Elements of Algebra*, which was based on the translation of the popular Bourdon textbook. The students despised this book so much that they initiated a nighttime funeral procession at the end of the semester in which a coffin was filled with copies of the book and carried around town, ending with a cremation ceremony accompanied by song and oration. A few years later a second coffin was added with copies of the required textbook from the English composition course. The authors of the cremated texts probably appreciated this innovative way of preventing sales of used copies.

Gilman at Johns Hopkins University. Daniel Coit Gilman began his term as president of the University of California, Berkeley, in 1872, one year before the start of classes on the Berkeley campus, but he soon found himself embroiled in politics.

His first years were especially challenging due to the college's severe fiscal problems; a cynic might say that things have not changed very much in the Golden State over the past 150 years. In addition, the president was beset from the outset with political interference from the regents, who engaged in fierce debates over the mission and structure of the university. Gilman soon grew weary of this state of affairs, so when he was offered the presidency of the new Johns Hopkins University (JHU) in 1875, he jumped at the opportunity to move from the left to the right coast. Gilman remained at Hopkins until his retirement in 1902 at age 70. His selection was one of two propitious presidential appointments for mathematics in the US, the other being Harvard President Charles Eliot, the longest-serving president in Harvard's long history and who transformed the provincial college into a preeminent research university, a byproduct of the radical change in higher education about to be described in this chapter.

Gilman's experiences as a fundraiser for the Sheffield Scientific School at Yale and his trips to Europe had imbued him with knowledge of the advantages of advanced training in the sciences and had stoked a strong desire to fashion them on American shores. Looking in his own backyard first, he scoured the country in search of the best professors he could find to head the various departments at Johns Hopkins. Gilman landed three altogether—Henry Rowland in physics, Ira Remsen in chemistry, and Basil Gildersleeve in classics. But the leading American mathematician, Benjamin Peirce, was comfortable at Harvard, so Gilman traveled throughout Europe in search of more experts, crossing England, France, Germany, Switzerland, and Austria before selecting three Englishmen, including Henry Newell Martin in biology and Charles D'Urban Morris as undergraduate professor of classics.

For mathematics, Gilman's leading candidate was **James Joseph Sylvester** (1814–1897), who accepted the offer despite two rather unpleasant experiences in the country 35 years earlier—with a recalcitrant student and an unrequited love. This appointment represents yet another English invasion of American mathematics, following on the coattails of Thomas Harriot and various English textbook authors of the colonial and early Republic days. At first sight, Sylvester might not seem like an appropriate choice for the position; in fact, in the photo of him shown here, the thought of returning to that unruly land seems to be giving him a headache. However, although he was a mathematician with truly impressive international credentials, he had not held a university position for the previous five years due to a forced early retirement from the Royal Military Academy at Woolwich. Besides, at age 61 he was thought to be past his prime as a mathematical researcher. To compound matters, Sylvester was not known to be a very good teacher, had never directed graduate students, had no experience as an administrator, and, perhaps worst of all, was poorly organized in almost all aspects of his life—personal and professional. Further, his prior experience in teaching in America ended unceremoniously after less than one year. On the other hand, Sylvester received strong endorsements from Benjamin Peirce and Joseph Henry (the secretary of the Smithsonian), both of whom urged Gilman to select him for this enviable position.

In retrospect, these seeming negatives indicate that Daniel Gilman deserves the bulk of credit for hiring J.J. Sylvester, despite his own initial reservations, because the appointment turned out to be perfect for Johns Hopkins and particularly beneficial for the history of mathematics in America. It exceeds Harvard's good fortune in hiring John Winthrop and John Farrar and Yale's in bringing Thomas Clap and Jeremiah Day

Figure 5.2. James Joseph "J.J." Sylvester

onboard. President Gilman was certainly a visionary, and his hiring of Sylvester was yet another instance when America profited from religious restrictions in Europe.

Naturally, Sylvester had misgivings about the position, regarding not only the lack of standing in the international community but also the economy. Regarding finance, Sylvester negotiated a contract that stipulated that his salary of $5000 a year, as well as a housing allowance of $1000 a year, be paid in gold, not currency. And, regarding the virtual absence of research-level mathematics in the US and Canada, his Parisian friend Charles Hermite wondered, "Is there really a mathematical future for the New World?"[7] Nonetheless, most of Sylvester's hesitations were based on self-doubts about his ability to create new mathematics. However, by the time he set sail on April 29, 1876, the mercurial Englishman was in good spirits and feeling confident that he could succeed. Upon arriving in New York, he boarded a train for Philadelphia to visit his brother and to view the Centennial Exhibition that had just opened on May 10. This World's Fair, the first major one held in the US, featured no mathematical component, unlike subsequent ones in Chicago and St. Louis, but Sylvester remained in Philadelphia until the end of the month. Then he rode the train to Baltimore, a city that felt familiar to him and was often termed "the Liverpool of the East Coast." However, he did not know one person among the 300,000 inhabitants, so initially the solitary visitor felt a keen social disconnect.

Nonetheless, Sylvester settled into rooms in the elegant Mount Vernon Hotel, located only a few blocks from the university and the city's seat of music, theater, and social life, Mount Vernon Place, and got right to work. The task at hand was daunting—President Gilman expected his professors not only to publish and train future researchers but to inaugurate specialized research journals in their fields, a task that only seemed to heighten his enthusiasm. In the beginning, the university consisted of only two dreary buildings; the trustees had determined to use funds from Hopkins' munificent will on scholars and not on buildings. There was no library, but fortunately the 60,000-volume Peabody Library was nearby and served immediate needs. Library practices were different then—the Peabody was a *reading* library, not a *lending* one.

After a few days of consulting with President Gilman over applications for fellowships from several seemingly competent students, Sylvester was on the train again, this time back to New York. The major reason was an obstacle encountered by many air travelers today—lost luggage. But in Sylvester's case, it was not shoes or socks, which could easily be replaced, but a box of manuscripts that were indispensable for his work. The box had failed to arrive in New York and he was promised it would come on a later ship. It didn't. So, Sylvester set out to track down its whereabouts. He was unsuccessful at the moment but, happily, the box ended up being delivered to Baltimore later in the summer. In the meantime, he had encountered a serious problem that would confront him repeatedly and did not admit an easy resolution.

Having experienced England's weather first hand, President Gilman had already warned his prized mathematics professor about the East Coast's summer heat and humidity, which had struck unseasonably early that year. Sylvester, however, was anxious to begin his duties, so he arrived amidst this suffocating weather instead of waiting until October, as Gilman had recommended. This then provides the second reason why the Englishman accustomed to cooler climes had to depart his adopted city soon after arriving. Now, there is no way New York would afford relief from such oppressive conditions, so Sylvester took to the rails again, this time to visit his old friend Benjamin Peirce in Cambridge. This meeting also allowed Sylvester to conduct his first matter of JHU business—to select an associate. He, along with President Gilman, was searching for someone well schooled in mathematics who could help him overcome his self-admitted administrative liabilities.

The decision for the highly desirable position of associate came down to two of Peirce's former students, William Edward Story and William Byerly, both Harvard graduates in the class of 1871. **William Elwood Byerly** (1849–1935) chose to remain in the US for his graduate education. He thereby earned Harvard's first doctorate in mathematics in 1873, and spent the next three years at Cornell University. The teaching load was rather heavy, however, so Byerly was anxious to land this new post. The section "Transition 1876" recorded that William Edward Story (1850–1930) traveled abroad for his graduate education, earned a doctorate at Leipzig in 1875, and spent the 1875–1876 academic year as a tutor at Harvard. He too was anxious to land the Johns Hopkins position. In fact, a whole generation of Americans was excited by this new university with its radically different mission. It is telling that Benjamin Peirce gave the edge to Story due to the advanced classes and doctorate he had taken in Germany, projecting that Story would become a more accomplished researcher than Byerly. Peirce then arranged for Sylvester to interview Story. Sylvester was duly impressed and recommended him to President Gilman. Story then went south to Baltimore, where his meeting with Daniel Gilman unexpectedly also went south.

Nonetheless, William Story was hired, which ends this part of our story, though certainly *not* the end of our Story.

It was August before Story was hired. Such a late date in the academic calendar seems incomprehensible today, when most faculty openings are filled shortly after the conclusion of the joint AMS-MAA meetings in early January. This, then, is another item in the long list of differences between academic practices now and then. The associate was charged with several tasks, including directing the undergraduate program, teaching all undergraduate courses and some graduate courses, and serving as managing editor of the journal yet to be established. However, not only was Story a

Figure 5.3. William Edward Story

propitious addition to JHU from this perspective, he turned out to play a pivotal role in the subsequent rise of mathematics in America in many ways.[8] In this sense, Sylvester's choice of Story was as good as Gilman's selection of Sylvester.

In the meantime, Sylvester continued his travels. Even the heat in the Boston area proved to be too draining, so he sought a more radical solution—sailing for England on July 22. But he only stayed five weeks before returning to the US for good, or at least until stifling summer weather would force him to vacation in London again. Classes commenced on October 3, but Sylvester did not teach any courses during that first term, preferring to work one-on-one with his new students instead. In fact, initially he thought his professorship required no classroom teaching at all, but President Gilman soon disabused him of that fanciful notion, and Sylvester taught every semester after that.

Sylvester's offspring. Just as Daniel Gilman had predicted, demand for higher education in America was on the rise, with a relatively large number of qualified applicants providing evidence of a groundswell of interest across the country for advanced training in the sciences, including mathematics. Yet, compared to today, the number of students was minuscule. How small? Well, the entering class at JHU numbered seven students in undergraduate classes in mathematics and eight in the graduate program.

Who were these young Americans who sought degrees beyond the bachelor's? What courses did they take? What backgrounds enabled them to succeed? What happened to them after graduation? Here I attempt to answer these questions by introducing those relatively unknown (today) graduate students who were awarded fellowships that covered tuition and fees and, moreover, who were paid the princely sum of $500 per year. (This is equivalent to about $11,400 in 2014. Generally, assistantships today pay $16,000 to $18,000 per year in addition to full tuition and sometimes medical benefits.) While such assistance might seem measly, this was the first time that American students were *paid* to study a subject they loved. What a concept—receiving remuneration for studying something you were passionate about to begin with. What could be better?

The mathematics department was initially allotted two fellowships. Gilman, in consultation with Sylvester and Story, chose Thomas Craig and George Halsted, both of whom turned out to be eminently successful. By the time classes began in late

September 1876, the department was given a third fellowship, which it awarded to Joshua Gore. Sylvester knew there was a jump discontinuity between his first two fellowship holders and Gore, and his doubts turned out to be prophetic, as Gore was the only one of the three fellows *not* to obtain a PhD. Overall, Sylvester seems to have been a very good evaluator of talent.

This was an exciting time to be an American graduate student, especially in the Sylvester School at Johns Hopkins, where students were exposed to research level mathematics on a daily basis. Moreover, the students themselves were inspired to engage in their own creative quest at once, were required to present their results in a seminar, and were expected to publish the best of these results in a journal that the university established. Such a three-pronged approach to academics—discover, present, publish—is standard today, but it was revolutionary in 1876.

First fellows. This section presents snippets of the lives of the first class of fellowship holders in America—Thomas Craig, George Halsted, and Joshua Gore. The online file "Web05-1stFellowsJHU" offers more details on their lives and careers.

Thomas Craig (1855–1900) can be regarded as the first true graduate student in mathematics in America. Although several PhDs in mathematics had been awarded before Johns Hopkins was even founded, not one of those recipients was enrolled in a legitimate *program* of study; rather, they were post-graduates who either worked entirely independently, and hence were awarded a degree for writing published papers, or they were mentored by a dissertation advisor. Craig was different, however.

Thomas Craig was born in Pittston, PA, of parents who hailed from Scotland. Craig attended the Pittston Seminary before entering Lafayette College in September 1871 at age 15. Four years later he earned a bachelor's degree in civil engineering. He was generally a model student, but on one occasion he was removed from campus by the registrar after a fraternity prank.

Craig never pursued a career in civil engineering. Instead, following graduation he moved to Newton, NJ, to become a mathematics teacher at a public high school. He taught mathematics classes to generally disinterested students by day and read mathematics books and journals by night. In this sense, he is also akin to Sylvester, who earlier in his career was an actuary by day but a mathematician by night. Craig did not merely read mathematics books and journals—he *studied* them, indeed, *devoured* them—becoming so accomplished with advanced mathematics subjects that in 1876, just one year removed from Lafayette, J.J. Sylvester chose him to be the first recipient of a graduate fellowship. What is more, President Gilman was so taken with his personal qualities that the new Johns Hopkins president loaned Craig money on a couple of occasions before classes started. Due to knowledge of advanced subjects he had acquired during his year of teaching, Craig taught graduate courses as soon as he entered Johns Hopkins even though he was just an entering graduate student at age 20.

Thomas Craig received his PhD in 1878, just two years after entering the graduate program. To further underscore his independence, he essentially wrote his dissertation himself, "The representation of one surface upon another, and some points in the theory of the curvature of surfaces." Since the subject of this work—differential geometry—was in Story's domain but not Sylvester's, it is likely that the Hopkins associate exerted more influence than the professor in this instance. In any event, Craig's fellowship was extended at Johns Hopkins for one year after graduation, and then he was hired as an instructor. He was only 24 at the time; Sylvester, on the other hand, was 65.

From 1879 to 1881 Craig worked part time at the US Coast and Geodetic Survey in Washington, DC, to supplement his Hopkins salary. During this period, he wrote two small books on hydrodynamics and one major work, *A Treatise on the Mathematical Theory of Perspective*, based on studies of projective geometry he had undertaken from the superintendent of the Coast Survey, Simon Newcomb.

However, Thomas Craig yearned to return to academic life full time. When offered an associate professorship at Hopkins in 1881, he quickly accepted and moved his family to Baltimore. He remained at JHU for the rest of his life, during which time he published numerous papers on the theory of functions (especially theta functions) and linear differential equations. Craig also served as editor of the *American Journal of Mathematics*. He took the position very seriously, making several trips abroad to guarantee a steady flow of articles from the very best European mathematicians, most notably his friend Henri Poincaré, who published two elaborate memoirs in the *AJM*.

Craig taught both undergraduate and graduate courses. Two graduate students he mentored ultimately became leaders in the emerging American mathematical research community. One of them, Edward Van Vleck, gave credit to his time in Craig's classes (1885–1887) with persuading him to pursue mathematics, while Luther P. Eisenhart, Craig's final doctoral student (PhD, 1900), enjoyed an illustrious career at Princeton.

During the year 1898, Thomas Craig began to suffer from health problems that forced him to resign as editor of the *AJM*. He continued to teach, however. On May 8, 1900, he spent the morning in his office at the university before walking home for lunch. Suffering from insomnia during his later years, he retired to his room to take a nap before dinner. His wife Louise found him in bed dead of a heart attack at age 44.

The second student to be awarded a fellowship for the inaugural year turned out to be very influential in the development of mathematics in America at the beginning of the twentieth century. **George Bruce Halsted** (1853–1922) was quite a character, someone you might call a wily Texan based on his deeds—and misdeeds—in Austin. Born in Newark, NJ, he went abroad to study in Berlin for a year after graduating from Princeton in 1875. He then returned to the US and enrolled at Johns Hopkins.

Figure 5.4. George Bruce Halsted

Figure 5.5. Charles Sanders Peirce

Halsted seems to have stirred up a tempest throughout his career, effectively tempering some otherwise significant influence he exerted on important mathematicians as well as on the development of non-Euclidean geometry. One regrettable incident occurred when Halsted, while preparing to teach a course in logic for the spring 1878 term, informed C.S. Peirce that he desired to devote a lecture to him, but claimed Sylvester advised him to refrain. Earlier, Peirce had made inquiries to Daniel Gilman about a permanent faculty position but had been turned down after the president's consultation with specialists within the university. Therefore, upon hearing Halsted's claims, Peirce penned a blistering letter to Sylvester complaining about this treatment. Sylvester instantly appealed to the good graces of President Gilman to mediate the affair since, as Sylvester protested, he had always held C.S. Peirce's work in the highest esteem. This unpleasant affair was completely resolved within a short time and C.S. Peirce became a part-time faculty member from 1879 to 1884.

Though known today primarily as a logician, **Charles Sanders Peirce** (1839–1914) made fundamental contributions to matrix algebra, the four-color problem, non-Euclidean geometry, and probability theory.[9] However, Peirce and Sylvester could both be prickly, and they had a major falling out from the fall of 1882 into the following spring.

The Halsted–Peirce affair did not end auspiciously for its instigator. Halsted delivered his series of lectures on logic during the spring 1878 as part of his doctoral requirements, but they were not well received. Worse, when submitted as a dissertation, they were rejected. It is rather curious, then, that President Gilman, and not Sylvester, determined what Halsted's requirements for the degree would be after that. Gilman gave him the opportunity to write a new thesis but without an extension of his fellowship for a third year. Therefore, Halsted left Johns Hopkins for Princeton, where he taught during the 1878–1879 academic year while writing a new dissertation on dual logic. This led to his being awarded a PhD from Johns Hopkins *in absentia* in 1879 for the dissertation "A basis for a dual logic." He remained at Princeton until 1884, when he was appointed professor and department head at the newly founded University of Texas at Austin. However, his tenure there ended in 1903 after he attacked the university's regents in print.

However, relations between Sylvester and Halsted remained cordial, in fact, symbiotic, both before and after this affair. For instance, Halsted took a reading course in modern algebra under Sylvester in the fall term of 1876, and their discussions sparked a revival of interest within Sylvester in earlier investigations he had carried out on invariant theory. In this case, the student inspired the professor, resulting in several papers in the French journal *Comptes rendus* over the next two years. Conversely, Sylvester exerted a lifetime influence on Halsted, who in the early 1890s recalled, "That the presence of such a man in America was epoch-making is not to be wondered at. His loss to us was a national misfortune."[10]

The remaining fellow from the first Johns Hopkins class did not fare as well as Craig and Halsted. **Joshua Walker Gore** (1852–1908) was born and raised in Frederick County, VA. He attended the University of Richmond (then Richmond College) 1871–1873 before transferring to the University of Virginia, where he obtained a bachelor's degree in civil engineering in 1875. He remained in the Johns Hopkins program until 1878, when he left without the desired degree. Next, Gore became a professor of physics and chemistry at Southwestern Baptist University in Jackson, TN, until 1881, when he was appointed assistant professor of mathematics at the University of Virginia. However, he left Charlottesville after only one year to accept a professorship in physics at the University of North Carolina. When UNC established its new department of applied sciences in 1901, Gore was appointed dean and the department enjoyed considerable growth under his direction.

Early doctorates. Sylvester began teaching his first course on February 19, 1877. He had never taught such a class—students who hungered for learning mathematics for its own sake. This list included the three mathematics fellows and another mathematics student, Fabian Franklin. However, the other four consisted of two engineering fellows and two students pursuing civil engineering. The latter quartet might have been less enthralled with the abstract nature of the course, and none of them obtained a PhD, but the subsequent careers of the two fellowship holders in engineering show just how hard it was to obtain a doctorate in those early years. **Daniel Webster Hering** (1850–1938; Sheffield Scientific College at Yale, class of 1872) was an engineer for railroads before becoming professor of mathematics at Western Maryland College 1880–1884, Western University of Pennsylvania (today the University of Pittsburgh) 1884–1885, and New York University 1885–1916, where he served as dean of the graduate school from 1902 until his retirement in 1916. **Erasmus Darwin Preston** (1851–1906; Cornell, class of 1875) remained at his *alma mater* as an instructor for a year after graduation before being awarded a Johns Hopkins fellowship in engineering for 1876–1878. Upon leaving Johns Hopkins, he moved to Paris for a year to study at the École des Ponts et Chaussées, and then returned to Baltimore to work with the US Coast and Geodetic Survey.

Officially Sylvester's course was titled "Determinants and Higher Algebra," but it was not a conventional class in any sense. His biographer Karen Parshall lightheartedly suggested that the course should have been called "My Current Researches."[11] Sylvester's manner of presentation was not fit for all tastes as he did not use a textbook and did not lecture in traditional style; rather, he went to the chalkboard, wrote out his most recent investigations, and then challenged the students to fill in some gaps and to prove other assertions themselves. This is called *doing* mathematics, not *watching* mathematics, and it is the essence of learning the subject. But this manner of instruction is not for everyone, as evidenced, in this case, by the fact that the two civil engineering students dropped out of the Johns Hopkins class at the end of that semester. And even within mathematics itself, Joshua Gore retained his fellowship the next year but then left without obtaining a degree. Likewise, the two other engineering fellows never received Johns Hopkins doctorates.

Overall, of the eight students that took Sylvester's course, only three were successful. Two have been introduced. The third, **Fabian Franklin** (1853–1939), was awarded a fellowship for his second year based on his outstanding performance in the course. Born in Hungary, his family immigrated to the US two years later, settling in Philadelphia but then moved to Washington, DC, in 1861 at a time when President Abraham Lincoln was dealing with the issue of secession. The precocious Franklin received his education in Washington, including a bachelor's degree from Columbian (now George Washington) University in 1869 at age 16. He worked as a civil engineer and surveyor for the next seven years before entering Johns Hopkins. Like Craig and Halsted, Franklin flourished under Sylvester's quirky teaching style, acting as a human calculator in carrying out various special cases of increasing complexity. In a paper that Sylvester submitted to *Comptes rendus* at the end of the spring semester of 1878 he acknowledged that "thanks to the intelligent cooperation and the great skill as a calculator of Mr. Franklin, one of my students in Baltimore, I am in the position to present to the Academy the table of fundamental invariants and covariants."[12] In a further show of appreciation, Sylvester extended Franklin's fellowship for 1878–1879, and the following year bestowed a PhD upon him for a dissertation titled "Bipunctual coordinates." Fabian Franklin became the first Johns Hopkins graduate in mathematics to have his dissertation published when it appeared in the second issue of the *American Journal of Mathematics*. He had joined the faculty as an associate professor in 1879, and he subsequently rose through the ranks to full professor. However, in a rare turn of events, he left the university in 1895. In fact, he left mathematics altogether for an entirely different life as a newspaper editor.

When the first Johns Hopkins school year ended in May 1877, Sylvester boarded a steamship to cross the Atlantic, having learned his meteorology lesson from the previous summer. When he returned to Baltimore that October, he was anxious to continue his course on determinants and modern algebra, which ran the entire year. This time he had five students, the four fellows (Craig, Halsted, Gore, and Franklin) and one newcomer who, like Franklin, was awarded a fellowship the next year and then continued through to the PhD.

Washington Irving Stringham (1847–1909), who never used his first name, was born in Delevan, NY (about 40 miles southeast of Buffalo). After the Civil War his family moved to Topeka, KS, where he established a house- and sign-painting business. He then attended nearby Washburn College, supporting himself by working in a drugstore and the college library, as well as teaching a course in penmanship. Then he enrolled as a freshman at Harvard in 1873 at age 25. He immediately came under the spell of Benjamin Peirce, who directed his senior honors thesis on applications of quaternions four years later. (That thesis was published the next year in the *Proceedings of the American Academy of Arts and Sciences*.) Stringham spent the summer of 1877 in Paris; I do not know if he had any contact with leading French mathematicians. He

Figure 5.6. Washington Irving Stringham

returned to Harvard in the fall for graduate school but lasted only one semester before transferring to the fledgling graduate program at Johns Hopkins in late 1877. His progress was so rapid that he published two papers in 1879 in the *American Journal of Mathematics*. The following year he became the fourth PhD recipient in mathematics at Johns Hopkins when he completed his thesis, "Regular figures in n-dimensional space," a work that dealt with the six regular figures, or polytopes, that can exist in four-dimensional space. This topic extends the fact, known from the ancient Greeks, that there are precisely five regular solid figures in three-dimensional space: cube, tetrahedron, octahedron, icosahedron, and dodecahedron. Stringham received his 1880 PhD officially under Sylvester but with considerable direction by William Story, whose help he gratefully acknowledged in the published version. His dissertation, like Franklin's, appeared in the *AJM*. There are no extant copies of the dissertations written by these first four Johns Hopkins doctoral students in mathematics, so the two versions that appeared in the *AJM* remain our only records from that time.

William Story too had studied under Benjamin Peirce at Harvard before earning his doctorate in Leipzig in 1875, so it is not coincidental that, upon graduation, Stringham would qualify for a Parker Fellowship to pursue additional studies in the same German city. Serendipitously, Felix Klein (compared with Story in the "Transition 1876" section) had just arrived in Leipzig. Stringham came under this rising star's gravitational pull, becoming Klein's first American doctoral student two years later in 1882. Altogether, four of the five students who took Sylvester's second-year course ultimately earned PhDs, Joshua Gore being the only disappointment. Moreover, the master certainly benefited as much as his students, his unconventional approach to teaching proving to be a fruitful strategy for carrying out his own research. By midsemester he was so enthralled with his progress after another productive night, one that probably began with rum and a cigar, that he posted a note outside the students' room the next morning announcing,

> The attendance of all members of the class of Higher Algebra is especially requested on Friday the 16th at one o'clock as Prof^r Sylvester will on that occasion give his newly-discovered Proof of the hitherto Undemonstrated Fundamental Theorem of Invariants.[13]

As historian of mathematics Judith Grabiner wrote, "Like a nineteenth-century British naturalist who wanted to name, classify, and describe the plant and animal kingdoms, Sylvester wanted to name, classify, and describe the invariants in the mathematical world."[14] Although his boast ultimately did not materialize, it does show how the English scholar benefited from his American students. (All too often mathematicians discover that, when vetting findings in the light of day to neutral observers, or even earlier to themselves, their heartfelt ardor comes crumbling down.)

The graduate program welcomed six new students in October 1878 at the start of the third year, yet the only person awarded a fellowship did not fare so well, which is rather surprising considering his achievements afterward. **Charles Ambrose Van Velzer** (1851–1945) graduated from Cornell in 1876 and remained on the faculty until entering Johns Hopkins. He remained in Sylvester's program for three years (1878–1881) but, for unknown reasons, left without a degree, though later he proved to be a capable and original mathematician. Hillsdale College conferred an honorary PhD on him in 1882. He was one of the first mathematicians to join the American Mathematical Society, having been elected a member of the New York Mathematical Society at its meeting on June 5, 1891.

Van Velzer went directly from Johns Hopkins to the University of Wisconsin, where he published several papers on determinants that most likely were initiated during his time as a graduate student. His contemporary, the mathematician/historian Florian Cajori, wrote, "To Professor Van Velzer belongs the credit of introducing the higher modern mathematics into the University of Wisconsin. ... [His] constant aim was to induce students to do independent work."[15] Van Velzer is thought to have been the thesis advisor for the first two doctoral dissertations at Wisconsin, written in 1897 and 1898.[16] However, in 1906 he developed a coal business interest that Wisconsin president Charles Van Hise charged was a conflict of interest. Van Hise insisted that Van Velzer choose between the business and academics. The mathematician chose to resign. Thereupon, he accepted a professorship at Carthage College (then located in Illinois) and stayed until retiring as professor emeritus. Imagine the scale of the events that transpired from the time Van Velzer was 10, when the Civil War broke out, until his death at the end of WWII.

Fellows up to 1883. So far, six graduate students who were awarded fellowships from the first three classes at Johns Hopkins have been introduced, four of whom earned PhDs under J.J. Sylvester. This section presents the remaining ten students whom Sylvester and Gilman chose for these highly competitive awards over the next four years; only one was awarded in 1880 and three each in the succeeding three years. (The reader might want to consult Table 5.1 on p. 261 to help keep names straight.) Two of the three fellows for 1879–1880 had attended the class on invariant theory that Sylvester had taught the previous year, Oscar Mitchell and Christine Ladd. They exemplify two important aspects of graduate-student life at the time: (1) personal hurdles that aspiring students had to overcome even to qualify to study under the British master, and (2) the hiring of visiting professors to teach specialized courses, in this case the logician C.S. Peirce. The third fellow for 1879–1880 embodies other aspects of collegiate life in mathematics at the turn of the twentieth century.

Oscar Howard Mitchell (1851–1889) played a small but important role in the emergent American mathematical research community. Undoubtedly, his name

would be better known today if pneumonia had not struck him down before he turned
38. Mitchell was born in Knox County, OH; it is tantalizing to ask if he was born in
the small town of Howard located in that county, because that was the first name he
bestowed upon his son, who became a respected mathematician at the University of
Pennsylvania. The elder Mitchell spent his boyhood on a farm, which forced him to
alternate seasons of school and farm chores. Many Americans at the time faced such
obstacles in attempts to do mathematics.[17] As a result, Mitchell was 20 years old by the
time he graduated from high school, though he had taught school in the meantime,
while independently carrying out a self-study program that helped him prepare for
college.

Oscar Mitchell enrolled in 1871 at Marietta College in Ohio and graduated four
years later as salutatorian of a class of 22 students. He remained in Marietta for the next
three years, but there are conflicting reports about whether he served as superintendent
of schools or as principal of Marietta High School. Both positions probably would have
required him to teach classes as well.

When Mitchell enrolled in the graduate program at Johns Hopkins, he performed
so well that he was offered a fellowship the next year. During his time at JHU (1878–
1882) the faculty grew to four—Sylvester (who occupied room 14 in University Building), Story (room 15), Franklin (room 16), and Craig (room 17). This expanded number
of faculty members was particularly sizable for the time; it meant that mathematically
rugged individualists were no longer isolated. In addition, new graduate students were
exposed to a much broader list of courses. Yet the course that had the greatest effect
on Mitchell was Sylvester's number theory class (1879–1880) an experience that served
as the inspiration for two papers he wrote under Sylvester's direction, one of which
resulted in his 1882 PhD thesis with the generic title, "Some theorems in numbers."

An analysis of Mitchell's work reveals a promising researcher with an optimistic
future who was mainly unfulfilled because the American educational landscape remained unprepared for researchers for another 20 years. Therefore, upon receiving
his doctorate, he returned to Marietta College, a select liberal arts institution, where
he endured a heavy teaching load that all but prevented serious research for the rest of
his life.

The person who best illustrates the two aspects of graduate-student life mentioned
above—overcoming personal hurdles and the hiring of visiting professors to teach specialized courses—was perhaps the best student Johns Hopkins produced during Sylvester's 7-1/2 years in Baltimore. Christine Ladd (1847–1930) appeared in Chapter 4 as
our last rugged individualist who contributed a couple of papers plus numerous problems and solutions to the journal *The Analyst*. Her persistence to achieve success in
science in the face of truly formidable obstacles can serve as an inspiration for twenty-
first-century female mathematics students.[18] Her mother was a women's rights activist
who often took her impressionable daughter with her to lectures on women's rights,
but her mother died when she was 12. Nonetheless, the family paid for her to attend
Wesleyan Academy and Vassar College, from which she graduated in 1869.

Christine Ladd taught in high schools for the next nine years, but by 1876 the
monotony of teaching lower-level subjects year after year had diminished her interest
in teaching, at which point she sought a radical change by applying to Harvard. She
was permitted to attend classes informally at the all-male bastion with the help of
two members of the mathematics faculty. One was William Byerly, who was known

to support women's causes at Radcliffe, then known as Harvard's "Female Annex;" indeed, Byerly Hall on the Radcliffe campus is named for him.[19] The other was **James Mills Peirce** (1834–1906), another of Benjamin Peirce's sons who—like his father—taught for 49 years at Harvard. Despite the support of such seemingly prominent figures, Ladd was unable to gain formal entry into the program, so after two years she decided to try her luck at Johns Hopkins with the hope of better success at a brand new university. She was wrong—partially.

In March 1878 Ladd wrote directly to Sylvester, laying out her credentials and inquiring if she could attend his lectures the next year: "Will you kindly tell me whether the Johns Hopkins University will refuse to permit it on account of my sex?"[20] Wanting to confirm her qualifications before presenting her case to President Gilman and the board of trustees, Sylvester directed Fabian Franklin, then a fellow, to do a literature search of her publications. The trustees deliberated throughout the next month, not only on Ladd's application but the broader issue of co-education. The trustees decided against admitting women but, as President Daniel Gilman informed her on April 26, "As this is an exceptional recognition of your mathematical scholarship, no charge will be made for tuition & your name will not be enrolled on the annual Register."[21] As one might imagine, Ladd excelled in the number theory course and, as a result, the restriction was loosened after that, allowing her to attend classes taught by both William Story and C.S. Peirce. Moreover, the next year Sylvester persuaded the university to award her a fellowship, which she held 1879–1882. During this entire period, however, she was neither entitled to be called a fellow nor was her name ever recorded on the list of students. In short, Christine Ladd was a Johns Hopkins student—formally—but an illegitimate one.

The enigmatic C.S. Peirce had joined the Hopkins faculty as a lecturer in logic in 1879 and offered a course in his specialty at once. Ladd was in his first class and took to his brand of logic. Within three years she completed a work that should have earned a PhD degree but since she had never been officially admitted to Hopkins, she was not entitled to graduate. Thus, she was not granted a doctorate at the time. That work, titled "On the algebra of logic," appeared as one of several essays in an 1883 book edited by C.S. Peirce but not in the university-sponsored *American Journal of Mathematics*. Ladd did publish three other papers in that journal, however.

The first time Fabian Franklin heard about Christine Ladd was when Sylvester assigned him the task of conducting a search of her work in 1878. His respect for her as a mathematician sprouted over the next year, but it blossomed in the number theory course that Sylvester taught to eight students, including Franklin and Ladd, during 1879–1880. It has been said that electricity crackled in Sylvester's classroom, but nonmathematical sparks seemed to fly in this one as well. One wonders if amicable numbers were studied, because the result was the marriage of Franklin and Ladd on August 24, 1882, after which time she adopted the name Christine Ladd Franklin in all her publications. The Franklins thus became America's first mathematical couple.

Ladd Franklin eventually was awarded a PhD in mathematics, but even that did not come easily. When Johns Hopkins was celebrating its semicentennial in 1926, she was offered an honorary degree. However, instead of feeling honored, she must have felt insulted, because the degree was offered because of her work in physiological optics. The persuasive Ladd Franklin argued that it was the PhD she had earned for the dissertation on logic 44 years earlier. Johns Hopkins agreed. And thus, although

Christine Ladd Franklin had been the first American woman to qualify for a PhD in mathematics, she became the oldest person ever to receive that degree—she was 78 at the time.

The remainder of this section presents the remaining eight fellows from Johns Hopkins while J.J. Sylvester was head of the mathematics department 1876–1883. (The online file "Web05-SylSons" contains additional information on all eight.) While the lives of Oscar Mitchell and Christine Ladd reveal some of the hurdles American students faced in order to pursue research-level mathematics in the last quarter of the twentieth century, other aspects of the life of the third member of their entering JHU class tell us something about the undergraduate experience a century ago.

Robert Woodworth Prentiss (1857–1913) graduated from Rutgers University in 1878. Just four years earlier, the university had initiated intercollegiate mathematics contests that were unlike any today, being conducted in a debating format. Prentiss's timing was propitious, as he took second place as a senior, one year before the university halted the competition. He was awarded a Johns Hopkins fellowship in 1879 but left two years later without obtaining a doctorate for a position at the US Nautical Almanac Office in Washington, DC. He joined the faculty at Rutgers in 1891. At the time, Rutgers undergraduate students could choose between two curricula—scientific or classical. But in 1907 all courses were organized into departments and became numbered for the first time.

The mathematics department, which had been formally organized a year in advance of the university-wide reorganization, offered eight courses in each of its three terms. The department remained in room 204 of the engineering building (Murray Hall today) until 1946, when it moved into the "Mathematics House." This remained its home until 1959, when it moved into today's DeWitt Hall, named after Rutgers' first math major, **Simeon DeWitt** (1756–1834), whose own education during the Revolutionary War was abruptly halted when British troops burned down the college buildings and dispersed the students.

Johns Hopkins awarded only one fellowship in 1880, to **Herbert Mills Perry** (1855–1898). Perry's father died when he was only seven years old, but his mother took charge of his early education. Consequently, Perry attended two prestigious preparatory schools in New Hampshire, Appleton Academy and Phillips Academy, before entering Harvard in 1876. An account of his class of 1880 stated:[22]

> At the close of his sophomore year he received "highest second-year honors" in mathematics, and he graduated "*magna cum laude*" with "final honors" in Mathematics.

The aging Benjamin Peirce then supported Perry's application for a fellowship at JHU. Perry held his fellowship 1880–1882 before an unstated illness forced him to withdraw. Despite this obstacle, he attempted to teach at two prestigious schools—the Cascadilla School in Ithaca, NY, and the University School at the University of Chicago—but he was limited to one year at each because of his illness and never held another position.

My information on two of the six Johns Hopkins fellows from 1881 and 1882 is also similarly incomplete. **George Stetson Ely** (1856–1917) was born in Fredonia, NY, where he attended the Normal School. Then he attended Amherst College in MA, where he "took every possible Mathematics prize and he was often called 'Mathematics Ely'."[23] He graduated from Amherst in 1878 and held his Johns Hopkins fellowship for two years, 1881–1883. George Ely took Sylvester's course in number theory during his first year along with Oscar Mitchell and Christine Ladd, becoming so smitten with the material that, by his second year, he had compiled a complete bibliography of nineteenth century works on Bernoulli's numbers. Preparation of this impressive document suggested an unsolved problem that became the topic for his dissertation. Titled simply "Bernoulli's numbers," the solution of the main problem has been judged "to be little more than a clever exercise."[24] Upon receiving his PhD in 1883, Ely accepted a professorship of mathematics at Buchtel College (now the University of Akron in Ohio) where he stayed for only one year before moving to the US Patent Office for the rest of his life. His older brother **Richard Theodore Ely** (1854–1943) was a renowned economist who was professor and head of the department of political economy at Johns Hopkins 1881–1892.

My information about **Gustav Bissing** (1862–1929) is even more sparse. He is the only student to transition from the undergraduate to the graduate program at Johns Hopkins, receiving his bachelor's degree in 1882 and then being awarded a fellowship for two years, 1882–1884. Although the university was virtually constructed as a graduate school, the expectation was that the undergraduate program would serve as a feeder for the advanced program. Obviously, that did not happen in mathematics. However, the one student who did successfully navigate the transition, Gustav Bissing, was able to earn a PhD, though he did not complete his dissertation until 1885, after Sylvester's departure from the university (and the US). It was titled "Notes on Gauss's coordinates and Steiner's quartic surface." Like Richard Ely, Gustav Bissing became an assistant examiner in the US Patent Office, 1884–1897. In 1912, he returned from Europe to join his brother in a New York City law practice specializing in patent law.

Altogether four of the six students awarded fellowships starting in 1881 and 1882 received PhDs at Johns Hopkins, another earned a DSc at an unnamed institution, while the sixth was certainly capable enough to have gotten one based on his accomplishments during and after his student days. More information is available on two of the other fellowship holders to paint pictures of their careers.

William Pitt Durfee (1855–1941) was born in Livonia, MI, and graduated from the University of Michigan at age 21. He then taught at the Berkeley Gymnasium 1877–1881 before being awarded a Johns Hopkins fellowship in the fall of 1881. Durfee took Sylvester's one-semester number theory class, whose seven students included also Oscar Mitchell, Christine Ladd, and George Ely. Durfee's description of Sylvester's inimitable manner of teaching serves to reinforce the essence of an unconventional yet extremely effective style:[25]

> Sylvester began to lecture on the Theory of Numbers, and promised
> to follow Lejeune–Dirichlet's book; he did so for, perhaps, six or eight
> lectures, when some discussion which came up led him off, and ... af-
> ter some further interpolations [he] was led to the consideration of his
> Universal Algebra, and never finished any of the previous subjects.

Durfee added that Sylvester's students "never received a systematic course of lectures," but he concluded that they "were greatly the gainers thereby." Durfee surely benefited, receiving a PhD in 1883 for a dissertation titled simply "Symmetric functions." One year later, he was appointed professor and chair of mathematics at Hobart College in Geneva, NY, a position he held until his retirement in 1929. After only four years at Hobart, he was appointed dean of the faculty, the first person to hold that position at an American liberal arts college. He served in this administrative position until 1925, and was the college's acting president on four different occasions.

His son, **Walter Hetherington Durfee** (1889–1974) received his Cornell PhD in 1930 and, like his father, became an important figure at Hobart College. Between them, *père* William taught mathematics for 45 years, was dean for 37 years, and served as acting president on four occasions, while *frère* Walter taught mathematics for 37 years, served as dean for many years, and filled in as acting president. Moreover, Walter's son **William Hetherington Durfee** (1915–2001) also obtained a PhD from Cornell in 1944. He taught at Dartmouth College and then was a mathematician at the US Bureau of Standards and the Operations Research Office in Washington, DC, before becoming professor of mathematics at Mt. Holyoke College in Massachusetts from 1955 until his retirement in 1980. Finally, to extend this line to a fourth generation, his son, **Alan Hetherington Durfee** became the third member of the family to receive a PhD from Cornell (in 1971).

Ellery Williams Davis (1857–1918) was another able mathematician/administrator to graduate from Johns Hopkins with a PhD. Born in Oconomowoc, WI, Davis received his BS degree from the University of Wisconsin in 1879, and entered Johns Hopkins two years later. Like several other students in the department, he was awarded a fellowship after gaining distinction in class that year and held it 1882–1884, culminating in his doctorate based on the dissertation, "Parametric representations of curves." Apparently, the dissertation was directed mainly by William Story; the degree was awarded six months after Sylvester's departure for England. Upon obtaining his doctorate, Davis accepted a professorship of mathematics and military tactics at the University of Florida (then Lake City Agricultural College) and later moved to the University of South Carolina, but he became disillusioned here too when unable to initiate a graduate program. Finally, in 1893, Davis landed at the University of Nebraska, where he spent the rest of his career. In 1901 he became dean of the college of arts and sciences, combining that position with the one in mathematics until his death in 1918.

Neither of the two remaining Johns Hopkins fellows from 1881–1882 ever received a PhD in mathematics. However, **Archibald Lamont Daniels** (1849–1918) came close. Born in Hudson, MI, he graduated from the University of Michigan in 1876. Daniels then studied at Göttingen and Berlin for six years before returning to the US, enrolling at Johns Hopkins, and being awarded a fellowship for 1883–1884. Daniels left Hopkins in 1884 without obtaining a PhD, and taught at the Chicago Manual Technology School and then at Princeton, which awarded him a DSc degree in 1886. This degree must have been honorary because Princeton did not award PhD degrees in mathematics at that time. Daniels accepted a professorship in mathematics and physics at the University of Vermont in 1886 and remained in Burlington for the rest of his life. The eldest of his six children, Archibald Lamont Daniels, Jr., earned a Yale PhD in 1912 in mathematical astronomy with a dissertation directed by Ernest Brown.

Arthur Stafford Hathaway (1855–1934) was as competent as others who held Johns Hopkins fellowships and his subsequent publications were better than most, yet he never obtained a PhD. Why? The answer highlights an important distinction between the meaning of an academic degree up to about 1925 and today. Hathaway was born in Keeler, MI, and attended Cornell, where he left an indelible mark both during his undergraduate days and later as a professor. Apparently, he invented his own shorthand method which only he could read, a skill he put to good use on several different occasions. The first that is relevant here is when he served as the personal secretary for Cornell's first president, Andrew D. White. Hathaway was also a court stenographer and used this skill when lecturers came to the university. Moreover, he showed mathematical prowess while an undergraduate student by publishing one small paper and solving a few problems in *The Analyst.*

Arthur Hathaway married Susan Hoxie (the first woman agricultural student at Cornell) in December of his senior year. After he graduated from Cornell in 1879, the newlyweds moved to Baltimore, where he was an instructor at Friends High School while also attending lectures given by J.J. Sylvester. Sadly, his wife died in March 1880 during childbirth (the son died the next day), a misfortune that exacted a heavy toll on the fledgling mathematician. Based on his achievement in Sylvester's classes, he was awarded a fellowship for the following year and held it 1881–1883, making great strides the whole time. Nevertheless, the death of his wife prompted him to leave Johns Hopkins without a degree. In the meantime, Hathaway wrote up the results he had obtained while at Johns Hopkins and published them in an 1884 paper. The opening paragraph of that article, "Some papers on the theory of numbers," is quite touching:[26]

> The principles upon which the following papers are founded were developed while attending the Lectures of Professor Sylvester on the Theory of Numbers, in 1879–80; and were presented before the Mathematical Seminary in May, 1880, Jan., Feb., March, 1881. The death of my wife, who had shared with me in the work, and the professional duties of a stenographer, left neither opportunity nor inclination to make a more extended publication of them. Feeling confident, however, that the principles were of value in simplifying the Theory of Numbers, I returned to the subject in the beginning of the present year (1884). It is due to Professor Sylvester to state that the beginnings of my knowledge in the Theory of Numbers were obtained entirely from short-hand notes of his lectures; and that it was his suggestive presentation of the Theory of Congruences that led to the development of these principles.

That paper ended with "To be continued." Although Hathaway published another paper on number theory a few years later, it bore no relation to this one.

Hathaway's life turned a corner when he remarried on September 8, 1885, shortly before assuming duties as an instructor in mathematics at Cornell. He remained at his *alma mater* for six years before leaving in 1891 to become professor and head of the department at the Rose Polytechnic Institute (today Rose-Hulman Institute of Technology in IN). It seems clear that he went to Rose through his direct connection back in Cornell with Henry Eddy, who had become president of the Institute. Hathaway retired from Rose-Hulman in 1920 upon reaching age 65.

This completes coverage of the 16 Americans awarded fellowships at Johns Hopkins (1876–1883) while J.J. Sylvester was head of the mathematics department. Table 5.1 assembles relevant information on all the fellows during this period. "Year" refers to the first year when the fellowship was awarded, not when the fellow first began taking courses in the program. Names of the eleven fellowship holders who received PhDs at Johns Hopkins are designated in bold print. These were the only mathematics students to receive doctorates during this period; I do not know when the first degree was awarded to a student who did not have a university fellowship.

Table 5.1. Johns Hopkins fellows 1876–1882

Year	Fellows		
1876	**Craig**	**Halsted**	Gore
1877	**Franklin**		
1878	**Stringham**	Van Velzer	
1879	**Mitchell**	**Ladd**	Prentiss
1880	Perry		
1881	**Durfee**	Hathaway	**Ely**
1882	Daniels	**Davis**	**Bissing**

Curriculum. Sylvester and Story constructed a program of study for undergraduate and graduate students that was the very best the country had ever seen. Story did the bulk of the teaching, three courses a semester, a load that itself was considerably lighter than the standard 15–20 hours per week at most American colleges and universities. Sylvester taught only one course per semester, rather light even by today's standards. The first fellow, Thomas Craig, also taught graduate courses because of the breadth and depth of his personal reading program. Fabian Franklin was similar. All other fellows except Ladd taught undergraduate courses.

The graduate program at Johns Hopkins ran deep, though it lacked breadth. Except for the first semester during Sylvester's Johns Hopkins tenure (1876–1883), he taught a graduate course each semester, always in his inimitable style. These included one-year offerings on the theory of numbers, and on determinants and higher algebra, as well as one-semester courses on the theory of substitutions, the theory of partitions, and the algebra of multiple quantities. Story, Craig, and Franklin taught the remaining graduate courses. Additional expertise at the graduate level was supplied by three visiting professors: Charles Sanders Peirce from the US Coast and Geodetic Survey, Simon Newcomb from the Nautical Almanac Office, and Sylvester's "invariant twin," Arthur Cayley, from England.[27]

Story oversaw an undergraduate program that included Story, Franklin, Craig, and the fellows. The program was totally prescribed, with students taking a one-year course each year according to this schedule: freshman (conic sections), sophomore (differential and integral calculus), junior (differential equations), senior (theory of equations and determinants). Two one-semester courses supplemented these offerings: higher plane curves and solid analytic geometry. This curriculum was far superior to any mathematics program offered in the country up to that point. With very few modifications, it became the standard for most American colleges up to about 1950.

Seminar. Sylvester did much more than write papers and train graduate students. One of his most notable contributions to American mathematics was the introduction of a monthly seminar modeled after the German prototype whereby graduate students reported on recent works written by other internationally recognized experts or, on special occasions, described progress on their own research. These meetings were held in a room with the quaint name "Seminary," which served as part lounge (where mathematicians and students gathered informally to discuss whatever topic happened to be fresh on their mind) and part departmental library (the walls were lined with journals, books, and physical models of surfaces). The university did not own its own library, so this room was vital for supplying students with appropriate books and journals.

The Seminary was also where seminars were conducted; the term "seminary" referred both to the room and to the lecture. It was in such seminars held in the Seminary that graduate students were expected to deliver lectures on their research, a practice that kept them quite busy. They printed synopses of their results in *University Circulars* that were established in December 1879. President Daniel Coit Gilman described the history and evolving nature of this publication as follows:[28]

> Since the opening in 1876, it has been found convenient to issue frequent statements in respect to the development of our plans. … Some of the features of the *University Reporter* of Cambridge, England, were introduced, particularly current announcements of lectures and courses of instruction. These bulletins were originally designed for the information of those who were connected with this university, or who wished to enter it; but they soon proved to be a convenient channel for acquainting the public at large with the condition of the foundation, and for spreading information in respect to scientific and literary investigations which were here in progress. Gradually the reports of society meetings, and the synopses of scientific papers claimed more and more space, and a demand for the successive circulars was indicated far beyond the company for which the original publication had been designed; and even to a limited extent in foreign countries. It is now thought best to modify in some particulars the plan of the circulars, so as to give still more space to the original work of the professors in the university and their more advanced students, without omitting the current information in respect to the plans of instruction, the rolls of attendance, and other purely local matters.

President Gilman added that all Johns Hopkins publications were "offered to subscribers, and are also the basis of important exchanges with foreign institutions and journals, as well as with many in this country." In short, the aim of the *Circulars* was to publicize the research accomplishments of the Johns Hopkins faculty and graduate students both at home and abroad.

Overall, Johns Hopkins published on average five *Circulars* a year. The practice of having graduate students present their findings and then writing them up for publication in such an unofficial outlet as the *Circulars* was new to America. Ultimately, this two-step procedure trained graduate students in the vital twofold skills of lecturing and expository writing.

To illustrate Gilman's plan to provide space for the original work of John Hopkins professors and advanced students, the December 1884 issue of the *Circulars* reprinted a notice from the popular journal *Science* about a series of lectures given at Johns Hopkins two months earlier by Sir William Thompson titled "Molecular Dynamics." The 60-year old Irish mathematical physicist, better known as Lord Kelvin (1824–1907), was regarded as one of the best scientists in the world, so his series of lectures on American soil were highly anticipated, drawing an audience of 21 listeners for 20 lectures delivered over a three-week period. But from our perspective, the announcement that 300 copies of Lord Kelvin's lectures were offered for sale by the university are equally important for what they indicate about the far-reaching abilities of Sylvester's mathematical "sons":[29]

> Stenographic notes were taken by Mr. A.S. Hathaway, ... lately a Mathematical Fellow of the Johns Hopkins University, and these notes (with additions subsequently made by the lecturer) have been carefully reproduced by the Papyrograph Plate Process. A bibliography of the subjects considered will also be given with the lectures. In all there will be about 360 pages.

The announcement then states that copies could be purchased at Johns Hopkins or from publishers in Berlin, Paris, and London.

What is a papyrograph plate process? Fortunately, Arthur Hathaway provided an explanation for those of us whose memories do not extend that far back in time:[30]

> As the papyrograph process which was employed for printing the lectures does not seem to be understood, a brief explanation of it seems desirable. A transparent paper plate is used; the copyist writes upon this with a chemical ink which is almost colorless, using a very fine steel pen. The work of the copyist must be revised and corrected before the ink eats through so as to form a stencil of the writing. When the stencil is formed, a prepared purple ink which flows steadily is forced by means of an impression pad through the stencil. In printing the lectures, about 300 impressions were obtained from each stencil before it began to "break" so as to be useless. The nature of the process, and the fact that the work had to go through the hands of a copyist, will explain, I think, why so many errors have occurred which to a mathematician are virtually *typographical* errors. The reader of the lectures will also see how pages 210, 252 were thus readily traced from Sir Wm. Thomson's handwriting by means of the transparent stencil sheet. [Signed] A.S.H.

How was Hathaway able to spend so much time with this project, having already resigned his fellowship? Once again, the *Circulars* provide an explanation, listing him as a "Fellow by Courtesy" for 1884–1885, meaning he was essentially what is called a visiting professor today. Former classmates Robert Prentiss and Gustav Bissing were also included on the list, but the latter resigned after the first semester to accept a position at the US Patent Office.

The importance of Lord Kelvin's lecture series was underscored when one of the two July 1885 issues of the *Circulars* reprinted a three-part report on the series by the English physicist George Forbes in the journal *Nature*. The initial and final parts of the report shed light on the marvelous work carried out by former Johns Hopkins fellow Arthur Hathaway and highlight the continuing interaction among American and British scientists:[31]

> [The lectures] show how much the author's attention was directed to the subject during his three months' sojourn in America. The audience at the Baltimore lectures consisted chiefly of American professors, and a few English men of science attended a larger or smaller

> number of the lectures. ... The course of twenty lectures was con-
> fined to the wave theory of light, largely dealing with the difficul-
> ties of that theory. The published lectures are not printed, but "jel-
> ligraphed," as Sir William Thomson would say. The number of copies
> is extremely limited, and are of unique interest, being reproduced
> from the short-hand notes taken at the lectures. Every one who knows
> how suggestive Sir William's lectures are and how fertile his mind is
> in bringing illustrative digressions to bear on the topic in hand, will
> expect these verbatim notes to be a rare treasure. Nor will he be disap-
> pointed. Mr. Hathaway, the reporter, has the unusual combination
> of being an expert stenographist, a skillful mathematician, and a clear
> and distinct calligraphist. His notes contain numerous errors, such as
> are unavoidable in such an undertaking, but, viewed as a whole, his
> work is almost a marvel.

Lord Kelvin's method of lecturing resembled the way J.J. Sylvester moved about his mathematics classroom to actively involve the audience:[32]

> Sir William went among his audience and had some conversation
> with them. It was ever his object to discard the professorial attitude
> and give his lectures the aspect of conferences. Discussion did not
> end in the lecture-room, and the three weeks at Baltimore were like
> one long conference guided by the master mind. It is not surprising
> that at the end of that time there was a genuine feeling of sadness at
> parting on the part of teacher and taught alike.

To emphasize the growing bond on both sides of the Atlantic, Forbes quoted Lord Kelvin's final remarks.[33]

> I am exceedingly sorry that our twenty-one coefficients [audience
> members] are to be scattered, but, though scattered far and wide, I
> hope we will still be coefficients working together for the great cause
> we are all so much interested in. I would be most happy to look for-
> ward to another conference, and the one damper to that happiness is
> that this one is now to end, and we shall be compelled to look forward
> for a time—I hope only for a time, and that we shall all meet again
> in some such way. I would say to those whose homes are on this
> side of the Atlantic, "Come on the other side and I will welcome you
> heartily, and we may have more conferences." Whether we have such
> a conference on this side or on the other side of the Atlantic again it
> will be a thing to look forward to—as this is to look back upon—as
> one of the most precious incidents I can possibly have. I suppose we
> must say farewell!

American Journal of Mathematics. In an 1876 article comparing the dreary state of mathematics in America with advances in Germany, France, and England, Simon Newcomb bemoaned the lack of an appropriate vehicle for publishing results in the field:[34]

> Not only is our scientific literature of every kind meagre in the ex-
> treme, but the facilities for the publication of any kind are extremely

> restricted, and have increased but little during the last fifty years. Sil-
> liman's Journal is now, as it was half a century ago, the solitary stan-
> dard journal of pure science published in the country.

Newcomb's lament was addressed with the founding of a new publication in 1878, the *American Journal of Mathematics*. This enterprise was the first to be sponsored by a university (Johns Hopkins), and it was the first journal in America to publish high-quality research results. Recall that president Daniel Gilman expected his professors to inaugurate specialized research journals in their fields, and the mathematics initiative was the first and, arguably, has been the most successful. The following year, 1879, saw the establishment of journals in two other JHU departments, the *American Chemical Journal* (edited by chemistry professor Ira Remsen, who ultimately succeeded Gilman as president) and *Studies from the Biological Laboratory* (biology professor Henry Newell Martin edited the latter, but it folded in 1893; the graduate program that Martin instituted was the first in biology in the country). After that, Johns Hopkins inaugurated the *American Journal of Philology* (edited by classics professor Basil Gildersleeve) in 1880, *Studies in Historical and Political Science* (edited by history professor Henry B. Adams) in 1882, *Memoirs from the Biological Laboratory* in 1887, and *Contributions to Assyriology and Comparative Semitic Philology* in 1889. The university did *not* sponsor a physics journal, perhaps because physics professor Henry Rowland published his findings in the mathematics journal.

The *American Journal of Mathematics* was the first serious American periodical aimed specifically at disseminating mathematical research, although initially it included a few expository articles. Fully aware of America's pitiful record for sustaining such publications, Sylvester entertained two serious reservations initially about initiating it: whether enough Americans would be able to contribute articles, and whether a sufficient number of individuals and libraries would subscribe. The response overwhelmed him on both accounts.

The first issue carried the expanded title *American Journal of Mathematics: Pure and Applied*, with J.J. Sylvester as editor in chief and William Story as associate editor in charge. The front cover added, "With the co-operation of Benjamin Peirce (in mechanics), Simon Newcomb (in astronomy), and Henry Rowland (in physics)." The first issue (formally called a *number*) of the *American Journal of Mathematics* (**AJM**) was printed in Baltimore, with distributors in New York, Boston, Philadelphia, Berlin, and London. This was quite an undertaking for the fledgling university.

That the *AJM* was aimed at being a very high-level publication was made clear by the "Notice to the reader." Signed by William Story, it began,

> In presenting to the Public this first number of the American Journal
> of Mathematics, the editors think it advisable, in order to prevent
> disappointment on the part of subscribers and contributors, to state
> briefly the principles by which they will be guided in its management.
> Although in the first instance designed to supply a want, as a medium
> of communication between American mathematicians, its pages will
> always be open to contributions from abroad, and promises of support
> from various foreign mathematicians of eminence have already been
> received. The publication of original investigations is the primary
> object of the Journal. ... The Journal will be published in volumes of

> about 384 quarto pages, each volume appearing in four numbers at
> periods not absolutely fixed, but ... by intervals of three months.

To underscore the difference between the mission of the *AJM* and others then in existence, Story cautioned, "It is to be understood that there will be no problem department in the Journal, but important remarks, however brief, may be inserted as notes." To emphasize this point, he advised "Persons desirous of offering to the Public mathematical problems for solution ... to send them to 'The Analyst' ... or to 'The Mathematical Visitor'."

The new Johns Hopkins journal was revolutionary. It was different than anything the country had ever seen. Volume 1, Number 1 exhibited all aspects of Story's notice. It began with a four-page note by Simon Newcomb on transformations that surfaces could undergo in space of more than three dimensions. It also included articles by two foreign mathematicians, Guido Weichold (who would earn a doctorate under Felix Klein in 1883) and Arthur Cayley, who exposed his American audience to group theory at an early stage in the area destined to become America's first specialty. The rest of the issue consisted of papers by Americans. The first installment of George Hill's pioneering work on lunar theory elicited praise from Henri Poincaré and thus helped to validate the *AJM* right from the start. In addition, the first issue contained a review by C.S. Peirce of an Italian book on the method of least squares, and articles on physics by Henry Eddy and Henry Rowland. There was also an important paper by J.J. Sylvester, "On an application of the new atomic theory to the graphical representation of the invariants and covariants of binary quantics," which constituted 40 of the 104 pages.

As Story wrote, four numbers would constitute each volume of the *AJM*, but this arrangement only lasted until Volume 4. The title page for this volume lists only Sylvester as editor but not Story, the result of a deep rift that occurred between the professor and his associate. Only in Volume 6 was a replacement for Story brought on board—Thomas Craig—but that volume consisted of just one number, as had Volumes 4 and 5. Only with Volume 7 did the journal reappear with four issues, and it has been published quarterly since then, right on up to the present time. In the year 2016 the *American Journal of Mathematics* published its 138th volume and was still headquartered at Johns Hopkins, making it the very first in what is now a long line of successful American periodicals devoted entirely to research in mathematics. This is quite a distinction!

The list of authors for the three remaining issues of Volume 1 during 1878–1879 reflect the spread seen in Number 1. Most of the 48 entries were short, consisting of seven or fewer pages, with major works by the French number theorist Eduard Lucas, the American George Hill, and Englishman J.J. Sylvester. It included brief notes by William Story, Henry Rowland, C.S. Peirce, and Simon Newcomb, all of whom were (or would soon be) on the Johns Hopkins faculty.

What about Sylvester's academic offspring? Three fellows contributed one article each to the remaining issues in Volume 1. Fabian Franklin published his doctoral dissertation in Number 2, George Halsted provided a bibliography of works on hyperspace and non-Euclidean geometry in Number 3, and Thomas Craig submitted his first paper, a six-page note on the motion of a point on the surface of an ellipsoid inspired by earlier work due to the eminent German mathematician, Karl Jacobi, in Number 4. Two omissions from the pages of the *AJM* are that Craig's dissertation never appeared, nor did any work by the only initial fellow not destined to earn a PhD, Joshua Gore.

Overall, 12 of the 16 fellows published a total of 73 papers in the *AJM* in Volumes 1–20, which appeared 1878–1899. Thomas Craig and Fabian Franklin accounted for most of that production—45 of the 73 papers (61%)—since both remained on the faculty throughout most of this period. After that came Irving Stringham (5), Ellery Davis (4), George Halsted, Christine Ladd, George Ely, William Durfee, and Archibald Daniels (3 each), Arthur Hathaway (2), and Charles Van Velzer and Oscar Mitchell (1 each). Those fellows who never published a paper in the *AJM* were Joshua Gore, Robert Prentiss, Herbert Perry, and Gustav Bissing; only the last in this quartet received a Johns Hopkins PhD.

Hopkins graduate students generally worked in those areas that held interest for J.J. Sylvester, but the expanded faculty enabled them to branch out as well. For instance, Sylvester continued to work on invariant theory with Arthur Cayley, before, during, and after Cayley's official visit for the spring 1882 semester, but Fabian Franklin proved to be a boon to Sylvester by carrying out many of the intricate computations necessary for the British computational approach to invariant theory (which differed dramatically from the theoretical approach preferred in Germany by Paul Gordan and Max Noether.) On the other hand, William Story's interest in geometry influenced fellows Thomas Craig (who became a specialist in differential geometry) and George Halsted (who became a specialist in non-Euclidean geometry).

Sylvester's main interest in number theory centered on partitions. As was the case with invariants, Fabian Franklin was a major contributor, as can be seen in two long papers on tables of generating functions "by J.J. Sylvester, assisted by F. Franklin."[35] Franklin also published a paper, "Sur le Développement du Produit infini," in the prized French journal, *Comptes rendus*. The degree of creativity in this paper impressed Charles Hermite, who presented it to the French *Académie des Sciences*. Afterward Hermite wrote his old friend Sylvester, "It certainly will not be unpleasant for you to hear that I was not the only person to be very interested in Mr. Franklin's very original and ingenious proof. ... Please tell Mr. Franklin that his talent is appreciated, as it deserves to be, by the mathematicians of the old world."[36] Just three years earlier, Hermite had cautioned Sylvester about going to America because it held no mathematical future. Franklin's paper surely vindicated Sylvester's passage across the Atlantic. In fact, Sylvester tried to induce Hermite to come to Baltimore, but the eminent French mathematician declined due to, as he wrote, his absolute inability to lecture in English.

Almost every issue of the early years of the *AJM* contained a paper on partitions by a Hopkins faculty member or graduate student. Perhaps the most interesting one of all, and certainly the one with the most distinct title, was "A constructive theory of partitions, arranged in three acts, an interact and an exodion," which was authored "by J.J. Sylvester, with insertions by F. Franklin."[37] The online file "Web05-Partitions" describes this 80-page work and the role the fellow William Durfee played in it. Today, the powerful software program *Mathematica* contains a command called "DurfeeSquare."

While the names of Franklin and Durfee occur repeatedly in the paper, Sylvester's correspondence with his friend Arthur Cayley indicates that the entire Johns Hopkins *community* was at work on this project.[38]

> I sent to the *Comptes rendus* two or three days ago my proof of the
> wonderful theorem [on] partitions of n into odd numbers and its par-
> titions into unequal numbers. Franklin, Mrs. Franklin, Story, Hath-
> away, Ely, and Durfee were all at work trying to find the proof—but

> I was fortunately beforehand with the theory and the only one in at
> the death.

The Mrs. Franklin mentioned here is Christine Ladd Franklin. This comment by Sylvester attests to the fact that the faculty members and graduate students that he brought to Johns Hopkins had evolved into America's first *community* of mathematical scholars who sometimes engaged in collaboration and sometimes in friendly, but fierce, competition, yet always worked as a unit enraptured with the thrill of discovery.

This was a revolutionary development!

The importance of a *community* of mathematicians is at the heart of the Parshall–Rowe book on mathematics in America 1876–1900 and the central part of its title.[39] As the Spanish mathematician Guillermo Curbera wrote, "Deep in the heart of mathematics there is a strong impulse towards communication and an intense sentiment of building a community."[40] The intertwining of community (especially mathematical organizations) and communication (journals and meetings) will prevail throughout the rest of this book.

Sylvester's community of scholars also worked on matrix algebra during the last semester of Sylvester's tenure at Johns Hopkins. The inspiration for this work was Hamilton's discovery of quaternions almost 40 years earlier. While examining conditions under which two $n \times n$ matrices commute, Sylvester was led to the concepts of, first, the characteristic polynomial of a matrix, and later, to eliminate repeated roots, the minimal polynomial. His study of linear transformations dovetailed nicely with another positive development to come out of the Sylvester School when C.S. Peirce revised his father's incipient work, *Linear Associative Algebras*, as a 133-page paper in Volume 4 (1881) of the *AJM*. The paper's title indicates that this was more than a mere copy: "Linear associative algebra. With notes and addenda by C.S. Peirce, son of the author." The republication is reminiscent of Benjamin Peirce's role in the Nathaniel Bowditch translation of Laplace, but C.S. Peirce went beyond explication, extending the material to new dimensions, and proving that finite-dimensional linear algebra is isomorphic to matrix theory. It is interesting to note that whereas publication of the original work gained Benjamin Peirce hardly any repute at all, its republication in the *AJM* corralled him an enduring, albeit posthumous, reputation.

The paper on the four-color problem, published in the September 1879 issue of the *AJM*, caused quite a controversy. Written by the Englishman, Sir Alfred Bray Kempe, a former student of Cayley at Cambridge, the introduction set out the history of the problem:[41]

> If, then, we take a simply connected surface divided in any manner
> into districts, and proceed to colour these districts so that no two ad-
> jacent districts shall be of the same colour, and if we go to work at
> random, first colouring as many districts as we can with one colour
> and then proceeding to another colour, we shall find that we require a
> good many different colours; but, by the use of a little care, the num-
> ber may be reduced. Now, it has been stated somewhere by Professor
> De Morgan that it has long been known to map-makers as a matter
> of experience—an experience however probably confined to compar-
> atively simple cases—that four colours will suffice in any case. That
> four colours may be necessary will be at once obvious ... but that
> four colours will suffice in all cases is a fact which is by no means

> obvious, and has rested hitherto, as far as I know, on the experience I have mentioned, and on the statement of Professor De Morgan, that the fact was no doubt true. Whether that statement was one merely of belief, or whether Professor De Morgan, or any one else, ever gave a proof of it, or a way of colouring any given map, is, I believe, unknown; at all events, no answer has been given to the query to that effect put by Professor Cayley to the London Mathematical Society … Professor Cayley, while indicating wherein the difficulty of the question consisted, states that he had not then obtained a solution. Some inkling of the nature of the difficulty of the question … may be derived from the fact that a very small alteration in one part of a map may render it necessary to recolour it throughout. After a somewhat arduous search, I have succeeded, suddenly, as might be expected, in hitting upon the weak point, which proved an easy one to attack. The result is, that the experience of the map-makers has not deceived them, the maps they had to deal with … can, in every case, be painted with four colours. How this can be done I will endeavour—at the request of the Editor-in-Chief—to explain.

While Sylvester, as editor in chief, may have requested Kempe's paper, the associate editor in charge, William Story, pursued an unusual protocol by adding a cautionary note at the end of the paper:

> In the foregoing valuable paper on the "Geographical Problem of the Four Colours," Mr. Kempe has substantially proved the fundamental theorem … by a very ingenious method; but it seems desirable, to make the proof absolutely rigorous, that certain cases which are liable to occur, and whose occurrence will render a change in the formulae, as well as some modification of the method of proof, necessary, should be considered separately, and I have endeavoured to do this in the following note.

In short, Story felt that the proof was not airtight, yet he was unable to pinpoint an error. It took another ten years before Percy John Heawood found that mistake. Unfortunately, Kempe is probably best known today for this fallacious "proof," yet it turned out to form the basis for a computer-aided proof in 1976, almost 100 years later.

There seems always to have been tension between J.J. Sylvester and his managing editors during the whole time he was editor in chief, 1878–1884. William Story was associate editor in charge of day-to-day operations from the journal's inception until the summer of 1880, when he had a falling out with Sylvester over allowing Henry Rowland to exceed the page limit imposed by Sylvester. (Sylvester was in London at the time a decision had to be made. Receiving a response via "snail-mail" would have taken weeks.) So, when Volume 4 appeared in the fall of 1881, no associate was listed, even though Fabian Franklin served as managing editor. But here too the two editors experienced a serious difference, and Franklin was fired. Similarly, Sylvester was listed as the only editor for Volume 5 (1882), though he had replaced Franklin with Mitchell by then. Only with Volume 6 was assistance noted formally, with Thomas Craig appearing as assistant editor and, subsequently, associate editor. Apparently, neither

Mitchell nor Craig had enough time to clash with Sylvester, because Mitchell returned to Marietta College and Sylvester departed JHU shortly after Craig came onboard.

It is somewhat curious that when Simon Newcomb succeeded J.J. Sylvester as editor in chief with Volume 7, three papers were published in French (including one by Henri Poincaré) and one in German, whereas, during the six volumes under Sylvester's editorship, only five appeared in French (two each by Faà de Bruno and Lucas and one by Hermite) and one in Italian.

Johns Hopkins University, 1884–1900. J.J. Sylvester escaped all but one East Coast summer (1882) by returning to London, yet the 1883 holiday was not as productive as usual due to the social swirl surrounding one overarching development—the vacancy for the Savilian Professor of Geometry at Oxford University. The Universities Tests Act of 1871 had removed the religious discrimination that had earlier denied Sylvester academic degrees and professorships at English universities because he was Jewish. So now he could be considered for that chair, one of the most prestigious in England. He formally applied in April through his friend William Spottiswoode, then president of the Royal Society of London but, more importantly, a member of the selection committee. However, matters did not progress smoothly for Sylvester that summer, as Spottiswoode's death in July from typhoid fever postponed the election indefinitely. He vacillated between remaining in England until the matter was resolved or returning to Baltimore. Finally, in September he informed President Gilman that he would resign his position at Johns Hopkins effective January 1, 1884, regardless of the outcome at Oxford. Gilman had dealt with Sylvester's mood swings before, so initially he tried to dissuade the head of his successful mathematics department, but when it became clear that the 68-year-old Englishman yearned to return to England, Gilman began discussions with him about a replacement.

That fall, Sylvester returned to Baltimore and taught his last class, on matrix theory. On December 5 he was notified by his good friend Arthur Cayley, a member of the selection committee, that he had been elected to the desired Savilian chair. The Hopkins Board of Trustees countered with a very attractive financial offer, but Sylvester was firm, so he made arrangements to sail on December 22. Two days before that, the university held a grand reception in his honor in Hopkins Hall at which the entire university turned out. After all, he was the first of the original six professors to leave. Furthermore, over the previous seven years, Sylvester had created the very first community of mathematical scholars in America, engaging his colleagues and students with a research mission never before witnessed in the US or Canada. It was revolutionary—and perfectly in step with the revolution begun by the philanthropist Johns Hopkins and carried out with marked success by Daniel Coit Gilman.

That fall, President Gilman set out to recruit a new mathematics professor for the thriving university. Sylvester, with an eye for young talent, recommended a rising star, Felix Klein, who was then in Leipzig. Gilman contacted Klein and negotiations commenced at once, but ultimately they broke down over financial differences. Next Gilman offered the position to Arthur Cayley, but he too rejected it. Finally, Gilman settled on his third choice, the Canadian Simon Newcomb. In 1875 Newcomb had been offered the directorship of the Harvard Observatory, but declined it in favor of continuing his work on mathematical theories and computation instead of collecting observational data. Two years later, he was appointed director of the Nautical Almanac Office, which had moved to Washington in the meantime.

On the surface, Simon Newcomb seemed like the perfect successor, but two aspects of his appointment militated against his ability to sustain the gains that the Sylvester School had crafted. For one, Newcomb's appointment was only half time, which allowed him to continue his work at the Nautical Almanac Office. Second, his primary interest was astronomy, not mathematics, so every course he offered at Johns Hopkins was in astronomy. However, Newcomb was elected as the fourth president of the American Mathematical Society for two terms (in 1897 and 1898), so he was certainly regarded as a leading mathematician by his peers. Perhaps most importantly, however, Newcomb lacked Sylvester's inspiring personality.

Despite these drawbacks, the business of mathematics continued under Newcomb right on up to the appointment of Frank Morley in 1900. William Story, Thomas Craig, and Fabian Franklin maintained the usual teaching duties, carried out research, and directed dissertations, with Story serving as the managing editor of the *AJM*. Johns Hopkins awarded 24 PhD degrees in mathematics during this period, including one to the University of Toronto's John Charles Fields in 1887 and another to Princeton University differential geometer Luther Eisenhart in 1900. Fields later established the internationally famous, quadrennial award named in his honor.

During this time, Johns Hopkins became the first American university to admit an African American student into its graduate program, **Kelly Miller** (1863–1939), the son of a slave. He graduated from the Fairfield Institute in 1880, enrolled at Howard University, and graduated four years later. While there, he obtained a part-time government job where he worked with Simon Newcomb, who recognized Miller's potential and determination, and recommended tutors for advanced study. Miller was admitted to Johns Hopkins, and enrolled in 1887 as a graduate student in mathematics. However, Johns Hopkins "later adopted a segregationist policy."[42] Unfortunately, an economic crisis caused JHU to raise tuition 25%, and the resulting financial situation forced Miller to leave after two years without earning a degree. Gilman and Newcomb then recommended him for a faculty position at Howard. Kelly Miller was appointed professor of mathematics at Howard in 1890 and, beginning five years later, taught courses in sociology a well. In addition, he served as the dean of the college of arts and sciences. Miller remained at Howard until his death. The online file "Web05-Miller" contains more information on Kelly Miller.

From about the time that Kelly Miller left Johns Hopkins in 1889, the fledgling community that had flocked to Sylvester began to disintegrate, with Story, Craig, and Franklin leaving Johns Hopkins by 1900. William Story had accepted a position as head of the department at the new Clark University in 1889. Thomas Craig died at age 44 in 1900. But Fabian Franklin made perhaps the most radical career change of all in 1895, after having been at Johns Hopkins as a student and professor for 18 years. At age 42, he left academia altogether to enter the publishing world, initially accepting an offer as associate editor of the *New York Evening Post*. Franklin developed a firm opposition to radical politics and to socialism in particular. During WWI he founded a periodical called *The Review*, which merged in 1922 with another paper, *The Independent*. He authored several books, with *Plain Talks on Economics* being perhaps his most successful. He died suddenly in January of 1939.

Figure 5.7. Kelly Miller

The defections of the Franklins—Christine Ladd Franklin had moved to biology—illustrate the plight of mathematics at Johns Hopkins. They also show that existing academic positions throughout America were insufficiently enticing for many Americans. Therefore, by the turn of the twentieth century, the vital element, the *community* that had formed around Sylvester, had dissipated.

What happened to the Savilian Professor himself? J.J. Sylvester died in 1897, a little over 13 years after leaving Johns Hopkins. Did he keep in touch with the members of the school he had founded there? Yes and no. He did exchange letters regularly with Gilman, continuing his role as a trusted advisor for the president and team player for the university, a testament to the close working relationship and educational philosophy the two forged during the university's first teetering steps. Though Sylvester did not correspond with fellow professors Story, Craig, and Franklin, when any of them visited England, he always invited them to tea. What could be more British? Also, Sylvester followed the progress of his academic sons and daughter, occasionally commenting about them to Gilman. At first it might seem surprising that no American students went to Oxford University to work with Sylvester, but Oxford did not grant PhDs or have real graduate education during Sylvester's tenure there.[43]

Where were the other 13 figures (besides Thomas Craig, Fabian Franklin, and Christine Ladd Franklin) who were awarded fellowships under Sylvester's watch located in 1900? Two were teaching at small colleges (Oscar Mitchell at Marietta and William Durfee at Hobart) while one was probably teaching high school (Herbert Perry). Another two were employed at the patent office (George Ely and Gustav Bissing). The remaining eight held professorial positions: George Halsted (at Texas), Joshua Gore (physics at North Carolina), Irving Stringham (Berkeley), Charles Van Velzer (Wisconsin), Robert Prentiss (Rutgers), Ellery Davis (Nebraska), Archibald Daniels (Vermont), and Arthur Hathaway (Rose-Hulman). This is a particularly impressive list of universities, and the placements must have made Sylvester proud. All would be regarded as research universities today, yet few of the fellows published. Why? Because American institutions of higher learning were slow to adopt the path blazed by Johns Hopkins. Professors were still teachers first, with most teaching loads consisting of five courses a semester, which afforded little time, let alone incentive, to engage in research.

What was an aspiring student in mathematics to do?

Figure 5.8. Felix Klein

Klein Klub

Despite the decline of Sylvester's mathematical community at Johns Hopkins, the experiment carried out in Baltimore demonstrated that an increasing number of American students possessed the interest and ability to pursue doctoral degrees in mathematics and even to seek postdoctoral training. This new generation of Americans was symptomatic of a different time for mathematics in America, in terms of both quantity and quality. What were they to do? After all, Johns Hopkins offered the only viable graduate *program* in mathematics, but the person responsible for starting it was gone. One possibility was for the inspired student to work under an isolated individual at one of the (mostly) East Coast universities. Yet up to 1884, the only universities that awarded PhDs in mathematics were Yale (six, but none since 1877), Harvard (four), and Columbia, Cornell, Michigan, and Syracuse (one each). Even among these schools, however, some of the dissertation topics would not be considered mathematics proper today.

As a result a general feeling arose by the mid-1880s that aspiring students who wanted to be *somebody* in American mathematics had to obtain advanced training in Germany. So pervasive did the German influence become during the period 1884–1904 that about 20% of the American Mathematical Society membership had obtained doctoral degrees from a German university or had studied there after obtaining an American PhD. From 1884 until the turn of the twentieth century, the most gifted and dedicated American graduate students gravitated to one leading figure, Felix Klein (1849–1925). The section "Transition 1876" presented an overview of his meteoric rise to prominence among mathematicians in the world. J.J. Sylvester had recommended Klein as his successor and Johns Hopkins President Gilman had negotiated with Klein about the position, but the German chose to remain in his native land instead, a pattern that repeated itself a decade later with a different president at a different university. As a result, most—though definitely not all—of his influence on American mathematics would come from abroad.

Leipzig. The flood of American students studying under Felix Klein began as a slow trickle when Irving Stringham traveled to Leipzig, thus forging a direct link from the Sylvester School to the Klein Klub. Upon graduation from Johns Hopkins, Stringham's two-year Parker Fellowship permitted him to continue his studies in Germany at exactly the time when Klein was beginning his tenure at Leipzig. While there, Stringham attended Klein's one-year course on function theory, a topic Klein presented from Riemann's geometric point of view. More importantly, Stringham delivered two lectures in Klein's higher-level seminar on geometry and function theory based on results from his Johns Hopkins PhD dissertation. He spoke in German, of course. Undoubtedly, his success in Sylvester's seminar had armed him with the confidence needed to deliver lectures in front of someone with Klein's exacting standards. Stringham did not seek a degree at Göttingen, but he spent two very intense years immersed in cutting-edge research and, more generally, in the milieu of collaborative and individualistic investigations of original material.

The German innovation of seminars played a central role in shaping American research mathematicians. A manifestation of this development was conducted by J.J. Sylvester at Johns Hopkins, but under the meticulous Klein, seminars were brought to their maximal benefit by transporting students to the threshold of research activity. Klein chose the subject with an eye toward correlating it with his lecture courses, and announced each seminar's goals in advance. Participation was by permission only, and Klein assigned a specific topic for each student to research and to report on formally, with a practice session conducted privately with Klein before the public lecture. The experience did not end there, either: The student had to record a synopsis in a notebook that Klein called a protocol book. The *Protokollbuch für das Seminar* from 1880–1881 records two lectures by Irving Stringham, "On regular bodies in 4-dimensional space," and "Groups of motions of 4-dimensional bodies." No other entry for an American would appear for another five years. (Although these lectures were delivered in German, the titles have been translated into English.) Overall, Klein's seminars "constituted one of the most important proving grounds for young American mathematical talent."[44]

Irving Stringham was certainly an attractive candidate for a professorship when he returned to the US in 1882. The University of California, Berkeley, had summarily dismissed William Welcker the year before, along with the president of the university, in order to elevate the school's standing. The mathematics professorship remained vacant for another year until the university's earlier president, Daniel Gilman, then at Johns Hopkins, lured Irving Stringham to Berkeley from Michigan to upgrade the department. Stringham thus became the first mathematics faculty member on the West Coast to hold a PhD degree.

It took three years after Stringham left Leipzig in late 1881 before another American crossed the Atlantic to study with Klein. This time there were two and, unlike Stringham and several other professors and postdoctoral students, they sought not just higher studies but advanced degrees. In 1884, Henry Fine and Frank Cole attended the second semester of Klein's course on elliptic function theory without benefit of the first part, a path that required strenuous effort to catch up to the class. Neither American was invited to attend the concurrent seminar that Klein conducted. The next semester, both took Klein's sequel course on higher curves and surfaces, yet once again neither was chosen to participate in the coexisting seminar. Both mathematicians possess celebrated names today. The building now affectionately known as "old Fine Hall" formerly housed Princeton University's department of mathematics, and the American Mathematical Society awards annual Cole prizes in algebra and number theory.

Henry Burchard Fine (1858–1928) was identified with Princeton University almost his whole life, beginning in 1875, when he and his mother settled in the town of Princeton. He was born and raised near Gettysburg, PA, but after his father died when Henry was eleven, his mother moved the family (including another son and two daughters) to New York and then to Minnesota. As an undergraduate, Fine served as a member of the editorial board of the student newspaper, the *Princetonian*, along with his lifelong friend Woodrow Wilson; Fine was managing editor in his senior year. He specialized in the classics before coming under the spell of George Halsted, who recognized Fine's impressive talent and subsequently convinced him to pursue mathematics. Halsted taught at Princeton from the time he left Johns Hopkins in 1878 until 1884. Fine remained at Princeton in two capacities upon graduation in 1880—as a fellow in experimental science for a year and then as tutor in mathematics for three.

Figure 5.9. Henry Burchard Fine

In March 1884 Henry Fine sailed to Germany even though, "According to his own account, he knew very little German and almost no mathematics."[45] He headed straight to Leipzig to work with Felix Klein, who agreed to advise him. Klein proposed a dissertation problem on enumerative geometry that turned out to be unsuitable so, with the advisor's consent, Fine switched to another problem proposed by fellow graduate student Eduard Study. This turned out to be successful, and Fine completed his *Doktorarbeit* in one year. Surprisingly this dissertation was written in English, the only time Klein ever approved a language other than German. Why did the demanding Klein approve a thesis written in English? The answer remains a mystery. In any event, Fine's dissertation "On the singularities of curves of double curvature" appeared in the *AJM* in 1886. (Notice how this research journal continued to play a vital role not just for Johns Hopkins students, but also for other Americans with no connection to the university.) Overall, then, Harry Fine, as he was generally called,[46] was not only tied to Klein directly but to the Sylvester School as well, in two ways, via the Johns Hopkins graduate George Halsted and the journal initiated there just eight years earlier.

Irving Stringham had studied under Felix Klein with funds from a Parker Fellowship, which provided one of the few opportunities for American students to continue their studies abroad. John Parker, Jr. (1783–1844) was a Harvard benefactor who established the fellowships for meritorious Harvard students in the natural sciences. (See the section "Transition 1876" for details.) Like Stringham, **Frank Nelson Cole** (1861–1926) was such a student, graduating second and with highest honors in mathematics in a class of 189 from Harvard University in 1882. (Today's entering undergraduate classes at Harvard College number about 1600.) In 1883 he was awarded a Parker Fellowship to study abroad with the intention of pursuing the mathematical side of physics, but one year later he came under the spell of Felix Klein and changed his major interest to pure mathematics. Frank Cole ended up spending the next year under Klein's tutelage, but his fellowship ended before he could complete the program. Although Cole returned to Harvard to finish his dissertation on sixth-degree equations, he taught courses on two topics that initiated a new era in graduate education in the US: 1) complex function theory à la Riemann, and 2) the theory of substitutions. One of Cole's promising graduate students, William Osgood, called these courses "truly inspiring."[47] Every member of the Harvard mathematics faculty attended the courses, yet none could seriously discuss Cole's dissertation, so he was forced to work in isolation, thus serving yet another example of someone who could be tagged a rugged individualist. Harvard awarded Cole the PhD in 1886 for his work on the problem Klein suggested, "A contribution to the theory of the general equation of the sixth degree," which was published that year in the *AJM*.

Figure 5.10. Frank Nelson Cole

Göttingen. In 1886, when dissertations written by the two Klein protégés appeared in the *AJM*, Klein's career took a giant leap forward—a professorship at the University of Göttingen, the Mecca of mathematics. Is it any wonder that any young, aspiring, talented student would want to matriculate there? The operative word in that question is *talented*, for Klein gave his time only to those passionately devoted students with the capability and perseverance to master his universal approach to mathematics. The achievements of Klein's students from 1886 to 1900 document his fundamental influence on a generation of American mathematicians who ultimately played vital roles in advancing not only research, but in establishing true graduate programs at universities scattered across the US.

To understand Klein's contributions at Göttingen, it is instructive to put in historical perspective the intense rivalry that existed between Göttingen and the University of Berlin, the two major German centers for mathematical research in the nineteenth century.[48] There is no record of any American student getting caught in the crossfire, but the rivalry does point out some fundamental differences in these two very different schools of mathematical thought, and it is just those differences which confronted all students, German and foreign. The differences can be summarized in a contrast drawn in 1875 by the leading mathematician, Gösta Mittag-Leffler, who compared Weierstrass's completely analytical system (Berlin) to the geometric function theory of Riemann (Göttingen).

Despite Gauss being associated with Göttingen from 1807 until his death in 1855, Berlin dominated not only the national, but also the international, mathematical scene throughout most of the nineteenth century, even though Gauss, the "Prince of Mathematics," was succeeded in turn by Dirichlet, Riemann, and Clebsch. From the mid-1850s until the early 1890s, the faculty at Berlin had a nucleus of Kummer, Weierstrass, Kronecker, and Fuchs. All four were giants in the field. In fact, during the 1860s and 1870s, Berlin graduates occupied practically all chairs in mathematics in Germany. Moreover, the Franco-Prussian war caused many German mathematicians to feel that the capital city should play a role analogous to Paris. Felix Klein remained aloof from this rivalry, even when the 20-year-old student undertook postdoctoral studies at Berlin in 1869–1870. The first overt sign of the struggle between Klein and Berlin occurred when he engaged in a nasty priority dispute with the Berliner Lazarus Fuchs while Klein resided in Leipzig. Matters worsened when Klein accepted the Göttingen post in 1886, especially when he arranged for the Norwegian—and thus foreigner—Sophus Lie to succeed him in Leipzig. Many German mathematicians, especially those in the Berlin School, were upset with this turn of events. However, three actions in the early 1890s allowed Göttingen to surpass Berlin: Kronecker's death in 1891, Weierstrass's retirement in 1892, and the appointment at Göttingen of David Hilbert in 1894. Together Klein and Hilbert attracted international mathematical talent to Göttingen in droves, but Klein's magnetism for American students manifested itself the moment he arrived in Göttingen. (The Hilbert Colony in America is described in the section "Transition 1900.")

The first American to study under Klein at Göttingen was **Mellen Woodman Haskell** (1863–1948), a Salem, MA, native who had received a bachelor's degree from Harvard in 1883. He applied for a Parker Fellowship after receiving a master's degree two years later but, by this time, competition for these coveted awards among Harvard graduates had become intense, with only one or two being available each year. However, with Frank Cole's firm backing, Haskell became the next recipient in line with Story, Stringham, and Cole himself.

When Mellen Haskell arrived in Göttingen in the summer of 1886, he signed up for two of Klein's courses. Indeed, Haskell became somewhat of a Göttingen fixture over the next four years, enrolling in nearly every course Klein offered. Right from the start, Haskell participated in Klein's seminars too. Klein's *Protokollbuchs* record that Haskell lectured on "Resolvents of the fourth degree" and "Linear differential equations" in his very first seminar, which gives an indication of the broad and deep background Haskell had received at Harvard under Frank Cole before his trans-Atlantic voyage. But two lectures presented in two different seminars in 1887 determined his future direction, resulting in his dissertation *"Über die zu der Kurve $\lambda^3\mu + \mu^3\nu + \nu^3\lambda = 0$ im projektiven sinne gehörende mehrfache Überdeckung der Ebene."* This work contained an elaborate construction of the projective Riemann surface associated with a certain normal curve that Klein had studied in conjunction with his work on the geometric Galois theory of a modular equation. Haskell's thesis "came closer than the work of any of Klein's American students to capturing that peculiar mix of ideas from group theory, algebraic geometry, and complex function theory that lay at the heart of Klein's own mathematical research."[49] Moreover, the mathematics was impressively deep—at a level never before seen in America. It showed just how far American students had come. To complete the circle, the dissertation was published in the *AJM* the year it was completed, 1890.

There are two important distinctions between higher degrees in America and many European countries, including Germany. Haskell's dissertation was formally called a *Doktorarbeit*, and is roughly equivalent to an American thesis. However, the German system had a "higher doctorate," the *Habilitation*, a completely independent work that culminates in a written account called the *Habilitationsschrift*, or Habilitation thesis. German universities required this higher degree to become a *privatdozent*, a position roughly equivalent to an American associate professor but without any salary. Of course, *privatdozents* aspired to become full professors, a promotion based entirely upon subsequent research.

The second important distinction between the degrees is that in America, graduate programs generally require courses and qualifying exams that are written and/or oral. The German system had no such requirements, only the presentation of research ability as demonstrated by the *Doktorarbeit* and/or *Habilitation*.

Mellen Haskell had accepted an instructorship at the University of Michigan the previous year, joining Frank Cole in Ann Arbor. Although the standard teaching load of 17–18 hours per semester militated against his project of translating Klein's famous *Erlanger Program*, he was finally able to include it as part of his paper "A comparative review of recent researches in geometry," which he published in the second volume of the *Bulletin of the New York Mathematical Society* in 1893. However, Haskell was no longer at the University of Michigan by then. Three years earlier he was induced to join Irving Stringham at the University of California as an assistant professor, thus becoming the second mathematics PhD on the faculty there. The two men became stalwarts at Berkeley and remained there for the rest of their lives. Haskell's accomplishments as chair of the mathematics department will be elaborated in Volume 2.

Mellen Haskell had studied under former Klein student Frank Cole before setting sail for Germany. The next American was similar. **Henry Dallas Thompson** (1864–1927) was prepared by Harry Fine at Princeton for advanced studies. Thompson lectured on the zeros of Θ-functions in the *"Seminar über hyperelliptische Funktionen"* during his very first semester at Göttingen in 1887. And then he lectured on a similar topic in the ensuing semester. In the fall semester of 1888, Thompson returned to Princeton, where he remained as a faculty member for the rest of his life. He obtained his doctorate under Klein in 1892 for the dissertation *"Hyperelliptische Schnittsysteme und Zusammenordnung der algebraischen und transzendenten Thetacharakteristiken."*

Next came two Harvard students later known as the "great twin brethren."[50] Both had received a solid foundation in Klein's brand of mathematics under Frank Cole and both returned to Harvard afterward to spend the rest of their careers in the "American

Figure 5.11. William Fogg Osgood

Cambridge." Nonetheless, vast differences separate these emerging scholars. Their role in the emerging American mathematical research community has been described as follows:[51]

> It was the moment of the great awakening in American mathematics, when a number of able and enthusiastic young men, largely trained in Germany, set about raising the science as pursued in this country [the US] to the same plane on which it was pursued in Europe. ... It was the beginning of the era ... of Cole at Columbia, of Fine at Princeton. ... Osgood and Bôcher, early presidents of the Society, were in the very middle of it. They introduced into Harvard new and advanced courses, largely dealing with what one might call "Göttingen mathematics," [and] gathered around them graduate students whom they prepared for the doctorate.

Except for excursions to Germany, **William Fogg Osgood** (1864–1943) spent his whole life in the Boston area.[52] He graduated from the historic Boston Latin School[53] in 1882 with highest honors in mathematics and then moved to contiguous Cambridge to attend Harvard, where he obtained his bachelor's degree in 1886, placing second in a class of 286. Frank Cole knew full well of the lack of a genuine research spirit at Harvard, despite faculty members James Mills Peirce and William Byerly in mathematics, as well as Benjamin O. Peirce in mathematical physics. Therefore, Cole recommended that Osgood travel abroad for graduate education. Because Osgood had been deeply inspired by Cole's course on function theory, it was easy for Cole to steer his star student to Felix Klein. First, Osgood remained on campus another year to prepare for the rigors he would experience in Göttingen. Like Cole, Osgood applied for a Parker Fellowship but was awarded the lesser Harris traveling fellowship instead. Osgood's father had died just two years before his autumn 1887 departure for Germany, so he needed financial support to pursue his PhD abroad. The Harris fellowship supplied partial relief for 1887–1888, but the next year he was upgraded to the more lucrative Parker Fellowship.

William Osgood already possessed an impressive command of German before making the arduous trans-Atlantic voyage. Indeed, "Osgood was a known Germanophile,"[54] and he wrote his most influential book in that language, *Lehrbuch der Funktionentheorie* (1907), which ran to five editions and was regarded as the classical treatise on the theory of functions of a real variable. Furthermore, he married a Göttingen woman while studying there, and one of his daughters even married a professor of German at Harvard.

Osgood was an excellent mathematics student with a solid command of German, yet even he failed to meet Klein's demanding standards, and so after two years at Göttingen, he moved to Erlangen, where he received his doctorate in 1890 for a thesis on Abelian integrals. He returned to Harvard for the fall semester of 1890 and remained there until retirement in 1933 at age 69.

One of Osgood's most important papers appeared in the first volume of the *Transactions of the AMS* in 1900, when he proved the general theorem that every simply connected open region in a plane with more than one boundary point can be mapped conformally onto the interior of a circle. His proof relied on the existence of a certain Green's function. Osgood submitted this paper while sailing on one of his many trips across the Atlantic.[55]

William Osgood was one of the very few mathematicians at the time who owned their living quarters. During his tenure, he introduced a standard of rigor unknown at Harvard beforehand. He also served as the eighth president of the AMS. In the Cambridge community, he was widely known for touring in motor cars and smoking cigars. After retiring, he taught for two years at the National University in Beijing.

Maxime Bôcher (1867–1918) came from a highly cultured Boston family with strong roots in both France and New England.[56] His mother, a teacher, traced her ancestry to the Pilgrims in Plymouth Rock. His father was the first professor of modern languages at MIT before moving to Harvard as professor of foreign languages for three decades, so the son encountered no problems with the German language. Bôcher attended public and private schools before graduating from the Cambridge Latin School at age 16. He then attended Harvard, graduating in 1888 *summa cum laude* with highest honors in mathematics based on an honors thesis. Upon graduation, Bôcher intended to remain at Harvard if he did not win a traveling fellowship to study under Klein in Göttingen but, like Osgood the year before, he was awarded a Harris fellowship that was upgraded to a Parker Fellowship after one year.

Maxime Bôcher too married a German woman while at Göttingen. Unlike Osgood, Bôcher completed a dissertation under Klein in 1891 on potential theory, a topic he had studied at Harvard with William Byerly and Benjamin O. Peirce. (This is not *the* Benjamin Peirce, but a distant relative. B.O. Peirce had linked mathematics with physics in a way that was "frequently lacking at Harvard especially from the time of his death in 1915 to the appointment of John Van Vleck in 1934."[57])

Like Osgood, Bôcher returned to Harvard for the rest of his career. The standard teaching load at the time consisted of three courses: one elementary, one intermediate, and the third advanced. In his intermediate course on modern geometry, he revitalized a syllabus that had been developed by William Byerly; it was so successful that he repeated it almost every year for the rest of his life. Among some colleagues there seems to have been some question about the efficacy of the lecture method of teaching he employed, but his students expressed no such reservations; besides, the course was a free elective. The topic of his advanced course was modern—curvilinear coordinates and functions defined by differential equations—but the term "seminary" used to describe it suggests a throwback to J.J. Sylvester at Johns Hopkins or, more recently,

Figure 5.12. Maxime Bôcher

William Story at Clark. Two years later he offered an introductory course on the theory of functions of a complex variable. The previous decade saw American students flee abroad, chiefly to Germany, for further instruction and guidance. Bôcher's example shows how those attributes were being brought to these shores in short order.

While an undergraduate student at Harvard, Maxime Bôcher earned a reputation as an excellent debater and an expert in rebuttal. This characteristic is exemplified by an incident that occurred at a meeting of a club for anyone interested in mathematics and physics from Harvard and MIT shortly after his return from Göttingen. The speaker, Frederick Woods, illustrated his lecture on surfaces with models from the Brill collection at MIT. A physician with an interest in mathematics rose to ask about intersecting lines, and Woods, a fellow Klein Klub member, answered that they could "meet at infinity." The perplexed questioner expressed some doubt about this seeming impossibility, causing a professor from a third university to interject with a haughty attitude that annoyed Bôcher. The Harvard professor interjected, "That is not true in the geometry of inversion." When the professor replied, "Oh, that is not geometry," Bôcher rejoined, "It is what Klein calls geometry," thus casting his adversary in the untenable role of intimating that Felix Klein was not a geometer.

In 1894 Bôcher published a book on potential theory, in German, that G.D. Birkhoff regarded as the "first work of importance" produced by any of Klein's American doctoral students.[58] *Die Reihenentwickelungen der Potentialtheorie* provided what Bôcher regarded as an adequate treatment of the subject that he had been unable to publish in his dissertation due to space limitations. In 1912 he presented an invited lecture "Boundary problems in one dimension" at the International Congress of Mathematicians in Cambridge, England. He delivered a series of AMS Colloquium Lectures, six in all, in 1906 titled "Linear differential equations and their applications." He was active with the AMS in other ways as well, serving as president 1908–1910. He also was an editor of the *Transactions of the AMS* who strived mightily to enforce clarity in exposition, urging authors not to forget the early difficulties and underlying ideas that became obscured through familiarity. Unfortunately, he suffered from ill health that intensified after 1913, and he died six years later, just two weeks after turning 51.

A Puritan, Bôcher held no place for human weakness, respecting only results. Therefore, many people found him cold and unsympathetic. Beginning students in research sometimes thought he was unappreciative. Yet during his Harvard tenure, he directed 18 doctoral dissertations, paying forward his Klein ancestry to a new generation of American mathematicians that included James Glover (1895), Milton Porter (1897), Otto Dunkel (1902), Griffith Evans (PhD, 1910), Lester Ford (1917), and, posthumously, Joseph Walsh (1920). Osgood, on the other hand, numbered but four, of whom Lewis Darwin Ames was the most prominent.

Osgood and Bôcher were the quintessential Klein students. They prospered at Harvard despite initially laboring on one-year contracts that required heavy teaching loads. A relevant quotation from Osgood's memorial to his longtime friend bespeaks of Klein's legacy in America, in general, and the Osgood–Bôcher symbiotic relationship in particular:[59]

> In the early years of our professional lives ... each of us was seeking to clarify and simplify his subject. Neither of us regarded the theory of functions of a real or of a complex variable as an end in itself, for each had his own ulterior uses for the theory—by Bôcher, his differential equations, both complex and real. In fact, for each of us the theory of functions was applied mathematics, and in presenting its subject matter and its methods to our students, our aim was to show them great problems of analysis, of geometry, and of mathematical physics which can be solved by the aid of that theory.
>
> Bôcher was quick to grasp the large ideas of the mathematics that unfolded itself before our eyes in those early years. His attitude toward mathematics helped me to have the courage of my convictions. The *Funktionentheorie* is largely Bôcher's work, less through the specific contributions cited on its pages than through the influence he had exerted prior to 1897—long before a line of the book had been written. We worked together, not as collaborators, but as those who hold the same ideals and try to attain them by the same methods. It was constructive work, and in such Bôcher was ever eager to engage.

Wesleyan. In addition to Bôcher, three graduates of one small but distinguished college, Wesleyan University in the southern New England town of Middletown, CT, also received doctorates under Klein, though there seems to be no apparent direct connection to Klein beforehand. All three would achieve distinction at different American universities. Credit for their solid foundation in mathematics devolves from one of their academic fathers, Wesleyan professor **John Monroe Van Vleck** (1833–1912), who worked at the Nautical Almanac Office from the time of his own graduation from Wesleyan in 1850 at age 17 until 1853, when he returned to his *alma mater*. Not only did Van Vleck teach at Wesleyan until his death nearly 60 years later, he served as acting president of the university on three different occasions. Whatever the source of John Van Vleck's link to Klein, it is truly impressive that three of his best students could survive the rigors of the Klein seminar program to write dissertations. None went directly from Wesleyan to Göttingen; all spent time as Van Vleck's assistant in the astronomical observatory, thus adding substantial maturity that undoubtedly prepared them for the challenges ahead.

Figure 5.13. Henry Seely White

The first was **Henry Seely White** (1861–1943), who had attended the Cazenovia Seminary (in New York) where his father taught mathematics and surveying. Then he enrolled at Wesleyan University to study mathematics and astronomy under John Van Vleck, earning an AB degree in 1882. Henry White stayed on campus the next year as an assistant in the observatory, taught mathematics and chemistry for a year at the Centenary Collegiate Institute in Hackettstown, NJ, and then returned to Wesleyan as a tutor. Following his mentor Van Vleck's advice, he set sail in 1887 for Europe to pursue higher mathematics while also experiencing an atmosphere of collaborative/competitive research. Initially, he went to Leipzig with the aim of studying under Sophus Lie and Eduard Study but, once there, was advised by an American friend, William J. James, to proceed to Göttingen to study with Klein, whom James described as "not only the leading research mathematician but also ... a magazine of driving power."[60] So White ended up spending most of his one semester in Leipzig on improving his fluency in German.

At Göttingen, White thrived under Felix Klein's tutelage. He presented two lectures at the 1888 summer session, one on applications of Abel's theorem and the other on bitangents of C_4. Henry White obtained his doctorate under Felix Klein in 1891 for a dissertation on Abelian integrals. Later in White's career, he was instrumental in establishing successful departments of mathematics at Northwestern and Vassar. An excellent researcher who worked on invariant theory, the geometry of curves and surfaces, and algebraic curves, he served as president of the American Mathematical Society in 1907 and 1908. His most vital contribution to American mathematics might have been the Evanston Colloquium he conducted at Northwestern in 1893. Later in life, White recalled the ordeal of working under Klein[61]:

> Klein received me kindly and admitted me to his seminar course, then just beginning, in Abelian Functions. Other Americans working with him at the same time were ... friendly and helpful to the less experienced neophyte. ... Klein expected hard work, and soon had in succession Haskell, Tyler, Osgood, and myself working up the official *Heft* or record of his lectures, always kept for reference in the mathematical *Lesezimmer*. This gave the fortunate student extra tuition, since what Klein gave in one day's lecture (two hours) must be edited

and elaborated and submitted for Klein's own correction and revision
within 48 hours.

A relevant term in this quotation is the reading room (*Lesezimmer*), which was typical
of life in a German university at that time, but virtually unknown in America until J.J.
Sylvester imported his seminary to Johns Hopkins, one that William Story transported
to Clark in its second year.

The second Wesleyan graduate to obtain a doctorate under Klein was **Edward
Burr Van Vleck** (1863–1943), the son of John Van Vleck. After obtaining his bachelor's degree in 1884, he worked in his father's physics lab for one year before studying
at Johns Hopkins for two more. He gave credit to Thomas Craig for persuading him to
concentrate on mathematics during his time in Baltimore (1885–1887). Edward Van
Vleck then went to Germany for five semesters, an effort that culminated in his 1893
doctorate for a *Doktorarbeit* on Lamé functions. Although he taught at Wesleyan after receiving his degree, he ended up spending most of his career at the University of
Wisconsin, teaching there from 1906 until his retirement in 1929.

Figure 5.14. Edward Burr Van Vleck

Like Harry Fine, Edward Van Vleck had a mathematics building named in his
honor, with Van Vleck Hall being dedicated at Wisconsin in May 1963. Van Vleck Hall
was *not* the site of the bombing that occurred at Wisconsin in 1970, during a protest
against the university's research carried out under contract with the US military during
the Vietnam War. The bombing, which killed physicist Robert Fassnacht, occurred in
Sterling Hall, where such applied investigations were carried out. Sadly, Fassnacht was
neither involved with nor employed by the Army Mathematics Research Center.

Frederick Shenstone Woods (1864–1950) had been a fellow undergraduate student with Henry White and Edward Van Vleck at Wesleyan, receiving his bachelor's
degree in 1885. He was awarded a master's degree from Wesleyan three years later
and then taught at an academy in New York until 1890, when he accepted an instructorship at the Massachusetts Institute of Technology (MIT), known as the Institute of
Technology in Boston until moving to Cambridge in 1916. Harry Tyler, introduced below, had just joined the faculty. The two of them remained MIT fixtures for the next 40
years, with Tyler heading the mathematics department (1902–1930) and Woods (1930–
1934). Shortly after arriving at MIT in 1890, Woods travelled to Göttingen to study

under Felix Klein. He delivered three lectures in Klein's seminars during 1892–1894 on pseudo-minimal surfaces and then minimal surfaces. Frederick Woods submitted a dissertation on pseudo-minimal surfaces in 1894, and was awarded his doctorate the following year.

In the first quarter of the twentieth century, Frederick Woods wrote a series of calculus and advanced calculus textbooks with MIT colleague Frederick Harold Bailey that provided strong links in the chain of especially successful calculus books written by MIT professors, including George B. Thomas (1914–2006).

Other universities. Texts written by Frederick Woods were not Klein's most enduring gift to MIT, however. That honor is bestowed upon **Harry Walter Tyler** (1863–1938), whom Dirk Struik credited with playing an instrumental role in transforming the mission of MIT's mathematics department from a service department into one of the leading research centers in America. Tyler graduated from MIT in 1884 and then joined the faculty as an assistant. Three years later he traveled to Göttingen to study with Felix Klein, but soon found he could not abide Klein's dictatorial style, so he left after only one year. However, he landed on his feet in Erlangen, armed with a thesis topic suggested by Klein on Abelian integrals. Like William Osgood before him, Harry Tyler ended up completing the dissertation at Erlangen instead of Göttingen, receiving his doctorate in 1889. After that, he returned to MIT, where he remained for the rest of his life, heading the department from 1902 until his retirement in 1930. Frederick Woods succeeded him as head.

Cornell became another sphere of Klein's influence in America. Cornell professor **James Edward Oliver** (1829–1895) took a 14-month sabbatical leave to study in Europe. First, he attended lectures by Arthur Cayley at Cambridge, but the Englishman was ailing at the time. Next, he traveled to Germany to study with Klein; that time was much more productive. Oliver wrote, "My work here is likely to be of great service to me, including the trains of thought and plans it suggests, no very radically new plans, only as to the spirit, the aims, and the details of my Cornell work."[62] Subsequently James Oliver advised his best student, Virgil Snyder, to pursue doctoral studies in Göttingen, and Snyder earned his doctorate under Klein in 1895. Thereafter, Snyder returned to his *alma mater* for the rest of his life.

Even Frank Cole himself returned for a second tour of duty under Klein, this time at Göttingen instead of Leipzig. Although Cole had become disillusioned with the lack of mathematical expertise at Harvard in the mid-1880s, he continued to work so hard that he suffered a nervous breakdown. This caused him to resign his Harvard position to take a job as an assistant to a railroad engineer. The outdoor air seems to have cured his condition, and shortly thereafter he set sail to attend Klein's seminars to recharge his mathematical battery. Upon his return to the US, Cole taught at Michigan from 1888 until 1895. He then moved to Columbia University, where he served as the secretary of the American Mathematical Society from 1896 to 1920. This explains why the Society was headquartered on the Morningside Heights campus in New York City throughout this period, a setup that continued until 1951, when AMS headquarters were moved to Providence, RI, where they have remained ever since.

Second woman pioneer. The final American to write a doctoral dissertation under Felix Klein illustrates an important turning point in educational opportunities for American women on German soil. The first American woman pioneer, Christine

Ladd, was not granted her PhD degree from Johns Hopkins in 1882, even though she had completed all requirements for the degree. In 1891 Ladd Franklin accompanied her husband, Hopkins professor Fabian Franklin, on his sabbatical at Göttingen. She petitioned Klein to attend his classes; he in turn entered a motion at a faculty meeting to permit her to be a regularly matriculated student. It was denied: she could attend as an auditor or not at all. This was not her first brush with discrimination. After all, she had encountered this same discrimination at Harvard and Johns Hopkins. Now it was occurring again at Göttingen. Unfortunately, Ladd Franklin opted out of mathematics and instead decided to perform experimental work in a laboratory on vision.

Two years later, however, in April 1893, Klein received a letter from one of his former German students, Heinrich Maschke, then teaching at Chicago, recommending one of his undergraduate students for admission to Klein's program. Klein had the opportunity to meet her when he spoke in Chicago that summer. He was so impressed that he recommended that the student come see him that fall.

She did.

Mary Francis "May" Winston (1869–1959) came from a highly educated family in which all seven children were college graduates. Winston studied classical languages and mathematics at Wisconsin, graduating at age 20. After two years of teaching, she earned a fellowship to Bryn Mawr College. However, at the end of the year she moved back home with her parents to attend the newly opened University of Chicago, which awarded her an honorary fellowship that paid all tuition. One year later, in the fall of 1893, she embarked on a journey to Göttingen to take Felix Klein up on his offer. Another woman came to Göttingen to study under him at the same time, the English-woman Grace Chisholm (later Grace Chisholm-Young). Once again Klein petitioned for full matriculation, and this time it was accepted.

Figure 5.15. Mary Francis "May" Winston New-son

It is curious that German women were not permitted to matriculate at German universities until 1908. Both Chisholm and Winston earned doctorates in mathematics before then. May Winston became the first woman to speak at one of Klein's seminars in December 1893, just two months after arriving in Germany. Her lecture, "The connectivity formulas of the principal branches of the p-functions," was delivered at the seminar on linear differential equations and the p-function. She was awarded an

Table 5.2. Twenty leading members of the Klein Klub

	Seminars	PhD	University	AMS	
				Pres.	VP
Bôcher	1889–1890 (2)	1891	Harvard	1909–1910	1902
Bolza			Chicago		1904
Cole			Michigan, Columbia		1921
Fine		1885	Princeton	1911–1912	1892–1893
Franklin	1891 (1)		Johns Hopkins		
Haskell	1886–1887 (4)	1890	Berkeley		1913
MacKinnon	1894–1896 (5)				
Maddison	1894–1895 (2)				
Maschke			Chicago		1907
Noble	1894–1895 (2)				
Osgood	1888 (1)		Harvard	1905–1906	1903
Snyder	1893–1894 (2)	1895	Cornell	1927–1928	1916
Stringham	1880 (2)		Berkeley		1906
Thompson	1887–1888 (2)	1892	Princeton		
Tyler	1888 (1)		MIT		1923
Van Vleck	1891 (2)	1893	Wisconsin	1913–1914	1909
White	1888–1889 (3)	1891	Northwestern, Vassar	1907–1908	1901
Winston	1893–1895 (4)	1897			
Woods	1892–1893 (3)	1895	MIT		
Ziwet			Michigan		1903

Association for Collegiate Alumnae (ACA) Fellowship for study abroad during 1895–1896, her final year at Göttingen. Her dissertation on differential equations, "*Über den Hermiteschen Fall der Laméshen Differentialgleichung*" was completed by the middle of that year, and she passed her examination in July 1896.

May Winston received her doctorate *magna cum laude* in 1897 after finally finding a publisher in Germany for her dissertation. Two years earlier, Grace Chisholm and another American, Margaret Maltby, received their PhDs, the former in mathematics and the latter in physics, so Winston became the second woman, and the only American woman, to receive a doctorate in mathematics from a European university up to 1900. Before returning to the US, Winston wrote another paper that Klein felt was good enough to be published in the leading German journal, *Mathematische Annalen*, which was much more prestigious than the *AJM* at that time.

Table 5.2 summarizes some facets of 20 of the leading figures in the Klein Klub. The second column lists the years when they presented seminars (with the number of presentations in parentheses). Column 3 lists the dates when Klein's nine (successful) American-born doctoral students received their degrees, while the next column indicates the university (or universities) with which that person became associated. The final two columns illustrate the dominating influence that the Klein Klub had on the emerging American Mathematical Society (AMS).

Not all Americans who desired advanced training in mathematics beyond the bachelor's degree gravitated to Germany. And not all went abroad in search of a doctorate—some aimed to study under one or more specific individuals on American soil.

American Mathematical Society (AMS)

This section describes the American Mathematical Society's founding and initial officers. Today the AMS is one of the most important organizations for research in mathematics in the world, with a membership numbering about 30,000.

What was New York City's role in the early development of mathematics in America? Considering that New York replaced Philadelphia as the most populous city in the New World in 1810 and has been the heart of finance in the US since then, it comes as a surprise that today's "Big Apple" played such a minimal role in the first nine decades of the nineteenth century. To this point, the city was mentioned only in connection with two failed attempts to form mathematics organizations—by George Baron in 1804 and William Marrat in 1817—as well as four failed attempts to launch mathematics journals by Baron (*The Mathematical Correspondent*), Robert Adrain (the *Analyst; or Mathematical Museum*), Marrat (*Monthly Scientific Journal*), and Charles Gill (*Mathematical Miscellany*). In this sense the US differed markedly from Germany and France, where Berlin and Paris stood at the center of national mathematical traditions, though London did not quite play such a dominating role in Great Britain.

Founding. It was a young American's exhilarating trip to Cambridge, England, that paid organizational dividends for the US. While many American students sailed to Europe to pursue higher mathematics, from the mid-1880s through the end of the first decade of the twentieth century, most went to Germany. However, a few went elsewhere, including Thomas Fiske, whose six-month study tour in England inspired a significant step in the development of mathematics in America.

Thomas Fiske (1865–1944) earned his AB degree in 1885 at Columbia, where he remained for graduate classes, receiving an AM the following year. In the spring of 1887, when he was assistant in mathematics as well as graduate student, Columbia professor John Van Amringe suggested that he study for six months in Cambridge, so that fall the dutiful Fiske set sail across the Atlantic. Fiske was armed with letters of introduction to English mathematicians Cayley, Glaisher, Forsyth, and Darwin written by Columbia trustee George L. Reeves, who had been fifth wrangler on the 1872 Mathematical Tripos at Cambridge University (see p. 150). Fiske spent the fall 1887 and winter 1888 semesters in England with the intention of attending Cayley's lectures, but a seeming misfortune struck when Cayley slipped on an icy pavement and fractured his leg, thus ending his series of lectures abruptly. However, it was James Glaisher who became a more important figure, taking Fiske under his wing and allowing him to attend meetings of the London Mathematical Society (**LMS**). The invigorating experience of soaking up mathematics in the company of other like-minded specialists while being entertained "with gossip about scores of contemporary and earlier mathematicians" made an immediate and deep impression on Fiske.[63] He continued:[64]

Figure 5.16. Thomas Scott Fiske

> On my return to New York I was filled with the thought that there
> should be a stronger feeling of comradeship among those interested
> in mathematics, and I proposed to my classmates and friendly rivals,
> Jacoby and Stabler, that we should try to organize a local mathemat-
> ical Society.

In late 1888 the trio of graduate students—Thomas Fiske, Harold Jacoby, and Edward Stabler, all born in 1865—distributed a circular announcing an organizational
meeting for 10 o'clock on Thanksgiving Day, November 24. Yet only three people
besides the organizers showed up—professors Van Amringe and Rees as well as a
fourth graduate student, James Maclay, who had just obtained his bachelor's degree
in civil engineering at Columbia. Undeterred, the group elected officers and spent
the next month writing a constitution. When they met again on December 29, 1888,
they adopted the constitution, so that New Year's gathering became the first official
meeting of the organization they called the New York Mathematical Society (**NYMS**).
The membership planned monthly meetings from October to June, with the December
gathering devoted to business instead of pleasure. As Fiske had envisioned, the Society
remained local, holding all meetings at Columbia College, then located on East 49th
Street and Madison Avenue.[65] It also remained small, enlisting only ten more members
during 1889. The 16 members who joined the NYMS by the end of 1889 became charter
members of the Society.

The initial NYMS meetings were so vigorous that, despite adding just seven more
members in 1890, by the end of that year a deep need was felt to establish a periodical
containing announcements of mathematical activities throughout the western world
and disseminating papers read at those meetings. The idea for such an undertaking
was suggested formally by Harold Jacoby at the December 1890 meeting, inspired by
"a letter from an American who had attended a meeting of the London Mathematical Society ... [that] made reference to its *Proceedings*."[66] (The unnamed American
was not identified. Jacoby later became the head of the department of astronomy at
Columbia.) Thus once again, the LMS served as a model for the fledgling NYMS.

At the meeting held in January 1891, the officers presented a report recommending publication of a periodical of limited scope if sufficient funds could be found to pay
for the enterprise. Thomas Fiske was authorized to seek the views of principal mathematicians throughout the country on the matter. Within a month, he was able to

report on positive responses he had received from several influential figures, including Woolsey Johnson at the Naval Academy in Annapolis, Simon Newcomb and Thomas Craig at Johns Hopkins, and Harry Fine at Princeton. Because of these favorable views, approval was given for Secretary Fiske to seek funds. The membership was electrified. As usual, Thomas Fiske got right to work, initiating a campaign to increase membership to cover the costs of publishing a periodical. Once again, the Londoner James Glaisher played a pivotal role, from England, encouraging his friend Woolsey Johnson to join; the Naval Academy professor thus became the first member residing outside the New York circle. Fiske then mailed a prospectus of the *Bulletin of the NYMS*, and an invitation to join, to various potential members. His campaign was a monumental success, with membership expanding to 89 by April and including such leading figures as Harry Fine, William Byerly, George Halsted, John Van Vleck, and Alexander Ziwet. The geographical diversity—for instance, Ziwet was at Michigan—attested to the growing national interest in the Society.

The 24th AMS member was **Carl Proteus Steinmetz** (1865–1923), a German immigrant who had fled his native country to escape arrest as a socialist, having written articles for a local socialist newspaper that had been banned by Otto von Bismarck. Steinmetz came to the US in June 1889 and gained employment as an electrical engineer, ultimately ascending to chief consulting engineer of General Electric Company. When Thomas Fiske saw a paper by Steinmetz in a German journal, with residence listed as New York, he looked him up and encouraged him to join the embryonic Society. Later Steinmetz revealed that he always wished to devote his life to pure mathematics, but that economics forced him to become an engineer.

Another new member was Frank Cole, who became a vital cog in all aspects of the workings of the Society from the time he accepted an appointment at Columbia in 1895. Because of Fiske's campaign, the NYMS roll expanded from 23 living in the New York City area to 210 located throughout the East in just one year. Moreover, the *Bulletin of the New York Mathematical Society* met the need of these surging numbers by supplying an outlet for the rapidly increasing number of papers read at the meetings.

Initially the scope of the *Bulletin* was restricted to critical and historical articles, to avoid competing directly with the two existing publications that published research in mathematics: the *American Journal* and the *Annals of Mathematics*. The first *Bulletin* issue numbered 32 pages and appeared in October 1891. Raymond Archibald stated:[67]

> 2000 copies were printed for wide free distribution in the continuation for members. By the end of 1891 the membership totaled 210—an increase of 187 in this single year, when the Society blossomed forth from a purely local into a truly national organization; indeed by 1892 it had become international since, apart from Canadians, such men as Cayley, Glaisher, and Mindizábel-Tamborrel, were also members.

The author of the first *Bulletin* article was the Naval Academy's Woolsey Johnson, who had contributed many articles to *The Analyst*. Altogether, the ten issues that appeared from October to July constituted the first volume; it numbered 246 pages.

The year 1894 was a watershed for the Society due to four special events. The first occurred in April when the Chicago mathematician E.H. Moore approached the leadership about the possibility of the Society publishing the proceedings of the International Congress held the previous August. The editors of the proposed book wrote in the Preface:[68]

> Neither the management of the [World's Columbian] Exposition nor the government of the United States had made provision for the publishing of the proceedings of any of the Chicago Congresses. No publisher was found willing to issue the papers at his own risk. ... At last a guaranty fund of $1000 in all was subscribed, $600 by the AMS, and $400 by members of that Society and other mathematicians.

Eventually the Society paid $773.76.

In light of subsequent developments, the most important development in 1894 was the expansion of the AMS beyond the New York City area. With a membership roll including mathematicians located in all parts of the country, it seemed natural for the Society to assume a truly national character. On July 1, 1895, it became officially the *American* Mathematical Society when a new constitution was written and approved. The *Bulletin* was renamed accordingly. Given the English roots of the AMS, it is not surprising that the LMS also published a *Bulletin*,[69] though the physical appearance of the American version resembled more closely another English journal, *Messenger of Mathematics*, which was published by Fiske's mentor in London, James Glaisher.

A third notable change occurred when the summer 1894 meeting was held in conjunction with the American Association for the Advancement of Science. Although 22 attended the Section A meeting and ten delivered papers, only two of the papers were in mathematics proper; the rest were in astronomy. Beginning in 1898, summer meetings were held on a university campus, where mathematicians could be housed in dormitories with their families as part of an affordable vacation. The final significant change in 1894 occurred when AMS President John Emory McClintock, Thomas Fiske's successor, delivered a retiring address. From that time forward Society bylaws required the retiring president to deliver such an address at the annual meeting.

Figure 5.17. John Howard Van Amringe

Officers. The AMS had five presidents in the nineteenth century, all of whom were elected annually (until 1901, when E.H. Moore served a two-year term). The first was **John Howard Van Amringe** (1835–1915; usually called "Van Am"), who held the position for three years, 1888–1890. Of Dutch descent, Van Amringe was basically home-schooled until attending Yale for two years, 1854–1856. After teaching for two years, he enrolled at Columbia, earning a bachelor's degree in 1860. Despite the Civil War demanding the most attention in the country, he was a tutor at his *alma mater* for three years before being appointed to the faculty. Van Amringe remained at Columbia for the rest of his life, serving as department head 1892–1910. He also served as dean of two colleges at Columbia.

John Emory McClintock (1840–1916) served as NYMS president for four one-year terms, 1891–1894. He attended Dickinson College in his hometown of Carlisle, PA, at age 14 for two years; his father was initially a professor of mathematics at Dickinson, but then turned to classics. Our McClintock then transferred to Yale for a year before matriculating at Columbia, where he graduated in 1859 at age 18. Like his predecessor, he became a tutor before sailing to Europe, where his father was in charge of signal service to the Union during the Civil War. While abroad, John McClintock pursued higher studies both in Paris and Göttingen. He returned to the US in 1867 and held actuarial positions at three insurance companies for the rest of his life. McClintock "was universally recognized as the foremost actuary America had produced."[70] In addition to publishing numerous papers on actuarial science, he wrote 22 papers on pure mathematics.

Figure 5.18. John Emory McClintock

The next two presidents might be considered astronomers today. George Hill served as AMS president 1895–1896 and Simon Newcomb 1897–1898. The fifth AMS president, Michigan native **Robert Simpson Woodward** (1849–1924), who served 1899–1900, also did work in astronomy. Woodward graduated from the University of Michigan with a bachelor's degree in civil engineering in 1872. Over the next 21 years, he held engineering positions on the US Lake Survey, the US Venus Commission, the US Geological Survey, and the US Coast and Geodetic Survey. Only in 1893 did he accept an academic position as professor of mechanics at Columbia, adding mathematical physics to his title six years later. However, he left Columbia just two years after that for the newly founded Carnegie Institution of Washington, where the multitalented Woodward served as president 1901–1921. Even during his engineering years, his work evinced a strong mathematical flavor. For instance, he 1) determined deformations of the Earth's surface as a result of removing or adding a load over a large area, 2) studied field methods for primary and secondary triangulations, and 3) developed the iced-gar apparatus for measuring baselines and calibrating steel tapes for topographical mapping.

Figure 5.19. Robert Simpson Woodward

The fact that four of the first five presidents of the AMS were specialists in applied mathematics reflects ongoing specialization within the sciences, as two other professional societies were formed in 1899—the American Astronomical Society and the American Physical Society. The latter association of physicists held its organizational meeting in conjunction with the October meeting of the AMS. Two months later "the two societies met in joint session for the purpose of listening to the presidential address of Professor R.W Woodward."[71] What a nice way to ring in a new century!

At the time of the founding of the NYMS in 1888 there were only two officers, president and secretary. A vice president was added two years later, and a second in 1897. **Charlotte Angas Scott** (1858–1931) was the first woman vice president, elected in 1906; no woman would hold the position for another seventy years. Even that late occurrence predated a female president, Julia Bowman Robinson who held the office 1983–1984. Although there have been several women vice presidents since Mary Gray served in that position 1976–1977, there has only been one other female president, Cathleen Synge Morawetz in 1995–1996.[72]

The AMS established a series of Colloquium Lectures in 1896. The impetus came from Henry White, who had hosted Felix Klein's eminently successful Evanston Lectures three years earlier. Thirteen members, including two women, attended the first Colloquium, in which Yale's James Pierpont[73] presented a series of six lectures on "Galois theory of equations" and Harvard's Maxime Bôcher delivered six on "Linear differential equations and their applications." That year's October meeting was held in Princeton as part of the university's sesquicentennial celebration, at which time its name was switched from the College of New Jersey. One eminent speaker at those festivities was Felix Klein, who delivered his final address in the US.

What is more, activity within the Society increased so rapidly that the seventh volume of the *Bulletin of the NMYS* (1897–1898) contained 580 pages. Early issues provide evidence of the increasing number of Americans interested in research, but they also serve as an outstanding repository of AMS activity, allowing us, for example, to read about Felix Klein's special lecture to the Society in September 1893 during his six-week American tour. The *Bulletin* also cited another famous German mathematician, Edouard Study, for an unexpected appearance at the AMS meeting a month later.

The next two chapters provide details of the increasing number of mathematical activities that took place in the US throughout the decade of the 1890s.

6

Chicago and Clark

Mathematics in America grew by leaps and bounds in the final quarter of the nineteenth century, first domestically at Johns Hopkins under J.J. Sylvester, and then abroad, principally under Felix Klein. Another domestic leap forward was the founding of two universities harboring grand visions based on the research of their faculties—Chicago and Clark. This chapter provides details of their successes as well as their failures. Serendipitously, the University of Chicago was founded at the time of the famous Worlds Fair in the "Windy City," and its mathematics department head E.H. Moore took full advantage of the opportunity by sponsoring a meeting sometimes called the "zeroth" International Congress of Mathematicians. The Chicago Congress was followed by the two-week Evanston Colloquium conducted by Felix Klein. Following those successful celebrations of the emerging community of American research mathematicians, Klein undertook a tour of Eastern universities that "raised the bar" at several where his former students were establishing thriving programs. Along the way, a singular student under Klein at Göttingen was Mary Winston, arguably the most talented American woman mathematician of her era.

American colleges in 1890

What was the state of American higher education in 1890? Numerous universities and colleges had been founded up to that point but, unlike their European counterparts, they did not adhere to one overarching philosophy determined by a central government. For better or worse, in the US institutions of higher learning arose haphazardly, whether at large state universities benefiting from funds from the Morrill Act or at smaller colleges founded by wealthy financiers or religious groups. The best source for information on mathematics at the leading colleges of the time is the 1890 report by Florian Cajori for the US Bureau of Education called *The Teaching and History of Mathematics in the United States*.[1] The second half of my copy of the book contains an analogous report filed the next year on instruction in chemistry and physics, but I restrict our attention to mathematics.

In addition to colleges covered in the preceding chapters, Cajori's *Teaching and History* also supplies information about state universities in Virginia, North Carolina,

South Carolina, Alabama, Mississippi, Kentucky, Tennessee, Texas, Michigan, and Wisconsin, as well as the private schools Dartmouth, Bowdoin, Georgetown, Virginia Military Institute, Tulane, and Washington University. Not one of these institutions had a faculty capable of carrying out original research in 1890. Two universities not mentioned by Cajori are examined because they were just opening when he submitted his report to the Bureau of Education.

Stanford University. The name Leland Stanford Junior University is somewhat misleading because it sounds like the name of a junior college, so today it is generally referred to as Stanford University, but there is a historical reason for the expanded title. The university was founded in 1885 by Leland and Jane Lathrop Stanford as a memorial to their only son, Leland Stanford Jr., who died of typhoid fever in Europe a few weeks before his sixteenth birthday. Therefore, the "Junior" in the university's name refers to their only child. The founders used their farmlands to create Stanford as the first major research university in the West, co-educational from the start and with no tuition (initially!).

Unlike Johns Hopkins, **Amasa Leland Stanford** (1824–1893) was well educated, enrolling in the Clinton Liberal Institute (in New York) before studying law at Cazenovia Seminary (1841–1845). He practiced law in Albany until 1848, when he moved to Wisconsin, where he married the daughter of a Milwaukee merchant two years later. Their only child was Leland Stanford, Jr. (1868–1884). Stanford Sr. served as district attorney before the lure of the gold rush impelled him to seek his fortune in California in 1852. He went into business with five brothers who had preceded him to the "Golden State." Four years later he settled in San Francisco, and by 1861 he was one of the "Big Four" railroad magnates who founded the Central Pacific Railroad. As president of the company, he administered the construction of the first transcontinental railway. Over the years, the Central Pacific merged with Wells Fargo, and Stanford served as director of the merged company from 1870 until his death in 1893. He was elected the first Republican governor of California in 1861 and US senator in 1885. He also owned two wineries and a horse farm in the Palo Alto region. The Palo Alto Stock Farm became famous for its thoroughbred horses, which explains why the Stanford campus is sometimes called "The Farm." His family wealth was estimated at $50 million at the time of his death from a heart attack in 1893 (about $1.4 billion today).[2]

Stanford University officially opened on October 1, 1891, with 15 faculty members and 559 students (including the future president, Herbert Hoover). Initially, a cap was established on female enrollment because of the concern of **Jane Eliza (Lathrop) Stanford** (1828–1905) that the school might become an all-female institution, which she deemed inappropriate for the memory of her son. She supervised the university's affairs from the time of her husband's death until her own 12 years later. Unfortunately, some of her edicts were detrimental, such as building memorials to Stanford family members while university faculty members were living in borderline poverty.

Leland Stanford had attempted unsuccessfully to lure an Ivy League professor to head his new university, so he settled for **David Starr Jordan** (1851–1931), who had been president of Indiana University since 1885 when, at age 34, he was the youngest university president in the US. Jordan ended up being the president of Stanford for 25 years, from its opening in 1891 until his retirement in 1916. Nine years later he was an expert witness for the defense in the Scopes Trial.

Figure 6.1. Amasa Leland Stanford

I know only two facts about Stanford's first mathematics professor, **Joseph Swain**. First, he played in a baseball game in November 1891, six weeks after the university opened, between teams of students and faculty members (including President Jordan). How talented were they? "Unfortunately, there's no box score for that inaugural contest, so hitting ability (or lack of it) cannot be determined."[3] Second, a memo sent to President Jordan in November 1896 from Mrs. Stanford indicates that she was "unable to make Swain vice president now."[4]

Joseph Swain was not President Jordan's first choice to head the mathematics department in the new university. Jordan had negotiated with the Scottish mathematician Thomas Muir (1844–1934), who accepted a position instead as Superintendent General of Education for South Africa after being appointed by Cecil Rhodes, then prime minister of Cape Colony. Muir might have accepted the offer except his wife was in poor health, having been advised to move to a warmer climate, though one should also take into account Rhodes's reputation for the power of persuasion.

Apparently, Swain remained at Stanford but vacated his position when another Scottish mathematician, **Robert Edgar Allardice** (1862–1928) accepted the professorship of mathematics in 1892. He had been assistant professor of mathematics at the University of Edinburgh and a founding member of the Edinburgh Mathematical Society before being recommended for the job by the chair of his department, George Chrystal. Once settled in California, Allardice was no longer able to maintain close ties with mathematicians in his native Scotland, although he visited several times. He remained at Stanford until retirement at age 65, a year before his death. During his tenure, he published many papers, mostly on topics in geometry, in the *Bulletin* (and later the *Transactions*) of the AMS, as well as the *Annals*, yet he also managed to submit papers to the *Proceedings of the Edinburgh Mathematical Society*. He served as the second chair of the San Francisco Section of the AMS (1903–1904), a two-year position he held four more times, the last one in 1926–1927.

In a memorial address, Allardice's colleague Douglas Houghton Campbell wrote that he "had marked social gifts, was an admirable talker with a fund of interesting and amusing anecdotes, and he had a keen interest in games of various kinds, in some of which he excelled."[5] Who was Campbell? The Stanford Historical Society's "House and Garden Tour" in April 2006 featured the following description of his home which also indicates his role in importing the national sport of Scotland to the US (and perhaps underlines why the golfer Tiger Woods attended Stanford):

> A 1914 Italian Renaissance Eclectic style stucco house built for Douglas Houghton Campbell, professor of botany in Stanford's pioneer faculty, and Robert Edgar Allardice, professor of mathematics, who brought golf to Stanford in the early days. ... After a checkered history as a student rental, flood, earthquake, and house fires, the home has been beautifully restored.

Another mathematician soon joined Robert Allardice on the faculty, **Rufus Lot Green** (1862–1932). Born and raised in Indiana, Green studied under Daniel Kirkwood and David Starr Jordan at Indiana University (1879–1881). Green then transferred to Cornell, where he stayed for two years before returning to his home state. He earned a BS degree at Indiana in 1885 and was immediately hired as an instructor by university president David Starr Jordan. Over the next three years, Green prepared for the MS degree he received in 1888. He spent the year 1887–1888 at Johns Hopkins, but departed without receiving a degree. Two years after becoming president of Stanford, Jordan hired Green as associate professor of mathematics (in 1893). Green remained at Stanford for 34 years until his retirement in 1927, at which time he was executive head of the mathematics department. He introduced several new courses to Stanford over his long tenure, but the one that particularly interested him was statistical mathematics.

A recent article details Green's volunteer work in assisting victims of the 1906 San Francisco earthquake when he marshaled scores of students to help him overcome the difficult problem of bread supply; apparently Green was responsible for dubbing the relief station he supervised "Camp Stanford."[6] One of Green's former students summarized his professional life as follows:[7]

> The name of Professor Green does not stand out in the annals of science, nor will historians of science record any great achievements of his in mathematical research or published monographs—but in the hearts of his students are indelibly impressed the sterling and modest qualities of a quiet and unassuming teacher, endowed with high ideals of true scholarship. For a period of over 40 years, teaching mathematics was his one great task.

Swain, Allardice, and Green show the humble beginnings of what is now, some 100 years later, one of the top research departments of mathematics in America. The first Stanford mathematician of renown was Hans Frederick Blichfeldt, who arrived on campus as an undergraduate in 1894 and joined the faculty three years later.

The Clark story

The founders of Johns Hopkins and Clark Universities were similar in two ways relevant to our account: little formal education and the founding of eponymous institutions. Yet it was the well-educated Leland Stanford who inspired Jonas Clark to establish his own university. The two successful businessmen had struck up a friendship that resulted in Clark founding a university with such alacrity that its classes started in the fall of 1889, two years before Stanford opened its doors.

Jonas Gilman Clark (1815–1900) was born and raised about 20 miles northwest of Worcester, MA, where he came to appreciate education—a love of reading, books, and libraries—from his mother. Clark ran successful manufacturing businesses initially in furniture and later tin-ware. He married Susan Wright in 1836; childless, the couple became inseparable for the next 64 years. Both were ardent supporters of the antislavery movement and helped raise money to support the Civil War. Like Stanford, Clark moved to San Francisco by sea in 1853, becoming the largest furniture wholesaler west of the Rockies. He also purchased considerable land in the Bay Area and became a partner in both an insurance company and a water supply firm. However, ill health forced him to sell his West Coast enterprises, whereupon he moved to New York City in 1868 and began collecting rare books and works of art. During this time the Clarks made four trips to Europe, where visiting European universities inspired a passion for higher education. In 1878 they liquidated their California and New York area holdings, and moved to Worcester. By the fall of 1885 Clark had purchased nearly a full block in Worcester with the idea of establishing a university along the lines of Johns Hopkins. He also assembled a group of prominent citizens to form a board of trustees to oversee the new university. Legislation enacting Clark University in Worcester was passed and signed into law in March 1887.

Figure 6.2. Jonas Gilman Clark

Clark intended to offer an undergraduate division first and add graduate and professional programs later. His immediate task was to hire a president and faculty and to build a library. He pledged roughly $1.5 million (about $45 million today) for the construction of, and equipment for, buildings, an endowment fund to support purchases of books for the library, and funds to cover operating costs. However, several board members had associations with Harvard, and feared that a first-class undergraduate college might detract from the long-established college, compelling them to pursue a hidden agenda to found Clark as a post-graduate institution only.

These trustees, with Clark as president of the board, hired **Granville Stanley Hall** (1844–1924) to be the university's first president. Stanley Hall possessed impressive credentials, having studied psychology in Europe for several years before earning a PhD from Harvard in 1878. (His degree was in philosophy, since there was no psychology department at the time.) He had been a professor of psychology at Johns Hopkins since 1881, meaning that his tenure overlapped with Sylvester's. Initially, Hall expressed no interest in heading a new university until he realized he might be able to create a prestigious one catering only to graduate students. In his letter of acceptance, he stated this view quite bluntly, asserting he had no interest in "organizing another College of the old New England type, or even the attempt to duplicate those that are best among established institutions old or new."[8]

Hall got his way: Clark would train future researchers instead of first establishing an undergraduate college to serve as a feeder for the graduate program. Thus, when Clark University opened its doors in October 1889, it was hailed as the first all-graduate university in America. In the meantime, the board had sent Hall on an eight-month "pedagogic tour" of Europe to study the most recent educational practices, to purchase apparatus and books, and to recruit distinguished faculty members in the five departments that would constitute the university: mathematics, physics, chemistry, biology, and (given Hall's specialty) psychology. For mathematics, the obvious choice to head the department was Felix Klein, so Hall, like Gilman at Johns Hopkins, entered negotiations. However, among other reservations that Klein harbored, Hall reported to Clark from Göttingen that "his wife is so opposed to going to America ... that even if called he could not leave Germany."[9] Hall also conducted unsuccessful negotiations with Vito Volterra. Consequently, the Clark president, unlike Gilman at Johns Hopkins, returned from Europe empty-handed.

Faculty. Who were the qualified American candidates to head the mathematics department at the fledgling Clark University in 1888? The short list might include three faculty members at Johns Hopkins (William Story, Thomas Craig, and Fabian Franklin), two at Yale (Hubert Newton and Willard Gibbs), William Byerly (at Harvard), George Halsted (at Texas), and the nonacademician George Hill. Hall had his eyes on his former Johns Hopkins colleague William Story, but what would compel Story to leave Baltimore, where he was directing a doctoral dissertation almost every year? It didn't take Hall long to determine that Story harbored several reasons for considering a move to a new institution. Mainly, Story felt he had never been fully appreciated. Worse, he felt mistreated on several occasions during his 13-year tenure at Hopkins, being dismissed as editor of the *American Journal of Mathematics*, passed over in favor of the "hybrid" (astronomy and mathematics) Simon Newcomb as head of the department, and having remained associate professor since that position was established in 1883. On the flip side, going to Clark would enable him to develop the kind of graduate department he had envisioned when he first accepted appointment at Johns Hopkins. Besides, the promised lighter teaching load would enable him to pursue research more completely.

Therefore, William Story accepted the offer to be the first head of the mathematics department at Clark University. This was certainly a feather in the fledgling university's cap, but in the process Johns Hopkins lost much of its pedagogical and organizational focus.

Hall and Story set about hiring faculty members at once, settling on two young mathematicians, one foreign and the other domestic. The latter was drafted from Johns Hopkins, adding to the subtraction from the mathematics department in Baltimore. **Henry Taber** (1860–1936) had gone to Johns Hopkins as a graduate student in 1882 after having earned his bachelor's degree from Yale's Sheffield Scientific School. Although J.J. Sylvester chaired the department for the next three semesters, Taber did not take courses from him, instead electing higher plane curves and conic sections with William Story, elliptic functions with Thomas Craig, as well as logic and probability with C.S. Peirce. After Sylvester departed, Taber took virtually every course that Story offered, including number theory, synthetic geometry, advanced analytic geometry, quaternions, and linear associative algebras. In the meantime, chronic illness forced Taber to withdraw from the program for a year and a half, so it was June 1888 by the time he was awarded the PhD with the dissertation, "On Clifford's n-fold algebras." Although no thesis advisor is listed, all evidence points to Story. Taber was then appointed as "Assistant in Mathematics" at Johns Hopkins, but by the time his dissertation appeared in the *AJS* in 1890, he had accepted Story's offer to join him on the faculty at Clark.

Taber was appointed a docent, a position he held for three years before being promoted to assistant professor in 1892. At the time, that position, which no longer exists, was defined as:[10]

> The highest annual appointment is that of docent. These positions are primarily honors, and are reserved for a few men whose work has already marked a distinct advance beyond the Doctorate who wish to engage in research.

In the spring of Taber's only year on the Johns Hopkins faculty, he attended two lecture courses, one on Abelian functions taught by Thomas Craig and the other "Theory of substitutions" by **Oskar Bolza** (1857–1942). Bolza was born and raised in southern Germany, had enrolled at the University of Berlin at age 18 and had studied physics for two years (1876–1878) under leading physicists Hermann Helmholtz and Gustav Kirchhoff. However, he did not find experimental work to his liking, so he changed to mathematics, taking courses from Karl Weierstrass at Berlin, Hermann Schwarz at Göttingen, and Elvin Christoffel at Strasbourg. Weierstrass's 1879 course on the calculus of variations had an effect on him for the rest of his life. However, Bolza did not find an appropriate research topic for a dissertation, so he took examinations to become a secondary school teacher in 1882. Nonetheless, he soon discovered that a life of research was more attractive, so he began working independently on hyperelliptic functions. But then a nightmare occurred, the kind that nags every advanced graduate student—he ascertained that not only had his results already been anticipated by the Frenchman Edouard Goursat, but Goursat's methods were considerably more elegant. Some graduate students end their pursuit of an academic career upon such a discovery because dissertations must present *original* contributions. Bolza, on the other hand, was emboldened to compete with world-class mathematicians, and that inspired him to return to Berlin to work with Leopold Kronecker and Lazarus Fuchs. However,

Figure 6.3. Oskar Bolza

correspondence with Felix Klein on his generalization of Goursat's results compelled Bolza to return to Göttingen to be supervised by Klein. This proved successful, resulting in his doctorate in 1886.

However, while in Berlin, Bolza had developed a very close friendship with Heinrich Maschke that would last a lifetime. Upon graduation, Maschke joined Bolza in Göttingen as a postdoctoral research assistant for Klein, but after a year Bolza left to spend a year in England learning English to migrate to the US. He had been encouraged to blaze such a trail by a physicist friend who had found gainful employment at the Thomas Edison laboratory in New Jersey and by two of Klein's students from the American expatriate colony in Göttingen, Mellen Haskell and Frank Cole. Thus, the spring of 1888 found Bolza in Baltimore to present a strong letter of recommendation on his behalf written by Felix Klein to the department head at Johns Hopkins, Simon Newcomb. Newcomb, however, was discouraged about the state of American education. He wrote to Klein to dissuade him from recommending other German mathematicians to come to the US, stating, "We have indeed several hundred so-called colleges but I doubt that if one half of the professors of mathematics in them could tell what a determinant is. All they [administrators] want in their professors is an elementary knowledge of the branches they teach and the practical ability to manage a class of boys, among whom many will be unruly."[11]

However, Newcomb's message was too late for Bolza, who was hired for the spring 1889 semester to teach his classic course on substitutions, which exposed Americans to Galois theory for the first time. The course met five hours a week for four weeks. Ten audited the course, including William Story, who was duly impressed. Besides, Felix Klein had introduced President Hall to Bolza during Hall's European tour. At the end of the four weeks, Hall and Story extended to Bolza the kind of offer he had envisioned when he left Germany. He accepted on the spot.

In addition to the three Johns Hopkins hires—Story, Bolza, and Taber—the spanking new university hired renowned physicist **Albert Abraham Michelson** (1852–1931) in its first year. In fact, Michelson commanded the highest salary among faculty members, a sign that you do not need advanced degrees (he earned none) when your work is superior. Michelson had been born in East Prussia (in present-day Poland) but came to the US with his family at age two. He was raised in mining towns in Nevada and California before graduating from high school in San Francisco in 1869. Michelson was then appointed to the Naval Academy by US president, Ulysses S. Grant. He graduated four years later, spent two years at sea, and then returned to the Naval Academy in 1875 as an instructor in physics and chemistry. During his four years in this position, Michelson refined his measurements of the speed of light. In 1879 he was posted to the National Almanac Office in Washington to work with Simon Newcomb. But after only one year, he took a leave of absence to study in Berlin, Heidelberg, and Paris for two years. He resigned from the Navy upon his return to the US in 1881, and two years later was appointed professor of physics at the Case School of Applied Science in Cleveland. (The Michelson House is located on what is now the Case Western Reserve campus.) While there, he and Edward Morley conducted the famous Michelson–Morley experiment negating the existence of ether. Michelson remained at Case until his appointment at Clark in 1889.

The three competent mathematicians (Story, Bolza, and Taber) and one world-class physicist (Michelson) thrust Clark into the forefront of higher educational institutions at once. The second year bore witness to yet three more impressive appointments, two in mathematics (Joseph de Perott and Henry White) and one in mathematical physics (Arthur G. Webster).

Henry White played an important role in the emergence of American mathematics. Felix Klein introduced Stanley Hall to the young White when the Clark president was in Europe; apparently, Klein gave "flattering testimonials."[12] White returned to the US to teach in a prep school associated with Northwestern for the spring 1890 semester. Then President Hall offered him the position of assistant at Clark, which he accepted even though "the salary was hardly adequate for subsistence."[13] White's first year was quite productive (he published two papers) and successful (he was awarded his Göttingen doctorate in 1891 for a dissertation on Abelian integrals).

Joseph de Perott (1854–1924) was born and raised in Saint Petersburg of a Polish father and Russian mother. He left Russia at age 15 and probably lived in Portugal and Spain. Little is known about his whereabouts in the crucial period of the 1870s, so it remains a mystery where he acquired knowledge of advanced mathematics. He was in Paris in 1881, however, and that might explain why he pronounced his name "Pair-oh," as if it were French. The "de" in the name was granted by Napoleon because one of Perott's grandfathers saved the life of an important French general. Curiously, he always signed his first name Joseph, whether writing in French, German, or English.

Perott studied in Paris (1877–1878) and Berlin (1878–1880), yet he never earned a degree. During this time, he maintained a strong personal relationship with one of the world's leading mathematicians, Sophia Kovalevskaya, in both mathematical and literary activities.[14] It is not known when he immigrated to the US, but an unusual exchange of letters between him and the president of Johns Hopkins at the end of 1889 and beginning of 1890 reveal a somewhat disagreeable disposition and an unusual occurrence. Perott read in a Hopkins *Circular* that he had been awarded a Fellowship by Courtesy, a position he felt was significantly below his credentials based on his published work. He balked that he had not been a student for nine years, and deserved a faculty position. Therefore, he resigned his fellowship before ever holding it. Perott's output was somehow more appreciated at Clark, where he was appointed docent in 1890, which required only two hours a week. Although the salary was low, the frugal Perott managed to live on it without need of additional compensation.

Joseph Perott became quite a controversial character in the town of Worcester because of his bohemian lifestyle. This enigmatic figure came to the US with long hair, yet never cut it after that, and he generally tucked his mass of shoulder-length hair under a derby he wore jammed tightly over the top of his head. Perott remained at Clark for the rest of his career, though he never became a citizen.

The other figure who joined the Clark faculty in the fall of 1890 was the mathematical physicist **Arthur Gordon Webster** (1863–1923), who had graduated at the top of his Harvard class of 1885 and had remained there the next year as instructor in mathematics and physics. He then was awarded a Parker Fellowship that he used to study in Berlin, Paris, and Stockholm, culminating in a PhD in physics from the University of Berlin in 1890 based on a dissertation written under Hermann von Helmholtz. Albert Michelson was also studying in Berlin, so undoubtedly the two Americans became acquainted there. Upon graduation, Webster returned to the US to accept appointment as docent in mathematical physics at the upstart Clark University. Surprisingly, it was not Michelson who brought Webster to President Hall's attention, but a Massachusetts senator on the Clark Board of Trustees who had been a Harvard classmate of Webster's father.

Arthur Webster was a proficient mathematician and quite competent in experimental physics, and his lectures on mathematical physics became classics. Delivered at blinding speed, these lectures were more detailed and comprehensive than anywhere in the country. During his 33 years on the Clark faculty, he directed 27 doctoral students. He also wrote two classic textbooks in theoretical physics that were important in advancing physics education in America: *The Theory of Electricity and Magnetism* (1897) and *The Dynamics of Particles and of Rigid, Elastic and Fluid Bodies* (1904). Moreover, his posthumous *Partial Differential Equations of Mathematical Physics* (1927) was translated into German by Gabor Szegö before the translator immigrated to the US. Webster was also a founder of the American Physical Society, which he helped launch in May 1899 (almost 11 years after the AMS was formed).

Arthur Webster was a brilliant classical scientist and leader in the American physics community, yet his ultimate demise was quite sad. Early in his Clark career, he relished the many startling discoveries that took place in the 1890s: Röntgen's discovery of X-rays, the detection of radioactivity and subsequent work on atomic transmutation by Marie and Pierre Curie, and Thomson's characterization of the electron. However, by the early 1920s, he was distressed by the growing acceptance of relativity and quantum theory. Adding to his woes were problems at Clark. Thoroughly depressed, Webster took his own life by pistol in May 1923, six months before reaching the age of 60.

Matters at Clark were much more positive at the time of its founding, however, especially in mathematics. The department that William Story assembled was young and vigorous. Story himself was only 40 when he moved to Clark in 1889. By the next fall Bolza was 33, Taber 30, White 29, and Perott 36. Moreover, the physics department included Michelson 38 and Webster 27. This faculty was arguably the best in the US for graduate and postgraduate education. Moreover, Clark University had the ability to retain its standing among the tops in the country for decades, yet several faculty members departed in 1892. Because the department thrived at such a high level during its first three years, the period 1889–1892 can be called the "golden age" of mathematics at Clark University.

A panoply of attractive courses was available to advanced students from the first year (1889–1890) when the mathematics faculty numbered only three. William Story taught seven hours a week, slightly more than half the 13-hour load he had carried at Johns Hopkins. He used this time to lecture on nine different topics, four of which could easily fit into a twenty-first-century program (modern algebra, number theory, finite differences, and probability theory) and five others that were modern at the time but are no longer fashionable (plane curves, surfaces and twisted curves, quaternions, synthetic geometry, and analytical mechanics). Oskar Bolza offered three courses that are sophisticated even by today's standards: calculus of variations, elliptic functions, and theory of functions. In addition, Henry Taber taught a course on the theory of matrices based on Benjamin Peirce's *Linear Associative Algebras.* Moreover, graduate student **Rollin Arthur Harris** (1863–1918), holding PhB (1885) and PhD (1888) degrees from Cornell, lectured on the use of analytic function theory in the construction of mappings.[15] Harris had published three papers in the *Annals* before even matriculating at Clark on a fellowship; two of them were cited in the Smith–Ginsburg classic, *The History of Mathematics in America before 1900.*[16] Upon leaving Clark at the end of the year, he spent his entire career at the US Coast and Geodetic Survey as a computer/mathematician. His 1200-page book *Manual of Tides* was regarded as indispensable for predicting tides.

There were no new faculty appointments in Clark's third year, 1891–1892, yet the department prospered, awarding its first PhD at the end of the year. Unfortunately, however, problems brewing below the surface beset the faculty with growing discontent. It turns out that President Hall had made numerous extravagant promises to faculty members for lab equipment (for mathematics this meant halving the number of journals in the departmental library), had forbidden the faculty from convening to discuss university affairs, and had disciplined some faculty members without consulting the heads of their departments. Hall cast blame on either the trustees or Jonas Clark over these matters. Moreover, he continually misled Jonas Clark concerning the financial health of the university and kept delaying the opening of the undergraduate division that Clark favored. Before long, Clark University had two warring camps, the founder and most of the faculty in one group aligned against the president, and most trustees in the other. Relations became so strained that several trustees refused to ante up pledged donations, thus sending the university's finances tumbling. In the spring of 1891, Clark himself resigned as treasurer of the board of trustees but retained his position as president. From then on, he communicated with Hall and the board mainly by letter.

What did this mean for mathematics? In a word, *trouble*, not because of animosities within the department (it appears that there were none), but the sour mood pervading the campus impelled some to investigate other opportunities. And change did not take long to occur: Henry White and Oskar Bolza, as well as the physicist Albert Michelson, left Clark at the end of the spring 1892 semester; White went to Northwestern, and Bolza and Michelson went to Chicago. In fact, Michelson became the first head of the physics department at Chicago when it opened in 1892 and, 15 years later, became the first American to win a Nobel Prize.

While these understandable defections did not decimate mathematics and physics, the vacancies were not filled, leaving all mathematics instruction in the hands of Story, Taber, and, to a much lesser extent, Perott. These three loyal faculty members constituted the department until 1921, with two exceptions: **Reginald Bryant Allen** (1872–1938), who earned a PhD under Henry Taber in 1905, was an instructor the next year, and then became a longtime professor at Kenyon College; and **Frank Blair Williams** (1871–1933), who earned his PhD in 1900 under William Story, returned to the department as an instructor in 1910 and remained at Clark for the rest of his career. Nonetheless, Clark's golden age was clearly at an end, only three years after it had commenced.

Inglorious end. In the fall of 1892, Jonas Clark strongly urged the board to let most contracts expire and to reallocate resources toward opening the undergraduate program, which would increase tuition revenue and potentially attract donors. Hall viewed this as dismantling the university and argued that additional scientists who had come for special opportunities in research and advanced instruction would leave if asked to teach undergraduates. The board, guided as always by Hall, rejected Clark's proposals and voted in December 1892 to continue down the same path they had blazed. This vote was the final straw for Clark. After scrupulously fulfilling his pledge for 1892–1893, he gave the university no more money during his lifetime. Nevertheless, he was a crafty character who would eventually get his way, even after his death in May 1900.

Ironically, the previous year Clark University had commemorated its first ten years in style with a celebration that included five eminent foreign scientists who lectured on their recent work, including Émile Picard in mathematics and Ludwig Boltzmann in physics. Their papers, including three by Picard running to almost 60 pages, were included in a mammoth tome on the meeting that included a historical sketch of the university.[17] Of more relevance here, however, are the ten-year "state of the department" accounts, the one for mathematics being written by the chair, William Story. This 24-page report, replete with lists of faculty and students over the decade, demonstrates Story's prescience and his vision for a department that all twenty-first-century mathematicians will recognize.[18]

> What I have in mind as a model mathematical department for post-graduate work would have, say, four professors and assistant professors, each having his personal assistant, and at least two instructors of lower grade for the more elementary work, and would be provided with a complete mathematical library with all the apparatus that is now possible to procure, with suitable provision for the purchase of new books and apparatus as they appear on the market.

Story's paragon is for a smaller college, so a modern one would increase the numbers he suggested, but otherwise, most present-day research departments are similar. By "personal assistant," Story referred to what are called postdocs today, writing:[19]

> We ought to have the means of retaining our best graduates [PhDs] for a year or two as personal assistants to the instructors, during which period they might also be gaining experience in the class-room by teaching a few hours a week under the supervision of one of the regular instructors."

Story indicated that the library must shelve not only books but complete sets of journals and transactions of learned societies. By "apparatus" he meant physical models—string as well as plaster of Paris; if one substitutes computers instead, you visualize a modern department. There is one element that Story did not foresee, however: buildings for mathematics with private offices for the faculty. His library contained space for holding seminars, and today these two items are separate, but other than that, he envisioned what would take place at Princeton and Chicago in the late 1920s.

Despite the enthusiasm that the decennial celebration generated, deep-seated administrative ruptures ultimately ruined the great idea of a new university. In 1900 the will of Jonas Clark stipulated that his estate would be divided among relatives (the Clarks had no children) unless the university met his conditions, the most important being that the undergraduate Clark College would operate independently of Stanley Hall under a separate president who would, like Hall, report directly to the board of trustees. In addition, Clark created separate endowments to keep Hall and the trustees from diverting college and library funds to graduate programs. Even more pointed was a provision preventing the merger of Clark College into Clark University until Hall resigned or retired. And that is exactly what happened, mainly because Clark's bequest totaled \$2,915,000, (almost \$90 million today). The college and university merged in 1920 only when President Hall retired.

Wallace W. Atwood, the new president of the combined institutions called Clark University, was no friend of mathematics.[20] In 1921, this geographer discontinued graduate work in mathematics and forced the department's three professors to retire. By then, Story was 71, Perott 67, and Taber 61. Story and Taber had been active researchers and mentors throughout their lives. Atwood's move put an inglorious end to an idea that had been started with such promise but whose potential went unrealized. After Story retired, he served as president of the Omar Khayyam Club (1924–1927), indicating a longtime interest in the history of mathematics. He died of pneumonia in April 1930 after a brief illness. Henry Taber had raised his three daughters alone after his wife's untimely death in 1892, after only six years of marriage. He devoted his retirement to a wide range of interests, including chemistry, history, literature, music, and dancing; he also belonged to a boat club and was an excellent tennis player. I know nothing about Perott's ultimate demise except that he died in Worcester in May 1924.

Some of Clark's advanced students reflect the mentoring roles that Story and Taber played while carrying out their own research programs. When the university opened its doors in the fall of 1889, 35 graduate students were awarded fellowships and scholarships. Story's decennial account does not describe the difference between fellows and scholars; I suspect that fellowships contained better benefits. It will be helpful to consult Table 6.1 (on p. 315) when reading accounts of these Clark fellows and scholars.

Classes. The initial Clark class in mathematics included five students with fellowships and one with a scholarship, all possessing strong backgrounds. The scholar L.P. Cravens had been professor of mathematics at Carthage College in IL, while Rollin Harris had already earned a Cornell PhD four years earlier. Two others had earned master's degrees, Henry Benner (Michigan) and Joseph McCulloch (Adrian College, MI). In addition, William Metzler had received a bachelor's degree at the University of Toronto and Jacob William Albert Young at Bucknell. Young, however, had spent the previous year studying at the University of Berlin. The faculty engaged in research and taught courses while emphasizing the centrality of original research. Moreover, they taught modern courses at a very high level. In fact, standards might have been a little too rigorous even for such seemingly qualified students, as only two returned the second year; Young and Metzler earned Clark's first PhDs in mathematics.

The next two classes brought in ten more individuals, amounting to 16 fellows and scholars who entered during Clark's golden age. Although only three of the 16 earned doctorates, this trio was quite successful. Jacob Young was the first to obtain a PhD in mathematics when he defended his thesis in September 1892. He was appointed at once as an instructor at the newly founded University of Chicago, where he taught for the rest of his long career.

The Canadian **William Henry Metzler** (1863–1943) received his PhD in June 1893. After teaching at MIT and Genesee Wesleyan (NY) 1892–1895, he accepted a professorship at Syracuse University. Initially, he produced an impressive array of papers on determinants and quaternions; two others appeared in the *Proceedings of the London Mathematical Society.*[21] William Metzler then switched gears to education, becoming a prominent researcher in the field and establishing *The Mathematics Teacher* as a quarterly publication of the Association of Mathematics Teachers of the Middle States and Maryland in 1908. Later he negotiated with the newly formed National Council of Teachers of Mathematics (**NCTM**) in 1920 to make it the NCTM's official journal. It has continued in this role from the first issue published by the NCTM in January 1921 up to today. Metzler remained at Syracuse until 1923, having also served as dean of the graduate school, but he then moved to the New York State College of Teachers at Albany, where he was professor of mathematics and dean for ten years until retiring at age 70.

Thomas Franklin Holgate (1859–1945), a Canadian and a University of Toronto graduate like Metzler, was a mathematics instructor at Albert College in Ontario before accepting his Clark fellowship. He moved to Northwestern University in 1893, two years before officially receiving his Clark doctorate. He was a particularly active contributor to mathematics on the national scene, one of five professors who made Chicago one of the focal points of the emerging AMS along with New York City. This quintet—which also included Holgate's Northwestern colleague Henry White, his former Clark professor Bolza, and two of Bolza's new Chicago colleagues, E.H. Moore and Heinrich Maschke—was responsible for organizing and conducting the 1893 Chicago Congress and Evanston Colloquium. In fact, Clark connections accounted for eight of the 23 AMS members who attended the meetings, counting two faculty members (Story and Taber), two ex-members (Bolza and White), and four students (Holgate, Hulburt, Keppel, and Loud). Forty years afterward, Holgate reminisced about these historic meetings in an invited lecture before an AMS audience. Earlier, in 1896–1897, he played an active role in founding the Chicago Section of the AMS; he served as the section's first secretary 1897–1905 and was elected to its initial three-member Program Committee. During this period, he served on the AMS Council 1901–1903. Thomas Holgate was devoted to Northwestern throughout his career, serving as interim president on two occasions and as dean of the faculty. In the latter position, he spent the year 1921–1922 at the University of Nanking in China lecturing on mathematics but, more importantly on the international scene, assisting in the general organization of that university. Holgate's text *Projective Pure Geometry* (1930) was highly regarded.

William Story directed the doctoral dissertations of Thomas Holgate and Jacob Young, while Henry Taber served as William Metzler's mentor.

Recall that early in 1891 Thomas Fiske initiated a campaign to increase membership in the NYMS to support its first journal, the *Bulletin*. The success of that campaign can be seen at Clark, where six new members joined the society between May 1891 and April 1892. This group included two faculty members (Bolza and Taber) and four fellows (Hall, Hulbert, Metzler, and Young). It is somewhat surprising that Thomas Holgate did not join the Society until 1895, a time by which the NYMS had morphed into the AMS. It is downright astonishing that department chair, William Story, did not join for another two years after that.

PhDs. Overall Clark awarded ten PhDs in mathematics before 1900, with only Johns Hopkins (30) and Yale (23) producing more. William Story directed eight dissertations, Taber one, and the two codirected Metzler's dissertation. However, none of the remaining seven recipients could match the careers of the three who entered during Clark's golden age. Even so, three of the seven landed at research universities: **Warren Gardner Bullard** (1867–1927, PhD 1896) was at Vermont (1896–1900) before joining William Metzler at Syracuse for the rest of his life; **Linnaeus Wayland Dowling** (1867–1928, PhD 1895) spent a long career at Wisconsin after earning his doctorate, and he directed five PhD dissertations, including one by Charlotte Pengra; and **Ernest William Rettger** (1871–1938, PhD 1898), the only one in this group to write his dissertation under Henry Taber, initially taught at Indiana University before moving to Cornell in 1908 as assistant professor of applied mechanics. Two other graduates began their professional careers at prestigious small schools, **Frederick Carlos Ferry** (1868–1956, PhD 1898) at Williams College and **Thomas Flint Nichols** (born 1870, PhD 1895) at Hamilton College (1896–1908, after a year at Wisconsin). Whereas Nichols left academia in 1908 for the New York State Engineer's Office and by 1915 was in Phoenix, Ferry became dean of Williams (1902–1917) and then president of Hamilton College (1917–1938). Frederick Ferry also served on the Council of the AMS 1913–1915. **John Shaw French** (born 1873, PhD 1899) was professor of mathematics at the tony Jacob Tome Institute in Port Deposit (1898–1908) in Maryland, and then the principal and headmaster at several New England private schools, while **John Ethan Hill** (1865–1941, PhD 1895) initially taught science and mathematics at private academies before becoming a sanitary engineer in New York City. In September 1895, just three months removed from receiving his doctorate, John Hill was one of 13 mathematicians who attended the initial AMS Colloquium Lectures, then a one-week affair following the Society's annual summer meeting.

Other Clark fellows and scholars. This section presents snippets of the lives and careers of the 13 other Clark fellows/scholars. The online file "Web06-ClarkFellows" provides more details, beginning with three who were awarded fellowships but earned doctorates elsewhere, or who were at Clark after 1900. **Henry Benner** (d. 1901), a member of the first class during 1889–1890, had earned a BS and an MS (1889) from Michigan before enrolling at Clark. Benner became a mathematics instructor at Northwestern University Preparatory School at the end of 1890. He left two years later to accept the same position at the Chicago Manual Training School, one of the two most influential institutions of that type in the country. Like many Americans up to 1910, Benner traveled abroad for an advanced degree, obtaining a doctorate at

Erlangen in 1897 for a dissertation titled, *"Bestimmung der Coefficienten, welche bei der Berechnung der Integrale $\int \frac{x^n dx}{\sqrt{1+ax+bx^2}}$ und $\int \frac{x^n dx}{\sqrt{1+ax+bx^2+cx^3}}$ auftreten."* Upon his return, he was appointed the Brockway Professor of Mathematics and acting Bostwick Professor of Astronomy at Albion College, MI. Tragically, at age 38 he drowned on Lake Orion in Michigan in the summer of 1901, thus cutting short what might have been a productive career.

John Van der Vries and Herbert Keppel earned PhDs at Clark in 1901. **John Nicholas Van der Vries** (1876–1936), who matriculated at Clark 1897–1901, wrote his dissertation under William Story. Upon graduation, he joined the faculty at the University of Kansas, and he served as chair of the Southwestern Section of the AMS on two occasions. He also became active in the fledgling Mathematical Association of America, yet left academia in 1918 to become the manager of the US Chamber of Commerce. **Herbert Govert Keppel** (1866–1918) had entered the Clark graduate program three years after graduating from Hope College in MI in 1889. He remained at Hope (1892–1895) but left without a degree to join the faculty at Northwestern. He took a leave to return to Clark in the fall of 1900 and received his PhD the following year (with a dissertation probably directed by William Story). With degree in hand, Herbert Keppel returned to Northwestern but, in 1905, became chair of the mathematics department at the University of Florida. He was the only professor of mathematics during his tenure in Gainesville; in fact, it was not until 1934 that the cardinality of Florida mathematics professors exceeded one. Keppel died at age 54, one year after marrying, upon contracting Spanish influenza while teaching mathematics for the YMCA at military camps during World War I.[22]

Besides Clark, the only other thriving doctoral program in America during 1889–1892 was at Johns Hopkins, where the sole requirement for the PhD degree was writing a dissertation. Even though William Story had been an integral part of this program under J.J. Sylvester, when he moved to Clark in 1889 he established requirements almost identical to those enforced by most American universities today. Story was brilliant. He stated that the basic goal of a doctoral program was to train advanced students for careers as professors, hence, "The requirements for this degree have been determined by our conception of the ideal teacher."[23] Being precise, he defined the latter term as follows: "the ideal teacher is a master of his subject, not only conversant with the general principles of all its more important branches ... but ... actively engaged in scientific research."[24] Having defined the ideal professor and stating that the goal of the program was to produce more of the same, Story listed the requirements for attaining the degree generally needed to enter the profession:[25]

> The candidate is expected to attend, during his first two years, specified courses of lectures on the general principles, methods, and results of all the more important branches of pure mathematics, to supplement these lectures by private reading, and to take an active part in the seminary. ... Advanced courses of lectures on special subjects that vary from year to year are also given, and each candidate for the degree is expected to attend a number of such courses. The student spends the greater part of his third year in the original investigation, under the constant personal guidance of one of the instructors, of a

> topic of his own selection. ... The results of the investigation, em-
> bodied in a dissertation suitable for printing, must be submitted to
> the instructor ... and must receive his approval before the candidate
> will be admitted to the final examination for the degree. ... The re-
> sults [must be] sufficiently novel and important to constitute a real
> contribution to science.

The one item missing from this list of requirements, which is found in most modern departments, is the written examination at the end of the second year.

Consequently, it seems appropriate to call William Story the father of PhD programs, a revolutionary development that would become the international standard throughout the twentieth century and into the twenty-first century. This advance represents perhaps America's first legitimate contribution to worldwide mathematics.

Story's use of the word "his" is purposeful, as Clark was a male bastion throughout its early history. However, Story was certainly not a sexist, as illustrated by a case involving **Leona May Peirce** (1863–1954), whose example also demonstrates the hurdles that women had to overcome to earn a PhD at that time. Peirce graduated from the all-women's Smith College in 1886. Over the next eight years she taught at the Springfield Collegiate Institute in MA and studied at Cornell. An agreement was then reached for her to study with William Story, but because Clark did not formally admit women, she had to take private lessons with him. She did this for three years, culminating in a dissertation completed in 1897. However, an administrative brouhaha over a final examination in one of her minor subjects prevented her from receiving her Cornell degree. Following Story's advice, she enrolled at Yale a year later and was awarded her PhD in 1899, after having taken a year of coursework there. Upon receiving the Yale doctorate, Peirce returned to her hometown of Springfield to help run the family music business. Nevertheless, she returned to teaching in her mid-sixties, though at high schools, before retiring in 1937.[26]

Clark played another positive role for a woman student about ten years later, when two new graduate students joined three others already in the program. One was **Alice Berg Hayes**, who became the first woman awarded a fellowship in the department for 1910–1911, and the first to earn a master's degree in mathematics. That year the other entering graduate student was **Solomon Lefschetz**, who recalled in 1970:[27]

> At Clark there was fortunately ... a well-kept library. Just two of us
> enjoyed it—my fellow graduate student in mathematics and future
> wife, and myself.

They married shortly after Lefschetz obtained his American citizenship in the summer of 1913. Lefschetz was undoubtedly the very best PhD to come out of Clark in the university's history, but he was well known for being obstreperous. One of his biographers said of Alice:[28]

> [she was] a pillar of strength for Lefschetz throughout the rest of his
> life, helping him to rise above his handicap and encouraging him in
> his work.

Although the "handicap" referred to here was the loss of both of his hands, it could just as easily have meant his combative nature, which she moderated over the course of their lives together.

Since dissertations had to be suitable for publication, Clark students up to 1900 published their results in two basic journals, with Young, Metzler, Holgate, and Rettger appearing in the *American Journal of Mathematics*, and Hill, Dowling, Nichols, and Bullard in *Mathematical Review*. Establishing the *AJM* was an initiative that William Story undertook while at Johns Hopkins. In 1896 he obtained funding from a Clark board of trustees member to found *Mathematical Review* which, unfortunately, ceased publication after one year. These two journals account for eight of the ten Clark dissertations. Ferry published his in *Archiv for Mathematik og Naturvidenskab*; I have been unable to determine if John French found an outlet willing to publish his dissertation.

Clark admitted 41 fellows and scholars 1889–1899, with ten earning PhDs. The 31 who left without the degree contain several who earned a modicum of success. Nine of them are cited briefly to indicate the high standards enforced by the graduate program conducted by William Story and Henry Taber. The online file "Web06-ClarkFellows" provides fuller accounts.

Levi Leonard Conant (1857–1916; Clark scholar, 1890–1891) received two degrees from Dartmouth (BA, 1879; AM, 1887). He then accepted the professorship of mathematics at the newly opened (South) Dakota School of Mines, where he stayed for three years before enrolling at Clark for a year. He left Clark without a degree to accept a professorship at Worcester Polytechnic Institute (WPI). Two years later, in 1893, he earned a PhD from Syracuse; neither the title of his dissertation nor the advisor were known. Levi Conant taught at WPI for the rest of his life, heading the mathematics department (1908–1916) and serving as interim president (1911–1913). His 1896 book *The Number Concept: Its Origin and Development* was a seminal work in the anthropological and psychological study of numerals, particularly Native American number systems, though many of his conclusions are not accepted today. He was struck and killed (at age 59) by a truck in front of his home in October 1916.

The father of the Canadian **Alfred Tennyson DeLury** (1864–1951; Clark fellow 1890–1891) was a shoemaker who emigrated from Ireland to Canada. His bent for poetry is reflected in the name of his son, another mathematician, Ralph Emerson DeLury. Upon graduation from high school, Alfred DeLury became a high-school teacher before attending the University of Toronto, where he graduated in 1890. Classmates included Thomas Holgate and William Metzler. He matriculated at Clark the next year. Over the next two years, DeLury taught high school in Vancouver and Toronto and then studied in Paris. Upon his return to Canada in 1892, he was appointed lecturer at Toronto, where he received an MA ten years later. He was promoted to professor before becoming chair of the mathematics department in 1919, a position he held until his retirement in 1934. He also served Toronto as dean of the faculty of arts (1922–1934). A nephew, Daniel Bertrand DeLury, received his Toronto PhD in 1936 under William J. Webber and then taught at the university for the rest of his career, directing two doctoral dissertations.

James Norris Hart (1861–1959; scholar, 1890–1891) graduated from the University of Maine in 1885, taught high school for two years, and was then appointed an instructor in mathematics and drawing. He earned a civil engineering degree in 1890, whereupon he was promoted to professor of mathematics. He was also awarded a Clark scholarship, which he held for only one year. He returned to Maine at the end of that year, 1891, as professor of mathematics and astronomy, and later served as dean. James Hart remained in Orono (ME) until retiring in 1937. He earned a master's degree from the University of Chicago in 1897 based upon a thesis on the perturbation of Jupiter upon a minor planet. Hart was awarded a DSc degree from Maine in 1908. Hart Hall on the Maine campus is named for him.

The next two Clark graduate students attended the historic Evanston Colloquium in 1893. **Lorrain Sherman Hulburt** (1858–1943; Clark fellow, 1891–1892) graduated from Wisconsin in 1883 and was appointed professor of mathematics and astronomy at the University of South Dakota in 1887. While remaining on the faculty there until 1891, Hulburt was awarded an AM at Wisconsin in 1888 and spent the academic year (1889–1890) studying with Felix Klein in Göttingen. After that one year at Clark, he moved to Johns Hopkins, where he earned his PhD in 1894 with the dissertation "A class of transcendental functions." Hulburt remained on the Hopkins faculty until retiring in 1926. **Frank Herbert Loud** (1852–1927; Clark scholar, 1890–1891) graduated from Amherst College in 1873 and then joined the faculty as instructor for the next three years. In 1877, he moved west as the first head of the department at Colorado College, where he became a mainstay of mathematical life until his retirement 30 years later. Initially, the willingness to perform many duties at the college, including serving as the first librarian, prevented him from pursuing his first love, meteorology, but the appointment of Florian Cajori to the faculty in 1889 freed him to write several research papers on the topic. Desiring a doctorate, he took leave from Colorado College for one year to enroll at Clark, but he left without a degree. He did obtain a PhD in 1900 under Frank Morley at Haverford College, but I am unaware of the circumstances of awarding this degree.

This chapter will reference **Albert Harry Wheeler** (1873–1950; Clark scholar, 1896–1899) several times, mainly to illustrate the stratification within mathematics instruction among 1) professors who were researchers, 2) college instructors who engaged in little research, and 3) high-school teachers. Harry Wheeler was a Worcester "lifer," graduating from the public high school and then from Worcester Polytechnic Institute in 1894, teaching for two years, and then entering Clark in 1896.[29] He left the university after three years without obtaining a doctorate. The chair of mathematics, William Story, insisted that the departmental library exhibit string and plaster models. Wheeler became fixated with constructing all known polyhedra during that time, a practice he carried throughout 1899–1920 while teaching high school. Wheeler then re-enrolled at Clark, but his timing could not have been worse because that was precisely when new president Wallace Atwood eliminated the graduate program in mathematics. Nonetheless, Wheeler petitioned the administration to work toward his PhD under Harvard geometer Julian Coolidge. Yet Wheeler apparently never presented the outline of his proposed dissertation that the administration requested. Instead, he returned to high-school teaching for the rest of his life, gaining a modicum of fame among the general population for dazzling lectures and demonstrations, and among mathematicians for the paper he presented at the 1924 International Congress of Mathematicians (**ICM**) in Toronto.

It is instructive to interrupt this chronology to put Wheeler's education and career in perspective and to gain an appreciation for changes that were taking place in the early part of the twentieth century that have ramifications for today. In his address at the Toronto ICM, he lectured on a group of triangles and certain of their subgroups in a loose way, not using the terms *group* and *subgroup* in the specific way adopted by mathematicians, but in place of *set* and *subset*. His misuse of terminology and general misunderstanding of abstract concepts was thus exposed, preventing him from ever gaining full respect within the mathematical research community. Wheeler did, however, engage in lively correspondence with the well-known geometer **Harold Scott MacDonald Coxeter** (1907–2003). Whereas Coxeter married abstract structures with classical geometry, as illustrated by such modern structures as Coxeter groups and Coxeter diagrams, Wheeler never seemed to grasp the underlying goal of modern mathematics to classify structures without reference to physical models. This failing on Wheeler's part might account for the fact that he never earned his PhD at Clark in the first place during a time when the abstract structure of a group was in full fashion.

Three of the remaining Clark scholarship students provide a direct transition from this section to the next, by transferring to the University of Chicago for the fall semester of 1892. Recall that Oskar Bolza and Jacob Young were both on the Chicago faculty by then; in fact, two of these Clark students moved to the Windy City expressly to continue working with Bolza. One was **Napoleon Bonaparte Heller** (b. 1863; Clark fellow, 1891–1892), who never received a doctorate, yet subsequently returned to his hometown of Philadelphia and established mathematics departments at two embryonic institutions, Drexel Institute and Temple University. Regrettably, "Heller [was] impoverished in retirement, and [his] pleas to [university president] Beury for part-time work or a larger pension went unheeded."[30]

The other two Clark transfers received PhDs at Chicago, but neither had such historical names. **John Irwin Hutchinson** (Clark fellow, 1890–1892) entered Clark after graduating from Bates College to study with Oskar Bolza, so he transferred with him when the University of Chicago opened for the fall semester of 1892. John Hutchinson left Chicago two years later to accept an instructorship at Cornell, where he was teaching when he completed his dissertation under Bolza in 1896. This placed him in Chicago's first graduation class alongside Leonard Dickson. Hutchinson ended up spending 41 years at Cornell. His early publications were concerned mostly with theta functions and automorphic functions, with two cited in the Smith–Ginsburg history.[31] so Hutchinson quickly gained a reputation that accounts for his 1902 inclusion among the first group of seven cooperating editors for the *Transactions of the AMS*. He also coauthored two calculus textbooks with his Cornell colleague and former Klein student, Virgil Snyder. Hutchinson served the AMS as one of two vice presidents for 1910; Dickson was the other. However, a nervous breakdown in 1912 interrupted these endeavors for several years, although he continued editorial duties with the *Transactions* until 1915. When he resumed activity, his main interest switched to analytic number theory, particularly the generalized zeta function.[32]

Ernest Brown Skinner (Clark scholar, 1891–1892) graduated from Ohio University in 1888 and then taught for three years at Amity College in Iowa. He accepted the Clark scholarship in 1891 so he could study under Story and Bolza. Unlike Heller and Hutchinson, however, he did not accompany Bolza to Chicago; instead, he accepted an instructorship at the University of Wisconsin, where he remained until retiring in 1934 after 43 years. A leave of absence for 1899–1900, combined with the two summers that sandwiched that leave, allowed him to enroll in the graduate program at Chicago and complete requirements for a PhD by the end of the year. Ironically, his dissertation on group theory was written under Heinrich Maschke and not his former Clark professor, Oskar Bolza. Ernest Skinner was so highly valued as an educator that he

Table 6.1. Clark fellows and scholars 1889–1899

Name	Years	PhD	Primary Institution
Benner, Henry	1889–1890	1899, Erlangen	Chicago Manual Training School
Bullard, Warren	1893–1896	1896, Clark	Syracuse
Conant, Levi	1890–1891	1893, Syracuse	Worcester Polytechnic Institute
DeLury, Alfred	1890–1891		Toronto
Dowling, Wayland	1892–1895	1895, Clark	Wisconsin
Ferry, Frederick	1895–1898	1898, Clark	Williams College
French, John	1895–1899	1899, Clark	Jacob Tome Institute
Harris, Rollin	1889–1890	1888, Cornell	US Coast and Geodetic Survey
Hart, James	1890–1891		Maine
Heller, Napoleon	1891–1892		Drexel, Temple
Hill, John	1892–1895	1895, Clark	Brooklyn Manual Training School
Holgate, Thomas	1890–1893	1895, Clark	Northwestern
Hulburt, Lorrain	1891–1892	1894, Johns Hopkins	Johns Hopkins
Hutchinson, John	1890–1892	1896, Chicago	Cornell
Keppel, Herbert	1892–1895; 1900–1901	1901, Clark	Florida
Loud, Frank	1890–1891	1900, Haverford	Colorado College
Metzler, William	1889–1892	1893, Clark	Syracuse
Nichols, Thomas	1892–1895	1895, Clark	Hamilton College
Rettger, Ernest	1895–1898	1898, Clark	Cornell
Skinner, Ernest	1891–1892	1900, Chicago	Wisconsin
Van der Vries, John	1897–1801	1901, Clark	Kansas
Wheeler, Harry	1896–1899		Worcester High School
Young, Jacob	1889–1892	1892, Clark	Chicago

was invited to lecture at Chicago during its summer 1894 series "On teaching mathematics"; one of the three other lecturers in the series was Henry Benner. Beginning in 1909, Ernest Skinner accepted the onerous task of coordinating and supervising the work in freshman mathematics at Wisconsin, which resulted in two popular textbooks on algebra, trigonometry, and analytic geometry written for the famous Series of Mathematical Texts edited by Earle Hedrick. However, Skinner's foremost text was arguably *The Mathematical Theory of Investment* (1913), which met a pressing need and became a model for other texts on the same subject. All the while he continued research in group theory, and thus was chosen to write the article on this topic for the 1929 edition of *Encyclopædia Britannica*.[33] Skinner was also involved in the founding of the Mathematical Association of America in 1915.

Table 6.1 summarizes some facts about 22 of the graduate students who enrolled at Clark 1889–1899. The column of primary institutions associated with these mathematicians shows the extent to which Americans who earned a PhD in the US in the last decade of the nineteenth century were afforded many more opportunities at research-level institutions than the Sylvester School 20 years earlier.

E.H. Moore at Chicago

During the 1890s, the wave of Americans returning from Europe with advanced knowledge of mathematics, plus the enthusiasm shown in America for research, helped swell the ranks of qualified professors to numbers never dreamed of when Sylvester came to the US in 1876. These activities included meetings (both local and national) publications, and a radical revision of graduate and undergraduate curricula. A new private university was founded during the early part of this decade that would elevate America to yet a new level. That university was not Cornell but Johns Hopkins, which is somewhat ironic considering that Andrew White, the president of the self-proclaimed "first American university," was the only one of three presidents who served as advisors to the board of trustees charged with carrying out Johns Hopkins's will, to favor the idea of a university based almost entirely upon scholarship. The historian Gary Cochell argues persuasively that the most important reason why Cornell was unable to step into the leadership role assumed by Johns Hopkins was due to undergraduate teaching loads of 15–18 hours per week.[34]

Instead, the University of Chicago filled the void.

Oskar Bolza and Albert Michelson left the fledgling Clark University for the University of Chicago just three years after Clark had opened in 1889 with great fanfare and promise. Bolza is the more important figure for our story, partly because three promising graduate students followed him: Napoleon Heller, John Hutchinson, and Ernest Skinner. Jacob Young, the first recipient of a Clark PhD in mathematics (in 1892) also migrated to Chicago, which quickly fulfilled the promise once expected of Clark.

In some ways, Chicago was different from Johns Hopkins and Clark. For instance, whereas the latter two universities were named for the founders, Standard Oil magnate **John D. Rockefeller** (1839–1937) forbid that practice even though his munificent gifts, amounting to $35 million over the years, bankrolled the new university. Unlike Johns Hopkins and Jonas Clark before him, Rockefeller was robust and at the height of his powers. His wealth had soared—he was the world's richest man and America's first billionaire. Another telling difference separating Rockefeller from Hopkins and Clark was that he maintained no overriding educational vision; his involvement came in response to appeals from the American Baptist Educational Society to establish a major university in Chicago. To underscore that point, Rockefeller visited campus only twice during the first ten years of its existence.

Once again it was the first president who performed the spadework of recruiting faculty, and Chicago's William Harper turned out to be at least as effective as Hopkins' Daniel Coit Gilman, both of whom were more successful than Clark's Stanley Hall. **William Rainey Harper** (1856–1906) was labeled a prodigy very early in life. At age eight, he began preparing for college-level courses and two years later enrolled at Muskingum College in his Ohio hometown, earning a bachelor's degree at age 14. In 1872, Harper matriculated at Yale, earning a PhD in comparative philology in four years. Upon graduation, he married the daughter of the Muskingum College president and took up teaching positions in Tennessee and Ohio before landing at the Morgan Park Academy in Chicago in 1879. Soon he began offering summer school classes at Morgan Park, and then he created correspondence courses for students who lacked the

Figure 6.4. William Rainey Harper

time or money to attend classes during the normal academic year. He also taught summer courses at the Chautauqua Institution, in upstate New York, for aspiring ministers. These endeavors earned him an appointment at Yale in 1886.

The Yale position and presidency of the Chautauqua Institution brought Harper into close association with John D. Rockefeller, who in September 1890 selected the 35-year-old Harper as president of an educational institution in the Midwest that could rival Yale and other leading Eastern universities. Harper didn't accept the offer for another six months, but once he did, he set out at once to match established European universities by attracting world-class educators. Right from the start, he established very high standards for professors, and students as well, seeking to attract the best professors by securing Rockefeller funds to elevate compensation beyond school teachers. The University of Chicago opened its doors in the fall of 1892, and within 18 months Harper managed to recruit 120 faculty members (including leading scientists Albert Michelson and Oskar Bolza from Clark) to a campus with ten buildings. His visit to Clark University in the spring of 1892, when the faculty was up in arms in protest, is sometimes called "Harper's raid."

The University of Chicago was founded primarily as a graduate university, not just a college with a graduate school attached, thus continuing the metamorphosis of the American universities from being undergraduate colleges. How would this bipartite structure affect the mathematics department? Who should head the department? After all, the university's first president mandated that all faculty members—in the graduate school and the undergraduate college—were expected to be researchers, although those in the college would have higher teaching loads. President Harper wrote that professors must "seek not to stock the student's mind with knowledge of what has already been accomplished in a given field, but rather so to train him that he himself may be able to push out along new lines of investigation."[35]

What American mathematicians possessed the tools Harper sought?

Faculty. A Who's Who list of American mathematics professors in 1892 would feature Thomas Craig and Fabian Franklin (Johns Hopkins); William Story (Clark); William Byerly, William Osgood, and Maxime Bôcher (Harvard); Henry Fine (Princeton); Irving Stringham and Mellen Haskell (Berkeley); Henry White (Northwestern); George Halsted (Texas); and Hubert Newton (Yale), as well as Willard Gibbs and

George Hill, two individuals who had ascended to high ranks in the scientific community, but who remained peripheral players on the academic scene. Story held a PhD from Leipzig while, Osgood, Bôcher, Fine, Haskell, and White had all studied under the estimable Felix Klein at Göttingen. Byerly, Halsted, and Newton were homegrown, while the outlier, Stringham, was a product of domestic and Klein programs.

Harper, it seems, must have had a keen eye for latent brilliance, because when he found himself unable to lure established professors from New England universities, he gambled on a virtual unknown quantity, E. Hastings Moore. By the turn of the century, Moore would alter indelibly the perceived need for the most talented American students to travel across the Atlantic for advanced training in mathematics. Chicago adopted the same goals as Johns Hopkins but heeded the lessons learned to create a community that has endured as a world-class department to this day. The two institutions shared the same dual goals of teaching and research for faculty and graduate students, yet they staffed their mathematics departments in different ways. Whereas Hopkins hired an established foreigner to head its mathematics department, one who in turn hired Americans to fill the faculty positions, Chicago did the reverse by foisting their hopes and aspirations on the shoulders of a young, aspiring, American mathematician.

Eliakim Hastings Moore (1862–1932) attended public schools in Cincinnati, where he came under the sway of Ormond Stone, the director of the Cincinnati Observatory. In 1884, Stone would found the second-oldest mathematical journal in the country, the *Annals of Mathematics*, but at the time when Moore worked under him, Stone had not yet moved to the University of Virginia, the journal's first home. (The *Annals* moved to Harvard in 1899 and has been at Princeton since 1911.)

Moore entered Yale in 1879 and quickly earned the nickname "Plus Moore" for achieving distinction in every class. He graduated four years later with a bachelor's degree in mathematics and then remained at Yale two more years, receiving his PhD in 1885 for the dissertation "Extensions of certain theorems of Clifford and Cayley in the geometry of n dimensions" written under Hubert Newton, the grandfather of American mathematics. Moore's focus on the British school of mathematics was altered dramatically the next year during postdoctorate study in Germany that began with a summer of language immersion in Göttingen followed by two semesters in Berlin, where he attended lectures by Karl Weierstrass and Leopold Kronecker. This experience exposed Moore to an entirely different level of research and to topics then unknown in the US. The highly successful German university tradition imbued him with a research ethic that would fundamentally affect him and, ultimately, American university education.

Upon returning to the US, Moore was appointed instructor at Northwestern for a year, but a heavy teaching load sent him scurrying back to Yale as a tutor (instructor today) for two years. Although his duties at Yale were equally taxing, he felt more comfortable in the company of Hubert Newton and the renowned scientist Willard Gibbs. More importantly, it is probably during this time that Moore met Harper, who then held a chair in the divinity school at Yale. Yet in 1889 Moore returned to Northwestern, this time as an assistant professor; he was promoted to associate professor just two years later.

Moore published but four papers during this period, not counting his 18-page dissertation. The first three papers dealt with subjects related to his dissertation and appeared in the *American Journal of Mathematics*. The fourth, Moore's first work in

Figure 6.5. Eliakim Hastings (E.H.) Moore

analysis, appeared in the fledgling Italian journal *Rendiconti del Circolo Matematico di Palermo*. These works present minor contributions to the mathematical literature, providing no clues about the genius who penned them.

This was Moore's situation when President Harper chose him as the first mathematics professor and acting head of the department in 1892. (He was promoted to full head four years later.) Moore set out to form his faculty with gusto. He proposed hiring two professors and two tutors, but he had to negotiate with Harper intensely to get his way. For his first appointment, Moore proposed Henry White, whom he knew from Northwestern. Moore was aware of White's unhappiness at Clark, but White accepted the offer to succeed Moore at Northwestern instead of joining him at Chicago, located just twelve miles away. Next, Moore sought to hire one of Felix Klein's students. He sounded out Maxime Bôcher and William Osgood from Harvard but was unable to budge either one.

Only then did Moore approach Oskar Bolza. Although Bolza was somewhat unhappy at Clark, mainly because of the halving of its journal subscriptions, he was not personally involved with its internal political shenanigans. Therefore, even these negotiations did not go easy, because Bolza insisted on a professorship for his friend Heinrich Maschke. This appointment presented a much riskier proposition to Harper, because Maschke had left mathematics for engineering after coming to America. But Bolza insisted. So Moore persisted. The department head sensed the worldwide shift of mathematical power to Germany, so he ultimately persuaded President Harper to hire both figures for his nascent department.

Heinrich Maschke (1853–1907) attended the gymnasium in his hometown of Breslau (then Germany, now Wrocław, Poland) before enrolling at the University of Heidelberg in 1872. After a year of compulsory military service, he continued his studies at the University of Berlin under Weierstrass, Kummer, and Kronecker. Initially, Maschke took examinations for certification for secondary-school teaching, but he wanted to become a university professor. Therefore, he took courses at different German universities until completing his doctorate at Göttingen in 1880. But obtaining a faculty position in Germany at the time was nigh impossible, so Maschke taught in secondary schools for the next six years before being granted a sabbatical for 1886–1887. He spent that year in Göttingen working with Felix Klein. In the meantime,

Figure 6.6. Heinrich Maschke

Maschke and Bolza became close friends and were now reunited under Klein, whom they worked with in evenings at Klein's house. This led to Maschke's first paper, published in 1887, on substitution groups. The next year, while back in Berlin, Maschke published a second paper, this one showing that a particular sixth-degree equation could be solved using hyperelliptic functions. (It was soon proved by Brioschi that an arbitrary sixth-degree algebraic equation can be reduced to Maschke's equation and, therefore, can be solved in the same way.)

Even these high-quality papers were unable to snare a university position for Maschke, so he wrote to Bolza at Clark in 1889 asking about the possibility of landing a professorship in the US. Bolza warned that academic positions were equally elusive in the New World, whereupon Maschke began studying electronics with the hope of pursuing a different career path. He immigrated to the US in 1891 and landed a job at an electrical company in Newark, NJ.

Once Moore convinced Harper to appoint Maschke along with Bolza, the University of Chicago began classes on October 1, 1892, with three professors forming the nucleus of the mathematics department. There was no way to predict that a chair with a meager research record (Moore), an immigrant with only three years' teaching experience (Bolza), and an even more recent immigrant who had been employed the previous two years as an electrical engineer (Maschke), would end up forming the basis for what would become the best mathematics department in America by the first decade of the twentieth century. Each one seemed to bring out the best in the other two despite strikingly contrasting personalities and modes of teaching.

> They supplemented one another notably: Moore, a fiery enthusiast, brilliant, and keenly interested in the popular research movements of his day; Bolza, product of the meticulous German school of analysis

> led by Weierstrass, able and widely-read scholar; Maschke, deliberate,
> sagacious, brilliant in research, delightful lecturer on geometry.[36]

By 1900 the trio had produced several influential American doctorates whose names are generally recognized today: Dickson, Slaught, Lehmer, and Bliss. That list would swell during 1903–1907, headed by the "Big 3" of Oswald Veblen, R.L. Moore, and G.D. Birkhoff.

As E.H. Moore had planned, Chicago's mathematics department did not consist of Bolza, Maschke, and him alone. In addition, he hired Jacob Young and Harris Hancock to teach undergraduate courses. Although this caused a two-tiered faculty, graduate and undergraduate, teaching loads were kept low enough to encourage—indeed, to demand—original research and publications from both tiers. Tellingly, that research had to be carried out at the very highest levels too, a policy that created two entirely different experiences for the two lower-level faculty members.

Jacob William Albert Young (1865–1948) graduated from Bucknell University in Pennsylvania in 1887 and then joined its faculty for a year before going abroad with the intention of earning a doctorate in Berlin.[37] However, he found himself woefully unprepared regarding coursework, a different method of instruction, and the contrasting thought processes, so he returned to the US after one year. He then enrolled at Clark, which awarded him a fellowship. Jacob Young was much more successful in these surroundings, holding the fellowship for three years until earning Clark's first PhD in mathematics (under Oskar Bolza) for a dissertation on determining all groups of order p^n, where p is a prime. When Chicago opened that year, 1892, E.H. Moore appointed him as an associate based on the support of Bolza and strong letters of support from William Story, Henry White, and George Hill. This position was lower than an instructor, but Young was promoted to instructor in 1894. His main interest soon changed from group theory to mathematical pedagogy, whereby he was promoted to associate professor of that specialty in 1897 and to professor in 1908. In the meantime, he studied pedagogy in Germany 1897–1898, France 1901, France and England 1904–1905, Austria 1906, and Italy 1908. Clearly, Jacob Young understood where Chicago ranked relative to the major European universities by then. He retired as professor emeritus of mathematical pedagogy in 1926.

Harris Hancock (1867–1944) serves as an example of Moore's exacting standards. The blue-blooded Hancock, whose parents were prominent horse breeders in Virginia, graduated from Johns Hopkins in 1888 before pursuing graduate studies abroad, first at Cambridge and then Berlin, where he studied under Weierstrass, Frobenius, and H.A. Schwartz. His appointment at Chicago was contingent upon the completion of his doctorate by the time classes started. When it became clear that this would not happen, he requested a leave of absence to write a dissertation. His fiery department head was in no mood to compromise standards—he not only refused the request, he lowered Hancock's position from associate to assistant. Hancock took an approved leave the next year, earning a Berlin PhD under Lazarus Fuchs in 1894. But that alone was not sufficient to satisfy Moore, who did not place high value on Hancock's presentations at departmental seminars. As a result, even though Hancock was promoted to associate in 1895, he continued to feel undervalued by Moore, who was also unhappy with Hancock's teaching. After receiving no encouragement from President Harper over the next five years, Hancock left Chicago in 1900 for the University of Cincinnati, where he enjoyed a productive career up to his retirement in 1937, producing five doctoral students and writing advanced books on elliptic functions (1910), elliptic integrals (1917), algebraic numbers (1931–1932), and Minkowski geometry of numbers (1939). Future Chicago graduate and Harvard great G.D. Birkhoff valued Hancock's work highly: "All that is required [to thrust a department into national prominence] in many cases is that mathematicians in a position of influence take the proper steps. As instances in point, I would cite what was done by Fine at Princeton and by Harris Hancock at Cincinnati.[38]

Unlike Harris Hancock, both Jacob Young and Herbert Slaught prospered at Chicago, not so much for their mathematical research but for their devotion to, and success in, administration and pedagogy. Slaught became the "father of the MAA." He was a 31-year-old high-school principal when President Harper awarded him one of Chicago's first three fellowships. But Slaught's training was insufficient for the vigorous graduate program, so he moved his family (wife and young daughter) to Chicago in early summer to devote every waking moment filling holes in his mathematical background. Moreover, the paternal Moore did not assign him any classes to teach the first year, enabling Slaught to devote all his time to graduate courses. What a contrast with Moore's treatment of Hancock!

Herbert Slaught did teach during the second year of his fellowship, when E.H. Moore came to appreciate his outstanding teaching ability. Thus, Moore assigned Young to a faculty position. President Harper believed in giving young people with special skills frequent encouragement, though in rather small doses, so between 1894 and 1897, he appointed Slaught successively reader, associate, assistant, and instructor. During this whole time, Slaught managed to complete his doctorate (under E.H. Moore in 1898) despite a very heavy teaching load; the title of his dissertation was "The cross ratio group of 120 quadratic Cremona transformations of the plane." Slaught remained at Chicago for the rest of his life, retiring in 1931 as professor emeritus. He played major roles with the three major organizations of mathematics teachers in the country: the AMS, the MAA (established in 1915), and the NCTM (National Council of Teachers of Mathematics, established in 1920).[39]

Program. The Chicago program that Slaught encountered when classes started on October 1, 1892, was impressively comprehensive from top to bottom. Unlike many other leading American institutions, both the undergraduate and graduate programs were co-educational from the outset. At the undergraduate level, all 242 entering students were required to take what is called a General Education course today; its topics included algebra, plane trigonometry, and coordinate geometry. Those students interested in pursuing mathematics were then required to take courses in algebra, analytic geometry, calculus, differential equations, applications of calculus to geometry, analytical mechanics, projective geometry, the theory of functions, and elliptic functions. If one regards the last two courses as advanced calculus, this program is not unlike those at most American universities today, except for linear algebra and modern algebra. For its time, this level of sophistication was unprecedented in the history of higher education in America. Clearly, the major aim was to prepare majors for graduate study, whence for research in mathematics.

The "modern" algebra course (groups, rings, fields) could not have been offered in 1893, because it did not take hold in American undergraduate programs until the

1950s, except at a few elite universities. Yet matters were quite different at the graduate level, where Oskar Bolza offered a graduate course on the theory of substitutions and algebraic equations, while Heinrich Maschke taught one on finite groups. The graduate program had to be even more serious and sophisticated if it hoped to compete with German universities in attracting the very best students. To accomplish this goal, the trio of Moore, Bolza, and Maschke offered courses that rivaled foreign institutions, almost ensuring Chicago's success right from the start. In addition to the two algebra courses, the list of nine graduate courses given in 1892–1893 included the theory of functions and elliptic functions (both taught by Moore), the theory of hyperelliptic functions (Bolza), line geometry and potential theory (both by Maschke), and number theory and the theory of invariants (both by Young). This was a serious program indeed. Within six years the department was offering two dozen graduate courses annually.

Five of Chicago's initial entering group of 170 graduate students concentrated on mathematics. President Harper laid down specific guidelines for departments in their use of fellowships, directing that 5/6 of their time was to be spent on scholarship and only 1/6 on duties related to teaching. Clearly, the object of the program was to train students to become independent investigators. Four members of the entering class spent successful careers in mathematics—Herbert Slaught, May Winston, and two who came with Oskar Bolza from Clark (Napoleon Heller and John Hutchinson). The fifth, James Archy Smith, dropped out of the program when his two-year fellowship ended; he did not pursue mathematics afterward.

The University of Chicago built upon the other aspect of a graduate education that lifts students to cutting-edge research—what was called a seminary at Johns Hopkins, a seminar at Clark (and notably at Göttingen), and the Mathematical Club at Chicago. These meetings were held weekly from January 5, 1893, through the end of the academic year on June 8. During this time, all faculty members and all but one student (Slaught) gave presentations that were generally reviews of papers or books by others or, more importantly, talks that vetted their own results before a supportive but critical audience. For instance, on January 19 there were two presentations on group theory, a review by Jacob Young of the enumeration of all simple groups up to order 200, and a proof of the existence of a simple group of order 168 by E.H. Moore. Four weeks later, John Hutchinson became the first graduate student to talk when he presented his research on hyperelliptic functions begun at Clark under Bolza. He was followed the next week by May Winston, who discussed a theorem in linear differential equations. On March 23, the astronomer T.J.J. (Thomas Jefferson Jackson) See, an 1889 Missouri graduate who came to Chicago with a newly minted PhD from Berlin, discussed the secular action of tidal friction. See was the only speaker that year from a different department; only one came from outside the university.

The very last Mathematical Club lecture of the year was notable, not for its topic, but for future developments. On June 8 Henry White discussed "Hesse's enumeration of the bitangents of the plane quartic curve," a topic which does not seem to have materialized in any of his subsequently printed works. White's lecture highlighted the collegiality between his university, Northwestern, and Chicago that would expand four years later into the Chicago Section of the AMS.

Many of the lectures before the Mathematical Club that year dealt with algebra, which evolved into a Chicago specialty during the 1890s, an area that has remained a strength to this day. One of the later sessions, for instance, was devoted to the fundamental theorem of algebra. The graduate student Napoleon Heller presented Gauss's first proof, and then Oskar Bolza sketched later proofs by Gauss (his third proof), Argand, Cauchy, and Weierstrass.

The Chicago calendar was divided into four terms, each running twelve weeks with a one-week vacation in between. The faculty taught three terms a year. Of special interest is the summer quarter, which attracted numerous high-school and college teachers in search of higher degrees. Both Jacob Young and Herbert Slaught took special interest in these students, which probably accounts for their decision to devote their lives to the promotion and improvement of teaching mathematics instead of researching mathematics. Slaught oversaw conferences for secondary-school teachers held on campus during summers; this led to his active participation in the Central Association of Teachers of Science and Mathematics and, ultimately, the NCTM. Some first-rate work emerged from these sessions. For instance, the 22-year old Norwegian expatriate Nels Lennes enrolled as an undergraduate in 1896, graduated in two years, became a public-school teacher in Chicago, and took summer courses until receiving his doctorate in 1907. During this time, he became the first person in the world to define the notion of a connected set (in topology) in the way it is used today. In addition to this, Lennes joined with Slaught to form a two-person authoring team that wrote two books on algebra for high schools in 1906; their later collaborations included several such projects over the next three decades.

The mathematics department also made use of summer quarters for special courses taught by visiting professors. For the first summer quarter in 1893, President Harper attempted to recruit Felix Klein to teach such a course. However, negotiations came to naught, just as they had done when President Gilman attempted to bring Klein to Johns Hopkins ten years earlier. Yet Felix Klein, the world's reigning mathematics champion, did come to Chicago in the summer of 1893 for an outside event that brought the new university to the spotlight.

Chicago Congress

E.H. Moore took advantage of a serendipitous event that took place the next summer which would elevate Chicago's stock even higher when the city sponsored a World's Fair that ran from May to October in 1893. One hardly associates academic gatherings with World's Fairs, which had generally attracted huge throngs who were entertained by amusements and awed by exhibits featuring the latest technologies since the first one that was held in London in 1851. But some of these gatherings were high-minded as well as commercial, allowing people to explore unknown worlds that transcended their everyday experience—different cultures, new scientific advancements, and new inventions. The germ of the idea to include congresses began with the Paris Exposition of 1867. The first attempt at adding an academic component to amusement was uneven, as meetings were unrelated and organization was somewhat muddled. Not much is known about the first World's Fair held in America in New York in 1853, except that it attracted paltry crowds.

However, a much bigger fair took place in Philadelphia in 1876, when the Centennial Exposition was held to celebrate the founding of the United States. (The most recent fair in North America was held in 1986 in Vancouver to commemorate that city's centennial.) The centerpiece of the Centennial Exposition was the display of the Liberty Torch that served as a preview of the Statue of Liberty. Perhaps the most popular exhibit of all, however, was the Corliss steam engine, the largest ever built at 70 feet tall, which powered virtually every exhibit at the Exposition after being switched on by President Ulysses Grant and Emperor Dom Pedro of Brazil. Many of the other exhibits are still on display at the Smithsonian Institution's Arts and Industries Building. While the Philadelphia Centennial Exposition was wildly successful in honoring the country's birth 100 years earlier, it did not include an academic component amongst the amusements and exhibits in what is now Fairmount Park.

It was in France again where the idea of congresses fully developed at the Paris Exposition of 1889. Although more popularly known for the construction of the Eiffel Tower, the *Exposition Universelle* included a series of 70 such gatherings that attracted an international audience and was universal in scope.[40] Therefore, when the next American World's Fair was planned for Chicago three years later, the organizers enthusiastically borrowed the idea of concurrently staging scientific congresses with an international component. The purpose of the resulting fair, formally called the Columbian Exposition, was to celebrate the 400th anniversary of the voyage of Christopher Columbus. Its ultimate success, with an attendance exceeding 27 million during its May 31–October 31 run with impressive commercial profits generated by amusements, other entertainments, and eateries that lined the 7/8-mile long Midway, earned the event the enduring reputation as arguably the greatest World's Fair ever held. Inspired by the success of the congresses at Paris, the organizers established a World's Congress Auxiliary to conduct a similar series of meetings devoted to specific themes. Some of the numbers are staggering; for instance, the World's Parliament of Religions attracted 700,000 spectators while the Congress of Representative Women attracted 150,000.

It was a fortunate stroke of luck for the University of Chicago that, due to delays, the World's Fair opened a year late, in 1893, because otherwise it would have been too early for its fledgling mathematics department to take advantage of the opportunities it offered. While the World's Fair was held in Jackson Park on the south side of Chicago—a monumental feat that required the reclamation of 633 acres of marshland with a verdant island surrounded by a lagoon—the venue for the scientific congresses, the Hall of Manufacturers and Liberal Arts, was situated seven miles north of the fairgrounds, but directly adjacent to the budding university.

Chicago's president, William Harper, lent vocal support to the idea of having his embryonic university participate as actively as possible. As early as the summer of 1892, he negotiated with Felix Klein to offer a 12-week lecture course the following summer at a time that would overlap with the Chicago Congress. In the meantime, the acting head of his mathematics department, E.H. Moore, was busy making plans to insert a mathematical component into the Congress on Science and Philosophy held in conjunction with the World's Fair during the week of August 21. Within these congresses, the department of mathematics and astronomy named Moore and three other local mathematicians as an organizing committee: Henry White (who had moved one year earlier to nearby Northwestern) and Moore's Chicago colleagues, Oskar Bolza and Heinrich Maschke. What is more, the fledgling NYMS and its journal played a role in publicizing the Congress.

Moreover, the relationship between the Chicago Congress and the NYMS ultimately became symbiotic, with the Congress helping the Society expand in important

Figure 6.7. 1893 World's Fair at Chicago, the Columbian Exposition

ways. But before then, back in October 1892, just as the University of Chicago was opening its doors, E.H. Moore inserted an announcement of his plan in the *Bulletin of the NYMS*:[41]

> A SCIENTIFIC CONGRESS is being arranged in connection with the World's Fair at Chicago in 1893. The general committee on mathematics and astronomy has issued a preliminary address inviting the cooperation of all persons and societies interested.

The announcement listed Moore as the member of the general committee in special charge of pure mathematics. The chair of the committee on mathematics and astronomy was the astronomer George W. Hough, which not only reinforces the way these two subjects were connected at the time, but also suggests that astronomy was held in somewhat higher esteem. Indeed, of the ten NYMS members on the advisory council for this committee, seven were primarily astronomers, though all possessed strong mathematical backgrounds—Simon Newcomb, Ormond Stone, and five not yet introduced: Charles Sumner Howe (Case School of Applied Science, Cleveland), John Krom Rees (professor of astronomy at Columbia), Truman Henry Safford (professor of astronomy at Williams College), George Mary Searle (director of the observatory at Catholic University in Washington, DC), and Charles Augustus Young (professor of astronomy at Princeton). The three mathematicians on the advisory council were Arthur Cayley (who thus represented a slight international component), Thomas Craig (Johns Hopkins), and Robert Simpson Woodward (Columbia). Although Woodward was a professor of mechanics at the time and had previously worked at the US Geological Survey (1884–1890) as well as the US Coast and Geodetic Survey (1890–1893), he was editor of the *Annals of Mathematics* and served a two-year stint as president of the AMS (1899–1900), so I feel secure in labeling him a mathematician.

Despite this seemingly lower position for mathematics, the Chicago Congress turned out to be a stunning success, both for the fledgling university and for mathematics in America, mostly due to the star attraction, Felix Klein. Although negotiations between President Harper and Klein about the proposed course of lectures stalled over remuneration, by the spring of 1893, both parties were pursuing different but complementary courses of action. Harper was engaged in getting his university off the ground, lending full support for his prized mathematics department to sponsor the Congress that Moore announced in the *Bulletin*. Klein meanwhile had arranged with the Prussian Ministry of Culture to be its official representative on the bigger stage, the World's Fair itself.

However, it was early June by the time Klein informed President Harper of this development, which left Congress organizers little time to react. This served as the background for an almost breathtaking announcement made at a meeting of the NYMS just two months before the Congress was slated to begin: "Professor [John] Van Vleck read an extract of a recent letter from Professor Felix Klein, in which Professor Klein stated that he expected to visit America during the present summer and to attend the mathematical congress at Chicago."[42] Although Moore's only contact with Klein before the Congress was an exchange of letters, all three other members of the organizing committee had studied under him in Göttingen, with Bolza earning a doctorate in 1886 and White in 1891.

Matters moved quickly. In the July 1893 issue of the *Bulletin of the NYMS*, E.H. Moore announced that Felix Klein would attend as an imperial commissioner of the German government. In this role, the eminent *wunderkind* agreed to coordinate Germany's scientific display at the one-week meeting. He also promised to deliver lectures and coordinate activities at the Congress. Not only did his presence mark a seismic shift in influence on American mathematics toward the German and away from the British and French, it signaled the beginning of the passage of international supremacy from Europe to America that was completed 50 years later. Moore's announcement one month before the Congress singled out Klein in capital letters:[43]

> The international congress on mathematics, astronomy, and astrophysics at Chicago will be held in the new Art Institute building on the lake front during the week beginning August 21, 1893. There seems to be every indication of a most successful session. The programme in mathematics will include a series of reviews of the recent development of particular branches of the science. Every one interested in mathematics will appreciate the importance of attending the congress, although papers will be received from authors even when they cannot be present in person. ... PROFESSOR FELIX KLEIN of Göttingen at the request of the German government will attend the congress as the official representative of German mathematics. He will deliver several addresses, among them one on "The development of the theory of groups during the last twenty years." Professor Klein has many enthusiastic pupils, friends, and admirers in America (as everywhere), who will rejoice at the opportunity thus afforded of

meeting him. He will remain in Chicago during the month of September, and will hold regular conferences with his mathematical friends and any others interested in the recent development of mathematics who may wish to attend.

Imagine—until that point the only other professional mathematics conferences in America were one-day meetings of the embryonic NYMS, founded only four years earlier. Yet here, Midwest mathematicians were inviting everyone to attend a one-week affair, which Moore had the audacity to call "international." Could it possibly succeed? In a word: Yes.

Altogether 45 mathematicians attended the six-day Chicago Congress, which ran from Monday to Saturday, August 21–26. Now, one could justifiably ask how to assess an event as successful when it attracted only 45 people amongst the 3.5 million who filed through the Fair's turnstiles in August alone. However, one must also consider the size and geographical spread of American mathematicians at the time. For instance, NYMS secretary Thomas Fiske reported that, up to 1893, "The average attendance of members at the meetings had been 15 or 16, and the average total attendance 18."[44]

An analysis of the assemblage provides the following snapshot of the first group to attend a sustained conference in America. Four came from Europe—Eduard Study and Klein from Germany plus one each from Austria and Italy. While calling the meeting "international" seems like a euphemism by today's standards, it seems appropriate in context. Among the 41 from the US, six did not have to travel far (four at Chicago and two at Northwestern), but a listing of institutional affiliations provides evidence that attendees traveled from across the country to spend a week with Klein and his brand of mathematics. The educational attainment in the audience, even for this early period in America's mathematical history, was quite high, with 23/45 having earned doctorates (21 PhDs and two MDs); I know nothing about the medical doctors John Purdan (from Alabama) and Montague Severson (Virginia). Not all attendees were mathematicians; in addition to the two MDs, Alexander Macfarlane (Texas) was a physicist, Arthur Webster (Clark) a mathematical physicist, John Johnson (Mississippi) an astronomer/physicist, and Merriman Mansfield (Lehigh) a civil engineer.

Pioneering women. Herbert Keppel was one of three graduate students who attended the Chicago Congress. He was at Clark at the time. The other two were among three women who attended the Chicago Congress. May Winston met Felix Klein there; the program listed her as an honorary fellow at the University of Chicago. Winston then studied under Klein at Göttingen, where she remained until completing her dissertation in 1896. She taught in a high school the next year but, upon receiving her PhD in 1897, she was appointed professor of mathematics at Kansas State Agricultural College (now Kansas State University). She showed signs of serious interest in conducting research by joining the AMS at once and maintaining her membership for at least 45 years. Over the next three years she attended the three December meetings of the Chicago Section of the AMS, held at Northwestern and the University of Chicago, but she did not present any papers.

However, Mary Winston resigned her position at Kansas State at the end of the 1899–1900 academic year to marry **Henry Byron Newson** (1860–1910), a mathematician at the University of Kansas, in July 1900. Henry Newson was a graduate of Ohio Wesleyan University who pursued graduate studies at Johns Hopkins, Heidelberg, and

Leipzig (under Sophus Lie). AMS records indicate he had earned a PhD by 1896 but no details are known about his degree. The marriage seemed to exert little effect on Henry Newson's professional career, as he continued to attend and present papers at AMS summer and annual meetings, as well as meetings of the Chicago Section and the Southwestern Section. But life changed dramatically for his wife, **Mary Winston Newson** (1869–1959), because antinepotism regulations prevented her from obtaining a position at Kansas; curiously she was permitted to teach during summers. When she resigned her position, she expressed regret that a woman of her scientific attainments could not fill two positions at once—college professor and homemaker.

In between the births of her daughters in 1901 and 1903, Winston Newson managed to publish an authorized translation of David Hilbert's famous lecture on "Mathematical Problems" for the *Bulletin of the AMS*, with permission of the author, of course.[45] Her translation of Hilbert's opening statement has become a landmark exhortation:

> Who among us would not be glad to lift the veil behind which the future lies hidden; to cast a glance at the next advances of our science and at the secrets of its development during future centuries? ... For the close of a great epoch not only invites us to look back into the past but also directs our thoughts to the unknown future.

Winston Newson provided a greater service than merely translating the piece, poetic wording notwithstanding. She laced it with numerous footnotes supplying bibliographic details that demonstrate her broad knowledge of various fields and reflect the highest level of learning that Felix Klein could impart to his students. For instance, Hilbert stated, "A. Kneser in a very recently published work has treated the calculus of variations from the modern point of view."[46] Not only did the translator provide bibliographic details on Kneser's book, written in German and just published that same year, but she added a long explanation that began, "As an indication of the contents of this work...."[47]

Unfortunately, Winston Newson seems to have lost touch with the mathematical community after that, except for attending a meeting of the Southwestern Section meeting in 1908 held at the University of Kansas. Her husband submitted a paper for the next meeting of that AMS section, but illness prevented him from reading it. Tragically, he died in 1910 at age 49, yet even after his untimely death she was unable to join the Kansas faculty, once again due to antinepotism policies, this time because her younger sister had just accepted a position in the English department. Their son Henry Winston Newson (1909–1978), born a year earlier, became a distinguished experimental nuclear physicist at Duke University.

Two years after being widowed, Winston Newson attended a Southwestern Section meeting held again at the University of Kansas in November 1912. Although she did not present a paper, her attendance marks a revived interest in her professional career, because she was appointed assistant professor at Washburn College the next fall. Washburn's location in Topeka, Kansas, was close enough for the single mother to return to Lawrence on weekends to spend time with her three children, who lived with her father. In this position, she soon became active in the Kansas Association of Teachers of Mathematics. Serving as chair of that association in the fall of 1915, she quickly prepared for the expected establishment of the MAA that December. Thus, when that event did occur, the Kansas association was prepared to submit an application to become one of the first three MAA sections. Moreover, Kansas was the first MAA section to hold a meeting (in 1916). She became a charter member of the MAA upon its founding and maintained membership into the early 1940s.

Mary Winston Newson had developed into a leader of American mathematics by the time she left Washburn in 1921 to become head of the mathematics department at Eureka College in Illinois. The next year she spoke about early members of the Chicago Section of the AMS at a meeting held in Chicago to celebrate the twenty-fifth anniversary of the founding of the section. In 1937, the tables got turned at the summer AMS-MAA joint meeting during which "a special luncheon for women … was held in honor of the women who were pioneers in mathematical research in America. The guests of honor were Professor Mary Winston Newson … and Professors Clara E. Smith and Clara L. Bacon."[48] Ironically that was the year she was required to retire from full-time teaching when Eureka College instituted a mandatory retirement policy, yet she continued teaching on a part-time basis for another five years. She wrote her last piece during this time, a review of a pamphlet by the mathematics historian David E. Smith about Thomas Jefferson. Appearing in a 1940 issue of the forerunner of today's *Mathematics Magazine*, when Winston Newson was 71 years old, she ended her article with a statement that might well ring true today:[49]

> When we turn from these thoughts to this early leader of the Republic
> to the present rather precarious position of our science, we may well
> wish that more of our politicians and statesmen or even our leaders in
> **education** had as sound an understanding of the principles of math-
> ematics and as correct a judgment of its followers. [The emphasis is
> hers.]

In 1956 Mary Winston Newson moved to a nursing home in Maryland near her elder daughter, where she died in December 1959 at age 90. Her children endowed the Mary Newson lecture series on international relations at Eureka College upon her death.

The two other women who attended the 1893 Chicago Congress were Ida Schottenfels and Charlotte Barnum. **Ida May Schottenfels** (1869–1942) had graduated from Northwestern the previous year. She then taught high school in Chicago while taking courses at the university, resulting in a master's degree in 1896. Even though that was the highest degree she would ever earn, her productivity was so great that she is listed as the second most active woman mathematician in the period up to 1906.[50] In 1901 Schottenfels obtained her first position in higher education, as an instructor at the New York State Normal College (now the University of Albany), but she returned to Chicago in 1907 (in an unknown capacity). One source reported that she became the chair of mathematics at Adrian College in Michigan in 1913 but AMS records list her at various addresses in Chicago from 1907 through the early 1930s.

Recall the travails of Christine Ladd Franklin in attempting to study higher mathematics at Johns Hopkins during 1879–1882. Ten years later another woman with a similar background attempted the same feat and met the same fate. **Charlotte Cynthia Barnum** (1860–1934) came from a family whose educational attainments resembled May Winston's, as both of her brothers graduated from Yale while she and her sister graduated from Vassar at a time when very few Americans graduated from high school let alone college. Charlotte Barnum attended Vassar (1877–1881), and graduated 12 years after Ladd. From commencement until 1890 she was employed in a variety of positions: teaching at various academies, being a human computer at the Yale Observatory, computing angles of crystals for a book on mineralogy, and being an editor for *Webster's International Dictionary.*

After teaching astronomy at Smith College during 1889–1890, Barnum got a mathematical itch like Ladd's. In early 1890 she requested permission to take courses at Johns Hopkins University, which still did not officially admit women, even for graduate work. Department chair Simon Newcomb supported her request with a letter to President Gilman, requesting that she be admitted to classes taught by Thomas Craig and Fabian Franklin. The board of trustees voted to allow her to attend lectures without a charge for tuition, but without enrollment. Consequently, Charlotte Barnum studied mathematics, physics, and astronomy at Johns Hopkins (1890–1892). (Johns Hopkins did not admit women into its graduate school officially until 1907. Clara Bacon, who had spent several summers taking courses at the university, applied at once and became the university's first woman PhD in mathematics four years later.)

Like Christine Ladd, however, Charlotte Barnum sought a doctorate, so she transferred to Yale in 1892, when women were first admitted to its graduate school. She remained at Yale for three years, receiving her PhD in 1895 for the dissertation "Functions having lines or surfaces of discontinuity." No dissertation advisor is stated. The attendance record from the Chicago Congress of 1893 lists Barnum's address as New Haven without any affiliation, but she was a graduate student at Yale at the time.

Not surprisingly, a good portion of the rest of the audience at the Chicago Congress had a direct link to Felix Klein in Germany or J.J. Sylvester at Johns Hopkins. Counting Winston, there were six former Klein students: Oskar Bolza and Heinrich Maschke (both at Chicago), James Oliver (Cornell), Harry Tyler (MIT), and Henry White (Northwestern). Klein stayed with White at his home in Evanston, located some 12 miles away, during the one-week congress and the two-week colloquium that followed. Another person associated with the Göttingen star, John Van Vleck (Wesleyan), was surely an honorary member of the Klein Klub, having sent three of his students to study under, and ultimately to receive doctorates from, Klein.

Longtime Michigan faculty member **Wooster Woodruff Beman** (1850–1921) seems to have had no connection with Klein, yet he traveled from Ann Arbor for the meeting. Beman was known for coauthoring several elementary textbooks and for translating several famous monographs. His attendance contrasts with his colleague Frank Cole, whose absence (like former Klein student Irving Stringham of Berkeley) is somewhat surprising. Both Cole and Stringham submitted papers to the Chicago Congress, however. Although there is no mention of William Osgood or Maxime Bôcher from Harvard, Klein met up with these easterners during the next month. Several other members in the audience were associated with J.J. Sylvester during his time at Johns Hopkins, with three being awarded fellowships: Ellery Davis (Nebraska), George Halsted (Texas), and Charles Van Velzer (Wisconsin).

Papers. E.H. Moore's call for participation in the Chicago Congress produced a variety of submissions for the program, buffeted by a huge trove of papers from internationally known luminaries that Felix Klein brought with him. On the home front, papers were submitted by some of the best mathematicians in America. Clearly, the seed Moore planted when he wrote that "papers will be received from authors even when they cannot be present in person" fell on fertile soil. Altogether, 39 individuals submitted a total of 45 papers, with 10 of those in attendance reading a dozen of them.

The general sessions for the three scientific congresses—mathematics, astronomy, and astrophysics—convened on Monday, August 21, 1893. Mathematicians and astronomers were included in the section "Science and Philosophy," which convened in the Memorial Art Palace (today the Art Institute of Chicago) at 10:30 a.m. After welcoming remarks, including those from Felix Klein and his illustrious compatriot, Heinrich Helmholtz, the assembly dispersed to individual rooms for the diverse divisions.

Initially, mathematicians and astronomers reassembled at noon for an introductory address by Felix Klein titled, "The present state of mathematics." Delivered in English in about ten minutes, Klein deplored the way mathematics had evolved up to 1872 due to the growing trend toward specialization and away from integration with applications. However, he felt that two concepts developed since then had enabled mathematicians to connect seemingly disparate fields: groups and functions (notably analytic functions of a complex variable). Klein noted how the group concept had forged "geometry and the theory of numbers ... in many ways to appear as different aspects of one and the same theory."[51] He surely could have appealed to his own Erlangen Program of 1872, famous for using groups of transformations to characterize different types of geometry. It turns out that Klein was rather perspicacious in singling out these two basic concepts because they have, in fact, been central in mathematics since that time. But predicting the future direction of mathematical research can be dicey, even for the world's acknowledged leaders. David Hilbert is well known for the list of 23 problems he stated would form the basis for research in mathematics throughout the twentieth century. On the other hand, a recent study gave a mixed review on predictions made by Henri Poincaré at the 1908 International Congress of Mathematicians (Rome) in the lecture "The future of mathematics."[52]

Back in Chicago, at the end of Klein's opening remarks, the assembly voted to meet in two separate rooms for mathematicians and astronomers over the rest of the week. Thereupon, the 45 mathematicians moved to a different room for a business meeting chaired by E.H. Moore. A nominating committee was appointed and, after some deliberation, recommended officers: President William Story (Clark), Vice President E.H. Moore (Chicago), and Secretary Harry Tyler (MIT). The executive committee consisted of these three plus Felix Klein and Henry White (Northwestern). Recall that Tyler and White were doctoral graduates under Klein. A recess was then called, after which the executive committee returned with a program for the Congress for the remainder of the week. Each session was to begin at 9:30 a.m. and last for three hours, though discussions extended another hour or so on most days. In addition, on three of the afternoons the mathematics section adjourned to the German University Exhibit on the Exhibition grounds at 3 p.m. to view the items Klein had assembled and to hear his lectures on the books, papers, and physical models from his homeland.

The flood of submitted papers was too numerous and extensive for each paper to be presented in full during the remaining five days. Therefore, it was decided to offer one to four lectures each day and to have only the abstracts of the remaining papers read— by authors when present or by those designated by the executive committee otherwise. By the time the group reassembled on Tuesday morning at 9:30, printed programs were

Table 6.2. Daily themes at the Chicago Congress

Day	Theme
Tuesday	Arithmetic, algebra, and multiple algebra
Wednesday	Algebraic curve-theory and the theory of functions of a real variable
Thursday	Theory of functions of a complex variable
Friday	Theory of groups
Saturday	Geometry

distributed. This was no small feat at a time when copiers or, of even earlier vintage, mimeograph machines had not yet been invented.

An overview of the papers read and lectures delivered each day at the Congress, drawn from the official proceedings of the Congress, indicate its historical importance and its significance in the development of mathematics in America. I emphasize the contributions of American mathematicians which, admittedly, were a step below the Europeans, except perhaps for those by the Chicago trio of Moore, Bolza, and Maschke. Congress secretary Harry Tyler provided brief summaries in the *Bulletin of the NYMS*.[53] The outstanding 1994 book by Karen Parshall and David Rowe on this crucial period in the emergence of an American mathematical research community supplies deeper accounts of some of the papers.[54] Table 6.2 summarizes the themes of the papers for the remaining five days of the Congress.

Tuesday began with the reading of abstracts of papers by David Hilbert, Heinrich Weber, Eugen Netto, Adolf Hurwitz, and Artemas Martin. Now, hearing the summary of a paper by the great David Hilbert was not the same as being in the same room with him, but the Congress took place a full seven years before Hilbert gained mathematical immortality with his list of 23 problems that would define the subject in the twentieth century. On the home front, Artemas Martin serves as an example of an isolated but talented and devoted amateur who contributed to mathematics more by establishing journals than carrying out original investigations. The seven-page paper he submitted to the Congress might portray him as a human calculator, as he provided several solutions for d as a sum of fifth powers for the equation $q^5 + d = p^5$, where p and q are not required to be relatively prime. As an example, for $q = 33$ and $p = 36$, he calculated that

$$d = 4^5 + 5^5 + 6^5 + 7^5 + 9^5 + 11^5 + 15^5 + 18^5 + 21^5 + 27^5.$$

Martin defended himself against critics who would carp that his methods were "tentative and not rigorous," and who then cited an 1814 book of tables he used for his calculations.[55]

Tuesday was devoted to "arithmetic, algebra, and multiple algebra." Two of the three lectures were delivered by Americans. The foreign speaker was the 31-year-old **Eduard Study** (stew'-dee), whose lecture in German may have prevented some audience members from complete understanding. Study had come to the US for the same reason as Oskar Bolza and Heinrich Maschke—professorial opportunities.[56] After delivering two papers at the Congress, he spent the 1893–1894 academic year based at Johns Hopkins, but he also lectured at several other universities before returning to his homeland without having landed a position as enticing as the one that awaited him with Klein at Göttingen.

The first lecture at the Congress was delivered by the Texas physicist **Alexander Macfarlane**, who read the first of his two papers; unfortunately, no information on either paper is available—except the titles—because he published his talks as separate pamphlets and not in the conference proceedings. The other American speaker, Henry Taber, submitted a five-page abstract of his lecture on two seemingly distinct topics he was able to tie together: generalizations of Stieltjes' theorem, and orthogonal transformations that arose from the interplay between linear transformations and matrices, based on earlier work by Arthur Cayley. Taber was one of only a handful of American mathematicians to have papers published abroad at this time; earlier findings related to his Congress presentation had appeared in an 1890 issue of the *Proceedings of the London Mathematical Society*, and a paper on automorphic linear transformations was printed in the *Mathematische Annalen* three years after the Congress. One other American, Wisconsin's Albert M. Sawin, submitted a paper that had appeared in the *Annals of Mathematics* the preceding June when he was at the University of Wyoming, "The algebraic solution of equations."[57] In this work Sawin applied Newton's identities for roots of a polynomial equation in terms of coefficients to derive various properties that resolvents of cubic and quartic equations must satisfy; he also briefly indicated extensions to fifth- and sixth-degree polynomial equations.

Figure 6.8. William Holding Echols, Jr.

The session on Wednesday was devoted to "algebraic curve-theory and the theory of functions of a real variable." Five of the six summaries read that day were submitted by foreign mathematicians, with the one by Hermann Minkowski on number theory and geometry being the deepest of all. Others were by the Germans Alfred Pringsheim and Max Noether, the Italian Salvatore Pincherle, and the Czech Matyáš Lerch. The only American to have a summary read was **William Holding Echols, Jr** (1859–1934); it was concerned with finding the curve of best fit passing through a finite number of given points using the determinant of infinite matrices. Echols had earned BS and CE degrees at the University of Virginia in 1882, after which he worked as an engineer for railroad and mining companies until accepting an engineering professorship at the Missouri School of Mines (now MUST—Missouri University of Science and Technology). He returned to his *alma mater* in 1891 and taught there for the rest of his life. Virginia established a scholars program in his honor in 1960. Known as "Reddy" for the color of his hair (not apparent in Figure 6.8), Echols was the oldest of three children of William Holding Echols, an 1859 graduate of West Point. Reddy's brother, Charles Patton Echols, attended the University of Virginia but left for West Point after three years, graduating third in his class of 1891. He returned to the Military Academy in 1904 as head of the mathematics department, a position he held until 1931.

The three papers read in their entirely that day were all by Americans, with Alexander Macfarlane providing a continuation of his first lecture. George Halsted's paper outlined the history of non-Euclidean geometry from an international perspective. It heaped effusive praise on the work of Lobachevski, whose work Halsted later translated into English and popularized, and cited several results due to Saccheri.

The third American to present a paper on Wednesday was **Henry Turner Eddy** (1844–1921), who surveyed recent work on graphical methods in applied mathematics, including his own. His paper evinced an impressive knowledge of descriptive geometry, especially graphical statics, an application of mathematics Felix Klein found particularly appealing. When Klein returned to Germany, he took with him Eddy's ideas on forging stronger ties between scientists and engineers. Eddy had received his bachelor's degree from Yale in 1867 and then pursued engineering studies at the Sheffield Scientific School. After a year at the University of Tennessee as professor of mathematics and Latin, he moved to Cornell as assistant professor of mathematics and civil engineering. While there, he was awarded a PhD in 1872 for work done on integrating square roots of quadratics, the first doctorate ever granted by Cornell in any subject. Eddy then spent 1873–1874 as adjunct professor of mathematics at Princeton before accepting a professorship of three fields—mathematics, astronomy, and civil engineering—at the University of Cincinnati, where he remained 1874–1890. Even though he also assumed the role of dean of the faculty, he somehow stole time to pursue research in mathematics and engineering, notably on reinforced concrete flat slabs. These investigations were aided by a sabbatical in 1879–1880 that took him to the University of Berlin in the fall of 1879 and to the Sorbonne and the Collège de France in the spring of 1880. While in Germany he wrote a book titled *Neue Construetionen aus der graphischen Statik*; it was published in Leipzig in 1880.

Eddy left Cincinnati in 1890 to become president of Rose Polytechnic Institute (now the Rose-Hulman Institute of Technology in Indiana), the position he held during the Chicago Congress. The next year he induced Arthur Hathaway to leave Cornell and become professor and head of his department of mathematics. However, after only four years Eddy moved again, this time to the University of Minnesota, where yet again he wore three hats—professor of engineering, professor of mathematics, and dean of the graduate school. In 1896 he was elected president of the Society for the Promotion of Engineering Education. Eddy Hall was constructed as a Mechanic Arts building and remains today the oldest structure on the Minnesota campus.

The session on Thursday was devoted to "Theory of functions of a complex variable." It began with the reading of summaries of works submitted by the Frenchman Charles Hermite and the American Irving Stringham. Both papers dealt with elliptic functions, with Stringham comparing the traditional approach espoused by Jacobi with newer techniques due to Weierstrass, showing certain advantages of the older theory for transformations of elliptic integrals. In this regard, Stringham demonstrated that he was one of the very few Americans able to extend the frontiers of research in analysis significantly. Summaries were also read of submissions by four German mathematicians: Martin Krause, Lothar Heffter, Heinrich Burkhardt, and Robert Fricke. The two papers read in their entirety included one on spherical trigonometry by Eduard Study (the second paper he had brought to the Congress) and the other on the theory of hyperelliptic integrals by Chicago's Oskar Bolza, who gave an exposition of Weierstrass's theory of Abelian functions based on Riemann's methods, the confluence of which was important in Klein's recent work.

Friday's session was devoted to the "Theory of groups," which became an American specialty within a decade. Summaries of papers submitted by mathematicians at two American universities supplemented three by Germans (Arthur Schoenflies, Franz Meyer, and Victor Schlegel) and one by a Czech (Edouard Weyr). The long-haired Joseph de Perott of Clark University constructed the Galois group of a group of order 660. Further, Frank Cole, then at Michigan, made use of a method he had learned from Felix Klein to discover a simple group G of order 504 (as a subgroup of S_9, the symmetric group on nine letters). The resulting group G had apparently eluded earlier lists, supposedly complete, of all simple groups of order ≤ 660. The introduction to Cole's paper turned out to be prophetic:

> Despite the great advances of the past fifty years, the Theory of Groups remains to-day in many respects in a very unfinished state. It is true that we possess an accurate system of general classification on the one hand and an elaborate knowledge of special types on the other. But between these two extremes lies a vast middle ground, the exploration of which is extremely slow and difficult.

The classification Cole envisaged was not resolved until the early 1980s, some 90 years later. Progress would indeed be slow and difficult.

The two substantial papers read by Chicago colleagues Heinrich Maschke and E.H. Moore that day showed a high level of research attainment that shortly made their department the best in the country for studying what is now called abstract (or modern) algebra. Indeed, the Chicago algebra tradition has continued to this day. The contributions by the three Chicago mathematicians illustrate the effectiveness of collaboration. Although Oskar Bolza had delivered a Congress paper on analysis, he taught a high-level graduate course on the theory of substitutions throughout the preceding year. Moreover, his Clark student and now Chicago colleague, Jacob Young, had written a dissertation on the determination of all groups of order p^n. Heinrich Maschke taught a similar course on finite groups; in fact, his first paper, published back in 1887, dealt with substitution groups. His Congress paper demonstrated an effect of the emphasis on group theory during the year, as it completely determined all absolute invariants of the quaternary group of 2520 substitutions occurring in line geometry (that is, A_7, the alternating group on seven letters), leading to the construction of a group of order 336 that had arisen in Klein's work on modular forms.

Up until the opening of the University of Chicago, E.H. Moore had mainly worked in algebraic geometry, but throughout his career he showed a keen aptitude for detecting the most important areas of research and immersing himself in them at once. That first year demonstrated such a facility for the first time. In the last communication of the day, Moore presented a momentous paper announcing significant results that cannot be ascertained directly from its title, "A doubly infinite system of simple groups." Serendipitously, Moore's paper provided a ready example of Cole's prediction that it would require new ideas of an abstract nature to mark progress toward the classification of simple groups. Moore began by listing all finite simple groups known up to that time. He then added a class of groups based on "the group of the modular equation for the transformation of elliptic functions of order q [a prime]."[58] This quotation reflects a common feature in the work of many great mathematicians, namely, the ability to weld together elements of seemingly unrelated parts of mathematics, in this case, group theory and elliptic functions. Moore's first major result was to extend the list of known simple groups of order $\frac{q(q^2-1)}{2}$ to those of order $\frac{q^n(q^{2n}-1)}{2}$, where q is a prime number greater than 2 (except the case where $q = 3$ and $n = 1$) and those of order $2^n \left(2^{2n} - 1\right)$ for $n > 1$. Notice that Cole's example is a particular case of the latter type for $n = 3$. The term "doubly infinite" in the title refers to the fact that both q and n assume infinitely many values.

As if the announcement of that result wasn't significant enough, the title of Moore's paper veils an entirely different part of algebra he investigated containing arguably his most important contribution to mathematics. He wrote, "Assuming now (nothing but) the existence of a field of order s ... I ... prove ... that *Every existent field* F[s] *is the abstract form of a Galois-field GF*[q^n], $s = q^n$."[59] In short, Moore proved that every finite field is a Galois field. Above and beyond being important in its own right, Moore's theorem formed the basis for the proof of the Wedderburn theorem some ten years later: every finite division ring is a (commutative) field.[60]

Two operative words in the statement of Moore's theorem are "abstract form." The primary dictum for Moore's approach to all of mathematics—to proceed from the abstract to the concrete—is encapsulated in the fundamental principle of "generalization by abstraction" that he enunciated at the beginning of his New Haven Colloquium Lectures of 1906: "The existence of analogies between central features of various theories implies the existence of a general theory which underlies the particular theories and unifies them with respect to those central features."[61] Although one of Moore's students recalled that Moore "emphasized the abstract and algebraic side of mathematics,"[62] another indicated that he took "delight ... in its concrete significance for the comfort and progress of the human race."[63]

The final session of the week on Saturday was devoted to geometry. It had been highly anticipated but did *not* live up to the excitement generated the previous day. The morning consisted of the reading of brief summaries of submissions by the Parisians Émile Lemoine (two papers) and Maurice D'Ocagne, the German Victor Schlegel (his second), and the Russian T.M. Pervouchine. After lunch, the imperious Felix Klein took center stage with heightened eagerness for his lecture "*Über die Entwicklung der Gruppentheorie der während der letzten zwanzig Jahre*" ("Concerning the development of group theory over the past twenty years"). In light of results announced by Cole, Maschke, and Moore on Friday, the audience undoubtedly anticipated a grand tour of progress from the time of Klein's Erlangen Program of 1872 (which had just been translated in the *Bulletin of the NYMS*) to perhaps a culmination of discoveries by the American trio. After all, this lecture by Klein was the one singled out by Moore in his NYMS announcement. Instead, it turned out to be somewhat anticlimactic, as Klein digressed wildly (one might say in Sylvester's mode) from what was announced beforehand, and he ended up publishing only a one-paragraph abstract (in German) in the Congress proceedings. Instead of focusing on group theory, he described a course on higher geometry developed at Göttingen and which he taught the prior summer. In this context, Klein discussed certain reforms in mathematics education that had been implemented in Göttingen. He also described a course aimed at exposing future teachers to applications in astronomy, physics, and technology. Klein dearly wanted to publicize this course beyond the 45 mathematicians who attended the Chicago Congress, so he arranged for a translation of a circular describing the course to appear in the *Bulletin*, coauthored with the five other directors of the seminar on mathematics and physics—the mathematicians Ernst Schering, Issai Schur, and Heinrich Weber, as well as the physicists Eduard Riecke and Woldemar Voigt.[64] In spite of the disappointment that some might have felt at not hearing about group theory, "On motion of Professor Moore it was voted unanimously [that] the thanks of this mathematical section be tendered to Professor Klein for his very valuable contributions to the proceedings of the Congress."[65]

Felix Klein bemoaned specialization. In the closing ceremony, his thoughts were echoed in remarks by the Clark University mathematical physicist Arthur G. Webster, who deprecated "the separation, in our educational curricula, of the different branches of mathematical and physical science."[66] Yet in this regard, it is relevant that the assembled mathematicians and astronomers had voted to split into distinct sections as soon as the Congress began.

The one-week Chicago Congress was concluded by its president, Clark's William Story, who "acknowledged its indebtedness to Professor Klein, and the indebtedness of American mathematics in general to the influence and inspiration of German universities and mathematicians."[67] That influence and inspiration had started with Story when he studied at Leipzig and returned home with a doctorate in the early 1870s. It would continue right through the end of World War II.

Table 6.3 lists the Congress papers submitted by Americans. Those who delivered lectures are in bold; the other submitted papers were read by title or by abstract.

Assessment. The success of the Chicago Mathematical Congress surprised almost everyone except its organizers. It was the first sustained mathematical meeting of note in America. Furthermore, it served as a prototype for today's quadrennial International Congresses of Mathematics, the first of which was held four years later in 1897. In this sense, the Chicago meeting is sometimes called the zeroth International Congress. The emergence of a sufficiently large number of mathematicians is the one vital element that separated the national situation in 1893 from the one that existed 17 years earlier, when Sylvester set foot in America. Undoubtedly, all 45 participants benefited from the mathematics presented at the Congress and the mutual interaction with a group of professionals sharing the same general enthusiasm, interests, and values.

The Congress also paid dividends for the AMS, as one of its three watershed moments in 1894 was the decision to publish the Congress proceedings. In the preface to that volume, which appeared in 1896, E.H. Moore acknowledged, "No publisher was found willing to issue the papers at his own risk. ... The Editors take this opportunity to express their grateful appreciation ... of the interest in the undertaking shown by the officers."[68] Those AMS officers arranged for the Society to publish the book in conjunction with the Macmillan Company. When Princeton's Harry Fine reviewed the volume, he brought "a number of claims to the attention of the mathematical readers besides the intrinsic merits of the papers which it contains,"[69] going on to mention

Table 6.3. Papers by American mathematicians at the 1893 Chicago Congress

Day	Author	Title of Paper
Tuesday	**Alexander Macfarlane**	On the definitions of the trigonometric functions
	Henry Taber	Concerning matrices and multiple algebra
	Artemas Martin	On fifth-power numbers whose sum is a fifth power
	Albert Sawin	On the algebraic solution of equations
Wednesday	**Henry Eddy**	Modern graphical methods
	George Bruce Halsted	Some salient points in the history of non-Euclidean and hyper-spaces
	Alexander Macfarlane	The principles of the elliptic and hyperbolic analysis
	William Echols	On interpolation formulae and their relation to infinite series
Thursday	**Oskar Bolza**	On Weierstrass' system of Abelian integrals of the first and second kinds
	Irving Stringham	A formulary for an introduction to elliptic functions
Friday	**Heinrich Maschke**	On a quaternary group of 2520 linear substitutions
	E.H. Moore	A doubly-infinite system of simple groups
	Joseph de Perott	A construction of Galois' group of 660 elements
	Frank Nelson Cole	Concerning the formation of groups

specifically its significance in being associated with a World's Fair and its international character.

E.H. Moore's motion tendering thanks to Felix Klein also included appreciation for his elucidation of the material in the German University Exhibit at the Exposition. On those occasions when Klein hosted receptions on Exhibition grounds on three of the afternoons, he took personal pleasure in 1) describing the workings of plaster and string models, as well as such instruments as harmonic analyzers, planimeters, and calculating machines, 2) expostulating on the telegraphic instrument designed by Gauss and Weber near the bust of Gauss, and 3) pointing to portraits of Jacobi, Dirichlet, and Riemann on the wall. Klein also derived great pleasure from highlighting various items lining bookshelves, including the collected works of notable German mathematicians (Dirichlet, Gauss, Jacobi, Möbius, Riemann, and Weber), the entire series of publications from Academies of Science in four German cities (Berlin, Göttingen, Leipzig, and Munich), complete sets of seven German journals devoted to research in mathematics, and copies of every doctoral dissertation written at a German university after 1850 (including, presumably, Harry Fine's thesis in English).

Clearly, Klein's lectures and demonstrations reinforced the fact that American mathematics paled by comparison with their German counterparts. But Klein was prepared to play the role of mentor for those who hailed from the US and Canada. And he began that program the very next week.

Evanston Colloquium

The upstart Midwesterners built an upside-down metaphorical "mathematical cake" at the zeroth International Congress of Mathematicians, wherein the sugary exterior lined the nutritious interior. The "icing" was concocted from various ingredients in Chicago, featuring a wide cast of lecturers and also a full range of topics from submitted papers. Two days after that meeting adjourned on Saturday, August 26, 1903, 25 mathematicians assembled at Northwestern University, roughly 12 miles north, to relish the "nutritious interior." In this case, the figurative cake was a two-week marathon of 12 lectures delivered by the indefatigable Felix Klein. Only one year earlier the host of the meeting, Henry White, moved from Clark to Northwestern, which provided meeting space and placed a collection of mathematics books from its library at the disposal of all attendees. Because this meeting took place at Northwestern in Evanston and was even more convenient for Klein, who bedded down in Henry White's house, it has become known as the Evanston Colloquium.

The NYMS had been sponsoring four or five meetings a year since its organizational event in late 1888, so such gatherings were not a novel concept by 1893. However, most were one-day affairs; the longest lasted two days. The historic Chicago Congress endured for six, but the Evanston Colloquium doubled that. Aside from its length, the Colloquium was very different in a vital way: there was only one speaker, Felix Klein. His aim was to present an overview of the major areas of mathematical research at the time. Although he had never attempted such a feat before, he was perhaps singularly positioned to offer such a panoramic view of a field that had exploded over the preceding 25 years. For many audience members, it was the only time they would be able to soak up not only Klein's mathematics, but his approach that eschewed almost all proofs, yet provided deep geometric intuition as an aid to finding relationships among supposedly different subfields.

Klein lectured daily, and in English. Of even more importance for the 24 listeners, he fielded numerous questions and followed the talks with lengthy conversations with participants. Instead of showing fatigue by the end of the second week, these gatherings lengthened to five hours. What endurance Klein had!

Table 6.4 lists the titles of the twelve lectures delivered from Monday August 28 to Saturday September 9, with only the intervening Sunday set aside as a day of rest. The titles of the lectures are taken directly from the thin tome published in January 1894.[70] Curiously, the proceedings from the Evanston Colloquium appeared almost three years before the proceedings from the Chicago Congress. They were highly anticipated, as the October 1893 *Bulletin of the NYMS* had already announced, "His addresses at these meetings are of course in publication by Macmillan & Co. in commission for Northwestern University."[71] The book's early publication was enhanced by the University of Michigan's Alexander Ziwet, who took notes during the lectures, submitted them for Klein's editing each day, made the changes, sent the resulting page proofs to Klein before he boarded his ship back to Germany four weeks later, and made all of Klein's suggested changes in subsequent page proofs to ensure the work would appear just four months after the Colloquium adjourned.[72] What a herculean task!

Table 6.4. Titles of the 12 lectures by Felix Klein at
the Evanston Colloquium

	Week 1	**Week 2**
Monday	Clebsch	The transcendency of the numbers e and π
Tuesday	Sophus Lie	Ideal numbers
Wednesday	Sophus Lie	The solution of higher algebraic equations
Thursday	On the real shape of algebraic curves and surfaces	On some recent advances in hyperelliptic and Abelian functions
Friday	Theory of functions and geometry	The most recent researches in non-Euclidean geometry
Saturday	On the mathematical character of space-intuition, and the relation of pure mathematics to the applied sciences	The study of mathematics at Göttingen

Some of the lecture titles do not provide a clue to their content.[73] For instance, Klein's opening talk divided mathematicians into logicians, formalists, and intuitionists. The next two dealt with his friend Sophus Lie's theory of partial differential equations. Lecture 4 on Thursday presented his own work on algebraic geometry, while the next one investigated hypergeometric functions. As the title indicates, week one ended with a discussion of two seemingly unrelated topics: one was the duality of naïve versus refined intuition that shaped most mathematical ideas; the other was the age-old question of the relationship between pure and applied mathematics. Lecture 6 was the only one of the twelve that Klein had reprinted in his collected works.

The second week began with the classical problem from antiquity of squaring the circle, a problem that had been resolved by Klein's doctoral student Ferdinand Lindemann. Lecture 8 on the second Tuesday applied Kummer's theory of ideal numbers to binary quadratic forms. Algebra was seemingly the main topic the next day too, when Klein provided an overview of the status of Galois theory, but in typical Kleinian style—joining algebra and geometry—he added an unexpected twist by showing that the group of rotations of an icosahedron is isomorphic to the Galois group of a 60-degree polynomial. The titles of the three remaining lectures describe their contents. It is regrettable that George Halsted was unable to extend his stay in Chicago onto Evanston, because he would have been a lively contributor to Klein's history of non-Euclidean geometry in Lecture 11. Finally, it is worthwhile to elaborate upon Lecture 12 because the title might suggest that it was related to the topic that concluded his work at the Chicago Congress two weeks earlier, when he discussed a course he had helped develop at Göttingen instead of lecturing on group theory. This lecture, however, was unrelated to that topic. Instead, he discussed mathematical prerequisites for American students desiring to pursue the doctoral program at Göttingen, and he encouraged only the best-prepared to undertake such a program. One vitally immersed listener was May Winston, who had attended every one of the events in Chicago and Evanston. As remarked earlier, she sailed to Germany shortly after the Colloquium ended, reached Göttingen even before Klein returned home, and within three years wrote the final dissertation by an American under Klein.

Roughly half (22 out of 45) of those who had attended the week-long meeting in Chicago were able to recharge mathematical and religious batteries on Sunday, August 27, before moving northward to Evanston. They were joined there by three new listeners, one of whom is rather curious. As noted above, **Albert Munroe Sawin** (1858–1917) submitted a paper to the Chicago Congress but did not read it there, even though he was in the midst of a two-year graduate program at Northwestern (1892–1894). Sawin had graduated from Wisconsin in 1883 at age 25 with BS and MS degrees before matriculating at Northwestern, where he began investigating algebraic solutions of polynomials, resulting in two papers in the *Annals*. He then taught at Minnesota and Wyoming, where he was located when he joined the NYMS in November 1891. In 1899 a note in the *Bulletin* announced his election to the AMS while he was at Clark University in Atlanta, GA, but not only did he let his membership lapse after joining the Society the first time, his name never appeared on AMS rolls again even after this *Bulletin* note.[74] Albert Sawin published three more papers in the *Annals* up to 1900. In 1899 he accepted a position at Syracuse, but he soon moved to Brooklyn, where in 1906 he was involved in some invention involving "the perfect steam engine."[75] In 1915 he accepted a temporary professorship at Rockford College in Illinois but one year later became sectional superintendent in Walla Walla, WA, across the state border from his residence in Oregon. That is where he died two years later. Albert Sawin is perhaps the most itinerant mathematician encountered so far.

Seven attendees had already had the advantage of having studied with Klein in Germany. Two others would soon do so, Ziwet and Winston, the latter being the only woman in the audience. Sylvester's former student Fabian Franklin attended Klein's lectures, but was unable to reach Chicago the previous week. And a genetic father-son connection took place when Edward Van Vleck joined his father John Van Vleck immediately upon returning from Göttingen with a doctorate under Klein in hand. Franklin, Sawin, and Edward Van Vleck were the three at Evanston but not Chicago.

Another Kleinian connection was **Clarence Abiathar Waldo** (1852–1926), perhaps best known today for his appeal to the Indiana state senate in 1897 urging the defeat of a bill to define $\pi = 3$. However, he holds an intriguing connection. Clarence Waldo graduated from Wesleyan University under John Van Vleck in 1875 and returned as a tutor and registrar (1877–1881). The next year, 1882–1883, he was one of the first Americans to go abroad to study mathematics, in his case at Munich and Leipzig, where he attended Klein's lectures. This causes us to wonder if he was the source for John Van Vleck sending three subsequent Wesleyan students to study with Klein (or if Van Vleck knew of Klein before Waldo's study tour). In any event, upon returning to the US, Waldo accepted a professorship at Rose Polytechnic Institute, where he stayed until 1891; he served as acting president on two different occasions during his eight-year tenure. Waldo was then appointed professor of mathematics at DePauw 1891–1895, when he attended the Chicago Congress and Evanston Colloquium. He was awarded a PhD by Syracuse University the next year, 1894. The following year he became head professor of mathematics at Purdue (1895–1908) before moving to Washington University at St. Louis, where he served as head professor (1910–1917).

Alexander Ziwet (1853–1928) played a heroic role in producing the book of the Evanston Colloquium lectures. Born in Breslau, Germany, his father died before he

Figure 6.9. Clarence Abiathar Waldo

graduated from gymnasium in 1873, whereupon he moved to Warsaw to join his sisters and his Polish mother while attending the university there. After three years, he transferred to the University of Moscow, but left that institution after only one year to attend the Polytechnic School at Karlsruhe, where he graduated with a CE degree in 1880. That fall, he immigrated to the US, landing in Detroit, where he worked for two years with the US Lake Survey followed by three with the US Coast and Geodetic Survey. Ziwet (pronounced ziv′-et) was appointed instructor of mathematics at the University of Michigan in 1885 and remained there until retirement in 1925. During his tenure as mathematics professor, he also served as the head of two units within Michigan's College of Engineering, the department of mathematics and the department of modern languages. He was very active with the nascent AMS, serving as vice president and editor of the *Bulletin*. Ziwet's publication record is slim, consisting of three short papers, two books on analytic geometry, and three books on mechanics. Nonetheless, "His familiarity with the literature of mathematics was amazing, covering long periods of time and journals in half a dozen languages."[76] Ziwet was a beloved figure on the Ann Arbor campus, being a founder of the Mathematical Club and leaving his personal library of over 5000 volumes to the university upon his death.

Ziwet was left with lots of editorial work to perform during the four weeks in between September 9, the final day of the Colloquium, and October 7, when Klein set sail for Bremen, Germany. In the meantime, the world-renowned German mathematician went on a whirlwind tour throughout the eastern part of the US to visit as many of his former students as possible before returning home. Such an exhausting trip by rail and coach might have drained lesser men but, on the contrary, Klein not only survived the grind, he relished it. And he was triumphant in his goal of surveying the state of mathematics in America.

Eastern tour

Klein's first stop was Ithaca, NY, where even today Cornell is regarded as equally inaccessible from everywhere (as the popular saying goes), yet he had little difficulty reaching that first destination. Cornell has been encountered on several occasions so far, mostly in connection with Johns Hopkins, beginning when Cornell president Andrew White served as one of three advisors during the founding of Johns Hopkins in 1876. Several changes had taken place at Cornell by the time of Klein's visit in 1893, the most important being the presence of its mathematics professor, James Oliver. While he attended the Chicago Congress, his student Virgil Snyder was at Göttingen in the midst of doctoral studies under Felix Klein. Oliver had traveled to Europe with his wife, initially to study with Arthur Cayley but, unhappy with the state of affairs in England, wrote to Klein to ask about the possibility of not only attending his lectures, but discussing Klein's method of teaching, courses of study as related to promising directions for research, and "to see something of your Seminary work."[77] Oliver then proceeded to Göttingen, where he became one of only five auditors for the second half of Klein's two-semester course on Lamé functions; two of the other four were the Americans Henry White and Maxime Bôcher. During that 1890 winter semester, the Olivers and Kleins struck up a very cordial relationship. When Klein returned the visit to Ithaca in 1893, three years after the American couple had returned home, the two men enjoyed long strolls along Ithaca's famed ravines.[78]

From Cornell, Klein traveled east to the Boston area. First, he visited Clark University, whose chair William Story had departed from Leipzig five years before Klein arrived, but he had advised Irving Stringham to attend Klein's lectures there. This occasion probably provided Henry Taber with his first opportunity to meet the visiting German. One can only wonder if Perott made his presence known.

From Worcester, Klein moved on to Cambridge, where Harvard housed three of his former American students (Maxime Bôcher, William Fogg Osgood, and Frank Cole) as well as William Byerly and Benjamin Osgood Peirce (the Hollis Professor of Mathematics and Natural Philosophy). Next, he traveled to Boston, where MIT was then located, to see his former student Harry Tyler. Another MIT faculty member, Frederick Woods, was in Göttingen at that time pursuing graduate studies under Klein, having already delivered two lectures at Klein's seminar. Woods received his doctorate two years later.

The next stop was Wesleyan, where Klein met up with the two Van Vlecks. John Van Vleck had attended both the Congress and the Colloquium. He was in his 40th year of teaching at Wesleyan; his son Edward had joined them at the Colloquium after returning home with a doctorate from Göttingen, and had just joined the Wesleyan faculty. John Van Vleck had also sent his students Henry White and Franklin Woods to Göttingen; the two earned doctorates under Klein in 1891 and 1895, respectively. John Van Vleck accompanied Klein on his visit to Yale after leaving Wesleyan. Here Klein met Willard Gibbs.

From New Haven, Klein moved on to New York City, where the NYMS held a special meeting in his honor. Thomas Fiske, who was soon to found the AMS, reported in the *Bulletin*:[79]

> A special meeting of the New York Mathematical Society for the purpose of giving the members an opportunity of meeting Professor Felix Klein of Göttingen was held Saturday afternoon, September 30,

> at half past three o'clock, the president, Dr. McClintock, in the chair.
> The president introduced Professor Klein in a brief address describing
> Professor Klein's mission to America and paying tribute to him in his
> different capacities as teacher, investigator, organizer, and editor.

Klein then delivered a lecture on recent advances by Schilling on non-Euclidean geometry and related them to other recent work on spherical trigonometry, using these
two topics to counter the opinion that elementary subjects offer no opportunities for
research. Fiske added:[80]

> This address was followed by a general discussion. Several questions
> were put to Professor Klein, to which he replied at length. After the
> adjournment of the meeting, those present were individually intro-
> duced to Professor Klein, and a general conversation ensued.

This incident reinforces the importance of personal acquaintances, made and renewed,
at professional meetings.

Yet even this stop did not end Klein's peregrinations, though his stay at Mayer's
Hotel in Hoboken, NJ, enabled him to mingle with other German guests at this favored spot located across the Hudson River from Manhattan. Klein wrote to Simon
Newcomb, so Klein ended up catching a train for Baltimore, where he not only met
one of the leaders of mathematics and astronomy in America but, through Newcomb,
he met Daniel Coit Gilman as well. Ten years earlier, President Gilman had negotiated
with Klein over succeeding J.J. Sylvester at Johns Hopkins. On the train back to New
York, Klein made one more stop, at Princeton, where he met up with faculty members
Harry Fine and Henry Thompson, the first having earned his doctorate under him in
1885 and the second in 1892, just one year earlier.

Finally, on October 7, Felix Klein boarded the *Saale* for the eleven-day trans-
Atlantic cruise to Bremen, Germany. Would he rest during the journey? No way.
Totally invigorated, Klein edited the book of his Evanston lectures and drafted a preliminary report for the German ministry assessing the current situation and prospects for
mathematics in the New World. Basically, he echoed J.J. Sylvester's view that America
possessed untapped potential that went beyond the usual assumption regarding the
proverbial "Yankee ingenuity" for experimentation and invention, asserting that it extended to the theoretical sciences as well. He criticized the country's secondary schools
and even its universities, which he found too absorbed with teaching and not enough
with research. Nonetheless, he concluded:

> Without question, at the present time and for the immediate future,
> America represents the richest and most promising object for scien-
> tific colonization. Already in past years, large numbers of Americans
> have studied in German universities. Up until now, however, this
> has happened without any special initiative on our part. My trip ev-
> idently represents a change in this policy, and I would like to add
> immediately that I have already made definite plans and provisional
> arrangements to return to America in the future.

Klein did return to the US just three years later to attend the Princeton sesquicentennial celebration at which he presented a series of four lectures, "The mathematical
theory of the top," that Harry Fine edited and published as a small booklet. An event
just two days later showed how quickly developments in mathematics were taking

place, as the AMS held a meeting in Princeton at which Klein lectured. The earlier meeting where he had spoken was sponsored by the NYMS, which had expanded to the AMS in 1894. Moreover, as a brief *Bulletin* report indicates, the meeting was held outside New York City, and Klein was not the only special speaker on the occasion:[81]

> A special meeting of the American Mathematical Society was held at Princeton University, on Saturday, October 17, at quarter past three, p.m. There were thirty-four members of the Society and thirteen visitors present. The President, Dr. G.W. HILL, occupied the chair, and introduced Professor Felix Klein and Professor J.J. Thomson, who addressed the Society. Professor Klein discussed the stability of a sleeping top. Professor Thomson spoke on mathematical questions connected with Röntgen rays and kindred phenomena. All business, including the election of new members, was postponed until the next meeting of the Society.

A four-page abstract of Klein's paper appeared within six months of that October 1896 meeting.[82]

Karen Parshall and David Rowe provided a succinct summary of the effect of the Chicago Congress and Evanston Colloquium on the development of mathematics in America:[83]

> Even if the gathering of mathematicians on the shores of Lake Michigan looked more like a regional meeting than an international congress, this event and the Evanston Colloquium that followed on its heels carried repercussions of immense significance for the crystallization and emergence of the American community of research mathematicians.

The quest to find a publisher for the Chicago Congress proceedings played an integral part in the NYMS expanding nationally to the AMS in 1894. In addition, Klein's former student Henry White was so impressed with the success of his mentor's Evanston Colloquium that, in February 1896, he wrote to Thomas Fiske suggesting that the AMS sponsor colloquium lectures. That idea was so timely that it was approved by the AMS council the next month and put into practice the same year with the creation of a series of colloquium lectures at annual meetings—a tradition that runs strong to this day. The first two series of colloquium lectures were delivered by James Pierpont and Maxime Bôcher at the annual summer meeting held September 1896 in Buffalo,[84] one month before the Princeton meeting.

The idea for a colloquium lecture series was one of many important contributions that the emergent group of Midwest mathematicians made to the development of mathematics in America. The establishment of sections within national organizations was another.

Mathematical pursuits

This section briefly examines the research by the University of Chicago's major figures, which landed its mathematics department atop national rankings by 1900, when it was poised to launch into the international arena. The leader, E.H. Moore, published papers in four distinct areas in four distinct time periods: algebraic geometry, algebra (group theory, finite fields, number theory), foundations (axiomatics and postulation theory), and functional analysis (integral equations, theory of functions, general analysis.) E.H. Moore's contributions to mathematics were compartmentalized into stages. Initially, he concentrated on algebraic geometry, the area of his doctoral dissertation. However, as evidenced by his paper at the Chicago Congress, his interest shifted to group theory at the beginning of the 1890s. Another change occurred at the beginning of the twentieth century based on David Hilbert's 1899 book *The Foundations of Geometry*. Shortly thereafter, E.H. Moore and his students Oswald Veblen, R.L. Moore, and Nels Lennes began investigating related foundational issues. Hilbert's work continued to exert an enormous influence on Moore. Not only did he shift interests in distinct periods, as did Hilbert, but midway through the first decade of the twentieth century, he changed his primary focus to integral equations, as reflected by the topic of George Birkhoff's 1907 dissertation. Moore spent the rest of his career in this area, which culminated in a field he called general analysis. Although this work exerted little influence either at home or abroad, it did include the Moore–Smith convergence, which he developed with his 1926 doctoral student, Herman Lyle Smith, who later became a leader at Louisiana State University.

E.H. Moore's major accomplishments represent several dominant developments that molded the American mathematical community in the closing decade of the nineteenth century and the opening decade of the twentieth. The first was the emergence of research universities, with Chicago exceeding the achievements that had been attained by Johns Hopkins and, to a lesser extent, by Clark. Moore was the acclaimed leader here.

A second major development from the 1890s was the emergence of the country's first research specialty—group theory. Altogether, Moore wrote eighteen papers from the time he arrived at Chicago in 1892 until 1899, and most of these were devoted to some aspect of group theory. The *Bulletin* was the outlet with the largest number, seven. However, an equal number appeared in foreign journals, with five in the prestigious *Mathematische Annalen*. The impetus behind publishing in the *Annalen* was a paper that had appeared there earlier in the year by the German algebraist Eugen Netto.[85] Moore then lectured on this topic at the Mathematical Club, was inspired to extend the results, and published them in the same journal under the title "Concerning triple systems." Two years later, he inserted a note in the Italian journal *Rendiconti* (of the Circolo Matematico di Palermo) with the same title and followed that up with a third paper in the *Bulletin* in 1897.

Moore presented his work on finite simple groups at the Chicago Congress, and announced the major results that year, 1893, as an abstract in the *Bulletin*. A footnote states, "This paper will be published in full in the Proceedings of the Congress. It will also appear as the first part of a paper, 'The sub-groups of the simple group whose order is $\frac{1}{2}q^n \left(q^{2n} - 1\right)$ if $q > 2$, or $q^n \left(q^{2n} - 1\right)$ if $q = 2$,' to be published in the *Mathematische Annalen*."[86] Although Moore did publish two more papers in that journal during the 1890s, this announced paper did not appear. Nor did it ever reach print. However, a similar type of classification was described in a paper he published in the *Proceedings of the London Mathematical Society* with the verbose title "Concerning the abstract groups of order $k!$ and $(k!/2)$ holoedrically isomorphic with the symmetric and the alternating substitution-groups on k letters."[87]

Moore also contributed indirectly to group theory via his doctoral students. In one of his lectures on groups during the spring 1894 semester at Chicago, he stated that, "in connection with the members of that course, Messrs. Brown, Dickson, Joffe, Slaught, [I] worked up the linear fractional configuration ... I take this opportunity to thank them for their cooperation, and especially Mr. Dickson."[88] This quotation addresses the benefit that individual mathematicians derived from the sense of a community with a common denominator of purpose. Moore acknowledged the value of a supportive community again a few years later in a paper that appeared in the inaugural issue of the *Transactions of the AMS*: "The basal notions of this paper were communicated to Chicago colleagues in February and March 1899."[89]

Herbert Slaught attested to the example that E.H. Moore set for courage and honesty in his memorial biography of his mentor and colleague:[90]

> A striking and critical test of [Moore's *intellectual* honesty occurred when he] was presenting a paper on a highly technical topic to a large gathering of faculty and graduate students from all parts of the country. When halfway through he discovered what seemed to be an error (though probably no one else in the room observed it). He stopped and re-examined the doubtful step for several minutes and then, convinced of the error, he abruptly dismissed the meeting— to the astonishment of most of the audience. It was an evidence of intellectual *courage* as well as *honesty* and doubtless won for him the supreme admiration of every person in the group—an admiration which was in no wise diminished but rather increased, when at a later meeting he announced that after all he had been able to prove the step to be correct.

Imagine—Moore dismissed the meeting. That takes guts!

Two other leading researchers in group theory during the 1890s were Frank Cole and Moore's colleague Heinrich Maschke, whose contributions appeared in the two halves of this decade. Cole's seven years at the University of Michigan, 1888–1895, proved to be the most productive period in his life, beginning with his translation of Eugen Netto's book, *The Theory of Substitutions and its Applications to Algebra*. Published in 1892, this translation was not only completed with the permission of Netto but was really a revision by both the author and the translator. That same year a paper appeared listing all simple groups of order 201–500. And then, during 1894–1896, Cole published another six papers on group theory, half on simple groups and half on substitution groups. Although no other paper on group theory by Cole would appear until 1904, he published several more over the next 20 years. Unfortunately, Cole's life was cut short in May 1926 by an infected tooth. Sadly, he was in the process of planning to retire to a farm in the Catskills of New York, where he anticipated long walks to study trees and wildflowers.

Maschke too died prematurely, at age 54. When he was called to the faculty position at Chicago, he had been away from research mathematics for several years. Therefore, he began his tenure at a decided disadvantage. Although he spoke about invariants of a group of 336 substitutions at the Chicago Congress, it was another four years before he contributed to the subject in a deep manner. During a talk to the Mathematical Club in May 1897, he announced a major result that appeared in *Mathematische*

Annalen the following year, the cyclotomic theorem.[91] At the end of that year he generalized this result; a particular case now bears the name Maschke's theorem. This work too appeared in 1899 in *Mathematische Annalen*. It reaffirms the positive effect of the Mathematical Club on the research environment at Chicago.

Group theory was surely not the only area pursued in depth in the 1890s. Oskar Bolza, for instance, spoke about elliptic and hyperelliptic function theory at the Chicago Congress. He pursued these topics throughout the remainder of the decade, and this was reflected in the dissertations written by John Hutchinson, William Gillespie, and John McDonald. However, by the turn of the century, Bolza's primary interest had turned to the calculus of variations, which would become his specialty for the rest of his life and would be the area pursued by his most famous student, Gilbert Ames Bliss.

Moore Mob, I. In this chapter and the next, I adopt the term "Moore Mob" to refer not just to E.H. Moore's PhD students, but also to all Chicago students who excelled in mathematics during his tenure as chair of the department, 1892–1927. Therefore, I include John Hutchinson, who moved to Cornell when his fellowship ended in 1894, but who completed his dissertation under Oskar Bolza two years later, thus becoming one of Chicago's first PhD recipients in mathematics. Two other graduate students who received their degrees up to 1900 achieved far greater renown, Leonard Dickson and Gilbert Bliss. Both ended up joining Herbert Slaught on a faculty that ultimately became criticized for inbreeding.

Figure 6.10. Leonard Eugene Dickson

Leonard Eugene Dickson (1874–1954) was a brilliant student at the University of Texas, where he was influenced by department chair George Halsted, a leading member of Sylvester's flock.[92] After graduating as valedictorian of his class in 1893, Dickson was awarded a teaching fellowship for the next year, resulting in a master's degree in 1894. During that time, he displayed early promise as a researcher by publishing a six-page paper, "Lowest integers representing sides of a right triangle," in the very first issue of the *American Mathematical Monthly*. Dickson desired to continue his study of higher mathematics, which was not possible at Texas, but there were few options available in the country for such young, aspiring students. In fact, he applied to only two graduate programs at quite distinct institutions. One, Harvard, was the nation's first college, and boasted a rich history that now featured two young researchers, Maxime Bôcher and William Osgood. The other, Chicago, had only been in existence for two years, though the Chicago Congress had elevated its reputation appreciably. And so, when the 20-year-old Dickson received a fellowship offer from Harvard, it seemed certain he would matriculate there, yet when E.H. Moore matched it, Dickson chose to enroll in his upstart program. I do not know why he chose the fledgling Chicago graduate program over the established Harvard, but Halsted undoubtedly played a role in the decision. Garrett Birkhoff has stated that the Texan Dickson arrived at Chicago with as "much of the dynamic energy and rugged individualism that we associate with that state."[93]

Leonard Dickson's decision paid dividends at once. Arriving at Chicago for the fall quarter of 1894, when E.H. Moore was lecturing on group theory, Dickson excelled at research right from the start, completing the program in only two years, thus joining John Hutchinson as the first two students awarded PhDs in mathematics (in 1896). The 22-year-old's dissertation, "The analytic representation of substitutions on a power of a prime number of letters with a discussion of the linear group," combined group theory with finite fields, inspired by Moore's paper from the Chicago Congress. Dickson's long and detailed dissertation, published in two parts, represented the first in a very long line of works on algebra and number theory, including papers, books, and monographs.[94] In 1901 he revised and expanded this work into book form, *Linear Groups with an Exposition of the Galois Field Theory*, published when he was still only 27.

Upon graduation, Dickson sailed to Europe to attend classes in Leipzig (by Sophus Lie) and Paris (by Camille Jordan). Upon returning to the US a year later, he taught at the University of California, Berkeley, for two years (joining Irving Stringham and Mellen Haskell) and the University of Texas for one (joining former mentor George Halsted) before returning to Chicago in 1900. The indefatigable Dickson flashed signs of a prolific output at once, publishing the classic book on linear groups mentioned above. That work was presented in two parts, one that gave "the first extensive presentation of the theory of finite fields," and the other that was devoted "to a general theory of linear groups over $GF[p^n]$, a generalization of Camille Jordan's work over the prime field $GF[p]$."[95] Moreover, the last part of this tome listed all simple groups known up to that time.

Dickson engaged the German publisher Teubner to print the work. Initially, he intended to contact American firms, but E.H. Moore convinced him to contact Felix Klein about acquiring a German company. This accomplishment impelled the modern historian of mathematics Karen Parshall to conclude,

> By thus securing the Teubner seal of approval, Dickson's work not only came before one of the most important segments of the international mathematical community at the turn of the [twentieth] century, but it also served to build its author's reputation at home and abroad.[96]

Over the next four decades, Leonard Dickson published 18 books and roughly 300 research articles. Overall, his contributions transformed Chicago into a world center for groups, rings, and algebras. Except for visiting positions back at Berkeley on three different occasions, Dickson spent the rest of his career at the University of Chicago.

Two burning questions arise from Leonard Dickson's behavior immediately before, and entirely after, his retirement in 1939. Ominously, upon retiring, he burned all his correspondence. Why? Probably because this intensely private person was aware of how one of his sisters used personal correspondence from their parents to write biographical accounts, and Dickson therefore sought to avoid risking public exposure and perhaps misunderstanding. Then he moved back to his home state of Texas and never pursued mathematics over the next 15 years. Why? Probably because he "felt he had seen too many world-class mathematicians do mediocre work long after they reached retirement."[97] He died five days shy of his eightieth birthday.

In addition to being a productive scholar, Leonard Dickson was one of America's most prolific doctoral advisors, directing 67 doctoral dissertations altogether. Saunders Mac Lane wrote, "In 1931, Dickson's best student, Adrian Albert, was appointed [to the Chicago faculty]."[98] Albert (PhD, 1928) succeeded his advisor in carrying on the Chicago algebra tradition. A recent *Notices of the AMS* article suggested that Ke-Chuen Yang (PhD, 1928) and Mina Rees (PhD, 1931) could also be considered among Dickson's "best" students, based on different criteria.[99] Yang was Dickson's only Chinese student. Other notable Dickson doctorates include Lloyd Williams (PhD, 1920; the founder of the Canadian Mathematical Society), Cyrus MacDuffee (PhD, 1921; directed 31 dissertations at Ohio State and Wisconsin), Burton Jones (PhD, 1928),[100] Arnold Ross (PhD, 1931; famous for nurturing young talent in a program at Ohio State from 1964 until his death at age 96 in 2002), and Ivan Niven (PhD, 1938; University of Oregon number theorist).

Numbered among Leonard Dickson's doctoral students were 18 women. Mina Rees was perhaps the most prominent member of this group. Two other Dickson graduates are notable, one for celebrating a long career and the other for suffering a distressingly short calling. Dickson's first woman student was **Mildred Leonora Sanderson** (1889–1914), who enrolled in the Chicago program on a Bardwell Memorial Fellowship awarded upon graduation from Mount Holyoke College in 1910. She received a master's degree one year later for a thesis Dickson felt could have served as her doctoral dissertation; it was published that same year in the *Annals of Mathematics*.[101] Instead, Sanderson undertook a new investigation that resulted in her PhD in 1913, thus making her the first of Dickson's 18 women doctoral students. Adopting the tradition of Dickson's own advisor, E.H. Moore, he had Sanderson outline her major results at a meeting of the Chicago Section of the AMS in March 1913; her dissertation was published in the *Transactions of the AMS* later that year.[102] Upon graduation, Mildred Sanderson joined Arnold Dresden as another Chicago graduate on the faculty at the University of Wisconsin. Her career got off to a propitious start when she presented a paper at an AMS meeting held in Madison that September but, sadly, she was forced to leave her position in February when she became ill with pulmonary tuberculosis. She died in October 1914 at age 25. E.T. Bell assessed her *Transactions* paper glowingly: "Miss Sanderson's single contribution (1913) to modular invariants has been rated by competent judges as one of the classics of the subject."[103]

While Mildred Sanderson was a math whiz early in life, **Mayme Irwin Logsdon** (1881–1967) came late to the subject. She got married shortly after graduating in 1900 from the Hardin Collegiate Institute in her home state of Kentucky, where she probably was enrolled in the "normal course" for teachers. Although only 19 at the time, she became a high-school teacher and principal, positions she held until 1911. Her husband, a widower with two children when they married, had died in 1909. At that point Mayme Logsdon enrolled at Chicago and earned her bachelor's degree in one year.

E.H. Moore's scintillating department enticed Logsdon to ascend directly into the graduate program, but she left after one year to become instructor and dean of students at Hastings College in Nebraska. She remained at Hastings 1913–1917, but had returned to Chicago during the 1914 summer quarter to earn a master's degree. In 1917 she was appointed instructor at Northwestern, where she remained for two years before returning to Chicago to resume full-time studies. It took but two years for her to complete requirements for the PhD, with a dissertation that described when a pair of Hermitian forms are equivalent. She was 40 when she received her doctorate in 1921, 16 years older than Sanderson had been.

Moore hired Mayme Logsdon as an instructor on the spot, and Logsdon remained on the Chicago faculty from that time until her retirement 25 years later. Early in her tenure, her lecture in Rome drew the famous number theorist André Weil's attention to L.J. Mordell's work on elliptic curves, an area that became important in Weil's subsequent career. Logsdon also directed four PhD dissertations during her Chicago years, yet she was never promoted beyond associate professor. She did not end her teaching career in 1946, however, instead moving to the University of Miami in Florida and remaining there until her second retirement at age 80 in 1961.

Figure 6.11. Gilbert Ames Bliss

In addition to Leonard Dickson, the other leading Moore Mob stalwart was Chicago-born **Gilbert Ames Bliss** (1876–1951), who entered the university in the fall of its second year, 1893, straight out of high school. Upon graduation four years later, he began studies in mathematical astronomy under Forest Ray Moulton, and his first published work was titled "The motion of a heavenly body in a resisting medium." However, two events aligned to encourage him to switch to mathematics after only one year. For one, he was not granted a fellowship in astronomy. Equally important was a course on the calculus of variations taught by Oskar Bolza referencing an 1879 paper on that topic by Karl Weierstrass. Bliss attacked this field with shock and awe, writing a master's thesis the next year (1898) and then expanding it into the dissertation, "The geodesic lines on the anchor ring," for his PhD under Bolza in 1900. The calculus of variations became his specialty throughout his career.

Bliss's doctoral work shows the ascendance of mathematics in America on the eve of the twentieth century. The famed mathematician Karl Jacobi had shown that all points on a surface of negative curvature are of the first kind, while all points on a closed surface of positive curvature are of the second kind.[104] Bliss showed that all points on the inner equator of an anchor ring are of the first kind, while all other points on it are of the second kind. His example, like Dickson's, shows American mathematicians educated entirely in the US entering the international arena on equal footing, thus marking an important turning point in American mathematics.

With degree in hand, Bliss was appointed instructor at the University of Minnesota, but he left there to spend 1902–1903 in Göttingen, not only to interact with Felix Klein, but also to get to know that program's rising star, David Hilbert. While there, Bliss also became quite friendly with Max Mason, an American student who would soon earn a doctorate under Hilbert. Once back in the US, Bliss taught at three different institutions over the next five years: Chicago (1903–1904), Missouri (1904–1905), and Princeton (1905–1908). The faculty at Princeton was young, vigorous, and up-and-coming; it would seem the perfect fit for someone of Bliss's ability and drive. But Chicago made him an offer he could not refuse—successor to Heinrich Maschke, who had just died.

Consequently, in the fall of 1908, Gilbert Bliss returned to his *alma mater*, where he remained for the rest of his life. The faculty was particularly robust, including E.H. Moore, Oskar Bolza, Herbert Slaught, Jacob Young, and Leonard Dickson, though Bolza returned to Germany two years later. Twenty-year old Theophil Henry Hildebrandt was also on the faculty at the time; he received his PhD in 1910, but one year earlier had accepted an appointment at Michigan, where he became a mainstay for the rest of his life. Other Chicago faculty members with a strong interest in mathematics included Forest Ray Moulton (astronomy), Arthur C. Lunn (applied mathematics), George W. Myers (the teaching of mathematics and astronomy), William D. Macmillan (mathematics and astronomy), and Charles R. Mann (physics).

Curiously, Bliss lectured on geometry during his first two years back on the Chicago faculty. But from the time Bolza left for Germany in 1910, Bliss happily took up his first love, which ranged broadly over the field of analysis with special focus on the calculus of variations. One year earlier, in 1909, Bliss delivered one of the two Colloquium Lectures at the AMS meeting in Princeton on "Fundamental existence theorems," which lie at the heart of the calculus of variations. During the 1920s, he solidified his international reputation with a series of papers on boundary-value problems with applications to physics, mainly in the newly emerging fields of quantum mechanics and relativity. But mathematics can change quickly, and Harvard mathematician Marston Morse would soon take the subject in a different direction, one that Bliss was unable to follow toward the end of his career. Nonetheless, Bliss's comprehensive 1946 book *Lectures on the Calculus of Variations* provided text beyond what had previously existed. Gilbert Bliss was also a prolific mentor of doctoral dissertations, advising 54 over the period 1910–1939. Among his most distinguished students were Lawrence Graves (PhD, 1924), William Duren, Jr. (PhD, 1930), Magnus Hestenes (PhD, 1932), and Alston Householder (PhD, 1937); Graves co-directed the thesis of Edward McShane with Bliss in 1930.

Subsequent chapters will document Bliss's contributions to the development of ballistics during World War I, the establishment of fellowships in mathematics from the National Research Council, and the construction of buildings for mathematics departments. Within the mathematics department, he succeeded E.H. Moore as chair in 1927, and held that position until his retirement in 1941. All the while, he remained

active with the Mathematical Club. Bliss served the AMS in various capacities, notably
as editor of the *Transactions* (1909–1916) and with a two-year term as president (1921–
1922), during which membership increased from 770 to 1127. He also wrote the first
Carus Monograph (on the calculus of variations, of course) and was awarded the first
Chauvenet Prize by the MAA in 1925.

Notably, Gilbert Bliss resisted all attempts to enact high admission standards in
order to restrict student enrollments, which placed him in opposition to the policy of
the highly selective University of Chicago at the time, one that today is sometimes
regarded as heresy at elite universities. His strong views on the matter were based on
his perception that some poorly trained students develop real intellectual power, while
conversely, some initially brilliant students gradually fade. Yet Bliss was very fair, and
listened to all sides of arguments.

PhD students. Table 6.5 lists the ten students who completed Chicago dissertations
up to 1900. The dissertations were written in three general areas of mathematics,
directed almost equally by the three professors (Bolza four, Moore and Maschke three
each). The six degrees awarded in the year 1900 was the highest in any single year for
quite a while, but the quality of these ten students fell far short of those who graduated
in the first eight years of the twentieth century.

The online file "Web06-ChiPhDs" contains snippets of the five not mentioned
above. Briefly, **John Anthony Miller** (1859–1946) earned a PhD in 1899 at age 39,
thus becoming Heinrich Maschke's first doctoral student. Seven years later, he moved
to Swarthmore College. In 1897, **George Lincoln Brown** (1869–1950) joined the faculty at the South Dakota State University, where he remained for the rest of his life.
While in this position, he completed his dissertation on a group of linear transformations under Heinrich Maschke in 1900. The Canadian-born **William Gillespie**
(1870–1947) enrolled in the Chicago graduate program in 1893 upon graduation from
the University of Toronto. He left after four years without a degree to accept an instructorship at Princeton, but completed his dissertation under Bolza while teaching
there. **Derrick Norman Lehmer** (1867–1938) entered the Chicago graduate program
in 1897 and received his PhD three years later under E.H. Moore. Lehmer went directly to the University of California, Berkeley, where he remained until retirement in
1937. Like William Gillespie, **John Hector McDonald** (1874–1953) was a Canadian

Table 6.5. Chicago PhDs up to 1900

PhD Recipient	Year	Advisor	Field
Leonard Dickson	1896	Moore	Algebra
John Hutchinson	1896	Bolza	Analysis
Herbert Slaught	1898	Moore	Geometry
John Miller	1899	Maschke	Analysis
Gilbert Bliss	1900	Bolza	Geometry
William Gillespie	1900	Bolza	Analysis
John McDonald	1900	Bolza	Algebra
George Brown	1900	Maschke	Algebra
Ernest Skinner	1900	Maschke	Algebra
Derrick Lehmer	1900	Moore	Analysis

and a University of Toronto graduate (1895). He entered Chicago the following year. His 1900 dissertation reads, "I feel under a deep obligation to all the teachers named, but particularly to Professors Moore and Bolza for the continued and varied assistance which they gave me throughout my whole term of graduate study." McDonald joined Lehmer at Berkeley in 1902 and remained there for the rest of his life.

The aim of the University of Chicago was to join the ranks of the best universities in the world. For mathematics, this meant being able to compete with Germany primarily, and France secondarily. Up to 1900 one could hardly conclude that the Chicago mathematics department had attained this level regarding research accomplishments, though it had made an excellent start with the faculty of Moore, Bolza, and Maschke and outstanding students Dickson and Bliss. A core of doctoral students who emerged in the first decade of the twentieth century vaulted Chicago mathematics onto the world stage.

7

The 1890s

Chapters 5 and 6 described the emergence of the American mathematical research community under the influence of three different figures—J.J. Sylvester at Johns Hopkins, Felix Klein in Göttingen, and E.H. Moore at Chicago. Another important development during the revolutionary period 1876–1900 was the founding of the American Mathematical Society in 1888. This chapter describes its activities over the remainder of the century. It also charts developments in graduate education in mathematics that took place at Stanford, Clark, Cornell, and Penn. Yet another central element from the 1890s was the school of American students who studied with Sophus Lie in Leipzig. Concurrently, American women achieved access to, and success in, graduate programs for which they qualified. This chapter introduces the ten who were awarded PhDs before 1900 and one who was prevented because of her sex. The chapter ends with an account of the emergence of statistics courses at some leading American universities over the last two decades of the nineteenth century.

American Mathematical Society (AMS)

Chapter 5 described the founding of the American Mathematical Society (AMS) in 1888 as the New York Mathematical Society and the establishment of its first journal, the *Bulletin*, three years later. Within the next three years, the AMS was transformed into a truly national organization, partly caused by the search for publishing support for the proceedings of the Chicago Congress. This section describes four developments that occurred within the newly minted AMS during the quinquennial period 1896–1900 that were related to the Evanston Colloquium, the Chicago group, and the *Bulletin*. The first was the initiation of a lecture series at summer AMS meetings in imitation of the Evanston Colloquium. Second, a different type of meeting was held every summer, starting with the Toronto meeting in 1897. The third was the founding of a section within the organization. The final development was the establishment of a second AMS journal, the *Transactions*, which assumed the primary role of publishing research papers written by AMS members. The section concludes with a summary of annual AMS meetings held during 1889–1899.

The year 1896 witnessed the historic establishment of the Colloquium Lectures, which have been mentioned several times so far. The impetus for this development came from Henry S. White, who had hosted Felix Klein's eminently successful lectures at the Evanston Colloquium in September 1893. White had therefore experienced first-hand the value of intensive and focused lectures. After attending the first two summer AMS meetings in 1894 and 1895, White wrote to Thomas Fiske in February 1896 to compliment him for the successful gatherings. White then requested an addition to the program, asking, "Would not a series of three or six lectures on nearly related topics ... prove attractive and useful to larger numbers?"[1] To be specific, White suggested five qualified candidates, sagaciously referencing individuals at five different universities: Bolza (Chicago), Craig (Clark), Cole (who had just arrived at Columbia from Michigan), Fine (Princeton), and Osgood (Harvard).

Fiske agreed with the idea and, as usual, acted with dispatch. He urged its acceptance to the AMS ruling body, the Council. The *Bulletin* reported that at its March meeting:[2]

> The Council of the American Mathematical Society has given its approval to a proposition for holding a colloquium in connection with the summer meeting at Buffalo next August. It is thought by many that the benefits arising from the summer meeting will be greatly increased if those attending have an opportunity of hearing brief accounts of the progress made in different branches of the higher mathematics. It is intended to have two or three short courses of lectures by eminent mathematicians, every lecture being followed by an informal discussion.

On April 16, 1896, within one month of that approval, Henry White mailed a circular to all AMS members announcing, "a meeting auxiliary to the Summer meeting, to continue for one week after the regular session, and to be designated as a Colloquium or Conference."[3] The circular provided grounds for the proposal and a plan for the colloquium. A subsequent mailing in July announced details of the historic event.

Thus, in early September, just seven months after Henry White's suggestion, 13 members, including two women, paid the \$3 fee to attend the first AMS Colloquium Lecture Series at the summer meeting in Buffalo, NY. The colloquium was held in a lecture hall in the Society of Natural Sciences building, the only time it did not occur on a university campus. (From this time on, summer meetings were named for the town in which the university was located, such as Ithaca for the meeting at Cornell.) in Buffalo, Yale's James Pierpont[4] presented a series of six one-hour lectures on "Galois theory of equations" and Harvard's Maxime Bôcher gave six on "Linear differential equations and their applications," with sessions meeting at 9 a.m., 11 a.m., and 3 p.m. on each of four days. The lecturers published abstracts of their talks as part of the AMS report on the meeting.[5]

The participants were delighted with the results. The AMS report on the meeting stated, "At the close of the last lecture ... a motion was adopted recommending to the Council that arrangements be made for a Colloquium in connection with the next Summer Meeting."[6] Apparently, the motion was carried unanimously. James Pierpont strongly agreed that the AMS should continue to sponsor such gatherings. He "regretted that time did not permit him to develop the theory of finite groups from [an] abstract standpoint and to touch upon some of the beautiful results obtained by

Frobenius, Holder, Cole, and others. The importance of these methods and theories not only for the Galoisian theory, but for many other branches of mathematics, makes it desirable that they be made the subject of a future colloquium."[7]

Despite these testaments, a colloquium was not held at the 1897 summer meeting in Toronto because the joint meeting with the British Association made such an arrangement impractical, so it was postponed until the next year. A note in the May 1898 issue of the *Bulletin* announced:[8]

> The fifth Summer Meeting of the American Mathematical Society will be held at the Institute of Technology, Boston, Mass., on Friday and Saturday, August 19th and 20th, thus immediately preceding the meeting of the American Association for the Advancement of Science. A colloquium will be held in connection with the Meeting. This will be conducted by Professor W.F. Osgood, who will give a course of lectures on some of the problems and methods of the modern theory of functions, and by Professor A.G. Webster, who will treat certain portions of mathematical physics. A circular giving a more definite announcement in regard to the colloquium and the regular sessions will shortly be issued by the Secretary.

The circular was mailed on June 20, 1898, about five weeks beforehand, and provided abstracts of the courses of lectures by the speakers.

This time, 26 AMS members attended the colloquium at the summer meeting, held in late August in Sever Hall on the Harvard campus, where William F. Osgood and Arthur G. Webster (who was at Cornell by then) each presented six lectures on "Selected topics in the theory of functions" and "The partial differential equations of wave propagation," respectively. The AMS report on this colloquium ended with a statement suggesting a need for the lectures to appear in print. "Professors Webster and Osgood have consented to prepare reports of their lectures for publication in the *Bulletin*."[9] Osgood's paper was printed immediately after the AMS report.

I do not know why no colloquium was offered at the summer 1899 meeting, but the famous 1900 ICM in Paris, at which David Hilbert proposed his list of 23 problems, precluded one in the first year of the new century as well. When the December 1900 issue of the *Bulletin* announced the third colloquium to accompany the summer 1901 meeting held in White Hall on the Cornell campus, it asserted that "conditions are favorable for extending the colloquium through a period of two or more weeks."[10] Such an extended period never materialized. The speakers were Oskar Bolza (Chicago) and Ernest W. Brown (Haverford College). Even after only two successful colloquia, it became a badge of honor to be selected to deliver such a series of lectures. Edward Kasner, though only 23 years old at the time, but holding a PhD from Columbia and having spent a postdoctoral period of study in Germany with Felix Klein and David Hilbert, asserted that "lecturers of high standing ... serve purely as a matter of honor, receiving no compensation, except a very moderate allowance for the travelling and other expenses provided for by the collection of a nominal fee [still $3] from each auditor."[11]

Ernest Brown, like William Osgood before him, published his lectures in the only AMS journal at the time, the *Bulletin*. But Oskar Bolza, who regarded his invitation as one of the two outstanding events in his career,[12] harbored greater plans. The February 1903 issue of the *Bulletin* announced that his lectures would "be published early in the spring as one of the supplementary volumes of the Decennial publications of the University of Chicago, under the title: 'Lectures on the calculus of variations'."[13] As most authors know, works do not always see light of day as quickly as anticipated; Bolza's work finally appeared the following July.

The perceived need for wider distribution of the Colloquium Lectures would finally take root at the fourth colloquium, held at MIT (then located in Boston) for four days following the two-day summer 1903 meeting. This time there were three Colloquium speakers, all former students of John Van Vleck and Felix Klein—Edward Van Vleck (Wisconsin), Henry White (Northwestern), and Franklin Woods (MIT). By this time the demand for publication of the lectures seemed ripe, so the following February the AMS reported, "A committee consisting of the President [Thomas Fiske] and Professor Osgood was appointed to arrange for the publication in book form of the courses of lectures delivered ... before the Boston Colloquium, in September, 1903."[14] Fiske and Osgood arranged for the Macmillan Company to issue 1000 copies of the book in April 1905. The contents consisted of White's three lectures titled "Linear systems of curves on algebraic surfaces" (pp. 1–30), Woods's three lectures "Forms of non-Euclidean space" (pp. 31–74), and Van Vleck's six lectures "Selected topics in the theory of divergent series and of continued fractions" (pp. 75–187). The three speakers dedicated this historic AMS volume to their mentor, in honor of his having completed 50 years of service as a professor at Wesleyan: "To Professor John Monroe Van Vleck, LLD, these lectures are affectionately inscribed by his former pupils." At $2 ($1.50 for AMS members), the book's price was roughly $55 in 2014.[15]

The next colloquia lectures took place at Yale the following year, 1906. All three speakers who delivered their lectures in Lampson Hall before an audience of 43 were associated with Chicago:

- E.H. Moore, "On the theory of bilinear functional operators,"
- Max Mason, "Selected topics in the theory of boundary value problems of differential equations,"
- E.J. Wilczynski, "Projective differential geometry."

Because Moore was such a successful Yale graduate, the university assumed responsibility for publishing 500 copies of a book containing all three sets of lectures. This prompted Cornell's Oliver Snyder to comment that "the colloquium has become a highly important element in the Society's activities."[16] In retrospect, this assessment appears off the mark. For one thing, the next four colloquia took place irregularly: 1909 (at Princeton), 1913 (Wisconsin), 1916 (Harvard), and 1920 (Chicago). The AMS managed to publish these lectures despite stringent financial difficulties.[17]

The AMS Colloquium Lecture Series was inspired by Felix Klein's 12 lectures at the Evanston Colloquium in 1893. Shortly after the first AMS version took place in 1896, another notable event occurred at the October AMS meeting in Princeton as part of that university's sesquicentennial celebration. During the ceremony, the elite institution's name was switched from the College of New Jersey to Princeton University. One eminent speaker was Felix Klein, who delivered his last address on American soil. This undoubtedly accounts for the large crowd that attended the meeting; for the year 1896 a *Bulletin* note commented, "The average attendance at the ordinary meetings during the year was 15; the attendance at the last annual meeting 25; at the summer meeting 30; at the colloquium 16; and at the Princeton meeting 34."[18] With Klein's final

American student Mary Winston having just completed her dissertation, this occasion would mark the German master's swan song from American mathematics.

Toronto meeting. When the New York Mathematical Society changed its name to the American Mathematical Society on July 1, 1894, the name of its official journal, the *Bulletin*, was switched accordingly. To reflect the expanding geography of the organization, the AMS decided to supplement the monthly gatherings, held during the academic year in New York City, with a summer meeting. An appropriate occasion presented itself—a meeting of the AAAS the next month. Accordingly, the AMS held its first summer meeting at the Brooklyn Polytechnic Institute on August 14–15, 1894, the two days before the AAAS meeting. The attendance of 22 AMS members would have seemed to have gotten the summer program off to an auspicious start, yet Thomas Fiske, in his official report, published as the very first article in the newly named *Bulletin of the American Mathematical Society*, griped, "[A]ttendance was not so large as had been anticipated."[19] However, the quality of the ten papers compelled the AMS founder to add that nonetheless, "the interest was well maintained throughout." Moreover, by holding the affair during summer recess, George Halsted and Alexander Macfarlane could attend from faraway Texas, and the indefatigable E.H. Moore could meet up with his mentor, Hubert Newton, who, considering his protégé's success at the University of Chicago, must have been a proud *Doktorvater* indeed.

The summer 1894 gathering was the first AMS meeting not held at Columbia University. The next summer meeting took place outside New York State—in a high school in Springfield, MA, in August 1895, once again the two days before the start of the AAAS meeting. The attendance of 33 AMS members, a 50% increase, included former Sylvester students William Durfee and Arthur Hathaway. The latter presented one of the 16 papers read (on quaternions) during the two sessions that were conducted on each day, one in the morning and the other the afternoon. The final session was devoted to two topics: (1) "A general subject catalogue or index of mathematical literature," and (2) "The mathematical curriculum of the college and scientific school." The AMS report on the meeting by Thomas Fiske contains abstracts of the papers and further details on the discussions.[20]

The third summer meeting was held even farther away from New York City, at the Society of Natural Sciences in Buffalo, NY. This time:[21]

> The Society continued for the present year the policy of affiliation, as regards the Summer Meeting, with the American Association for the Advancement of Science. In deference to the desire of the latter organization, the meeting of the Society was held the week following that of the Association. The register of attendance and the close and undivided attention given to the work of the meeting show that this arrangement was, on the whole, profitable for the Society.

That register recorded the presence of Sylvester's former students Ellery Davis and William Durfee, Clark graduates Alfred DeLury and Thomas Holgate, Chicago mentor E.H. Moore along with the university's first two PhDs in mathematics, Leonard E. Dickson and John Hutchinson, as well as a veritable who's-who from Klein students (Maxime Bôcher, Frank Cole, William Osgood, Virgil Snyder, Henry White, Mary Winston, and Franklin Woods). These 14 account for almost one half of the 34 AMS members in attendance.

The next summer meeting was historic for being held outside the US, as the AMS continued to break new ground in 1897 by sponsoring it in Toronto, sandwiched between the Detroit meeting of the American Association for the Advancement of Science (AAAS) and the annual meeting of the British Association for the Advancement of Science (BAAS). The Canadian Mathematical Congress was not formed until 1945, so the BAAS partially filled the needs of Canadian mathematicians at this point in the country's history.[22] The BAAS had previously met in Montreal in 1884, with the Germans Ferdinand Lindemann and Walther Dyck in attendance. The 1897 gathering attracted an estimated 1300 scientists, which was a drop of some 400 from the one held 13 years earlier. "For the Toronto meeting, the foreign (which here includes colonial) contributions came from Canada and the United States."[23]

Summer AMS meetings became a staple of the organization from this time until the end of the twentieth century. In 1973 the AMS added another set of Colloquium Lectures at the winter meetings.[24] The AMS president in 1897 was Nova Scotia's Simon Newcomb, which perhaps accounts for the fact that the meeting was held in Canada. It was an overwhelming success. AMS secretary Frank Cole gushed:[25]

> The actual outcome ... exceeded all anticipation. ... [T]his meeting will rank among the most important scientific gatherings that have taken place on this continent, and is a distinct landmark of the vigorous growth of the Society and the scientific activity of its individual members.

The attendance of 55 included 44 AMS members, perhaps the most famous being J.C. Fields, who interrupted his 1892–1900 study tour in Europe to present one of the 21 papers read. Other Canadian figures in attendance were three who had spent their undergraduate years at the University of Toronto (Alfred DeLury, Thomas Holgate, and William Metzler). However, two notable Toronto graduates were not listed: William Gillespie and John McDonald.

The summer 1897 AMS meeting reflected the growth of the department of mathematics and natural philosophy at the University of Toronto. The three professors in the department up to 1875 all hailed from England. Only J.B. Cherriman distinguished himself while chairing the department 1853–1875. In 1876, his successor, James Loudon, became the first native-born Canadian to hold the position, but he first had to overcome discrimination. The Ontario Legislature, which made all university appointments, opposed his professorship in 1875 because he had been born and educated in Canada. However, due to public protests, the Legislature caved in, and Loudon thus became "the first native-born Canadian on the faculty of an English-Canadian university."[26]

An 1862 graduate of Toronto, **James Loudon** (1841–1916), taught classes in the fall of 1863 until Cherriman returned from his usual summer vacation in England. The following year Loudon was appointed the first mathematics tutor at Toronto, a position he held until his appointment as professor. During his 11-year professorship, he published three papers in the *American Journal of Mathematics* and two textbooks—*The Elements and Practice of Algebra* (1873) and *Algebra for Beginners* (1876). However, Loudon's main interest was physics, so when the department of mathematics and natural philosophy was divided into two parts in 1887, he moved to the newly formed physics department. Five years later Loudon was named president of the university, a position he held until retiring in 1906.

Figure 7.1. James Loudon

In 1887, **Alfred Baker** (1847–1942) was appointed the first chair of the mathematics department. A distinguished student at Toronto, he graduated in 1869 and then taught in secondary schools for six years before becoming mathematical tutor in 1875. That was his standing for another 12 years until being promoted to professor. Baker served as chair until his retirement in 1919. His main interests were geometry (notably quaternions) and probability, but he also taught advanced courses in higher plane curves, modern algebra, and differential equations (adopting a well-known text by George Boole). Classes were so small at the time that, in the third and fourth years, they were essentially private affairs.

Baker, like his predecessors, was not a research mathematician. The first person at Toronto who might be considered a research mathematician was **George Paxton Young** (1819–1889), a graduate of the University of Edinburgh who immigrated to Canada in 1847 and became the chief superintendent of education in Canada in 1864. Young was appointed professor of logic, metaphysics, and ethics (generally called philosophy today) at the University of Toronto in 1871. Yet he published six papers in the *American Journal of Mathematics* and one in the *Quarterly Journal of Mathematics* in England. Regarding this situation, the later Toronto mathematician Gilbert Robinson wrote:[27]

> It is fascinating to compare these mathematical papers of Young with those of his contemporaries, Cherriman and Loudon. The implication is obvious—mathematicians were not abstractionists in those days, they had a practical role to play in solving the world's problems. Philosophers, on the other hand, having no such role were free to associate with like minded colleagues wherever they might be found.

One other Canadian who published original results at this time was **John Cadenhead Glashan** (1844–1932), who had no direct connection with Toronto. Rather, he was the superintendent of education for Canada with an office located in Ottawa; his book *Public School Arithmetic and Mensuration* was very influential in the country, and an Ottawa public school is named after him. While in this position he published eight papers in the *American Journal of Mathematics* from its first volume in 1878 to 1901, as well as four papers in the *Proceedings of the Royal Society of Canada*. John Glashan offered to sell his extensive library to the AMS in 1924, provided the Society would cover the cost of transportation, legal services, and traveling expenses for the librarian (Raymond Archibald). However, the AMS did not approve the deal.[28]

Modeled after the Royal Society of London but with the addition of literature and the social sciences, the Royal Society of Canada was the first professional organization of scientists in Canada. Also known as *La Société royale du Canada*, it was founded under the personal patronage of the Marquis of Lorne in 1882, and incorporated the next year. Serving as Canada's national academy of science, the RSC today 1) promotes Canadian research and scholarly accomplishment in both official languages, 2) recognizes academic and artistic excellence, and 3) advises governments, nongovernmental organizations, and Canadians on matters of public interest. One of the original sections of the RSC was "Mathematical, Physical and Chemical Sciences," which initiated two journals from its start in 1882, the *Proceedings of the RSC* and the *Memoirs of the RSC*.

Chicago Section. While the Toronto meeting was big and the establishment of the Fields Medal in the 1930s would become more famous, a smaller development took place in 1897 that had ramifications for the AMS over the next 30 years, and which played an even greater role when the MAA was founded in late 1915—establishing a section within a national, professional organization. E.H. Moore and his Chicago school were at the forefront of this development, having already been major cogs in the nationalization of the AMS and its publication initiatives.

Collaboration between the mathematics departments at Chicago and Northwestern had begun during Chicago's first academic year, which ended when Henry White delivered the final lecture before the Mathematical Club. After the close of the Chicago Congress and Evanston Colloquium, joint discussions and seminars took place involving Moore, Bolza, and Maschke (Chicago), and White and Holgate (Northwestern). The success of these endeavors encouraged this quintet to seek a larger umbrella under which to expand their small band of research mathematicians into a much bigger, and more broadly defined, geographically, community of scholars.

At some time during 1896, E.H. Moore informed the AMS Council about the group's desire to serve a larger population of researchers in the Midwest. Thus, he wrote and then distributed a circular in early December 1896 proposing this idea. The affixed signatures of 28 Midwest mathematicians reinforced its appeal. The "Call to a Conference in Chicago," as the circular was called, was sent to teachers and students of mathematics in neighboring institutions. Its purpose was stated plainly:[29]

> The regular monthly meetings of the Society afford ... opportunities
> to those who live in the vicinity of New York. By the organization
> of *sections* of the Society, can similar advantages be secured for other
> parts of the country? Shall, for instance, a Chicago section be orga-
> nized?

That question, among others, was answered when 18 AMS members (and at least two others) met that New Year's Eve and Day for two days of reading and discussing papers. However, the core group's plan was broader—to propose the formal establishment of an organization that would abet communication through direct contact and conversation in addition to the usual mode—publication. Minutes from the meeting reveal:[30]

> A resolution was adopted that, in the opinion of the Conference, it
> was desirable for the members of the American Mathematical Soci-
> ety to hold in Chicago at least two meetings a year for the reading and

discussion of mathematical papers, one during the Christmas vacation and one in the spring. Professor H.S. White, Dr. E.M. Blake, and Dr. J.W.A. Young were appointed a committee to make the necessary arrangements for a meeting to be held in the spring of 1897.

The appointed committee came from Northwestern, Purdue, and Chicago. To illustrate the wide geographical swath of the proposal, attendees came from surrounding states (Wisconsin, Ohio, and Michigan) as well as distant ones such as Nebraska (Ellery Davis), Kansas (Henry Byron Newson, who would wed Mary Winston in 1900), and Louisiana (Philetus Harvey Philbrick, chief engineer of the Kansas City, Watkins, & Gulf Railway). Altogether, 12 individuals delivered 14 papers, including Mary Winston, fresh from completing her dissertation under Felix Klein. Half of the presenters accounted for half of the papers, and hailed from the two core institutions, Chicago and Northwestern, but others found the meeting a convenient opportunity to vet their investigations, like Purdue's Edwin Mortimer Blake (PhD, Columbia 1893). As well, the meeting allowed aspiring mathematicians to introduce their incipient works before a learned audience, including Derrick Norman Lehmer, who was headmaster of the Worthington Military Academy (Lincoln, Nebraska) during 1896–1897 but matriculated at Chicago in the fall following this meeting.

At the close of the meeting, a committee was charged with formulating a plan for a permanent, local section of the AMS. The committee acted with dispatch:[31]

> The Council has authorized the organization of a Section of the American Mathematical Society by those members who may be present at the conference to be held in Chicago on April 24.

While that development surely sounded hopeful, the question arose: because neither the US nor Canada had ever witnessed a successful organization of mathematicians until 1888, less than ten years earlier, was there any reason to suspect that a section within that Society could endure?

In a word—Yes.

Why? In a second word—community. The Chicago Section boasted a critical mass of mathematicians to hold successful biannual meetings throughout its history, a one-day affair in April and a two-day gathering on the Friday and Saturday between Christmas and New Year's Day. The online file "Web07-ChiSec" presents an overview of the six subsequent meetings held in the nineteenth century. Table 7.1 summarizes those meetings from the organizational one in 1896 just described (here called the "zeroth") to the one on the eve of the twentieth century.

Subsequently, two other AMS groups banded together to form sections—the San Francisco Section in 1902 and the Southwestern Section in 1907. The first chair of the San Francisco Section was J.J. Sylvester's 1880 doctoral student at Johns Hopkins, W. Irving Stringham, then ensconced at the University of California at Berkeley. The third chair was his colleague Mellen Haskell, who had obtained his doctorate in 1890 under Felix Klein at Göttingen. The second chair of the Southwestern Section was J.J. Sylvester's final American doctoral student (1884) at Johns Hopkins, Ellery William Davis. Hilbert's Göttingen doctoral student Earle Hedrick was the initial chair of the Southwestern Section, serving for five years altogether, and then chairing the San Francisco Section (1924–1925).

Table 7.1. Meetings of the Chicago Section of the
AMS in the nineteenth century

Meeting	Year	Date	Location	Members	Papers
0	1896	Dec 31–Jan 1	Chicago	18	12
1	1897	April 24	Chicago	13	10
2	1897	December 30–31	Northwestern	21	21
3	1898	April 9	Chicago	11	6
4	1898	December 29–30	Chicago	26	13
5	1899	April 1	Northwestern	24	12
6	1899	December 28–29	Chicago	18	14

Transactions of the American Mathematical Society. E.H. Moore played
a prominent role in expanding the scope of AMS activities in several ways, including
his part in creating the Society's second journal, its first devoted strictly to cutting-edge
research. The establishment of the *Transactions of the AMS* serves to bring down the
curtain on American mathematics in the nineteenth century, while thrusting the research community into the twentieth. Owing to the success of the Colloquium Lectures
begun in 1896, some AMS members sought other means of disseminating the fruits
of their labor. This led AMS founder Thomas Fiske to suggest to Simon Newcomb
that the AMS acquire the *American Journal of Mathematics*. Their informal exchanges
started off amicably, so at the August 1898 Council meeting, Fiske was appointed chair
of a committee including four other nationally eminent mathematicians (Simon Newcomb, E.H. Moore, Maxime Bôcher, and James Pierpont) "to consider the question of
securing improved facilities for the publication of original mathematical articles in this
country."[32]

It sounds as if negotiations should have proceeded smoothly, especially since Newcomb was AMS president, cooperating editor of the *AJM*, and the former chair of the
department at Johns Hopkins which published the *AJM*. But they did not. In late October, the AMS Council approved the following motion.[33]

> It is recommended that the American Mathematical Society offer
> to The Johns Hopkins University the following plan of cooperation,
> for the purpose of enlarging and improving the *American Journal of
> Mathematics*, of extending its influence among the members of the
> American Mathematical Society; and of supplying in these ways the
> need which the Society feels for an organ in which meritorious origi-
> nal investigations may be promptly published.

The approved recommendation was followed by five bulleted points, the third of which
stated, "That the *Journal* shall have a board of seven editors, of whom two shall be
selected by The Johns Hopkins University, and five by the Council of the American
Mathematical Society."[34] This point proved to be unacceptable to Hopkins president
Daniel Coit Gilman, and six weeks later Newcomb wrote to Fiske, "The failure of our
negotiations will not lessen the desire of our University here to make the *American
Journal* all that its name implies."[35]

Thus, Johns Hopkins was willing to cooperate with the AMS in upgrading the
American Journal but on its own terms. To this end, Simon Newcomb proposed an

editorial corps to achieve this goal. But the suggested editors declined to serve, causing Newcomb to gripe vituperatively, "And now, after I have done all this, the Society runs away from me, as it were, and refuses to furnish the associate-editors, or to have anything to do with the arrangement."[36]

However, the breakdown in negotiations fostered a sense of solidarity and independence within AMS leadership. William Osgood talked about the many informal discussions on this issue conducted in *Rathskellars* at all AMS meetings, especially after James Pierpont returned from the December 1898 meeting of the Chicago Section, which he attended specifically to sound out its membership on the idea of establishing a research journal. He came away from the meeting heartened by the unbounded enthusiasm for the project and encouraged that its members would supply an adequate amount of papers needed to keep such a publication in existence.

Thomas Fiske acted expeditiously yet again, this time arranging a meeting in a New York restaurant with about 20 AMS leaders. John McClintock, among others, opposed forming a direct rival to the *American Journal of Mathematics*, fearing that such a project would be seen as an unfriendly act toward Johns Hopkins. Such a strongly held view seemed to scuttle the idea until Maxime Bôcher's famed skill at diplomacy came to the rescue. "Would Dr. McClintock feel it improper for the Society to publish its *Transactions*?" he asked. "No. Certainly not. Any society may publish its *Transactions*."[37] What a brilliant tact. By replacing the word "journal" with "transactions," the whole matter was settled.

A short while later, an unofficial but influential group of six leaders met at the home of Emory McClintock and agreed to recommend that the Society undertake publishing the journal. Their recommendation was adopted at an AMS Council meeting on February 25, 1899, that gave birth to the *Transactions of the AMS*.[38] (It is interesting to note the influence of informal meetings in the early days of the AMS, with the notion of a "Section" and the *Transactions* arising from such gatherings.) As the name *Transactions* implies, however, its papers had to have been read at an AMS meeting, either by a member or someone "presented by" a member; thus, Paul Gordan's "*Formentheoretische Entwicklung der in Herrn White's Abhandlung über Curven dritter Ordnung enthaltenen Sätze*" was presented by none other than Henry White.

Imagine—an American journal attracting manuscripts from abroad. Would John Farrar have thought such a development possible when he spearheaded the movement away from British textbooks to French translations? Would William Story have imagined that his groundbreaking study tour in Germany—as well as any one of the glut of Americans who flocked to Göttingen, Berlin, Leipzig, and Paris on similar educational quests—would have become unnecessary by the time the *Transactions* was inaugurated in 1899?

The first volume of the *Transactions* appeared in 1900. It ran to over 500 pages with 35 papers contributed by 27 different authors. The first editor-in-chief, E.H. Moore, maintained the position for eight years; the initial associate editors were Ernest W. Brown and Thomas Fiske. In something of an irony, the first submitted manuscript came from abroad, sent by the eminent French mathematician Édouard Goursat. Other foreign submissions were sent by the Englishman J.E. Campbell and the Germans Paul Gordan (two papers) and Martin Krause. As desired, almost all articles were contributed by American authors, including luminaries such as Maxime Bôcher, Oskar Bolza, Leonard Dickson, E.H. Moore, William Osgood, and Heinrich Maschke.

The initial entry was written by Henry White. Of the 35 papers in Volume 1, nine came from Chicago faculty members, four each from Cornell and Harvard (Osgood submitted his while "on the Atlantic"), and two each from Columbia, Northwestern, and Yale. None came from Johns Hopkins.

The *Transactions* was an instant success, reflecting the hard work and high standards of E.H. Moore and his associates. Nonetheless, several decades would pass before it gained the international stature it now maintains. Yet associate editor Ernest Brown emphasized the quest to print high-quality works from the start. Brown later recalled, in words sure to be trumpeted by editors to this day:[39]

> I like to think of the immense amount of trouble that we all took—
> and especially Moore—to get the best information, the best printing,
> the best editing, and the best papers before the first number appeared.
> And the work did not stop there. We wrestled with our younger con-
> tributors to try and get them to put their ideas into good form. The
> refereeing was a very serious business.

Of course, the matter of finances for the new journal still had to be resolved, but once again Thomas Fiske headed a committee that managed to secure dependable support. Over the next decade, 16 universities signed up to be subscribers, with five guaranteeing subscriptions for all ten years—Chicago, Columbia, Cornell, Northwestern, and Yale—and the 11 others subscribing for varying numbers of years, always at least five—Berkeley, Bryn Mawr, Harvard, Haverford, Illinois, Missouri, Princeton, Stanford, Vassar, Wesleyan, and Wisconsin.

Now I backtrack in time to put the establishment of the *Transactions* in historical perspective for American journals of mathematics. Chapter 4 described *The Analyst*, edited by Joel E. Hendricks (1874–1883). This periodical serves as a transition to the Period of Emergence, 1876–1900, but the term "transition" suggests a smooth development from one period to another. However, there were no precedents in the US or Canada before 1876 for what transpired in the history of mathematics in America over the next 25 years:

(1)   the founding of a *community* of scholars at Johns Hopkins led by J.J. Sylvester;
(2)   the expectation of *original research* at the cutting-edge by faculty members transplanted from Göttingen, especially by Felix Klein; and
(3)   the establishment of a graduate *program* at the University of Chicago under E.H. Moore.

These three changes were revolutionary, not evolutionary. Yet the tradition of periodicals that emphasized problem solving, small contributions to mathematics (regarding depth as well as breadth), and articles on astronomy and related applications continued to proliferate. For instance, the two journals established by Artemas Martin, *Mathematical Magazine* and *Mathematical Visitor*, were in the same vein as the *The Analyst*, while the *Mathematical Messenger* was devoted largely to problem solving. They lasted three, seventeen, and nine years, respectively. On a higher plane, the *Mathematical Review* was founded by William Story at Clark University in the same spirit as the Peirce/Lovering *Cambridge Miscellany*, but it too was short lived.

Table 7.2 lists nine American mathematics journals founded during 1876–1900. The *Annals of Mathematics* was established in 1884 by **Ormond Stone** (1847–1933) at the University of Virginia as a successor to *The Analyst*, which was described in

Table 7.2. Mathematical journals founded 1876–1900

Year	Journal	Founder
1877	*Mathematical Visitor*	Artemas Martin
1878	*American Journal*	J.J. Sylvester, William E. Story
1882	*Mathematical Magazine*	Artemas Martin
1884	*Annals of Mathematics*	Ormond Stone
1887	*Mathematical Messenger*	G.H. Harvill
1891	*Bulletin of the New York Mathematical Society*	Thomas Fiske
1894	*American Mathematical Monthly*	Benjamin Finkel
1896	*Mathematical Review*	William E. Story
1900	*Transactions of the American Mathematical Society*	E.H. Moore, Thomas Fiske

Chapter 4. Stone's chief interest was astronomy, and he was director of the Cincinnati Observatory before accepting the professorship of practical astronomy at the University of Virginia in 1882.

Some of the contents of the *Annals* are described in the online file "Web07-Annals." The *Annals* satisfied the needs of serious researchers, including Cornell's James Oliver and Sylvester's former students George Halsted and Oscar Mitchell. The journal improved steadily over the years, reflecting the growth in numbers and quality of the American mathematical research community. In 1899, the headquarters for the journal moved to Harvard,[40] and the switch in content was radical. The first volume in the Second Series was on a par with the *Transactions*, with most papers on higher ground and no problem sections included.

One author of *Annals* papers not yet introduced is **Alexander Pell** (1857–1921). Born Sergei Petrovich Degaev in Moscow, he assassinated the head of the Czar's secret police in 1883, and quickly fled with his first wife to London.[41] Three years later the couple immigrated to the US, where they became naturalized citizens, adopting the names Alexander and Emma Pell (for unknown reasons). After working a series of menial jobs over the next nine years, he enrolled in the graduate program at Johns Hopkins in 1895 and earned a PhD two years later. At the time the University of South Dakota (**USD**) sought to establish itself in mathematics, so the Clark graduate and Hopkins faculty member Lorrain Hulburt recommended a first-class mathematician who happened to speak with a strong Russian accent. USD replied, "Send your Russian mathematician along, brogue and all." Pell became exactly what the university desired: a successful teacher who published numerous research papers and participated actively in the AMS, especially the Chicago Section. One of the USD students that Pell mentored during 1899–1903 was Anna Johnson. Two years after his wife died in 1904, he joined his former student in Göttingen, where she was studying on a Palmer Fellowship from Wellesley College for 1906–1907. They married there that July. Pell left USD for the Armour Institute in Chicago in 1908, but a stroke three years later limited his mathematical output after that; he died in 1921.

Some other notable authors of papers in the *Annals* were Earle Hedrick (who, like Anna Johnson, studied at Göttingen under David Hilbert), George Miller (known

Table 7.3. First 11 annual meetings of the AMS

#	Date	Attendance		Papers	Membership
		Total	Members		
1	December 6, 1889	10	10?	0	16
2	December 5, 1890	10	10	0	23
3	December 30, 1891	10	10	0	210
4	December 29, 1892	14	10	1	234
5	December 28, 1893	24	14	0	239
6	December 28, 1894	17	17	2	251
7	December 27, 1895	25	25	1	268
8	December 30, 1896	24	24	3	275
9	December 29, 1897	38	38	11	291
10	December 28, 1898	28	22	9	308
11	December 28, 1899	80+	38	13	326

for contributions to group theory), and three who earned doctorates from European universities other than Göttingen: Charles Bouton, Edgar Lovett, and James Pierpont.

Annual meetings. The AMS initially held monthly meetings at Columbia University during the academic year from October to June, with the annual meeting held in December. The purpose of the annual meeting was to hear reports on four matters— finances, membership, publications, and the development of a library at Columbia— and then to hold elections of officers for the ensuing year. No papers were read at annual meetings until 1892. Intriguingly, the speaker was not identified. Table 7.3 provides the dates, attendance (of members), and the number of papers presented at the 11 annual meetings held from 1889 until the end of the century, as well as the Society's membership in each of these years. (By contrast, in 2012 the annual meeting in Boston attracted over 7000 people; moreover, over 2000 papers were delivered, a feature that has become a chief attraction for many attendees.)

Table 7.3 shows implicitly that initially all annual meetings were held on one day. Beginning with the thirteenth meeting in 1901, they stretched over two days. With few exceptions, annual meetings remained two-day affairs in December until the mid-1920s, when they were extended to three days from late December into the New Year. Table 7.3 also indicates that attendance at annual meetings remained steady and small until 1893, when it began to increase regularly (though not monotonically). This increase took place about two years after the dramatic uplift in AMS membership listed in the final column.

Study abroad

Between 1880 and 1900 many American students who sought a doctorate in mathematics felt obliged to study in Germany, particularly in Göttingen, the citadel of mathematics and physics. Felix Klein exerted a major influence on the development of American mathematics by his systematic program of taking students from the most elementary courses to the cutting edge of research, which extended beyond the Sylvester program. American students benefited enormously from this systematic approach and exported it to various universities. David Hilbert took Klein's place from 1900 up to 1910. (The "Transition 1900" section provides details of this group.) Klein's students (and visiting professors) helped fill German classes that might not have run otherwise due to low enrollments. However, his students encountered much less home-grown competition than the newer cadre under Hilbert. Yet the new gang also succeeded admirably.

Both Klein and Hilbert thus served as "cluster points" for American students, who were not afforded preferential treatment. However, some American students received doctorates elsewhere. For instance, James Pierpont earned his at Vienna. Also, Raymond Archibald, Edward Huntington, and Louis Karpinski all received their doctorates at Straßburg, but their dissertations were not supervised by a cluster point. Another important cluster point for an American school of students is introduced next.

Lie Lair. Felix Klein's successor at Leipzig, the Norwegian **Sophus Lie** (1842–1899), exerted the second greatest influence on the American mathematical research community, though on fewer individuals and universities, despite an approach diametrically opposed to Klein. Lie (pronounced lee) was encountered in the section "Transition 1876" when Klein and he went to Paris in the spring of 1870 for a short but intense study tour. Lie was born in the capital city Christiania (renamed Oslo in 1925). Unlike Klein, he was not a child prodigy, so although he attended lectures by famed Norwegian group theorist Ludwig Sylow at age 20 on the emerging concept of a group of substitutions, he did not initially thrust himself into the Norwegian algebraic tradition of Abel and Sylow. In fact, like many undergraduate students today, he vacillated about what subject to pursue even after his graduation in 1865 from the University of Christiana. Consequently, it was two years before his independent study produced his first original mathematical idea, which led him to a detailed examination of the geometry of Plücker in Germany and Poncelet in France. Undeterred by having his first paper rejected by the Norwegian Academy of Science due to poor exposition, he nonetheless pursued his research, and was rewarded when his first paper was accepted by the celebrated *Crelle's Journal* in 1869.

Sometimes it takes just one stroke of fortune to propel a career forward, and this paper enabled Lie to obtain a scholarship for further study in Europe. First, he went to Berlin, where he gained recognition for his lecture in Kummer's seminar as well as for the errors he corrected in some of this leading algebraist's work. More importantly, this is where the 27-year-old Lie met the 20-year-old Klein. During the subsequent spring of 1870, the two friends traveled together to Paris to meet Darboux, Chasles, and Camille Jordan. Their timing was impeccable, in a mathematical sense, because Jordan's classic book *Traité des Substitutions et des Équations algébriques* had just appeared; consequently, the visitors benefited enormously from lectures by, and discussions with, Jordan on substitution groups. Moreover, Lie and Klein collaborated on ideas relating group theory to geometry that initially led to joint papers, and ultimately resulted in Klein's famous Erlangen Program two years later and Lie's subsequent development of groups of transformations.

But their timing was less than impeccable from a political standpoint because France declared war on Prussia that July in 1870. This sent Klein, a Prussian patriot, scurrying home to join the army. Lie remained behind. The Parisian geometer Gaston Darboux, who inspired Lie and Klein as much in his field as Jordan had in algebra, reported:[42]

Figure 7.2. Sophus Lie

> Surprised at Paris by the declaration of war, [Lie] took refuge at Foun-
> tainebleau. Occupied incessantly by the ideas fermenting in his brain,
> he would go every day into the forest, loitering in places most remote
> from the beaten path, taking notes and drawing figures. It took little
> at this time to awaken suspicion. Arrested and imprisoned at Foun-
> tainebleau, under conditions otherwise very comfortable, he called
> for the aid of Chasles, Bertrand, and others; I made the trip to Foun-
> tainebleau and had no trouble in convincing the procureur imperial;
> all the notes which had been seized ... bore in no way upon the na-
> tional defenses. Lie was released.

In mid-September, however, the German army blockaded Paris, causing Lie to hike
back home to Christiania over a circuitous route that took him first to Italy and then
through Germany, where he caught up with Klein once again and continued their
collaboration. Sometimes even a raging war cannot prevent the spread of mathematical
ideas.

By that time, Christiana authorities realized that Lie was a remarkable mathematician. Accordingly, they created a chair for him in 1872 and awarded him a doctorate based on the ideas he had developed over the preceding five years. At the base of
this development were *contact transformations*—one-to-one correspondences between
lines and spheres such that intersecting lines are mapped onto tangent spheres. Working alone after that, Lie developed what Darboux described as a "masterful theory of
continuous transformation groups which constitutes his most important work."[43] The
French geometer then listed the two overarching elements that were wholly essential
to Lie's research:

- contact transformations that shed light on the integration of partial
  differential equations, and
- infinitesimal transformations.

Lie sought a theory for partial differential equations analogous to the Galois theory of
polynomial equations. This led him to what he called an *infinitesimal group*, called a
Lie algebra today. He also discovered what are now called Lie groups in his theory

of continuous transformation groups. This work placed him atop the pantheon of Norwegian mathematicians with groups named after them: Abel, Sylow, and Lie.

Although drawn to his native land, mathematical isolation compelled Sophus Lie to accept an offer to succeed his friend Felix Klein as professor of mathematics in 1886 at Leipzig, where Lie remained for 12 years. During that time the Lie Lair numbered 20 Americans, with six earning doctorates. Three subsequently became driving forces at leading American institutions: James Page at the University of Virginia, Edgar Lovett at Rice, and Hans Blichfeldt at Stanford. In addition, several American faculty members went abroad to study with him, perhaps the most famous of whom was the group theorist George Miller. Shortly after Lie's untimely death in February 1899 from pernicious anemia, Miller concluded an appreciation of his mentor based on his study tour with Lie during the summer of 1895:[44]

> He was an inspiring teacher but his lectures were not always well prepared. He … [was] bent on training [his students] to think according to his methods rather than on giving them a systematic treatise on any subject.

Lie's approach, therefore, was akin to Sylvester's and hence quite different from Klein's. The two friends were of entirely different temperaments, as reflected in their mode of dress, Klein always formal but Lie frequently throwing off his collar and tie upon entering the lecture hall. Klein was a micromanaging taskmaster with exceedingly high expectations; that is why students of the caliber of William Osgood and Henry Tyler ended up taking their doctorates outside Göttingen. By contrast, George Miller wrote that:

> Lie was regarded as one of the easiest men at Leipzig for the doctor thesis. He generally assigned easy subjects and he assisted the students very freely.

And for Americans, in particular, Miller added:

> He seemed to pay special attention to foreign students. … During the summers of 1895 and 1896 he returned from his summer vacation about one month before the opening of the university and lectured daily before the few American students of mathematics who happened to be in Leipzig, with a view to prepare them to follow his lectures better during the university year.

Americans responded in kind, with George Halsted calling Lie "the greatest mathematician in the world" at the time of Lie's death.[45]

The remainder of this section examines the six American students who received PhDs under Lie in the order of when they arrived in Leipzig. As with Klein's students, the six were interconnected, with one inspiring another to work with the master. The life and work of the first five are described, and their place in the development of mathematics at the turn of the twentieth century, is summarized, before the leader of the lair is introduced. The online file "Web07-LieLair" provides further details.

The first of Lie's American students was **James Morris Page** (1864–1936), a Virginian with roots in the state going back to 1650. Page entered Randolph-Macon College at age 17 and graduated four years later. He then went to Leipzig and completed his dissertation in only two years. To underscore Lie's practice of helping his students, Lie acknowledged that Page "was largely shown a way that offered good prospects for success," but quickly added that "the task of carrying the entire investigation through required not only considerable energy but also a significant amount of finesse in dealing with the computational techniques."[46] Lie concluded that the "work must therefore be regarded as a true scientific accomplishment." Page published his dissertation in the *American Journal* one year after receiving his degree.[47] Because Lie's methods were not generally known at the time, this 54-page work began with a long introduction. Page then classified and described the 11 primitive groups in a four-dimensional space.

Returning to the US with degree in hand, Page established the Keswick School and served as headmaster 1888–1895. During this time, he became involved with the mathematics department at the University of Virginia, publishing three papers on transformation groups in the *Annals*. He also worked feverishly on a book that would introduce Lie's method for solving differential equations, so when he closed his school in June 1895, he traveled to Europe, first to consult with his mentor in Leipzig and then to confer with other mathematicians in Paris. Upon returning to the US in the spring of 1896, Page lectured on this material as a visiting professor at Johns Hopkins. That fall he was appointed adjunct professor of mathematics at Virginia, where he remained until his retirement in 1934, having been promoted to professor in 1901. A dormitory at Virginia was named in his honor.

James Page published three more papers in the latter part of the nineteenth century, but halted research in the new century. His most enduring work is the 1897 book, *Ordinary Differential Equations: An Elementary Text Book with an Introduction to Lie's Theory of the Group of One Parameter*, the first text in English aimed at elucidating Lie's notoriously complicated theory. Two reviews of the text shed light on interrelationships among members of the Lie Lair. Leonard Dickson, who had studied with Lie during 1896–1897, wrote a very positive review based on a one-year course whose aim was "to introduce a class of mature students to the theory of ordinary and partial differential equations through the medium of continuous groups."[48] Apparently, one of the "mature students" in the class was J.J. Sylvester's former student Arthur Hathaway, who was credited with finding and correcting a critical error in the text.

A second American reviewer chafed at a review in *Nature* questioning whether the book was appropriate for beginners: "In the hands of an instructor who is alive to both sides of the subject, the book is susceptible of successful application to the needs of those studying the subject for the first time."[49] That *Bulletin* reviewer was *not* a disinterested observer, but another student of Lie. By the time this American reached Leipzig, however, a serious breach had occurred between Sophus Lie and Felix Klein. It seems that Lie took offense to two Klein works that Lie felt did not adequately credit his contributions. Initially, Lie petulantly attacked Klein in private correspondence, but later his gripes went public in the preface to one of his books. Klein attempted to patch up this dispute in his second and third lectures at the Evanston Colloquium, but to no effect. By this time, Lie had assumed Klein's position as the leading mentor for young Americans seeking a doctorate from a German university, even though Lie, unlike Klein, never visited the US.

The *Bulletin* reviewer was **Edgar Odell Lovett** (1871–1957), whose life intersected with James Page in a meaningful way. Both graduated from small colleges, Lovett from Bethany College in West Virginia in 1890. The 19-year old then became a mathematics instructor at West Kentucky College (in Mayfield). Lovett left West Kentucky in 1892

Figure 7.3. Edgar Odell Lovett

to work with Ormond Stone in the Leander McCormick Observatory at the University of Virginia, where he earned a PhD three years later. During his three years in Charlottesville, Lovett attended lectures on Lie's theory of infinitesimal transformations given to the Mathematical Club by James Page. Upon receiving his doctorate and being intrigued by this topic, Lovett moved to Christiana to study under the renowned Norwegian, resulting in his second doctorate, this one awarded at Leipzig in 1898. His dissertation combined his two major interests of celestial mechanics and Lie theory of contact transformations.[50]

In the spring of 1897, Lovett lectured at Johns Hopkins and the University of Virginia, but perhaps his biggest presentation occurred at the April AMS meeting in New York. In addition to his talk, Leonard Dickson and Henry Newson lectured on groups of transformations and discontinuous groups, thus making this gathering an all-Lie affair. That summer Lovett offered a course at the University of Chicago on "The geometry of Lie's transformation groups." In the fall, he accepted an instructorship in mathematics at Princeton. To top off the year, Lovett presented two papers at the annual AMS meeting in New York on December 29. Once again, the program featured recent work by Sophus Lie; George Miller and Charles Bouton also lectured on this material.

The proximity of Princeton to New York allowed Lovett to present many papers at AMS meetings at Columbia. He also spoke at the semicentennial meeting of the AAAS in Boston, where he was preceded by Arthur Hathaway and followed by Leonard Dickson. In addition, he presented a paper at the AAAS's British counterpart, the BAAS, where Irving Stringham preceded him. Lovett was granted a leave of absence from Princeton for 1900–1901, and began it by presenting a paper at the ICM in Paris.

Edgar Lovett's productivity fell off dramatically in the new century; his output paled in comparison to the period 1897–1900. In 1907, Princeton president Woodrow Wilson recommended him to become the first president of a new university to be established in Houston. Lovett accepted the following year. He served as president of Rice for 38 years up to his retirement in 1946, having enforced the highest standards for his faculty all the while.

Rice University has a fascinating history whose drama makes the establishment of Clark University pale by comparison. It all started when Massachusetts businessman

William Marsh Rice, who had made a fortune in real estate, railroad development, and cotton trading in Texas, died in his sleep in 1900. Nine years earlier he had written a will to found an eponymous, tuition-free college upon his death. Shortly after he died, a suspiciously large check made out to his New York City lawyer, Albert Patrick, drew the attention of an alert bank teller due to a misspelling in the recipient's name. Patrick asserted that Rice had changed his will to leave the bulk of his fortune to him, but a subsequent investigation by the New York district attorney resulted in the arrests of Patrick as well as Rice's butler and valet, who, it turns out, had administered chloroform to Rice in his sleep. However, it took almost ten years for the culprits to be brought to justice. Rice's Houston friend and lawyer helped to direct the fortune ($4.6 million in 1904, roughly $110 million today) toward founding what was initially called the Rice Institute. In 1907, the board of trustees heeded Woodrow Wilson's advice and selected Edgar Odell Lovett as the first president of Rice Institute, which was still in the planning stages. Like Daniel Coit Gilman and William Rainey Harper before him at Johns Hopkins and Chicago, respectively, Lovett set out to visit universities throughout the country and Europe to help him select the best possible faculty. By the time he was formally inaugurated in October 1912, he had decided to adopt a uniform architecture for campus buildings, similar to the University of Pennsylvania and the residential college system adopted by Cambridge (England) and Yale.

Charles Bouton was one of the two mathematicians who joined Edgar Lovett in presenting papers connected with Lie's work at the annual AMS meeting in December 1897.[51] The title of the lecture, "Some examples of differential invariants," provides no clue to his connection with Lie.

Born and raised in St. Louis, **Charles Leonard Bouton** (1869–1922) graduated from Washington University in 1891. He returned to Washington as an instructor after two years, but stayed for only one year before matriculating in Harvard's graduate school. At the end of his second year, Bouton used a Parker Traveling Fellowship to work in Leipzig 1896–1898 under Sophus Lie. Edgar Lovett lived in Christiana during the fall of Bouton's first year in Leipzig, so it is quite possible that the two American students met at that time. Charles Bouton earned his Leipzig doctorate in 1898, and revised his dissertation, "Invariants of the general linear differential equation and their relation to the theory of continuous groups," as a 60-page paper in the January 1899 issue of the *American Journal*. Upon his return, Bouton accepted an instructorship at Harvard and remained there the rest of his life. During 1899–1900, the Harvard mathematics department offered three "courses of reading and research: —By Professor [Asaph] Hall: Selected topics in celestial mechanics. —By Professor Bôcher: Linear differential equations. —By Dr. Bouton: Continuous groups."[52]

However, Bouton's research sputtered—I am aware of only four papers (besides those already noted) and one doctoral student. Yet it was an *Annals* paper unrelated to his study with Lie that has earned him an enduring reputation in the history of mathematics: "Nim, a game with a complete mathematical theory." In the introduction, Bouton stated:[53]

> The game here discussed has interested the writer on account of its seeming complexity, and its extremely simple and complete mathematical theory ... the name in the title is proposed for it.

A 2016 article in *Mathematics Magazine* affirmed the simplicity of Bouton's strategy for the game Nim:[54]

Charles Bouton used simple mathematics to describe a winning strategy. Following Bouton, mathematicians showed his ideas apply to an entire class of games called "impartial, perfect information games."

Two other American students arrived in Leipzig during Sophus Lie's last year, but managed to complete dissertations and received doctorates anyway. **David Andrew Rothrock** (1864–1949) graduated from the Northern Indiana Normal School in 1887 and entered Indiana University four years later. He was associated with Indiana for most of the rest of his life. Rothrock earned his AB in one year (1892) and joined the faculty as instructor in mathematics. He was promoted to assistant professor in 1895 after having enrolled in E.H. Moore's graduate program at the University of Chicago 1894–1895. He then went to Leipzig to study under the ailing Lie, and received his doctorate in 1899 for a dissertation whose revision appeared in the *Monthly*: "Point invariants for the finite continuous groups of the plane."[55] This appears to be Rothrock's only published paper on a topic in mathematics, though he later wrote an article on the work of American mathematicians in World War II. He was the head of a committee that helped establish the Indiana Association of Mathematics Teachers in 1906. He retired in 1938.

Rothrock's name is linked with Indiana's mathematics department to this day. Scholarships in mathematics in his name were established in 1956 through a gift from a former student who had graduated in 1905. Moreover, in 1990 Rothrock faculty teaching awards were endowed in his memory.

John Van Etten Westfall (1872–1944) graduated from Cornell in 1895, whose department chair Lucien Wait wrote to Felix Klein, "We shall send two of our graduate students to Europe for an extended course of mathematical study."[56] Since Klein was no longer taking American students, it is possible that Westfall was one of the unnamed students because he received his doctorate from Leipzig under Lie in 1898. Westfall returned to Cornell as an honorary fellow for 1898–1899, when he accepted an instructorship at Iowa State University. The Iowa State mathematics department was small, with only a head, Lienas Gifford Weld, and "professor of mechanics in the department of mathematics" Arthur G. Smith. Both were Iowa State grads, neither possessing a doctorate. Nonetheless, John Westfall attended many meetings of the AMS Chicago Section, where he met Indiana's David Rothrock. During Cornell's summer 1901 term, Westfall taught a course on the "Theory of functions of a complex variable, the elements of the theories of Cauchy, Riemann, and Weierstrass."[57] He remained in Ithaca to attend the summer AMS meeting at Cornell, where he met up with Leonard Dickson and George Miller, both of whom had studied with Lie.

Iowa State was beginning to establish a graduate program, allowing Westfall to teach a range of courses, from advanced undergraduate offerings to the graduate "Advanced course, including the theories of Sophus Lie." Surprisingly, two rather unusual advanced-undergraduate courses that Westfall offered turned out to be decisive in his career, with "Insurance" serving as a prerequisite for "The mathematical theory of life insurance." This explains why he left Iowa State in 1905 to become vice president of the Equitable Life Assurance Society in New York City the next year.

This section has described five American students who obtained doctorates in mathematics under Sophus Lie. James Page, William Bouton, and David Rothrock spent long careers at Virginia, Harvard, and Indiana, respectively, John Westfall was at Iowa State for only six years before leaving academia altogether, and Edgar Lovett

Figure 7.4. Hans Frederick Blichfeldt, (top row, middle, partly obscured by hat) at the 1932 ICM in Zürich

spent nearly a decade at Princeton before becoming the first president at Rice. Not one of this quintet published extensively, and none beyond the first decade of the twentieth century. Such lack of productivity might doom Lie's students to second-class status within the mathematical research community. But the remaining member of the Lie Lair provides a counterexample.

Hans Frederick Blichfeldt (1873–1945) was born in Denmark. A precocious student, Blichfeldt showed remarkable mathematical ability early, discovering the solutions to polynomial equations of the third and fourth degrees some time before age 15. He then passed the Danish state examination for university enrollment with honors. However, his parents could not afford tuition, so he was unable to attend. To improve their lot, the family immigrated to the US when he was 15, whereupon the teenager spent four years as a laborer on farms and in sawmills in Nebraska, Wyoming, Oregon, and Washington, followed by two years traveling about the country as a surveyor. Having saved a substantial part of his wages during six years of labor, the tall, sturdy Dane entered the embryonic Stanford University in 1894, earning a BA in two years (1896) and an MA in one (1897). Stanford professor Rufus Green loaned him sufficient funds to travel to Leipzig, where Blichfeldt studied under the fellow Scandinavian Sophus Lie. Blichfeldt earned a doctorate *summa cum laude* in one year. An abridgement of his dissertation was published under the title "On a certain class of groups of transformation in three dimensional space" in 1900.[58]

Blichfeldt returned to Palo Alto and spent the remainder of his life on the Stanford faculty. He served as head of the mathematics department from 1927 until his retirement in 1938 at age 65. He was active in the AMS, elected vice president in 1912 and chair of the San Francisco Section for three two-year terms. Blichfeldt delivered numerous lectures throughout the country, mainly on his specialties—group theory and the geometry of numbers. His contributions to group theory and group characteristics were of considerable importance because of their applications to Lie groups. In addition, he published the text *Finite Collineation Groups* and coauthored *Theory and Applications of Finite Groups* with George Miller and L. E. Dickson, both of whom had also studied under Sophus Lie. According to E.T. Bell, Blichfeldt's "mathematical output was not voluminous,"[59] yet he was elected to the National Academy of Sciences, due to the quality of these publications, and he represented the NAS at the 1932 ICM in Zürich. In that vein, he represented the US government and the AMS at the next ICM in Oslo four years later. Blichfeldt's achievements rank him as the shining light among Lie's graduates, the one who contributed most to the development of mathematics in America.

Elsewhere. The section "Transition 1876" describes the third and final European school of American students, to complement those who studied under and with Felix Klein and Sophus Lie. The new group under David Hilbert was located in Göttingen. However, not all American students who pursued higher mathematics in Germany went to Göttingen or Leipzig; Ernest Wilczynski is an example. Beyond that, not all American students who obtained European degrees went to Germany; James Pierpont serves as an example. Wilczynski and Pierpont played important roles at Chicago and Yale, respectively.

Arthur Webster and Harris Hancock obtained doctorates at the University of Berlin, in 1890 and 1894, respectively. **Ernest Julius Wilczynski** (1876–1932) received his Berlin degree in 1897, so the three American figures hardly constitute a "school." Born in Hamburg, Germany, Wilczynski's family immigrated to the US after he completed his second year at an elementary school, so he completed his early education in Chicago, graduating high school in three years at age 17. Instead of matriculating in the University of Chicago, which opened the previous year, he studied at the University of Berlin with the help of an uncle, obtaining his doctorate in 1897. Wilczynski's precocity continued apace. He published five papers before completing his dissertation on applications of hydrodynamics to the theory of rotating bodies (no advisor is stated). This reflects an early interest in mathematical astronomy, with differential equations serving as the bridge linking the two fields.

Upon returning to the US, Wilczynski was unable to find an academic position, reputedly because at age 20 he looked young and boyish, so he worked as a human computer at the Nautical Almanac Office in Washington, DC, for a year before accepting an instructorship at Berkeley. Apparently, a friend from Berlin, Armin Otto Leuschner, a mathematical astronomer on the California faculty, helped him obtain the position. The Berkeley department then numbered one member from each of the three dominant schools from the period 1896–1900: 1) Irving Stringham (PhD, Johns Hopkins) was head of the mathematics department and dean of the faculties, 2) Mellen Haskell (PhD, Göttingen), and 3) Leonard Dickson (PhD, Chicago). However, Dickson left for Texas at the end of Wilczynski's first year.

Ernest Wilczynski remained at Berkeley from 1898 to 1907. He then left for the University of Illinois but stayed only three years before succeeding Heinrich Maschke at Chicago in 1910 upon Maschke's death. Wilczynski remained at Chicago for the rest of his career, establishing a reputation as an excellent teacher among undergraduate and graduate students for his lucid, elegant exposition:[60]

> His friendly, kindly, informal attitude toward his students did much
> to win their affection and loyalty. He believed that a student could do
> his best work when he was thoroughly at ease in the presence of the
> instructor.

The number of his doctoral students is disputed, ranging from seven to 25; I accept the figure 25 based upon an investigation by Øystein Ore.

Wilczynski's output can be divided into three distinct periods. Overall, he published 15 papers in mathematical astronomy before switching to differential equations, a field in which he published eight. But it was in projective differential geometry that he became pre-eminent, producing 32 papers and an influential book. He also delivered one of the three series of AMS Colloquium Lectures (with E.H. Moore and Max Mason) at Yale in 1906 titled "Projective differential geometry." Wilczynski also wrote

four papers on functions of a complex variable, 15 on miscellaneous topics, and two
college textbooks. His articles appeared in 19 journals in the four languages in which
he was fluent: German, English, French, and Italian.

Unfortunately, Wilczynski was not granted a long career, and he ended it in a
manner one can only admire:[61]

> His health failed gradually after 1919 but he resolutely continued at
> his post until early in the Summer Quarter of 1923, when in the midst
> of a lecture he finally realized that he could go no further and, with a
> simple statement to that effect, walked from his class-room never to
> return, leaving his students amazed by the classic self-restraint with
> which he accepted his tragic fate.

Wilczynski was only 47 at the time. He retired officially three years later when it
became apparent that he would never be able to resume active duty. He died after
a lingering illness at age 55.

Wilczynski was fluent in the four languages used internationally in mathematics
around 1900—German, English, French, and Italian. There had been talk of making English the official language of the first International Congress, held in Zurich in
1897, because it was neutral ground between French, on the one hand, and German
and Italian, on the other. In this regard the modern historian of mathematics Jeremy
Gray noted, "This incident underscored the fact that there was no recognized *lingua
franca* in which mathematicians communicated."[62] Due to the "Tower of Babel" regarding languages a century ago, the importance of translation was clear, yet there
was a critical difference separating the US and Great Britain in this regard. Recall that
American Mary Winston Newson translated David Hilbert's famous lecture on "Mathematical Problems" into English; moreover, it was the Americans, not the British, who
translated Poincaré's books and essays. According to Gray, "It is quite striking how
much farther the American mathematical community went than the British in this
direction."

Regarding language, Latin was the *lingua franca* of mathematics for about 1600
years after the fall of the Roman Empire—think of Cardano's *Ars Magna* (1542) and
Newton's *Philosophiœ Naturalis Principia Mathematica* (1687). The change to national languages began in the seventeenth century, as exemplified by the *Discours de
la method* of Descartes and its important appendix *La géométrie* (1637). By the mid-
nineteenth century, the triumvirate of English, French, and German prevailed. A
steady decline of French language publication began around 1910. Nonetheless, German began to fall just after World War I, replaced by English from the 1920s onward.
The aftermath of World War II saw a spike in the use of Russian, and around 1970, the
AMS began cover-to-cover English translations of the leading Russian mathematics
journals. That practice became unnecessary after the fall of the Soviet Union in late
1991.[63]

Returning now to American students who obtained European degrees, in many
ways the career of **James Pelham Pierpont** (1866–1938) resembled that of Ernest
Wilczynski, the most notable differences occurring at the beginning and the end of
their lives. Pierpont was born into a wealthy family, and reached age 72 without the
kind of lingering illness that caused Wilczynski to leave the classroom at 47. In educational opportunity, Pierpont would be considered today the ultimate "legacy" student,

with a forefather of the same name who was one of the founders of Yale, where Pierpont spent most of his career. Except he did not attend Yale, instead graduating from Worcester Polytechnic Institute in 1886, having matriculated there initially to study mechanical engineering.

From that point onward, the careers of Pierpont and Wilczynski converge along similar paths. Upon graduation, Pierpont set sail for Europe on a prolonged study of mathematics at various universities. His first stop was the University of Berlin, where he gained knowledge of the algebraic works of Kronecker that dominated his initial forays into research. He was still in Europe seven years later when Wilczynski, ten years younger, arrived in Berlin to begin his university studies, but by then Pierpont had moved to the University of Vienna, where he obtained a doctorate the following year (1894) with a dissertation on the history of fifth-degree equations from 1858 to 1894. Pierpont enjoyed the academic and cultural milieu in the Austrian capital immensely, and he visited the city several times later in life.

When James Pierpont returned to the US in 1894 with doctorate in hand, he was immediately offered a position at Yale, though at the lowest possible position—lecturer. However, his research output and missionary zeal toward mathematics enabled him to climb the academic ladder quickly, and within four years he was promoted to professor (in 1898). This was the year when Wilczynski began his own quick ascent up the Berkeley ladder.

Pierpont delivered the first AMS Colloquium Lecture (along with Maxime Bôcher) on "Galois theory of equations" at the annual summer meeting held September 1896 in Buffalo; subsequently, he bemoaned the fact that he never published the lecture series. Two years later he served on an AMS committee that ultimately led to the establishment of the *Transactions*; he then served as associate editor of this prestigious journal. While Wilczynski was teaching all levels of courses at Illinois and Chicago, Pierpont mainly concentrated on upper-level undergraduate and graduate courses. His biographer, Øystein Ore, found that Pierpont had advised 25 doctoral dissertations by contacting the authors, a tactic that was necessary because Yale did not then list mentors formally.[64] At the head of the list is undoubtedly the Canadian-born Roland Richardson, but others worth mentioning are famous calculus textbook author William Anthony Granville, Carl Eben Stromquist (grandfather of MAA governor Walter Stromquist), and two Missouri mathematicians, Mary Shore (Walker) Hull and George E. Wahlin.

There is another reason why Pierpont was regarded as an intriguing instructor—an ambidexterity that might be the envy of all mathematics professors:[65]

> He seemed to have equal mastery over both hands. He could write with them both and it always used to impress the undergraduates when he drew illustrations on the blackboard using both arms simultaneously and independently.

Moreover, Pierpont could read and write upside down, a trait surely useful when helping students in one-on-one settings. Unlike the norm today, professors did not then generally have their own offices for providing such help.

Ernest Wilczynski and James Pierpont serve as examples of American students who became notable twentieth-century mathematicians after traveling abroad for advanced study beyond Göttingen and Leipzig. At the turn of the new century, Pierpont commented:[66]

> It is patent to everyone that our new university life is being moulded
> largely after German ideals. Ninety-nine per cent of our younger men
> have received their inspiration in Germany; it is only natural that we
> should have German methods before our eyes in shaping the course
> our graduate instruction shall take.

Yet he added:

> It is a question in my mind whether it is wise for us to imitate so freely
> German methods, and be so largely dominated by the German way of
> looking at things. America is not a New Germany. To counteract this
> excessive German influence, it seems desirable that we should have
> among us a respectable minority who have spent considerable time
> under French influences. It is well, so it seems to me, that a part
> of our gifted youth should have received their inspiration from men
> like Poincaré and Picard, Darboux and Koenigs, Painlevé, Borel and
> Hadamard."

Few Americans heeded Pierpont's advice, however. But with universities in the
US beginning to develop serious graduate programs during the 1890s, many aspiring
mathematicians stayed home for their advanced training.

American doctoral programs. **Julian Lowell Coolidge** (1873–1954) occupied
an especially advantageous perch for viewing the quest for higher education at the
turn of the twentieth century. After receiving his BA degree from Harvard in 1895,
he traveled abroad to Oxford and earned a BSc two years later. Oxford did not award
doctoral degrees at the time, so Coolidge returned to the US, where he taught at the
Groton School (1897–1899). One of his prize pupils, and a lifelong friend, was Franklin
D. Roosevelt. Coolidge joined the Harvard faculty in 1899, and three years later was
granted a two-year leave for study abroad. This time he spent the first year in Turin,
Italy, studying under Corrado Segre, and the second at Bonn, where he received his
doctorate in 1904 for a dissertation on dual-projective geometry in elliptical and spherical spaces. His advisor, Eduard Study, had returned to Germany after being unable
to land an appropriate position in the US. Therefore, by 1904, when Coolidge resumed
his career at Harvard, he had experienced mathematical culture at an impressionable
age in England, Germany, and Italy. He remained at Harvard until retirement in 1940.

Julian Coolidge was deeply influenced by his experience in Italy and, before leaving Turin, wrote an article for the *Bulletin* describing programs there in a manner similar to Pierpont's article on France mentioned above. After stating that most Americans equated *study abroad* with *study in Germany* because of the latter's "well deserved
renown for free research and profound scholarship, and the scientific standing of those
who teach there. ... And yet ... [o]ther universities have come forward in the last
decades."[67] Regarding his compatriots, Coolidge asserted:

> American students of mathematics, who intend to devote their lives
> to the subject, have generally in the past deemed it wise to spend
> more or less time at a European university, and though we may point
> with pride to the work that is being done to-day in our own graduate
> schools, it is safe to predict that for a long time to come, this migration
> will continue.

Table 7.4. Americans who earned doctorates in the nineteenth century

	before 1870	1870–1874	1875–1879	1880–1884	1885–1889	1890–1894	1895–1899
Home	3	3	7	9	23	28	56
Abroad	2	0	4	1	10	12	11
% Foreign	40	0	36	10	23	30	16

It is notoriously difficult to predict future trends in mathematics, and Coolidge's last-quoted statement confirms that. For evidence, consider the study of American doctorates in mathematics by Roland Richardson. Admitting delimitations due to the fluid definition of mathematics in the nineteenth century, his results are generally reliable.[68] Table 7.4 extracts figures for Americans who earned doctorates in the nineteenth century either at home or abroad. The high percentages of degrees earned abroad in the late 1870s and early 1890s, as tabulated in the bottom row, were due to J.J. Sylvester (36%) and Felix Klein (10%), and did not extend into the twentieth century the way Coolidge felt was "safe to predict." In fact, the 22% who earned PhDs during 1900–1904 was the first term in a decreasing sequence that converged to zero by 1940.

Table 7.5 lists the top degree-granting American universities where American students earned PhDs in mathematics. Not surprisingly, John Hopkins and Clark University ranked high on the list. Three universities not on the list nonetheless produced mathematicians who turned out to be notable in the twentieth century: Syracuse (the historian of mathematics David E. Smith, PhD, 1887), Purdue (James Byrnie Shaw, PhD, 1893), and Kansas (Arnold Emch, PhD, 1895).

Some small colleges awarded doctorates in mathematics for work done independently; Lafayette College, for instance, awarded two. Few mathematicians who earned such degrees left their mark on the history of the subject. **George Abram Miller** (1863–1951) was someone who succeeded admirably despite such a limited background.[69] Born on a farm to a poor family, Miller nevertheless graduated from high school and obtained a teaching job that allowed him to attend college, initially at the Franklin Academy (now Franklin & Marshall College) and then Muhlenberg College, where he received a BA degree in 1887. After serving as a school principal in Kansas for a year, he was appointed professor of mathematics at Eureka College in

Table 7.5. Leading American producers of PhDs in mathematics before 1900

	before 1890	1890–1899	Total
Johns Hopkins	14	16	30
Yale	10	13	23
Clark	0	11	11
Harvard	5	5	10
Penn	0	9	9
Columbia	4	4	8
Cornell	4	2	6
Virginia	1	3	4
Chicago	0	4	4

Figure 7.5. George Abram Miller

Illinois. During his five-year tenure at Eureka (1888–1893), Miller became involved in upper-level mathematics in two ways. For one, he joined the New York Mathematical Society in October 1891. He also took correspondence courses offered by Cumberland University, a small school located about 30 miles east of Nashville. One could obtain a PhD from Cumberland by examinations in advanced courses instead of a thesis, and Miller chose this option for the doctorate he was awarded in 1892.

The next year, George Miller was appointed instructor at Michigan, where he lived in Frank Cole's house. This contact afforded Miller an opportunity to interact with a research mathematician, and discussions about group theory turned out to be decisive in his career. After two years Miller, like Cole, went to Europe for advanced training, but he first went to Leipzig to study under Lie instead of Göttingen under Klein, as Cole had done. Lie apparently introduced Miller to commutator groups, a topic he pursued regularly. Upon his return to the US, Miller taught at Cornell (1897–1901), Stanford (1901–1906), and Illinois (1906–1931). While on the West Coast, he helped form the San Francisco Section of the AMS and served as its first secretary (1902–1906). After that he became active with the Chicago Section, chairing it in 1908 and 1909. He was also a charter member of the MAA in 1915.

George Miller was a specialist in group theory, producing about 400 papers (more on finite groups than any other American), mostly on commutator groups, characteristic groups, and groups of isomorphisms. His 1916 book *Finite Groups*, written with Leonard Dickson and Hans Blichfeldt, became a classic. Miller also wrote numerous papers on the history of mathematics that are generally discredited for over-reliance on unreliable secondary sources. Despite narrow contributions to mathematics, and disreputable articles on history, he stands as the leading American researcher among those who obtained doctorates at small institutions. Moreover, he was one of two dissertation advisors (along with Arthur B. Coble) that situated the University of Illinois as "the only public institution among those that granted more than ten degrees to women before 1940."[70] Upon his death, Miller bequeathed $2 million to the University of Illinois (about $18 M in 2014).

Larger universities generally accounted for progress in research mathematics in America. Graduates of Johns Hopkins, Clark University, and the University of Chicago during the 1890s, plus the wave of Americans returning with advanced degrees from

Europe, helped swell the ranks of qualified professors to numbers never dreamed of when Sylvester came to Hopkins in 1876. While Hopkins, Clark, and Chicago represented a radical advance in graduate education (and an uplifting of undergraduate curricula as well), new and old universities, public and private, reacted to the stimulus by upgrading their offerings and evolving their own graduate programs. Two of the more successful universities on the list in Table 7.5 are presented as examples of developing programs at one emerging university (Cornell) and one with an especially long history (Penn).

University of Pennsylvania. The University of Pennsylvania (Penn) traces its roots back to 1749, when the precollege Academy of Philadelphia consisted of three schools: Latin, English, and Mathematics. With Benjamin Franklin intimately involved in founding the institution, it is not surprising that a separate Charitable School was opened for children of poor people as well. Speaking of Franklin, his 1749 pamphlet *Proposals Relating to the Education of Youth in Pennsilvania* led directly to the founding of the Academy of Philadelphia, known since 1791 as the University of Pennsylvania:[71]

> As to their STUDIES, it would be well if they could be taught *every Thing* that is useful, and *every Thing* that is ornamental: But Art is long, and their Time is short. It is therefore propos'd that they learn those Things that are likely to be most useful and *most ornamental*. ... All should be taught to write a *fair Hand*, and swift, as that is useful to All. And with it may be learnt something of *Drawing*, by Imitation of Prints, and some of the first Principles of Perspective. *Arithmetick, Accounts*, and some of the first Principles of *Geometry* and *Astronomy*.

The trustees appointed the first master of mathematics in the mathematics school of the Academy of Philadelphia in 1750, so in this sense Penn's mathematics department is over 250 years old. Although the Academy did not award degrees, the College of Philadelphia, chartered in 1755, did. Despite these developments, the department of mathematics today dates its founding to 1899, when it was established as an independent entity.[72] The first master of mathematics was **Theophilus Grew** (d. 1759), who is known to have published almanacs in Annapolis before moving to Philadelphia in 1734 as a teacher of mathematics, surveying, navigation, and astronomy. He opened a mathematical school in the city eight years later, and throughout the 1740s was a commissioner in a protracted boundary dispute between Pennsylvania and Maryland. Grew was appointed master of mathematics at the Academy in 1750, and professor of mathematics when the College of Philadelphia was chartered five years later; he served in both capacities simultaneously until his death. His 1753 work *The Description and Use of the Globes, celestial and terrestrial; with variety of examples for the learner's exercise* was an early textbook for students at the school.

No appropriate candidate was found to succeed Grew, so Thomas Pratt, the master of the English school at the Academy, taught mathematics until the Reverend **Hugh Williamson** (1735–1819) was appointed professor of mathematics two years later (see p. 104 for his portrait). He had entered Penn's first class at age 16 and received its first bachelor of arts degree. Thereupon he studied divinity and was admitted to the presbytery but was never ordained and never assumed a pastorate. Instead, Williamson accepted the professorship at his *alma mater* in 1761, but resigned in 1764 to study medicine abroad, first in Edinburgh and then in Utrecht, where he earned a medical degree. He practiced medicine upon his return to Philadelphia and helped found what is now the University of Delaware. Later he moved to North Carolina, where he was elected to the Continental Congress (1784–1786), and served as a delegate to the convention that framed the Constitution of the United States in 1787. He moved to New York in 1793 and lived there the rest of his life, organizing (with DeWitt Clinton) the Literary and Philosophical Society.

Once again Thomas Pratt taught mathematics until a new professor was appointed in 1766—Thomas Dugan, who held the position only three years. From 1769–1773 there was apparently no professor of mathematics, leaving tutors to teach the subject. However, since 1773, when James Cannon was appointed, the University of Pennsylvania has always had a professor of mathematics. Altogether there were ten such office holders over the 144-year period 1755–1899, with the contributions of three having been covered in earlier chapters: Robert Patterson (professor 1782–1814), his son Robert Maskell Patterson (1814–1828), and Robert Adrain (1828–1834).

American colleges in the first half of the nineteenth century were quite different from what they are today. At Penn, for instance, there were generally 50 or fewer students enrolled at a time. Moreover, one is led to ask why some of them matriculated at a university. At Penn, discipline and lack of attendance were major issues when Robert Adrain taught there around 1830; almost all faculty meetings were devoted to disciplinary measures. Minutes from those meetings reveal such infractions as leaving the math room and not returning, disturbing the mathematics recitation by speaking so as to disturb the exercises, and "attempting to cast ridicule upon the direction of the Professor by not recalling his exercises from the Board."[73]

The modern era of the university began in 1855 when **Ezra Otis Kendall** (1818–1899) was appointed professor. In 1872, the Penn campus moved to its present location in West Philadelphia, but the big academic leap occurred in 1881 when the faculty of philosophy, a forerunner of the graduate school, was formed in response to the success of the graduate programs at Johns Hopkins. Ezra Kendall was named the dean of the faculty. A similarly signal event for mathematics took place that year when Pennsylvania Railroad magnate Thomas A. Scott (1823–1881) endowed a chair of mathematics (Penn's second in arts and sciences), and Ezra Kendall became the first holder of the Scott Chair. Kendall occupied the chair until his death in 1899, three years after his retirement.

Edwin Schofield Crawley (1862–1933) not only succeeded Kendall in the Scott Chair but became the head of the first official department of mathematics in 1899. Crawley received his bachelor's degree from Penn in 1882 and immediately joined the faculty as instructor of civil engineering, even though he was only 20 years old at the time. He moved to mathematics two years later, and in 1892 was awarded Penn's first doctorate in mathematics for the dissertation titled "The forms of quartics and one node and four real distinct asymptotes." The next year Crawley directed the PhD dissertation of the Russian **Isaac Joachim Schwatt** (1867–1934), who then became an instructor of mathematics. **George Egbert Fisher** (1863–1920) and **George Hervey Hallett** (1870–1941) were awarded PhDs in 1895 and 1896, respectively. No dissertation advisor is listed for either one; indeed, no other Penn dissertation listed an official advisor until 1903. Fisher had been on the faculty since earning his AB degree in 1889,

Figure 7.6. Ezra Otis Kendall

and Hallett since one year after earning his in 1893. All four of these doctoral recipients remained at Penn for the rest of their careers.

The work of Schwatt and Fisher reflect Benjamin Franklin's goal of combining "useful and ornamental" studies. Schwatt's "Geometrical treatment of curves" may have been ornamental in 1893, but over the next century the geometry of curved surfaces became the basis for the physical theories of relativity and string theory, as well as an essential component of robotics and DNA molecular structure. Similarly, Fisher's "Some points in the theory of invariants and covariants," regarded as a generalized study of symmetry, has found applications in quantum mechanics and operations research.

Ezra Kendall held both the Scott professorship of mathematics and the Flower professorship of astronomy, but in 1896 relinquished the latter title in favor of **Charles Leander Doolittle** (1843–1919), who had been professor of mathematics and astronomy at Lehigh for 20 years before moving to Penn in 1895 in the same position. Titles changed in 1899 when Doolittle dropped the mathematics part of his duties and Edwin Crawley became professor of mathematics, thus putting an end to the unofficial marriage of mathematics with astronomy (and earlier with natural philosophy and physics). This explains why the Penn mathematics faculty in 1999 decided to date their existence from a century before, but it also reflects the more general independence of mathematics as a separate entity that took place internationally at this time. At its creation, the department numbered five faculty members: Crawley (who had come in 1886), Schwatt (1893), Hallett (1894), Fisher (1889), and Henry Brown Evans (an instructor in mathematics and astronomy at the time, but who soon moved to astronomy).

Penn's mathematics department did not award a doctorate to a woman until 1901: Roxana H. Vivian. However, the department played an important racial role, awarding PhDs to the second and third African American students (Donald Woodward and William Claytor). The much younger Cornell University, unfettered by longstanding traditions, upstaged the older University of Pennsylvania on both counts.

Cornell University. Located in the Finger Lakes region of central New York, Cornell was established in 1865 by two state senators.[74] **Ezra Cornell** (1807–1874) became involved with the experimental telegraph line of Samuel Morse, and then supervised the laying of the first telegraph line in America in 1843. Fourteen years later, he helped form the Western Union Telegraph Company and quickly amassed a fortune. He was elected to the state senate in 1863. **Andrew Dickson White** (1832–1918), a Yale graduate (1853) who studied in Paris and Berlin over the next three years, accepted a history professorship at the fledgling University of Michigan in 1856. He traveled to Europe again (1862–1863), this time to England, in an attempt to cure an acute case of dyspepsia. Upon his return, he too was elected to the senate. The freshmen senators soon became close friends, with White admiring Cornell's philanthropy and Cornell appreciating White's vision of an ideal university.

Thus, White and Cornell sponsored a bill in February 1865 establishing Cornell University as New York's land-grant institution on land donated by the philanthropist. Andrew White served as the school's first president 1866–1885. He sought two characteristics for the new venture:

- embrace the utilitarian as well as the liberal arts (the equivalent of Franklin's ornamental and practical), and
- not restrict enrollment by sex or color.

The first characteristic was reflected in the appointment of the initial professor of mathematics, **Evan William Evans** (1827–1874), a Yale grad (1851) who had become active in mining engineering while teaching mathematics at Marietta College (1857–1864). (Evans departed from Marietta before Oscar Mitchell enrolled in 1871.) Evans's mix of pure and applied mathematics made him an appropriate choice to help achieve White's vision. Evans spent the next year scouring Europe for ideas about mathematics education before taking up his new post.

Three assistant professors were hired in the second year, 1866. The only important one from our perspective was Henry Eddy, an assistant professor of civil engineering as well, who was awarded a Cornell PhD in 1872. Eddy left the university the next year, but the department had made two crucial hires in the meantime. **Lucien Augustus Wait** (1846–1913) joined the staff in 1870 immediately upon receiving his undergraduate degree at Harvard. Although Wait never published original research, he remained very active in undergraduate affairs and became chair of the department in 1895.

A more important addition to the mathematics faculty was James Edward Oliver (1829–1895), appointed in 1871 as an assistant professor. Oliver turned out to be critical for the department's future. Born into a devout Quaker family whose roots were planted in Plymouth in 1639, his mother was a teacher who taught her frail son at home so well that he thrived at the Lynn Academy before entering Harvard in 1846, when he was placed in the sophomore class. (At that time college presidents at American colleges interviewed all prospective students and decided where to place them in the four-year curriculum.) Oliver's roommate for his final two years was Horace Davis, president of the University of California, Berkeley, 1887–1890 and president of Stanford's Board of Trustees 1885–1916. Davis summed up Oliver's character and the role the great Benjamin Peirce played in forging it:[75]

> [Oliver] was studious in his habits, especially in those branches he was fond of, and he pursued these with an avidity which made him neglect other studies, for he cared very little for his rank. His first love, of course, was mathematics, and next to that came ethics and moral philosophy. His forensic compositions were especially terse and clear, but mathematics he devoured with an eager appetite, and when, in his senior year, he was given the *Mécanique Céleste* to study he would often become so absorbed as to prolong his work into the small hours of the morning, and I have many times waked up from my first nap to see him still poring over the ponderous volume long after midnight.

Oliver's habit of working into the wee hours of the morning caused his mother to appeal to Horace Davis, "He has great respect for thee, and thee can give him good counsel."[76] But there was no way Davis could lessen Oliver's drive to master mathematics.

With the confluence of both his graduation in 1849 (with an AB degree from the Lawrence Scientific School at Harvard) and the establishment of the *American Ephemeris and Nautical Almanac* one year later, and with Peirce's offer of a position, it only seems natural that Oliver would work there. This federally funded organization compiled nautical almanacs for naval ships as well as commercial vehicles. But Oliver did not perform his duties at the Cambridge headquarters; instead, he worked in his little study located in his parents' home in Lynn, located about 15 miles away. Although he "found the endless repetitions of the same arithmetical processes extremely wearisome,"[77] he remained with the *Almanac* until 1867, because it was a haven for the country's leading mathematicians and astronomers. Therefore, when the Nautical Almanac Office moved to Washington, DC, in 1867, he resigned his position. Over the next three years, Oliver pursued studies in applied sciences at Columbia in New York City and chemistry in Philadelphia, where he earned money by offering private instruction in mathematics and physics. But by the spring of 1871, he was back in Lynn while delivering a course of lectures on thermodynamics at Harvard.

James Oliver had intended to continue the course that fall, but in the meantime, he was offered the position at Cornell. When Evan Evans had to step down as head of that department two years later due to failing health (he died the next year), Oliver succeeded him in the post, which he held until his death 22 years later. Although Cornell was geared toward undergraduate work, Oliver fought valiantly to orient his department toward graduate-level training. Files from the 1880s record numerous battles between Oliver and the administration over the hiring of more faculty members, the reduction of teaching loads, and the pursuit of original investigations. (His epic battles probably sound hauntingly familiar to many department chairs even today.) Initially, Oliver taught lower-level classes, but after 1889 he concentrated on graduate courses, sometimes offering seven a year. This workload left little time for research, yet he managed to publish several papers between 1882 and 1892.

Upon being appointed head of the department in 1873, Oliver set about changing the Cornell curriculum. An ally in his fight was William Byerly, a faculty member at Cornell (1873–1876) before moving to Harvard after having been passed over for the job as Sylvester's assistant that went to William Story. Cornell had offered the following courses when classes began in 1868.

- Freshman: algebra and geometry
- Sophomore: trigonometry, analytical geometry, differential calculus
- Junior/Senior: analytical geometry, modern higher geometry, integral calculus

However, beginning in 1874 the upper-level curriculum expanded to include courses in differential equations, finite differences, and quaternions. Moreover, Oliver offered a seminar for prospective teachers.

While at Cornell, Byerly was influenced by the way Evan Evans introduced modern methods in the analytic geometry course based on Salmon's book *Conic Sections*. Therefore, when Byerly returned to Harvard in 1876, he offered a course dealing with these matters that became the standard at Harvard until supplemented by topics in calculus 60 years later.

Byerly's position was filled in 1859 by Yale graduate **George William Jones** (1837–1911). The triumvirate of James Oliver, Lucien Wait, and George Jones formed the core of the mathematics department for the next 18 years, during which time they initiated a vigorous graduate program in mathematics. The first such offering was described in the Cornell catalogue for 1873–1874 as an "advanced course of study in Pure and Applied Mathematics ... for resident graduates, and for such undergraduates as may elect."

James Oliver envisioned his department filling the void left by the departure of J.J. Sylvester from Johns Hopkins in late 1883. Unfortunately, two overarching issues prevented Cornell from assuming leadership of the emerging American mathematical research community. One was a lack of devotion to research, which resulted in a meager publication record. The trio that led the department did not consist primarily of research mathematicians; even James Oliver, the most highly trained, seemed reluctant to publish his results. Another issue was that all faculty members were mainly involved with undergraduate teaching, bearing loads between 15 and 20 hours per week. Even the arrival of Arthur Hathaway in 1885 did little to help the cause; he departed six years later.

Yet Oliver's unflinching determination to create a true graduate program began to pay dividends. The first PhD was awarded to **Hiram John Messenger** (1855–1913) in 1886 based on a dissertation written under Oliver. Messenger became an actuary with the Metropolitan Life Insurance Company in New York City until 1897, when he moved to the Actuary of the Travelers Insurance Company in Hartford, CT. In 1887, a year after Messenger's graduation, the mathematics department was thriving with 11 graduate students. During that year, two more doctorates were awarded, one of which went to Rollin Harris, who was among the first fellows when Clark University opened.

However, two events around 1890 changed Oliver's life radically. First, the supposedly confirmed bachelor got married in 1888 at age 59. Oliver then took a sabbatical leave during 1889–1890 that allowed the newlyweds to travel through Europe for 14 months. More importantly for mathematics, his study tour turned out to set the stage for dramatic changes that would take place at Cornell. Initially, Oliver sought to attend Arthur Cayley's lectures, because his own research in algebra concerned invariants, the Englishman's specialty. Unfortunately, Cayley was infirm by then. Therefore, upon the advice of one of his former students, he wrote to Felix Klein to ask about the possibility of attending his lectures as well as discussing methods of teaching, courses of study as related to promising directions for research, and "to see something of your Seminary work."[78] Oliver then proceeded to Göttingen, where he became one of only five auditors for the second half of Klein's two-semester course on Lamé functions; two of the other four were Henry White and Maxime Bôcher. Oliver was in good company. Also, the Olivers and the Kleins struck up a cordial relationship during that 1890 winter semester. Later in life Oliver summarized Klein's lifelong gifts to him as "the spirit, the aims, and the details of my Cornell work."[79]

When James Oliver attended the Chicago Congress in 1893, his best student, Virgil Snyder, was at Göttingen in the midst of doctoral studies under Klein. The major reason for Oliver to attend the Congress was to renew his relationship with Klein, but he also took advantage of the opportunity professional meetings always afford—personal contacts with other experts having a similar passion. Along this line, Oliver engaged in brisk conversation with George Halsted concerning the paper the Texas professor presented at the Congress. Less than two years later, Halsted stated that Oliver "was a pronounced believer in the non-Euclidean geometry. I vividly recall how he came up after my lecture on Saccheri at Chicago and expressing his interest in the most charming fashion ... [on] what Cayley called 'the physical space of our experience'."[80]

When Klein returned the visit to Cornell, the two men enjoyed long strolls along Ithaca's famed ravines. Even though both Klein and Oliver were involved in providing opportunities for gifted females to pursue higher mathematics, I do not know if any of Oliver's three women graduate students at the time joined the men on these walks. Oliver had directed Messenger's doctoral dissertation in 1886, but the seeds planted by his switch to the graduate curriculum were just beginning to bear fruit during Klein's visit in 1893. The three women students were in different stages of programs that would result in doctorates: Ida Metcalf had just been awarded her PhD, Annie MacKinnon was working on the dissertation for the degree she would earn in 1894, and Agnes Baxter was just beginning the thesis she completed for her PhD just before Oliver's death in March 1895 at age 65.

Throughout the 1890s, Klein's enduring influence was so pervasive that Parshall and Rowe wrote, "Cornell emerged as a prime sphere of Klein's influence in the United States."[81] One result was the Mathematical Club that Oliver organized to involve faculty and advanced students in research in a way like Klein's seminars. Called the Oliver Club today, the seminar celebrated its centenary in 1991. Three active participants in the Club from the 1890s studied under Klein in Göttingen—Virgil Snyder, John Tanner, and Annie MacKinnon—with Snyder earning his doctorate there.

When **John Henry Tanner** (1861–1940) studied at Göttingen in 1895, he befriended **Ernst Ritter** (1867–1895), Klein's assistant who had earned his doctorate four years earlier. Tanner recruited Ritter to Cornell, and this addition would undoubtedly have propelled the department forward, but distressingly Ritter caught typhoid on the trip and died at age 28 before ever setting foot in Ithaca. To add to the department's woes, James Oliver died in 1895 too. The Cornell mathematics department was shaken by these events, but nonetheless, progress was not retarded. Lucien Wait succeeded Oliver as department head and embraced Oliver's ideals. In a letter to Felix Klein, Wait lamented Ritter's death and mentioned plans for continued involvement with Klein, whose affection he described in glowing terms:[82]

> I am desirous of bringing a leading German mathematician to Amer-
> ica to assist us here at Cornell. I desire to have ONE advanced course
> of mathematical lectures delivered in the German language so that
> our students will be able to take hold of work in Germany without so
> much delay. ... We shall send two of our graduate students to Eu-
> rope for an extended course of mathematical study next Summer. ...
> I desire to express my grateful appreciation for what you are doing for
> our students in Göttingen. They not only enjoy your work but they
> all speak of you with a great deal of affection personally.

Although Cornell was unable to hire a German, the mathematics department did
secure someone with training under Klein—his former student Virgil Snyder, who
completed his dissertation that year. He would quickly lead Cornell into the ranks
of the best mathematics departments in America at the beginning of the twentieth
century. The state of the department at the turn of the century can be seen in a *Bulletin*
announcement of advanced courses offered during 1898–1899.[83]

- Wait: advanced analytic geometry; advanced differential calculus
- Jones: higher algebra and trigonometry; probabilities
- McMahon: higher plane curves; quaternions; potential, and spherical har-
  monics; mathematical theory of sound
- Tanner: binary quantics; theoretical mechanics; German readings
- Murray: differential equations; finite differences; astronomy
- Hutchinson: advanced integral calculus; elliptic functions; surface and twisted
  curves
- Snyder: projective geometry; general function theory; line geometries
- Miller: substitution groups; continuous groups; theory of numbers

This lineup of courses is certainly impressive for both its breadth and depth.

All eight mathematicians on the list (but the two here) have been introduced.
James McMahon (1856–1922), born in Ireland, came to Cornell as an assistant professor in 1890 after having earned an AM from the University of Dublin. One year
later he became a member of the New York Mathematical Society. McMahon was very
active in the AAAS, serving as secretary and vice president. Born in Nova Scotia, where
he received his bachelor's degree at Dalhousie University, **Daniel Alexander Murray**
(1862–1934) came to Cornell as an instructor in 1894, one year after having earned his
Johns Hopkins PhD for a dissertation on differential equations. He left Cornell in 1901
for a professorship at Dalhousie, where he remained until 1907, when he moved to
McGill University. Murray retired from McGill in 1930.

The most important person on this list, however, is **Virgil Snyder** (1869–1950).[84]
Born and raised in Iowa, Snyder graduated from Iowa State University with a ScB
degree in 1889. He matriculated at Cornell the next year, receiving an MA in 1892.
Upon the urging of department head James Oliver, he traveled to Göttingen on an
Erastus Brooks Fellowship to study under Felix Klein, earning his doctorate in 1895
for a dissertation on linear complexes in Lie's sphere geometry. Thus, Snyder was
in Germany when Klein visited Oliver at Cornell two years earlier. Snyder returned
directly to Ithaca from Göttingen as an instructor in 1895, and remained at Cornell until
his retirement in 1938. During his 43-year tenure he published many papers on topics

Figure 7.7. Virgil Snyder

related to his thesis and on surfaces invariant under infinite discontinuous groups of birational transformations. The quality of his work was recently described as follows:[85]

> Up until the 1920s, Snyder's prolific output and his talents as a teacher made him, together with Frank Morley of Johns Hopkins, one of the most influential algebraic geometers in the nation. Together with Henry White, in fact, Snyder emerged as a principal heir to Klein's geometric legacy.

Virgil Snyder was especially active with the AMS, elected president for two years, 1927–1928. He presented papers at many local and national meetings. Moreover, he and his wife engaged in much travel to Europe, resulting in frequent AMS reports on meetings he attended. His personal acquaintance with mathematicians abroad enabled him to play a conciliatory role at the 1924 International Congress of Mathematicians in Toronto, where international tensions resulting from World War I were running high.

Virgil Snyder was certainly the most prolific member of the Cornell mathematics department, not only in terms of publications, but also in mentoring students, directing 40 doctoral dissertations from 1902 through 1937. Thirteen of his PhD recipients were women, a disproportionately high percentage. The emergence of women in the American mathematical research community and the role that Cornell played in this stunning development is described next.

Exceptional women

The education of women in the US and Canada lagged behind men by a century or more. It took until the first half of the nineteenth century before schools for women were established above the high-school level; they were generally called "seminaries" in the Northeast and "colleges" in the South. Graduates of these schools normally studied algebra, geometry, and trigonometry before pursuing careers as teachers in secondary schools. Just four years after it opened as America's first co-educational school in 1837, the Oberlin Collegiate Institute (Oberlin College after 1850) became the first institution to offer women an undergraduate education in mathematics equivalent to those of men. Ultimately two Oberlin women graduates would earn PhDs in mathematics, but not until the twentieth century: Mary E. Sinclair, Chicago, 1908; Josephine A. Robinson, Syracuse, 1918.

The first public institution in the US to admit women on an equal basis was the University of Iowa, in 1855, but the Morrill Act of 1862 helped establish several land-grant universities that were co-educational from the start (Cornell, for instance) even though the Act included no mention of women or co-education. Most women who pursued mathematics at these universities were enrolled in a "normal department" that trained prospective secondary-school teachers. For the most part, women in the late nineteenth century who were interested in the serious pursuit of mathematics had to attend a woman's college, because most of the PhD-granting schools were restricted to males. Even the president of the College for Women of Case Western Reserve worried about the effect of undergraduate men and women taking seats in the same classroom. The 1876 Harvard graduate **Charles Franklin Thwing** (1853–1937, pronounced "twing") president of Case 1890–1921, wrote in 1894:[86]

> Such objections as certain scholars and administrators hold against co-education do not apply with equal force to graduate as with undergraduate students ... the greater age of graduate students may entirely or largely remove difficulties which are found in the way of men and women mingling in the undergraduate department. The Faculty of Yale University knows very well that to admit women to its graduate school is quite unlike opening the doors of Yale College to girls of the age of eighteen.

However, women lagged behind men in access to graduate study as well. As examples, the two trailblazing American graduate schools, John Hopkins and Clark, erected barriers to prevent women from enrolling in their graduate programs. On the other hand, the University of Chicago was co-educational from its founding, a policy that permitted Mary Winston Newson to enroll in the school's initial class in 1892; she earned a Göttingen PhD in 1897. Yet several pioneering women preceded her. The historian of science Margaret Rossiter adopted militaristic terms to describe attempts by these women to obtain PhDs at the end of the nineteenth century:[87]

> When all the attempts by women to gain higher degrees at universities in the United States and Germany over three decades (1870 to 1900) are viewed together, they can be seen as a process of infiltration, a kind of educational "guerrilla warfare" or slow "war of attrition" against the universities. Under this almost military strategy, individual women sought to test the repressive system on as many fronts (departments and universities) as possible, probing for weak points and using what friends they had to help them evade the rules informally, and, when enough "exceptional" women had been admitted in this way and had surpassed their fellow students without the imagined disruption, to push for a change in policy, which then could be seen as harmless, "only fair," and long overdue, and could be enacted quietly.

Christine Ladd Franklin is an illustration of Rossiter's "guerrilla warfare" on the home front. After graduating from Vassar, she was permitted to take courses at Harvard but not to enroll. Then she applied to Johns Hopkins, which accepted her as an exceptional but unofficial student. By 1882 she had published three papers in the *American Journal* and had completed a dissertation on logic under C.S. Peirce that was highly praised by the department, yet university officials refused to award the degree because of her sex. These trustees thus erected a roadblock, not just a hurdle. By the time the university attempted to redress this wrong, 44 years later, Ladd Franklin was 78 years old and the torch had been passed to a new generation.

Mary Winston Newson illustrates Rossiter's "guerrilla warfare" abroad. In the US, matters changed dramatically for exceptional women when the brand new—and co-educational—University of Chicago opened in 1892 and admitted Winston into its graduate program. At the end of that year, Heinrich Maschke recommended her for advanced study under Felix Klein, even before Klein met her during the Chicago Congress and Evanston Colloquium at the end of her first year of graduate study. Just two years earlier Christine Ladd Franklin had accompanied her husband Fabian Franklin on a sabbatical to Germany, but was denied permission to enroll as a student at Göttingen despite Klein's attempt to admit her. Her probing for a weak point seems to have paid dividends when Klein received permission to sponsor Winston. Moreover, Ladd Franklin's role in establishing a fellowship program for women in 1888, called the Association of Collegiate Alumnae (ACA), surely aided Winston. Years later, Winston recalled:[88]

> It had been a dream of mine to go to Europe to study and when the announcement was made that the "Association of Collegiate Alumnae" was offering a scholarship for a year's study in Europe, I made up my mind to apply.

Ironically, Winston did *not* win the scholarship. So how was she going to finance her study abroad? This is where the remarkable Ladd Franklin came to the rescue by sending her $500, a considerable sum at the time, to defer expenses (about $13,000 in 2014). Winston was eminently successful at Göttingen, earning her PhD under Klein in 1897.

In between Christine Ladd Franklin and Mary Winston, however, American women had made great strides, as illustrated by the three Cornell women graduate students who met Felix Klein during his visit with James Oliver on campus shortly after Winston had made his acquaintance in Chicago. None of them needed to travel abroad to seek doctorates because conditions were ripening right at home, yet one traveled to Göttingen for post-doctoral study. Altogether, ten American women obtained PhDs in mathematics before 1900, nine from institutions in the US. The lives and careers of these nine can be partitioned into the three types of colleges where they received their doctorates: those that were co-educational (like Cornell), those that were historically male until the latter part of the nineteenth century (like Yale), and emerging women's colleges (like Bryn Mawr). The online file "Web07-AmerWomen" provides further details on their lives and careers.

Cornell. Three trailblazers earned doctorates at Cornell, which had been co-educational since its founding in 1865. Women did not attend classes at the Ithaca campus when it opened in 1868; instead they had to wait another four years until the first dormitory for women was erected. It took another 20 years before the first doctorate in mathematics was awarded to a woman

That honor goes to **Ida Martha Metcalf** (1856–1952), who was born and raised in Texas but moved to the Northeast with her mother and brother after her father died when she was a preteen.[89] She enrolled in the private, co-educational Boston University, as a special student (1883–1885), and then as a regular student the next year. Two years after earning a bachelor's degree in 1886, Ida Metcalf matriculated in the graduate program in mathematics at Cornell, obtaining a master's degree in 1889. She returned to Cornell three years later, and in 1893 became the first woman to earn a PhD in mathematics at Cornell for the dissertation "Geometric duality in space." James Oliver is credited as her advisor, but George Jones probably directed her study, as she had taken his course on projective geometry that year. Although Metcalf entered the work force with a PhD in 1893, she was unable to secure a university teaching position. She taught at various high schools, but by 1910 left the teaching profession to become a security analyst at a banking house in New York City. Shortly thereafter she won a job in the Comptroller's Office in the city when she became the first woman to pass a civil service examination. This enabled her to work in the city's finance department until her retirement in 1921 at age 65. Ida Metcalf was bitter about being unable to land a suitable teaching position despite earning a PhD from a prestigious university and, as a result, maintained a very cynical view of higher education for women throughout the rest of her life. Considering her accomplishments, the struggle to find a teaching position commensurate with her ability must have been a very hard pill to swallow.

The next two women to earn Cornell PhDs led eerily similar lives. They were born in different provinces in Canada, Annie MacKinnon in Woodstock, Ontario, and Agnes Baxter in Halifax, Nova Scotia. The family of **Annie Louise MacKinnon** (1868–1940) moved to Kansas in the US when she was only two years old, so she received her early education in the American Midwest. She received her bachelor's degree from the University of Kansas in 1889 and master's degree two years later. She taught in high school for a year, and then enrolled in the graduate program at Cornell, being awarded a Brooks Fellowship her second year. Annie MacKinnon earned her PhD in 1894 for a dissertation written under James Oliver, "Concomitant binary forms in terms of the roots." She published her research in the *Annals of Mathematics* in two parts. In the first part, which appeared the next academic year, she stated that her aim was to bring a direct connection "between root and coefficient symbolic expressions for Covariants and Invariants … into clearer light [to illustrate] the practical value of German Symbolism in Modern Algebra."[90] At the end of the paper MacKinnon promised, "The tables referred to in Part I, Chapter III, will be given in a future number of the *Annals*."[91] However, the second part would not appear for another four years.[92]

Perhaps part of the reason for the delay was that Annie MacKinnon had won an ACA Fellowship for European study. In July 1894 she wrote to Felix Klein to gently remind him of their conversation the previous year at Cornell, and to ask for permission to attend his courses. She ended up spending two years in Göttingen, presenting more lectures at his seminars than any other American student, male or female: three in the "Seminar on Number Theory"[93] and one each in the "Seminar on the Foundations of Analysis for Functions of a Single Variable" and the "Seminar on the Foundations of Analysis for Functions of Several Variables." Mary Winston was studying under Klein at the same time. They were soon joined by yet a third American woman, Isabel Maddison. All three presented lectures at the two seminars on foundations of analysis. Clearly, Felix Klein had become *the* cluster point for American women in Europe.

The third Cornell graduate, **Agnes Sime Baxter** (1870–1917), came from a family that had immigrated to Nova Scotia from Scotland. She obtained bachelor's and master's degrees from Dalhousie University in 1891 and 1892. Both MacKinnon and Baxter entered Cornell in October 1892.[94] MacKinnon graduated in two years, but it took Baxter three to complete her dissertation, also written under Oliver, "On Abelian integrals, a resume of Neumann's Abelsche integral with comments and applications." Baxter too was awarded an Erastus Brooks Fellowship for her final year of study.

The women both entered the work force in 1896, MacKinnon after a two-year study tour in Göttingen and Baxter after another year at Cornell editing the works of their advisor, James Oliver. MacKinnon had won an ACA fellowship for 1894–1895 and a Women's Education Association of Boston Fellowship for 1895–1896. Upon returning to the US, she became the only mathematics faculty member at Wells College, but left after five years to marry someone she had met in Göttingen. Annie MacKinnon Fitch neither taught after her marriage nor continued the research program she had begun with such promise. However, she joined the AMS in 1897, was a charter member of the MAA when it was established in 1915, and maintained both memberships until about 1930.

Agnes Baxter had met her husband while still an undergraduate. She and **Albert Ross Hill** (1869–1943), also from Nova Scotia and a Dalhousie graduate, received PhDs in 1895, his in philosophy. They married in 1896. Subsequently, the Hills moved to the University of Nebraska and the University of Missouri, where he served as dean of the school of education. Like MacKinnon Fitch, Baxter Hill's subsequent career was typical of the times, following her husband's career and staying home to raise children. When she died of pneumonia shortly after her 47th birthday, Albert Hill donated her large collection of books to Dalhousie and, in appreciation, the university opened the Agnes Baxter Reading Room in the department's reading room in 1988.[95]

These three exceptional women are prime examples of the philosophy that Ezra Cornell and Andrew White espoused. Table 7.6 lists the six PhD recipients in mathematics from Cornell during the 1890s, half of whom were women. This fraction is very high for universities at that time, but it is not surprising due to the significant clustering that took place for women around a few specific advisors, in this case James Oliver. Yet these three women were not the only ones with sufficient persistence and academic ability to excel in mathematics at the time. **Anna Helene Palmie** (1863–1946) obtained her Cornell bachelor's degree in 1890 and remained two years in the graduate program before accepting a teaching position at the Women's College of Western Reserve University in Cleveland (Case Western today). And **Estella Kate Wentz** (1866–1938) received a master's degree at Cornell in 1894 before moving to Indiana to teach high school. Both were active in the AMS.

Table 7.6 masks another aspect of the state of higher learning in mathematics around the turn of the century—its dependence on individuals rather than a community. In Cornell's case, Lucien Wait did not share James Oliver's views, feeling instead that a PhD should be granted only in exceptional cases and only after significant periods of study, preferably abroad. With Wait succeeding Oliver as chair in 1895, and putting his overarching philosophy into effect, Cornell experienced a temporary halt in the production of doctorates until 1901, when Virgil Snyder became the leading force in the department. Snyder became one of the leading advisors of women PhDs in the

Table 7.6. PhD recipients in mathematics from
Cornell during the 1890s

Name	Year PhD
Ida Metcalf	1893
Annie MacKinnon	1894
Lionel Marks	1894
Agnes Baxter	1895
Ernest Nichols	1897
Charles Comstock	1898

country, with 14 of his 40 students being females. Only Leonard Dickson of the University of Chicago has produced more women PhDs in mathematics.

All-male bastions. Cornell was not the first university to confer a doctorate in mathematics on a woman. Columbia awarded a PhD to **Winifred Haring Edgerton** (1862–1951) almost a decade earlier, at about the time that the board of trustees at Johns Hopkins refused to bestow the degree that Christine Ladd earned. Winifred Edgerton thus became the first American woman to successfully navigate the "process of infiltration." In fact, she was the first woman to earn a degree of any kind from Columbia, whose undergraduate college was all male. Her determination to obtain this degree in 1886 blazed a trail for those who followed, as illustrated by the motto inscribed below her portrait hung in Columbia University to celebrate the fiftieth anniversary of her graduation from Wellesley: "She opened the door."

Winifred Edgerton was born in Wisconsin, but at age two moved with her family to New York City, where she was schooled by private tutors. An increasing interest in astronomy and mathematics impelled her parents to have an observatory built in one of their New Jersey homes. Edgerton then entered Wellesley College, which had been opened in 1875. When she graduated in 1883 she moved back home with her parents and became a teacher at a boarding school for "Young Ladies."

Edgerton's calculation of the orbit of a comet that year further stoked her interest in mathematical astronomy, and led her to seek access to the telescope at Columbia. This in turn created a strong desire to continue her study of mathematics and astronomy so, with the support of her parents and Columbia President Frederick Barnard, she applied for admission even though Columbia had never granted a degree to a woman. The board of trustees opposed admitting women, but in a compromise with Barnard, agreed that women could be given detailed syllabi and, if they subsequently passed necessary exams, would be awarded appropriate degrees. However, women were forbidden from attending classes. In February 1884, the board granted permission for Edgerton to use the telescope and to study under Professor John Rees.

Yet Winifred Edgerton still had to leap hurdles in her quest to earn a doctorate. In one class the students, all males, requested that the professor assign one of the hardest textbooks in the course, but they had not counted on the fact that she had already read it in one of her Wellesley courses. Undaunted, she continued her work in the observatory and her study of mathematics alone, without any communication with other students, and she was awarded her PhD in June 1886 at age 23. The *New York Times* recorded the historical event: "She was greeted with a terrific round of

Figure 7.8. Winifred Haring Edgerton

applause."[96] Edgerton's dissertation bore the expansive but descriptive title "Multiple integrals: (1) Their geometrical interpretation in Cartesian geometry, in trilinears and triplanars, in tangential; in quaternions; and in modern geometry. (2) Their analytical interpretation in the theory of equations, using determinants, invariants and covariants as instruments in the investigation." Edgerton submitted another dissertation based on the computation of the orbit of the 1883 comet, but official records indicate that her degree was based on the one in mathematics. It was directed by department head John Van Amringe.

Upon graduation, Winifred Edgerton was offered a position at Wellesley but turned it down in favor of teaching at a boarding school. One year later, in September 1887, she married a highly educated man, **Frederick Merrill** (1861–1916). She delivered her first child the next year, 1888, when she was also asked to serve on a small committee to form a separate woman's college to be part of Columbia. However, because there were men on the committee and its meetings were conducted in a downtown Manhattan office, her husband objected, so she declined to serve. Nonetheless, when Barnard College was founded in 1889, named in honor of the campaigner for women's education, Edgerton Merrill's name was listed on the request to the Columbia trustees to establish the woman's college.

Financial vicissitudes caused the Merrills to separate in 1904. After that, Winifred Edgerton Merrill remained in New York City to raise their four children. Initially, she was the principal at a girls' school in Yonkers. In 1906, she founded the exclusive Oaksmere School for Girls in New Rochelle, which opened a branch in Paris in 1912. Two years later, Oaksmere moved to Mamaroneck, NY, after a devastating fire destroyed the recitation hall and gymnasium.[97] Once again, however, lavish spending on her schools caused financial difficulties despite exceedingly high tuition, and Oaksmere closed in 1926.

Table 7.7 lists the ten American women who received PhDs in mathematics before 1900. The word *received* is used purposely here because Christine Ladd Franklin was the first American woman to have completed all requirements for a PhD in 1882. Also, Table 7.7 indicates that Mary Winston Newson was the only woman during this time to receive a foreign doctorate. The only other American-born woman to receive a PhD

Table 7.7. American women who received PhDs in
mathematics before 1900

Name	Year	Institution
Winifred Haring Edgerton	1886	Columbia
Ida Metcalf	1893	Cornell
Annie MacKinnon Fitch	1894	Cornell
Ruth Gentry	1894	Bryn Mawr
Charlotte Barnum	1895	Yale
Agnes Baxter Hill	1895	Cornell
Elizabeth Dickerman	1896	Yale
Isabel Maddison	1896	Bryn Mawr
Mary Winston Newson	1897	Göttingen
Leona Peirce	1899	Yale

abroad before 1900 (Anne Lucy Bosworth Focke) will be introduced in the section
"Transition 1900."

Although Winifred Edgerton Merrill "opened the door" in 1886, the door was
barely ajar, as no other woman earned a doctorate in mathematics for another seven
years. Yet her example demonstrated that women could carry out original research
in mathematics, and nine more were successful over the next seven years. The five
mathematicians in Table 7.7 not yet introduced obtained their degrees at an established
university (Yale) and a new brand of women's college (Bryn Mawr). The lives of this
quintet are described and then conclusions about them are drawn.

Table 7.7 lists three recipients of Yale PhDs in this period, but Leona Peirce had
performed her research under William Story at Clark University in an agreement with
Cornell. An administrative "snafu" prevented her from receiving her Cornell degree,
causing her to matriculate at Yale, which awarded her the degree for the dissertation
"Chain-differentiants of a ternary quantic." What about the education and subsequent
careers of the two women who earned doctorates at Yale, a traditionally male institution that admitted women only into its graduate program beginning in 1892? This
Ivy League institution did not list dissertation advisors formally, a policy that makes it
problematic to assign credit for the primary figure responsible for directing research.

Both Christine Ladd Franklin and **Charlotte Cynthia Barnum** (1860–1934) were
unsuccessful in attempts to obtain PhDs at Johns Hopkins. Barnum graduated from
Vassar in 1881, taught high school for eight years, and then taught astronomy at Smith
College (1889–1890). During that year, she was granted permission to take courses
at Johns Hopkins, where she remained (1890–1892). But she sought a doctorate, so
she transferred to Yale, and three years later became the first woman to receive a PhD
degree at the formerly all-male bastion for her dissertation "Functions having lines
or surfaces of discontinuity." Except for an instructorship the next year at Carleton
College (1895–1896), Charlotte Barnum worked in applied areas of mathematics and
in editorial work most of the rest of her career, as an actuarial computer for insurance
companies, a computer for the US Naval Observatory and for the tidal division of the
US Coast and Geodetic Survey, an editorial assistant in a biological survey for the US
Department of Agriculture, and an editor for Yale University Press.

Table 7.8. Women who earned PhDs in mathematics at Yale

Name	BA	BA	PhD	Notes
Margaretta Palmer	Vassar	1887	1894	astronomy
Charlotte Barnum	Vassar	1881	1895	
Elizabeth Dickerman	Smith	1894	1896	
Leona Peirce	Smith	1886	1899	
Ruth Wood	Smith	1898	1901	*Annals*
Helen Merrill	Wellesley	1886	1903	*Trans.*; Wellesley 1903
Clara Smith	Mount Holyoke	1885	1904	*Trans.*; Wellesley 1908
Euphemia Worthington	Wellesley	1904	1908	Wellesley 1909–1918
Mary Walker	Missouri	1903	1909	Missouri 1909–1911
Ida Barney	Smith	1908	1911	*Amer. Journal*; colleges

The remaining woman to earn a Yale PhD before 1900 was **Elizabeth Street Dickerman** (1872–1965), whose life was similar to Charlotte Barnum's. Dickerman attended Smith College (1890–1894) and then entered the graduate program at Yale. Therefore, her first year in New Haven overlapped with Barnum's final year. Dickerman completed her dissertation in just two years, "Curves of the first and second degree in (x, y, z) where there are conics having two points in common." Like Barnum, Dickerman obtained one collegiate faculty appointment, as a one-year substitute at the College for Women of Western Reserve University, 1906–1907. She taught mathematics and psychology in several different private schools (1904–1913) but turned her attention to literary pursuits for the rest of her life. I am not aware whether she published her dissertation, let alone any other works in mathematics. Moreover, she joined neither the AMS nor the MAA.

In 1920 Yale published an account of its alumnae since 1894. The author of the mathematics section, Ernest Brown, provided thumbnail sketches of the education and professional activities of the ten women who had obtained PhDs in mathematics during this period.[98] Table 7.8 summarizes Brown's findings, including the first entry, **Margaretta Palmer** (1862–1924), even though she would probably be categorized as an astronomer today. In fact, her dissertation carried out a recalculation of the orbit of a comet that had been discovered in 1847 by her undergraduate mentor, Maria Mitchell. Margaretta Palmer and Charlotte Barnum were in the first group of women admitted to the Yale Graduate School in 1892.

The "Notes" column in Table 7.8 highlights two essential differences between the three women who graduated before 1900 (excluding Margaretta Palmer) and the remaining six. First, not one of the three published her dissertation, whereas four of the remaining six did. Second, not one in the first group snared an academic position at a college (except in two one-year isolated incidents) whereas three of the last six taught at Wellesley, and Mary Walker taught at Missouri until she married the physicist Albert Wallace Hull two years after obtaining her doctorate. Another vital difference separating the two groups, one not indicated in the table, was having a dissertation advisor. In this sense, Ruth Wood and Helen Merrill are aligned with the pre-1900 trio with unknown advisors. Merrill is an unusual case. Sometimes she is listed as a student of James Pierpont, probably because he was on her examining committee and signed the official form. However, the problem for her thesis was suggested by Milton Porter, who directed her study but had left Yale in 1902, the year before she received her degree. (Helen Merrill became the first woman officer of the MAA when elected vice president in 1920; no other woman would be an officer for another 43 years.) However, James Pierpont did serve as the advisor for Smith, Worthington, Walker, and Barney, underscoring his role as a significant cluster point for women at Yale. Although only Helen Merrill will appear subsequently, it should be mentioned that **Ruth Goulding Wood** (1875–1939) created a trust fund ensuring that at least one female mathematics professor would earn the highest salary on the Smith College faculty.

The mention of Smith College focuses attention on schools where this group of pioneering women obtained their bachelor's degrees. Apart from **Mary Shore Walker Hull** (1882–1952), every one of the others attended a women's college, with four having done their undergraduate work at Smith.

Women's colleges. Women's colleges can be divided into two distinct types: co-ordinate colleges (like Barnard and Radcliffe) that were officially separate but closely affiliated with a men's college (Columbia and Harvard, respectively), and those that stood alone, independent of other institutions. The first women's college explicitly founded with the same requirements as men's colleges was Elmira College, NY, in 1855. Although Elmira played no role in mathematics, those historically women's colleges, known as the Seven Sisters, did. Organized in 1927 to promote private women's colleges, this group consisted of Mount Holyoke, Radcliffe, Smith, and Wellesley in Massachusetts, Barnard and Vassar in New York, and Bryn Mawr in Pennsylvania. Four graduates of these schools earned PhDs in mathematics at Yale alone.

All seven colleges opened between 1865 and 1889. Two are now co-educational, Vassar (as of 1969) and Radcliffe (which merged with Harvard in 1977; full integration was completed in 1999). Bryn Mawr is the only one with a graduate program in mathematics; Radcliffe also awarded doctorates, but its recipients took classes at Harvard and worked under Harvard dissertation advisors.

The American Civil War contributed to the amassing of great fortunes and, per Karen Parshall, "Interestingly, a number of those newly rich chose to direct significant amounts of money to the cause of higher education."[99] For instance, Matthew Vassar turned over half of the fortune he had made in brewing, as well as more than 200 acres of land, to establish the eponymous women's college in 1865. Sophia Smith invested assets from her wealthy father in stocks so smartly that she donated enough funds to establish her eponymous college in 1871. Four years later, Wellesley was founded from a generous gift from the successful Boston lawyer, Henry Durant, and his wife, Pauline. "Bryn Mawr College, on the other hand was made possible by Joseph W. Taylor, a Quaker physician and successful businessman, who purchased land outside of Philadelphia for the site of the college and then left some $800,000 in his will as an endowment." As the only women's college to offer graduate degrees:[100]

> Bryn Mawr was thus unique among institutions of higher education in providing *both* an educational environment fully supportive of women's aspirations to earn the PhD *and* a place for them to pursue careers as active researchers. (Italics due to Parshall.)

The first of the Seven Sisters was Mount Holyoke, which was opened as the Mount Holyoke Female Seminary in 1837 by Mary Mason Lyon (1797–1849), a pioneer in women's education in America. Initially chartered as a teaching seminary, Mount Holyoke became a model for many of the others 30 years later, with both Vassar and Wellesley specifically patterned after it. The school received its collegiate charter in 1888 and was known as the Mount Holyoke Seminary and College until 1893, when it became Mount Holyoke College.

Vassar was the first of the Seven Sisters to be chartered as a college when it was founded in 1861 by the brewer Matthew Vassar. The first person appointed to the Vassar faculty (in 1865) was the astronomer Maria Mitchell, a mentor to Christine Ladd Franklin and Margaretta Palmer. Vassar became co-educational in 1969 after declining an offer to merge with Yale. Today Smith College is the largest privately endowed college for women in the US. Smith graduate Euphemia Lofton Haynes, class of 1914, became the first African American woman to receive a PhD in mathematics in 1943 (from Catholic University).

Wellesley College was founded in 1870 as the Wellesley Female Seminary, a name that was changed in 1873. Henry Durant, had emphasized mathematics and science at the college's inception, a principal mission that remained in place even when the prescribed curriculum gave way to electives. Up to World War II, the full-time mathematics faculty at Wellesley consisted entirely of women. Moreover, the college enrolled more women than any other school in America up to 1900, when it was surpassed by Smith (which was superseded by Hunter College in New York City by 1915).

Barnard became affiliated with Columbia officially in 1900 but continues to be independently governed. However, Barnard's only connection with mathematics is the PhD awarded to Winifred Edgerton Merrill in 1886. The other coordinate college, Radcliffe, will play no role here. Radcliffe was originally created in 1879 by the Harvard faculty as "The Harvard Annex" for women's instruction. Fifteen years later it was chartered as Radcliffe College.

The remaining member of the Seven Sisters, Bryn Mawr College (pronounced colloquially as "brin-mar" in English, as opposed to the correct Welsh pronunciation) ranked among the top producers of women PhDs in mathematics through the middle of the twentieth century. When the college was founded in 1885, it was affiliated with Quakers but became nondenominational just eight years later.

The first dean of Bryn Mawr College was as instrumental in determining the future course of mathematics at the school as Daniel Coit Gilman had been at Johns Hopkins. Both looked to England for a mathematician to thrust the mathematics department into a leading role in the US. **Martha Carey Thomas** (1857–1935) graduated from Cornell in 1877 despite the wishes of her father, a trustee at Johns Hopkins. Thomas then enrolled in the graduate program at Johns Hopkins in her hometown of Baltimore, but withdrew when offered only private tutoring rather than being permitted to attend classes despite a formal petition to president Gilman. Carey Thomas, as she preferred to be called, next traveled to Europe, initially pursuing graduate work in Leipzig, but then moving to Zurich because the University of Leipzig did not grant degrees to women. When she received her PhD in linguistics from the University of Zurich in 1882, she became the first American woman to obtain a doctorate in any field. She moved to Paris for additional study at the Sorbonne before returning to the US. In 1884, a year before Bryn Mawr College opened its doors, she was appointed chair of the

English department and the first female dean in the US. A suffragist, Carey Thomas became the second president of the college. During her tenure as dean (1885–1908) and president (1894–1922), she was instrumental in recruiting several notable faculty members. Perhaps none was as successful as her appointment in mathematics, C.A. Scott, despite Thomas being only one year older than Scott. The two had met in England as early as 1882.

Like William Rainey Harper at Chicago seven years later, Carey Thomas took a chance on a young, untested, but magnetic figure to head her mathematics department instead of settling for an older, established mathematician. That figure was **Charlotte Angas Scott** (1858–1931), who in 1876 had entered Girton College, the first residential college for women in England, on scholarship. She graduated with honors four years later. Most of the faculty (22 out of 34) at the prestigious Cambridge University, located nearby, permitted women to attend their lectures, although always with a chaperone. Therefore, Scott received an excellent education. In her final year, she was permitted to take the highly competitive Tripos exam (which allowed students to graduate with honors) on an informal basis. Just taking the exam would test one's fortitude, as it required 50 hours over nine days. Yet Scott placed eighth, an unprecedented achievement for a woman. Nonetheless, due to her sex, she was unable to receive the title "eighth wrangler," and was barred from attending the commencement ceremony where these awards were announced with great pomp and circumstance. Cambridge students interrupted the ceremony unceremoniously:[101]

> The man read out the names and when he came to "eighth," before he could say the name, all the undergraduates called out "Scott of Girton," and cheered tremendously, shouting her name over and over again with tremendous cheers and raising of hats.

However, Cambridge did not grant degrees to women until 1948, so Scott remained at Girton as resident lecturer in mathematics for the next four years. During that time, she passed examinations administered by the University of London, which became the first British college to award degrees to women in 1880. At that time, the university was an examining body only, not a teaching institution. Charlotte Scott was then awarded a bachelor's degree from London in 1882 and a PhD three years later for research conducted under Arthur Cayley. By 1885 Cayley's friend J.J. Sylvester had returned to England from Johns Hopkins with firm knowledge of the collegiate landscape in America, whereupon Cayley, who had been a visiting professor at Johns Hopkins, recommended Charlotte Scott for the opening at Bryn Mawr. The 27-year-old Scott accepted at once and sailed across the Atlantic to establish undergraduate and graduate programs in mathematics. Altogether, Charlotte Scott directed seven dissertations by American women over her 40-year tenure at Bryn Mawr (1885–1925), including two in the 1890s. Scott and, later, Anna Pell Wheeler were two important cluster points for women in the US. A third leading advisor for women PhDs was Aubrey Landry of Catholic University, but the bulk of her students graduated in the 1930s.

The program at Bryn Mawr has always been relatively small, though not to the extent of Scott's first three years, when only three undergraduates and one graduate student who were seriously interested in mathematics enrolled in her classes. But by 1894 the number of graduate students had quadrupled, most likely due to fellowships the college offered; two of those four ended up being successful doctoral candidates. Like J.J. Sylvester, Scott returned to England for many summers, mainly to avoid the

Figure 7.9. Charlotte Angas Scott

humidity that engulfs the Philadelphia area. She too went back to the motherland one year after retirement in 1924, remaining at Bryn Mawr that year to supervise the completion of her final student's doctoral dissertation (Marguerite Lehr).

Charlotte Scott was a tireless researcher who was especially active with the AMS; she was the only woman elected vice president of the AMS until 1976.[102] However, her role as dissertation advisor was critical, thus the education and career of the two doctoral students she directed before 1900 are discussed next. In her role as mentor, she initiated the Bryn Mawr College Mathematics Journal Club, the equivalent of the seminars that had been so successful for taking graduate students from learners to creators at Johns Hopkins and Chicago.

The first two Bryn Mawr fellows were not able to obtain PhDs. **Ella C. Williams** was awarded a fellowship for 1885–1886 after having graduated from the University of Michigan in 1880 and studied in Göttingen and Newnham College, Cambridge, in 1884 before coming to Bryn Mawr. Williams departed before the start of the 1886–1887 academic year, and became a teacher at the private Miss Moses's School in New York City for a year before teaching for two years at the State Normal School in Plymouth, NH. She then taught at Miss Spence's School in New York City, 1896–1898. Williams was one of the six women who were the first of their sex to join the AMS (in 1892). That October, she became the first woman to present a paper at an AMS meeting, "An orthomorphic transformation of the ellipsoid." The fellowship that Williams had at Bryn Mawr was next awarded to **Anne A. Stewart**, who also left after only one year. She was a native of Nova Scotia with a SB degree from Dalhousie University (1886), who also studied at University College, London (1880–1882). After leaving Bryn Mawr, Stewart taught at the Mary E. Stevens' School in Philadelphia (1887–1893). She then studied at Newnham College at the University of Cambridge (1893–1895) before returning to the Stevens School (1896–1898).

Charlotte Scott's two ultimately successful nineteenth-century graduate students were different in many ways. **Ruth Gentry** (1862–1917) came from an Indiana farming family. She graduated from Indiana State University (then a "normal" school, meaning that its mission was to train prospective teachers), and she then held various teaching positions during 1880–1890, including a two-year stint at Stetson University from 1886 to 1888 (founded as Deland Academy and College) to save money to pay for college. She also attended the University of Michigan for one year before and two years after her brief tenure at Stetson, resulting in a PhB degree in 1890. She then accepted a fellowship to Bryn Mawr College for 1890–1891; the next year she won a two-year ACA Fellowship for study in Europe. However, she was unsuccessful in attempts to attend courses as an auditor in Berlin despite support by Lazarus Fuchs. She then wrote to Felix Klein in Göttingen asking if he would provide private lessons, only to be totally rebuffed once again. Frustrated with Germany, she ended up moving to Paris to study at the Sorbonne. Bryn Mawr rewarded her effort with yet another fellowship upon her return for the next year, and she completed all requirements for the college's first PhD in mathematics in June 1894. Several sources date that degree to 1896, an error now ascribed to the two-year delay in printing her dissertation.[103] AMS membership records confirm the earlier date; she joined the Society in 1894 and maintained membership for the rest of her life. The topic of Gentry's dissertation was her advisor's specialty, algebraic geometry. Titled "On the forms of plane quartic curves," its aim was to characterize curves of order 4 in the plane.

Upon receiving her degree, Ruth Gentry accepted an instructorship at Vassar as the first faculty member to hold a PhD. However, she resigned after eight years, likely due to ill health. In 1902 she became associate principal and head teacher at a private school for girls in Pittsburgh, where she remained for only two years. From that time on, she mostly lived in her Indiana hometown, where she died of breast cancer at age 55.

Charlotte Scott's second PhD student, **Ada Isabel Maddison** (1869–1950), was, like her, born in Great Britain. Isabel Maddison attended schools in Wales that prepared her to pass the matriculation exam for the University of London at age 16. Like Scott, Maddison attended Girton College (1889–1892) and was given permission to attend Arthur Cayley's lectures at Cambridge. Maddison and her classmate, Grace Chisholm Young, both passed the Tripos exam with first-class honors in the spring of 1892.

Isabel Maddison was awarded a fellowship to Bryn Mawr College for 1892–1893, no doubt due to the Cayley–Scott connection. She also won the first Mary E. Garrett Fellowship for study abroad the following year, but used it to attend lectures by Felix Klein and David Hilbert at Göttingen instead of England. She lectured twice in Klein's seminars on the foundations of analysis during her year abroad, when she also published a paper in the respected *Quarterly Journal of Pure and Applied Mathematics*. Upon her return to Bryn Mawr, she accepted a position as secretary to President Carey Thomas and completed two major projects. One was a translation for the *Bulletin of the AMS* of a Felix Klein lecture presented to the Royal Academy of Sciences of Göttingen while she was there.[104] The other, much more significant, was her dissertation, "On singular solutions of differential equations of the first order in two variables and the geometrical properties of certain invariants and covariants of their complete primitives," published in the British *Quarterly Journal of Mathematics*. This work reeks of Klein much more than Scott, and it seems possible that, as with Frank Cole at Harvard in 1886, Klein assigned the topic and gave his imprimatur on the results, whence the home institution awarded the PhD in 1896.

Maddison remained at Bryn Mawr for the rest of her career. Although she, like Sylvester and Scott before her, generally took summer vacations (which they would have called "holidays") in England, she remained in Bryn Mawr even beyond retirement in 1926. She held two positions during 1896–1910, one administrative (first, secretary, and then assistant to President Thomas) and the other academic (reader and then associate in mathematics). However, she dropped the mathematics position after that, becoming a dean in the college and assistant to the president.

Isabel Maddison remained connected with the American mathematical community over her career, maintaining membership in the AMS and attending annual meetings at Columbia University up to 1910. The *Bulletin* carried her announcement of an April 1922 meeting at Bryn Mawr "in honor of Professor C.A. Scott, on the completion of her thirty-seventh year as Head of the Department of Mathematics."[105] The principal speaker at the conference was Scott's longtime friend, the philosopher Alfred North Whitehead.

At the beginning of her career, Isabel Maddison published three book reviews for the *Bulletin*. The first contrasted two texts on elementary geometry,[106] while the second evaluated a textbook on analytic geometry.[107] The third review provided a short sketch of the contents of an advanced book by Rudolf Sturm on Jacob Steiner's 1866 lectures on synthetic geometry; her review supplied details for an audience not fluent in German.[108] During this time, Maddison used both of her European experiences in a one-paragraph note on the four-color problem that added an early source on the problem, due to Möbius in 1840, to a Heawood paper that had appeared in the *Quarterly Journal of Mathematics*.[109] Given this interest in geometry, Maddison offered advanced courses at Bryn Mawr titled "Analytic geometry of space" and "Analytic geometry of three dimensions."

Characteristics. This section distills information from the lives and careers of the ten American women who were awarded PhDs in mathematics before 1900 (not counting Christine Ladd Franklin) regarding family background, education, and work history. One unstated aspect which all ten possess is that they were Caucasian. The mention of one possible missed opportunity precedes a discussion of other common characteristics.

Kelly Miller serves as an example of a male who might have earned a PhD in mathematics despite racism. On the other side of the gender divide, **Susan Johnson McAfee** (1889–1974) was an "African American who might have earned a doctorate in mathematics if she had not encountered blatant racism."[110] Susan Johnson was the sixth of 11 children whose parents were slaves before the Emancipation Proclamation freed them. All 11 children in this impressive family attended college. There were two black colleges in their town of Marshall, TX. Susan Johnson graduated from the historically black Wiley College in 1912 with the intention of becoming a mathematics teacher, but the spelling portion of her Texas teachers' examination was conveniently "lost," and her father refused to pay $50 to "find" it. (Roughly $1260 in 2014.) This was not necessarily due to racism; apparently, corruption was common at the time, regardless of race.

Consequently, Susan Johnson was unable to teach mathematics, and institutional racism prevented her from pursuing higher mathematics. Shortly after graduating, she married the farmer and businessman Luther Ford McAfee, a graduate of the historically black Texas College in Tyler, which had been founded in 1894. She convinced her husband to move to Marshall, home to Wiley College (founded in 1873) and the historically black Bishop College (founded in 1881, moved to Dallas in 1961, but closed in 1988). All nine of their children attended Wiley College. Five graduated as mathematics majors, including their third son Cecil McAfee, who became a high-school teacher and principal. In turn, all five of his daughters graduated in mathematics, and four of them became mathematics teachers.

While this genealogy is impressive, what moved Patricia Kenschaft to state that Susan Johnson McAfee might have earned a PhD in mathematics? For a comparison, she looked to Marjorie Lee Browne, who became one of the first African American women to earn a PhD in mathematics from an American college (in 1949). Browne too graduated in mathematics from a historically black college, Howard, and taught in high school afterward. Then she enrolled at Michigan and earned her master's degree. Interestingly, Marjorie Browne was then appointed to the faculty at Wiley College, where she remained 1939–1947. She then accepted a fellowship back at Michigan, which enabled her to earn her doctorate just two years later. Kenschaft concluded, "Denied her rightful career, Susie Johnson McAfee became, as far as I know, the outstanding mathematical parent."[111] I would add "and grandparent" to that assertion.

Of the ten American women who were awarded PhDs in mathematics before 1900, seven were born in the US, with only Isabel Maddison hailing from abroad. Annie MacKinnon Fitch and Agnes Baxter Hill were born in Canada, but the MacKinnon family soon moved to the US.

Even this small sample size of ten reveals a variety of family backgrounds. Six of the ten women were born into working-class families; Winifred Edgerton Merrill and Ruth Gentry were raised on farms. Generally, fathers tended to have more education than mothers in those times. The first two Yale graduates, Charlotte Barnum and Elizabeth Dickerman, came from particularly well-educated families, both fathers being Yale graduates. Leona Peirce's father was a Colby graduate. Mary Winston Newson's father was a physician, and her mother taught French, art, and mathematics in high school.

All but one member of this group attended public schools before entering college; Winifred Edgerton Merrill was schooled at home by private tutors. Regarding undergraduate education, Isabel Maddison attended Girton in her native England and Agnes Baxter Hill attended Dalhousie University in Canada, but the other eight attended a mix of colleges in the US. Half attended Seven Sisters institutions: Elizabeth Dickerman and Leona Peirce at Smith, Winifred Edgerton Merrill at Wellesley, and Charlotte Barnum at Vassar. Ida Metcalf also attended a private school, Boston University. The remaining three attended midwestern land-grant universities: Annie MacKinnon Fitch at Kansas, Ruth Gentry at Michigan, and Mary Winston Newson at Wisconsin. Only two of the ten moved directly from undergraduate studies into a graduate program: Dickerman from Smith to Yale, and Maddison from Girton to Bryn Mawr. All the rest taught in high schools for different amounts of time before enrolling in graduate programs. Four pursued studies abroad, not including the Englishwoman Maddison, with Peirce spending a year at Newnham College in England in the midst of her Cornell program. MacKinnon Fitch and Gentry interrupted their graduate studies to take courses at Göttingen, while Winston Newson received her doctorate there. Overall, these ten women with PhDs in mathematics had a variety of educational experiences in terms of undergraduate and graduate programs, both in the US and abroad.

While American men who received PhDs in the latter part of the nineteenth century could find academic positions at desirable schools upon graduation, sometimes at PhD-granting institutions, women did not fare as well. Three accepted positions at Seven Sisters colleges: MacKinnon Fitch at Wellesley for five years; Ruth Gentry at Vassar for eight years; and Isabel Maddison at Bryn Mawr for ten years, though she was primarily an administrator under the president. During their careers, two of these women accepted one-year positions at other colleges, Barnum at Carleton, and Dickerman at Case Western. All ten taught in high school at some point in their career.

Undoubtedly, the biggest impediments to professional careers for women mathematicians after receiving PhD degrees were antinepotism policies that allowed only one member of a family to be on the faculty at a time. Such a regulation was particularly harmful to Mary Winston Newson, who had been professor of mathematics at Kansas State for three years before marrying Henry Newson in 1901 at age 30. Antinepotism rules prevented her from obtaining a position at the University of Kansas, where her husband was on the faculty, though she was permitted to teach during summer terms. How curious! What happened when Henry Newson died ten years later? Mary Winston Newson still could not join the Kansas faculty because her younger sister had just accepted a position in the English department. What a double-edged sword! (Incidentally, all six of her siblings were college graduates, a rare feat in those times.) Thus, Winston Newson moved to Washburn College and later to Eureka College, schools of lesser prominence from what one might expect for someone holding a Göttingen doctorate under the esteemed Felix Klein.

Marriage also dramatically affected the careers of the three other women. Even at a women's college like Wellesley, female faculty members were expected to remain single. Therefore, Annie MacKinnon Fitch was forced to resign her Wellesley position after five years. She never returned to academia after marrying Edward Fitch in 1901 at age 33. Similarly, after remaining at Cornell for a year editing her mentor's works, Agnes Baxter Hill got married at age 26 and never taught again. Winifred Edgerton Merrill's case was a little different. She taught high school for a year upon receiving her PhD before marrying in 1887 at age 25. She also taught at a private school for one year between then and 1904, when she returned to teaching, the first two years at a private school and then the next 22 at the private school she founded.

Among the women who remained unmarried, Ruth Gentry taught at Vassar for eight years before ill health changed her career trajectory. Isabel Maddison was primarily an administrator, but also a part-time teacher at Bryn Mawr. The other four seemed to have experienced satisfying and productive careers in various capacities: Leona Peirce managed the family business (until "retirement" at age 65, after which she taught for 28 years); Ida Metcalf was a civil servant, Charlotte Barnum a human computer and editor; and Elizabeth Dickerman a private-school teacher.

It seems that Susan Johnson McAfee shared many characteristics in common with these ten white women. She came from a working-class family, attended public school before college, and married a man versed in mathematics. She also lived to age 85.

The lifespans of all 11 exceptional women who earned PhDs in mathematics before 1900 (a criterion that includes Ladd Franklin but excludes Johnson McAfee) seems particularly notable. Even a glance at Table 7.9 reveals remarkable longevity, with

Table 7.9. Lifespans of the 11 American women
who earned PhDs in mathematics before 1900

Name	Age
Christine Ladd Franklin	82
Winifred Edgerton Merrill	88
Ida Metcalf	96
Annie MacKinnon Fitch	72
Ruth Gentry	55
Charlotte Barnum	73
Agnes Baxter Hill	46
Elizabeth Dickerman	92
Isabel Maddison	81
Mary Winston Newson	90
Leona Peirce	91

only Baxter Hill (pneumonia) and Gentry (breast cancer) not surviving into their 70s.
Overall this group's median age was 82, and their mean age was 78.7. (In 1900, life
expectancy for *women* in the US was 50.9 years.) This is just one other way in which
these women were truly *exceptional*!

Statistics

The period 1840–1880 saw the emergence of several statisticians, the founding of five
statistical societies, and the offering of the first courses in the field. In this initial stage,
statistics was limited to the collection and description of data. This section describes
the second stage, when a sixth statistical society was formed (though short lived). However, the main development that took place in the last two decades of the nineteenth
century was the emergence of statistics courses at some leading universities. Although
Johns Hopkins and the University of Chicago played important roles in this advance,
neither had the same dominating influence on statistics as on mathematics. In fact,
there was no equivalent of an American statistician for either J.J. Sylvester or E.H.

Table 7.10. Some institutions that began offering
statistics courses 1880–1900

Institution	Year	Academic Unit	Instructor
Columbia	1880	Graduate School of Political Science	Richmond Mayo-Smith
Pennsylvania	1883	Wharton School	Roland Falkner
MIT	1886	Nonprofessional studies	Davis Dewey
Johns Hopkins	1887	History and politics	Elgin Gould
Michigan	1887	Political economy	Henry Adams
Wellesley	1891	Political economy	Katharine Coman
Chicago	1892	Political economy	Isaac Hourwich
Cornell	1892	History and political science	Walter Wilcox, James Oliver
Harvard	1895	History and political science	John Cummings
Minnesota	1897	Political Science	William Folwell

Moore. One aspect that mathematics and statistics did share, however, was the decisive influence of Germany in the last quarter of the nineteenth century.

Over the course of the first 50 years of the American Statistical Association (ASA), 1839–1889, the ASA held numerous meetings featuring papers that dealt with medical and social issues, but none of the papers was mathematical in nature. Overall, the ASA remained a local organization headquartered in Boston. That changed in the 1890s, when the ASA expanded to include many members in New York City and Washington.

Slightly before that, a small group in Washington attempted to form a second national statistical society. Initially called the Census Analytical Association, its name was later changed to the National Statistical Association. The fledgling organization sponsored monthly meetings and, like the AMS also founded in 1888, initiated a journal six years later whose title reinforced its aspirations, *The National Statistical Journal*. The first issue described its aim:[112]

> The purpose of this Journal is to give the public the benefit of carefully prepared analyses upon National economic matters from the study of men who were known and recognized as high authority. ... The labor set forth in this work is—mining in the great mountains of Official Statistics, separating and grouping together figures, and finding facts of greatest value; exhibiting the true condition of our country's affairs and throwing some searching light upon its prospects, whether in financial or social problems.

Despite such lofty aims, only one issue of the official publication was ever published, and the organization itself folded within eight years of its founding. Consequently, of the six statistical organizations founded in the US in the nineteenth century, only the ASA survived into the twentieth.

A more notable development during 1880–1900 was the spread of statistics courses in 16 leading universities. Table 7.10 lists ten of these institutions, the first year the course was offered, its academic unit (a department, unless specified otherwise), and the instructor.[113] It is notable that no mathematics department offered a statistics course at this time.

Next, several leading statisticians who taught these courses are introduced. The online file "Web07-AmerStat" provides more details on their lives and careers.

Columbia introduced the course "Social science: Statistics, methods, and results" when it opened its new graduate school of political science in the fall of 1880. The course description reveals that it covered the statistics of population, economic statistics, moral statistics, and methods of statistical observations. The instructor was **Richmond Mayo-Smith** (1854–1901), an 1875 graduate of Amherst College who spent two years studying at the universities of Berlin and Heidelberg before joining the Columbia faculty in 1877. He remained at Columbia for the rest of his life as professor of political economy and social science, publishing numerous articles related to statistics as well as three notable textbooks that were used at various American schools. An article on early American statisticians concluded that Mayo-Smith "was the most eminent American scientific statistician and one of the outstanding ones of the world in the late nineteenth century."[114]

In 1881, the entrepreneur and industrialist Joseph Wharton established the world's first collegiate school of business at the University of Pennsylvania. His pioneering vision for his eponymous school was to produce graduates who would become "pillars of the state, whether in private or in public life."[115] In 1883 the Wharton School of Finance and Commerce offered a course in two parts entitled "Legislation and Administration. Statistics." By 1888 the title was changed to "Statistics. General Theory. Statistics of Population," indicating a course of study independent of political science. The instructor for the course was **Roland Post Falkner** (1866–1940), a Penn graduate (at age 19) who then studied in Germany for three years, resulting in his doctorate from Halle in 1888. Falkner then returned to Penn, where he became associate professor of statistics, making him the first American professor to have the word "statistics" in a professorial title.[116] During the 1890s, Falkner taught a course that by the end of the decade was titled "Statistical history, theory, and practice" using a text he had translated from German.[117]

Another person whose professional title included the word "statistics" was **Davis Rich Dewey** (1858–1942). Upon receiving his PhD in economics at Johns Hopkins in 1886, he was appointed instructor in history and political science at MIT. Davis Dewey remained at MIT for the rest of his career. Starting in 1888, he offered two statistics courses: "Statistics of the United States and graphic methods" and "Statistics of sociology."

Both Johns Hopkins and the University of Michigan began offering statistics courses in 1887. Uncharacteristically, the Johns Hopkins *Register* did not list a title, course number, description of contents, or name of the instructor, as were generally found in catalogs at that time. However, starting in the fall of 1887, the course "Principles of the science of statistics" was offered by **Henry Carter Adams** (1851–1921) at Michigan. He graduated from Grinnell College in his native Iowa in 1874. Two years later he was among the initial group of ten fellows in the first class at Johns Hopkins and received his PhD in 1878. Hopkins president Daniel Coit Gilman obtained funds to support Adams for an additional year of study at Oxford, Paris, Berlin, and Heidelberg. Upon returning to the US, Adams was appointed to the faculty at Cornell. He taught one semester at Cornell and one at Johns Hopkins from 1879 to 1886, when he was appointed at Michigan.

Wellesley College was instrumental in the emergence of women in American mathematics in the nineteenth century, and it played a singular role in statistics. According to the catalog for the fall 1891 semester, Wellesley offered the course "Statistical study of economic problems" in the department of political economy, but I do not know if it was given. However, it is known that the following year, the course was taught by professor of history and political economy, **Katharine Coman** (1857–1915). Coman graduated from the University of Michigan in 1880 and joined the faculty at Wellesley three years later as professor of history and political economy. Therefore, she was the first American woman to teach statistics. In fact, Wellesley was the only woman's college to offer a course in statistics before 1900. The description of Coman's offering for 1891–1892 reads:[118]

> The course is introduced by lectures on the principles of statistical research. Each member of the class undertakes the investigation of a particular problem, and reports the results of her inquiry to the class.

Four years later, the catalog description shows that the course came to rely on a variety of materials:[119]

Figure 7.10. Katharine Coman

The graphic method of presenting statistical results is emphasized. No single authority can be recommended. The United States Census, the Census of Massachusetts, the statistical reports of the Treasury Department, the reports of the National and State Labor Bureau furnish the statistical data.

This course description was repeated for the rest of the century, indicating the extent of the Wellesley influence.

In 1901, Katharine Coman was appointed the first chair of the college's department of economics and sociology, a position she held until illness from cancer forced her retirement in 1913. She is known today for pioneering research in sociology, particularly working and living conditions of immigrants, women, and the poor. She lived with Katharine Lee Bates, the chair of the English department at Wellesley. Bates is known as the author of the anthem "America the Beautiful."

Another colorful figure in the history of statistics was **Isaac A. Hourwich** (1860–1924), who was born in Russia. He studied medicine and mathematics before being banished to Siberia (1881–1886) for engagement in radical political movements. Three years later he fled Russia because of related activities. He lived briefly in Paris before immigrating to the US. Isaac Hourwich was a docent in statistics at the University of Chicago in its inaugural year (1892–1893) when he taught two courses in the department of political economy at Chicago, "Statistics" and "Advanced Statistics." However, he returned to New York after that one year. From his arrival in the US, Hourwich enrolled at Columbia, where he earned a PhD in economics in 1893.

Elgin Ralston Lovell Gould (1860–1915) became the first American professor to have the full title "professor of statistics" when he succeeded Isaac Hourwich at Chicago. This distinction is yet another jewel in the crown of the university that played such a critical role in the development of higher education in America. Gould continued to teach the two statistics courses that Hourwich had initiated, but renamed them "Statistics—history, theory, technique and analysis" and "Statistical investigation." Born in Ottawa, Ontario, he graduated from the University of Toronto in 1881 and then won a fellowship to Johns Hopkins for 1882–1884. He received his PhD in

1886. Before his appointment at Chicago, Gould gave occasional short courses on economic and social statistics (using graphical methods) at Johns Hopkins, but he left academics about 1897. Tragically, he was killed in a horseback riding accident in Canada in 1915.

Historian of science Theodore Porter wrote that, at Harvard:[120]

> Benjamin Peirce ... made routine use of error theory in his work, and even published a noteworthy contribution to it, a probabilistic criterion for the rejection of "doubtful observations," or outliers.

Porter added that

> [Peirce's son C.S. Peirce] devoted much of his career to the sciences of observation and measurement, ... which had grown highly sophisticated by the late nineteenth century.

Indeed, Porter devoted an entire section of his book, on the history of statistics during 1820–1900, to "Peirce's rejection of necessity."[121]

As noted in Table 7.10, Harvard offered its first course in statistics relatively late, in 1895. The course "The theory of statistics," was taught by John Cummings, an instructor in political economy in the economics division of the department of history and political science. In the fall of 1897 the title of the course was changed to "Statistics."

However, Cornell was the first American university to offer a statistics course with the involvement of a mathematician, department head James Oliver. During the academic year 1892–1893, Oliver taught the course "Statistics" in the department of history and political science. Its catalog description reinforces the fact that the course contained some mathematical fundamentals but was mainly concerned with applications: "Mathematical methods applied to treatment of certain economic and social questions."[122] Oliver, ill at the time, did not teach the course again and died in 1895. A course on mathematical statistics was not given at any American university for another 30 years.

Cornell offered a second statistics course in 1895. Starting in 1892, and continuing for many years, professor of social science and political economy Walter Willcox initiated a course later named "Social Statistics." The catalog description reads:[123]

> An elementary course in statistical methods and results with a survey of the statistics of the United States. Much practical work in investigation and tabulation will be done by the students.

These two courses at Cornell illustrate two distinct faces of statistics, one devoted to mathematical issues and the other concentrating on applications, in Willcox's case mainly demographics.

Walter Francis Willcox (1861–1964) is regarded as "one of America's most distinguished statisticians."[124] He enrolled at Columbia after graduating from Amherst in 1884 and earned a PhD in 1891. His dissertation was revised and published in 1897 as the book *The Divorce Problem: A Study in Statistics*. In the meantime, Willcox spent the academic year 1889–1890 at the University of Berlin. He joined the Cornell faculty in 1891 as an instructor and was promoted to full professor of economics and statistics ten years later. He remained in Ithaca for the rest of his long and distinguished life that extended from the Civil War to J.F. Kennedy's national goal of placing an American on the Moon. Willcox taught his statistics course at Cornell until 1899, when he took a three-year leave to serve as one of five chief statisticians for the 1900 US census, charged

with analyzing and interpreting demographic data. The culmination of this effort was
the monumental, 1100-page work *Supplementary Analysis and Derivative Tables*. At
the Census Bureau, he enacted a policy of hiring PhD-trained statisticians, seven of
whom later became presidents of the American Statistical Association. In the first part
of the twentieth century, Willcox played a central role in the apportionment of seats in
the US House of Representatives.

Despite this proliferation of statistics courses, it should be borne in mind that not
one was mathematical. The 1888 article "Statistics in colleges," by Carroll D. Wright,
head of the Massachusetts Bureau of Statistics of Labor and ASA president, expressed
the dominant view of the purpose of statistics at the time:[125]

> The gathering of original data in the most complete and accurate
> manner; the tabulation of the information gathered by the most ap-
> proved methods, and the presentation of the results in compact and
> easily understood tables.

This section ends with two notable developments from the final years of the nineteenth century. One was a statistical laboratory established at George Washington University (then called Columbian College) under the direction of Richmond Mayo-Smith
from 1897 until his death four years later. This occurred shortly before E.H. Moore
became an advocate for the laboratory approach to teaching mathematics at the University of Chicago.

The other important advance in the final decade of the nineteenth century was the
graduate course "The science of statistics," initiated by **William Watts Folwell** (1833–
1929) at the University of Minnesota in 1897–1898. Folwell was an 1857 graduate of
Hobart College in NY who became adjunct professor of mathematics there two years
later. However, after only one year he studied philology at the University of Berlin.
Folwell became chair of mathematics at Kenyon College in Ohio in 1869, but was
appointed the first president of the University of Minnesota later that year. William
Folwell resigned in 1884 but remained as professor of political science and librarian
after that.

Publishing abroad. The dominant theme throughout this chapter has been the
birth and childhood of an American mathematical research community. This specialized group was delivered at Johns Hopkins, but experienced emptiness when Sylvester
returned to England, causing some children to run away to study abroad. Meanwhile, a
successor to Johns Hopkins at Clark was virtually stillborn. Finally, a healthy program
emerged at Chicago which, in turn, inspired several other universities to change their
mission and elevate standards.

Although a true community of mathematical scholars had emerged by 1900, evidence presented in this chapter has been from an internal, that is, domestic, viewpoint
only. An investigation of international submissions in British mathematics journals by
Americans during 1800–1900 provides an external perspective.[126] That investigation
found, as expected, that no American published a paper from 1800 to 1850, but someone published a paper in the 1850s and someone published two papers in the 1860s.
These numbers are recorded in the first two rows of Table 7.11, which lists by decade
the numbers of American authors and the total number of their contributions. This
table clearly demonstrates the remarkable growth that took place over the last three
decades of the nineteenth century and the exponential increase in the very last decade.

Table 7.11. American contributors to British journals by decade

Period	Contributors	Contributions
1850–1859	1	1
1860–1869	1	2
1870–1879	10	31
1880–1889	11	17
1890–1900	28	80

However, Table 7.11 conceals two relevant facts. First, more American mathematicians published papers in British journals in the 1870s and 1880s than any other foreign nationality (though admittedly, France, Germany, and Italy had their own journals). Second, during the pivotal period 1890–1900, the US outnumbered every other foreign country in contributors *and* contributions by such a large measure that the same result holds for the entire period 1800–1900.

A peek ahead

When J.J. Sylvester came to Johns Hopkins in 1876, there was no such thing as a graduate program in mathematics in America. There were no seminars, no reading rooms, no mathematics libraries, nor was there a research journal in mathematics. Yet by the time Leona Peirce received a Yale PhD in 1899, she had already been enrolled in a full-fledged graduate program at Cornell and had studied under Sylvester's former assistant William Story at Clark. By 1900, then, several institutions in the US offered legitimate programs in mathematics that were organized essentially as they are today. Whereas American students had gone to England or France primarily for cultural or linguistic reasons during the first eight decades of the nineteenth century, and a new wave set out for Leipzig, Göttingen, and Berlin over the last two decades, by the dawn of the twentieth century it was no longer necessary to go abroad at all.

The last decade of the nineteenth century witnessed the founding of the University of Chicago, the expansion of the American Mathematical Society into a national organization, and the establishment of another research journal, the *Transactions of the AMS*. However, it was not until the first decade of the twentieth century that these developments bore fruit. Chicago served as a model for faculty and for curricula at several other universities, the AMS spawned additional sections whose meetings served as gatherings for mathematical researchers throughout the US, and the *Transactions* emerged as the primary outlet for the highest-level original investigations. Part IV (in Volume 2) elucidates these developments.

Transition 1900:
Hilbert's American Colony

The major cluster point for American students after Felix Klein supervised the dissertation of his final American student, Edward Kasner, in 1899, was **David Hilbert** (1862–1943). Hilbert biographer **Constance Reid** (1918–2010) asserted:[1]

> The Americans at the University were sufficient in number and wealth to have their own letterhead: The American Colony of Göttingen.

Therefore, I refer to the group of Americans that Hilbert supervised between 1899 and 1910 as the "Hilbert Colony." Overall, Hilbert was a doctorate-producing machine who supervised almost two students a year over the 37-year period 1898–1934, including Felix and Serge Bernstein, Alfred Haar, Georg Hamel, Erich Hecke, Erhard Schmidt, Hugo Steinhaus, as well as four who were forced to flee their homeland, and later contributed vitally to American mathematics from the 1930s onward—Richard Courant, Max Dehn, Otto Neugebauer, and Hermann Weyl.

This section chronicles the lives and careers of the 13 American mathematicians in the Hilbert Colony. (The online file "Web-Tr1900" provides more biographical details.) During 1900–1910, enrollment in upper-level courses in Germany increased dramatically from what it had been under Klein, thereby lessening the need for foreigners to fill classrooms and heightening competition for the newcomers. E.T. Bell observed:[2]

> Hilbert was still on his way to the top and absorbed in his own researches. Moreover, he seems to have been somewhat unapproachable, especially to Americans.

Yet the Hilbert Colony proved to be hardy, resilient, and, ultimately, successful. Moreover, it included his first woman student of any nationality.

Pre-1900. The only American to complete requirements for a doctorate under Hilbert before 1900 was **Legh Wilber Reid** (1867–1961). For inexplicable reasons, he is sometimes omitted from listings of Hilbert's students, but the authoritative biography *Hilbert* by Constance Reid (no relation to Legh) asserts rather plainly:[3]

> When Legh Reid, one of his former American students, wrote a book on the subject, Hilbert endorsed it with enthusiasm.

Indeed, Legh Reid is an authentic American (born in Alexandria, VA) who obtained a doctorate under Hilbert in 1899. Earlier he had received bachelor's degrees from Virginia Military Institute (1887) and Johns Hopkins University (1889). He worked as a human computer for the US Bureau of the Census and the US Coast and

Geodetic Survey (1889–1893). Then Reid was appointed instructor at Princeton, where he obtained a master's degree in 1896. At that point, he sailed abroad to study in Göttingen, culminating in his doctorate in 1899 for the dissertation *"Tafel der Klassenanzahlen für kubische Zahlkörper."* It was published in English in 1901 in the *American Journal* as "A table of class numbers for cubic number fields." Its major results lean heavily on a technical lemma due to Hilbert's close friend and later colleague, Hermann Minkowski.

Legh Reid spent the rest of his career at Haverford College (all-male until 1980), from his appointment as instructor in 1900 to his retirement as professor emeritus in 1934. His major contribution to mathematics was his book *The Elements of the Theory of Algebraic Numbers*, published by Macmillan Company in 1910 and containing an introduction by David Hilbert.

Until recently, Anne Bosworth was unknown, even though she was a student of Hilbert. The account here has been pieced together from material discovered by former University of Rhode Island (**URI**) archivist Sarina Wyant and a graduate student, as well as from additional searches.

Anne Lucy Bosworth Focke (1868–1907) was born in Rhode Island and graduated from Wellesley College in 1890. Two other members of that class also earned PhDs in mathematics—Grace Andrews from Columbia in 1901 and Clara Bacon from Johns Hopkins in 1911. Upon graduation, Bosworth taught in high school for two years. In 1892, she was appointed the first professor of mathematics and physics at the fledgling URI. As the sole member of the department, she was charged with developing the curriculum, initiating a book collection, and teaching courses in algebra, geometry, calculus, and electricity.

Figure 7.11. Anne Lucy Bosworth Focke

Bosworth was one of the many college teachers who attended courses during summer quarters at the University of Chicago. In April 1898 she was granted a leave of absence from URI to study abroad, and she returned home with a doctorate from one of the world's leading figures. The genesis of the awarding of this degree is rather curious because Bosworth was caught completely unaware. When she sailed, her only goal was to attend courses at Göttingen. That summer, she joined two Bryn Mawr graduates in a series of lectures on mechanics by Felix Klein. In the fall, she attended David Hilbert's lectures on Euclidean geometry. The following spring, he summoned her to tea, and asked when she planned to take her doctoral exams. She replied that she had not even given thought to a dissertation topic, let alone a degree. In characteristic Hilbert manner, he blurted out, "But your dissertation is finished!" Apparently, Hilbert judged her solution to a special exercise he had posed in the class to be worthy of a thesis. Thereupon, Anne Bosworth dropped plans to travel throughout Europe that summer, in favor of writing the dissertation and taking her oral exam. When the degree was formally awarded in 1900, she became David Hilbert's first female doctoral student based on the dissertation *"Begründung einer vom Parallelenaziome unabhängigen Streckenrechnung,"* which George Halsted described as:[4]

> a beautiful piece of non-Euclidean geometry [that] is, so far as I know,
> the first feminine contribution to our fascinating subject.

Bosworth joined the AMS in February 1900, six months after returning from Germany, where she had been escorted by her mother the whole time. While there, she met **Theodore Moses Focke** (1871–1949), who had been a tutor in physics and chemistry at Oberlin College (1893–1896) before traveling abroad to pursue a graduate degree in physics at Göttingen. He received his doctorate in 1898, but then traveled throughout Europe for a year before returning to Göttingen, when the two met. As required of her leave, Bosworth returned to URI in the fall of 1899, but left the college after her marriage to Focke two years later. The first six years of marriage reflected the culture of the time, and were very productive for both newlyweds. But tragedy struck in 1907, when Anne Bosworth Focke contracted pneumonia and died at age 38, leaving behind children aged five, three, and one. She never did revise her dissertation for publication in the US.

Why has Anne Bosworth remained virtually unknown to historians of mathematics? After all, being the first woman to earn a doctorate for a dissertation directed by someone as internationally recognized as David Hilbert is a major accomplishment in any period, let alone one in which women were prevented from enrolling in most graduate programs. When searching for biographical material on scientists from this era, a generally reliable volume is *American Men of Science*, which first appeared in 1906. But it contains no entry under "Bosworth," and not just because of the second word in the title of this volume. Indeed, she can be found there. At that time, entries for married women were subsumed under the last name of her husband, hence her listing reads, "Focke, Mrs. Theo. M. (Anne Bosworth)." This might explain why she is missing from all histories of women mathematicians up to the groundbreaking book by Judy Green and Jeanne LaDuke in 2009, whose genealogical inquiry sparked the URI archival search.

Rather than introduce the remaining 11 members of the twentieth-century Hilbert Colony in chronological order, I present them according to rankings published in the first three volumes of *American Men of Science*. This valuable series provides vignettes of the careers of American scientists who were living at that time. The first three editions appeared in 1906, 1910, and 1921. In each edition, scientists generally regarded as the most eminent in the field were asked to rank others in that field, and each of the top 1000 was listed with a star beside the name. Eighty mathematicians were starred, and the numbers in the *American Men of Science* column in Table 7.12 designate the edition in which they were listed (if at all): 1 for 1906, 2 for 1910, and 3 for 1921. Earning a star on this list became a badge of honor and, more practically, a bargaining chip for a higher salary.

Table 7.12. Hilbert's American students

Name	Year	AMoS	Name	Year	AMoS
Legh Wilber Reid	1899		Wilhelmus David Westfall	1905	
Anne Lucy Bosworth	1900		David Clinton Gillespie	1906	
Edgar Jerome Townsend	1900	3	William De Weese Cairns	1907	
Earle Raymond Hedrick	1901	1, 2, 3	Arthur Robert Crathorne	1907	
Charles Albert Noble	1901		Charles Haseman	1907	
Oliver Dimon Kellogg	1902	2, 3	Wallie Hurwitz	1910	3
Charles Max Mason	1903	2, 3			

Starred entries. Five members of the Hilbert Colony were starred in the first three volumes, with **Earle Raymond Hedrick** (1876–1943) the only one listed in all three. Born in Indiana, Hedrick received an AB degree at Michigan in 1896. After teaching high school for a year, he enrolled in the graduate program at Harvard, where he excelled under Maxime Bôcher and William Osgood for two years. Hedrick was then awarded a Parker Fellowship for study abroad, which he used to go directly to Göttingen to study under Felix Klein and David Hilbert. Hedrick completed his dissertation on differential equations, *"Über den analytischen Charakter der Lösungen von Differentialgleichungen,"* in two years. Harvard then extended his scholarship to cover a third year, which he used for additional studies in Paris with Emile Picard, Édouard Goursat, and Jacques Hadamard, among others.

Figure 7.12. Earle Raymond Hedrick

Upon returning to the US, Earle Hedrick accepted a position at the Sheffield Scientific School at Yale but left in 1903 to become professor and head of the department of mathematics at the University of Missouri. He remained in this position until 1924, except for six months as director of the mathematical educational corps with the American expeditionary force in France during World War I. At Missouri, he helped found the Southwestern Section of the AMS in 1906. He also translated (with Otto Dunkel) Goursat's famous *Cours d'Analyse* into English, a book that was "widely used thereafter in all college circles."[5] In 1924, Hedrick left Missouri to become professor and department head at the University of California, Los Angeles (UCLA). Five years later he was elected to a two-year term as president of the AMS.

A signal change in Hedrick's life occurred in 1937, when he became one of two provosts and vice presidents for the entire University of California system. He retired five years later, at age 65, but did not abandon mathematics. Instead, he returned to his earlier interest in applied mathematics, publishing one paper on Hooke's law and another on the transmission of heat in boilers. Because of this experience, Hedrick was offered the opportunity to develop the Program of Advanced Instruction and Research in Mechanics at Brown University, where he became Visiting Professor and helped inaugurate the *Quarterly of Applied Mathematics*. However, Hedrick became ill shortly after arriving in Providence, and died the same year the *Quarterly* was launched.

Closely aligned with Earle Hedrick is **Oliver Dimon Kellogg** (1878–1932), who was strongly influenced by Henry Fine to study mathematics while an undergraduate at Princeton. Apparently, Fine's lectures stoked mathematical flames in Kellogg so strongly that, upon his 1899 graduation, he remained on campus another year to get a master's degree. That background was sufficient for a successful program at Göttingen under the direction of David Hilbert, with funds supplied by a John S. Kennedy Fellowship. Kellogg received his doctorate in January 1903 for a dissertation on integral equations and a particular case of the Dirichlet problem. By the time the degree was officially awarded, he had already returned to Princeton as an instructor. Over the next few years, he published two papers that corrected flaws he had found in his dissertation, "*Zur Theorie der Integralgleichungen und des Dirichlet'schen Prinzips.*"

Earle Hedrick recruited Oliver Kellogg to Missouri in 1905. I do not know why Kellogg departed Princeton at that time, because Dean Henry Fine had just begun the historic upgrading of the Princeton faculty with the addition of preceptors Oswald Veblen and Gilbert Bliss. In fact, Kellogg and Bliss essentially traded places. At Missouri, Kellogg prospered despite heavy teaching loads and administrative duties in what was then the southwest region of the country, initially publishing impressive papers on potential theory, three in 1908 alone. During this time, he also wrote a series of articles on the teaching of mathematics, and then in 1909 teamed up with Hedrick to write the textbook *Applications of Calculus to Mechanics*. But it was his 1912 *Transactions* paper "Harmonic functions and Green's integral" that brought him acclaim for what is called today the "Kellogg theorem" on harmonic analysis and Green's functions.

Like many mathematicians of his generation, Kellogg's career was disrupted by war service when he was assigned to work on the design of devices to detect submarines for the US Coast Guard Academy. In the fall of 1919, Kellogg was appointed to a permanent position at Harvard caused by the sudden death of Maxime Bôcher just before the start of the academic year 1918–1919. In 1926, Kellogg produced his only doctoral student, Arthur H. Copeland, with the dissertation "Studies on the gyroscope." Copeland then spent a distinguished career at the University of Michigan. Just three years earlier, Kellogg published a related paper in the *Transactions* on this very topic, studying the gyroscope as an application of properties of spherical curves.

Perhaps Kellogg's most important mathematical achievement, however, was the 1922 joint paper with George Birkhoff, "Invariant points in function space," which contains the celebrated Birkhoff–Kellogg theorem, a generalization of the famous Brouwer fixed point theorem. Birkhoff and Kellogg were working at a level unprecedented in the US up to that time. Kellogg's name has come down to us in another hyphenated result, the Kellogg–Evans lemma, which he proved and included for the first time in his classic 1929 text, *Foundations of Potential Theory*. An analysis of the book concluded:[6]

> Accessible to both advanced undergraduates and beginning graduate
> students, it was noteworthy for its rigour and felicitous style. While
> not specifically mentioned, many of the proofs in the volume—even
> of well-known results—are original and due to Kellogg himself.

Four years later, Griffith Evans proved the lemma in greater generality, hence the hyphenated form of its name. At the time of Kellogg's death in 1932, he was working on an advanced volume to supplement his already famous 1929 book.

Overall, Earle Hedrick and Oliver Kellogg are known for different types of contributions. For Hedrick, it was administrative leadership, especially at UCLA, but also with the AMS and MAA, as well as editorial positions with journals in both organizations. For Kellogg, however, it was research, either with published papers in the most important American journals of the day or with influential books.

Arguably the most accomplished American scientist in the Hilbert Colony was **Charles Max Mason** (1877–1961). Always called by his middle name, he was not even aware that it was *not* his first name until he saw it on his college diploma. During the year 1898–1899, Mason taught high school, and modern historian Patti W. Hunter summed up his versatility by noting that he also "coached the track team, led the school orchestra, and trained the debating team."[7] Mason had just graduated from the University of Wisconsin, where he was a champion high-jumper, but also sailed, canoed, skated, and played the violin. Yet mathematics became his passion, and, after one semester of what he later referred to as "so-called graduate work"[8] at Wisconsin, he went to Germany in 1900 to study with Hilbert at Göttingen, even though he was seemingly ill prepared for such study. By dint of hard work, Mason excelled, and Hilbert soon gave him a problem to solve. The good news is that the aspiring mathematician solved the problem in short order; the bad news is that his elegant solution required only two pages. While the length of dissertations in mathematics might be shorter than those in most other subjects, none is *that* short, so Hilbert handed Mason a second problem, one whose solution required several months of intense investigation—abetted by a productive dream one night—that led to the considerably expanded 1903 dissertation on differential equations, *"Randwertaufgaben bei gewöhnlichen Differentialgleichungen."* This work was so substantial and impressive that Mason graduated with highest distinction in a class that included the well-known Japanese mathematician Teiji Takagi (1875–1960).

Max Mason set about an academic career as soon as he returned from Europe, teaching for one year at MIT followed by four at Yale. He spent the next 17 years (1908–1925) at Wisconsin, where George Birkhoff was a colleague the first year and Arnold Dresden the entire time. He was especially productive between 1904 and 1910, publishing eight papers in pure mathematics. However, Mason's primary interest changed to physics, and he never wrote another paper in pure mathematics. Nonetheless, he was one of three speakers invited to deliver the fifth Colloquium at the 1906 summer meeting at Yale, along with E.H. Moore and Ernest Wilczynski. AMS historian Raymond Archibald stated that this colloquium:[9]

> ...set a high standard of excellence. Through the kind offices of Prof.
> Pierpont, Yale U. assumed the responsibility for the publication of
> these lectures. The fact that Moore was such a distinguished alumnus
> of Yale was doubtless a determining factor in effecting the arrangement.

Figure 7.13. Charles Max Mason

The series of three sets of lectures, including Mason's "Selected topics in the theory of boundary value problems of differential equations," was published in 1910.

Max Mason took a leave of absence from Wisconsin for two years during World War I to manage a large research team at the National Research Council working on submarine warfare. During 1917–1919, he invented a submarine detection device known as the "M-V tube" (for Multiple-Variable), which "allowed the crew to determine the location of a submarine or other ship producing the noise."[10] His experience of successfully administering a group of scientists led the University of Chicago to offer him the presidency in 1925; he thereby became the first president who was *not* one of the original faculty members from 1892. The entire Chicago faculty, not just the mathematics department, was delighted with his performance, and expected him to enjoy a long tenure, but when Mason's first wife died in 1928, he left academia for good, moving to New York as director of the Natural Sciences Division of the Rockefeller Foundation. This position too proved to be short lived, because he was promoted to president of the Foundation the next year. Mason held this position for a seven-year period that turned out to be critical for funding in mathematics. He left the Rockefeller Foundation in 1936 to direct a team that was constructing the Palomar Observatory.

Edgar Jerome Townsend (1864–1955) received a bachelor's degree at tiny Albion College in MI in 1890. He then enrolled at the University of Michigan for a year, obtaining a master's degree and accepting an instructorship at the Chicago Manual Training School. While attending the Chicago Congress in August 1893, he was offered a position at the University of Illinois, where he spent the rest of his career until retiring in 1929. Townsend was one of 28 mathematicians who signed the circular announcing a conference to form the Chicago Section of the AMS in 1896 and was one of 17 members who attended that gathering. He took a two-year leave of absence (1898–1900) to study under Hilbert at Göttingen, where he obtained his doctorate in 1900 for the dissertation *"Über den Begriff und die Anwendung des Doppellimes."* At Illinois, he directed two PhD dissertations and also served as dean of the college of science.

One of Edgar Townsend's greatest accomplishments was the collection of models he assembled after returning home from Göttingen in 1900. Even before going abroad, he had been exposed to the impressive display of models made by the famous Brill firm and exhibited at the Chicago Congress, so in 1911, as dean, he hired the mathematician

Arnold Emch to expand the collection, resulting in the largest such exhibition on public display in the country.

The influence of Hilbert was reflected in Townsend's 1902 translation of Hilbert's classic *Grundlagen der Geometrie* (1899). The resulting book, *The Foundations of Geometry*, is officially titled an "authorized translation," yet caustic reviewer George Bruce Halsted was definitely *not* a fan:[11]

> Readers of the *American Mathematical Monthly* may consult a technical review of this translation in *Science*, … where in the interest of merest justice are pointed out some few among the blemishes in what Professor Townsend puts forth as a translation of Hilbert's beautiful "Festschrift."

Townsend wrote three notable textbooks of his own: *First Course in Calculus* (1909), *Functions of a Complex Variable* (1917), and *Functions of Real Variables* (1929). All three were reviewed favorably.[12]

The final Hilbert American graduate starred in *American Man of Science*, **Wallie Abraham Hurwitz** (1886–1958), was the son of Jewish-German parents who immigrated to Missouri about 1870. Hurwitz graduated from the University of Missouri in 1906 with two bachelor's degrees, and with a master's degree based on the thesis "Definition of improper groups by means of axioms." Moreover, in his senior year he was an assistant in mathematics and, in a course taught by Gilbert Bliss, "was an extraordinarily able and precocious student."[13] Upon graduation, Wallie Hurwitz matriculated in the graduate program at Harvard (1906–1909) before traveling abroad on a Sheldon Fellowship to study with David Hilbert. He earned a doctorate in one year for a dissertation on linear partial differential equations of the first order. Upon receiving the degree, he joined the faculty at Cornell, where he remained for the rest of his career (except for one year 1941–1942 spent at the Institute for Advanced Study).

Figure 7.14. Wallie Abraham Hurwitz

Throughout his career, Hurwitz was active with the AMS, beginning as cooperating editor of the *Transactions* (1915–1926). He was also a member of the Council (1919–1921), and served on the Committee of Publication for the *Bulletin* (1921–1923). In 1921 when the AMS first began the practice of inviting speakers to deliver lectures at meetings, he was the first one so chosen; he also delivered an invited lecture 11 years later, both given at meetings held in New York City.[14] Moreover, he served as an AMS consultant on cryptanalysis for the War Preparedness Board before the outbreak of World War II. Hurwitz was a charter member of the MAA, and edited the *American Mathematical Monthly* column "Questions and Discussions" from 1919 through 1921. Curiously, he gave up research at age 52, although he continued to teach advanced courses and attend meetings after that. Two colleagues attempted unsuccessfully to engage him in collaborative efforts, but he resisted. As Mark Kac commented, "A shame.

He had an excellent mind and was very knowledgeable."[15] Hurwitz was quite successful with the stock market, and left his sizable estate to three universities—Missouri, Harvard, and Cornell.

Nonstarred. None of the remaining Hilbert figures was starred in *American Men of Science*, yet all had successful careers. **Wilhelmus David Allen Westfall** (1879–1951) graduated from Yale in 1901 and remained on campus for the next two years. He then went to Göttingen and completed his doctorate under David Hilbert in 1905 for a dissertation on integral equations. David Westfall then accepted a position at Missouri, joining the chair Earl Hedrick and Oliver Kellogg in forming *the* Hilbert Colony outpost in the US. Westfall remained at Missouri until retiring in 1949.

Charles Albert Noble (1867–1962) was born and raised in northern California. He was an 1889 graduate of Berkeley, where Irving Stringham imbued him with the primacy of research. Noble then taught in high schools for four years, before heading to Germany to study with Felix Klein and David Hilbert at Göttingen for another four years (1893–1896). A memorial article written upon his death some 66 years later records:[16]

> Like other American mathematicians of the period, he acquired an idiomatic knowledge of German, and a happy, intimate knowledge of German student life.

His fellow students in the English-speaking Hilbert Colony selected him as "patriarch,"[17] following in the footsteps of Earle Hedrick and Max Mason.

Although Charles Noble returned to California in 1896 without a degree, he published his first paper that year (in German!) on a boundary-value problem, and then spent the next year continuing his studies on a fellowship at Berkeley. By that time, the Berkeley faculty included another active researcher, the former Klein student Mellen Haskell. The following year, 1897, Noble was hired by department chair Irving Stringham, and he remained there until retiring in 1937.[18] During his 40-year tenure, Noble viewed the evolution of Berkeley from a small institution to one of national prominence. Fiscal problems had plagued the university up to 1899, when Benjamin Wheeler assumed the presidency. Over the next 20 years the classical scholar stabilized finances and attracted library and scholarship funds, research grants, and a distinguished faculty to Berkeley, so the campus grew in size and distinction as its reputation improved rapidly. The largess of benefactor Phoebe A. Hearst helped construct several elegant and stately structures, including Gilman Hall.[19]

Charles Noble became one of Berkeley's distinguished faculty members, returning to Göttingen in 1901 only to submit his dissertation on the calculus of variations, "*Eine neus Methode in der Variationsrechnung*." His enduring fame rests with his translation (with Earle Hedrick) of Felix Klein's popular book *Elementarmathematik vom hoheren Standpunkte aus*, based on lectures Klein gave at Göttingen in 1908–1909. The two volumes were published in 1932 and 1939 as *Elementary Mathematics from an Advanced Standpoint*. Raymond Archibald wrote that the two volumes "rendered real service to students and college teachers of mathematics in this country."[20] A review of the second volume by David Curtiss niggled about material not included by the translators, and wondered if a third volume was in the planning."[21] (If it was, it never appeared.)

Charles Noble was 70 years old when he retired in 1937, yet he returned to the Berkeley campus during World War II to donate his services to the university in place

Figure 7.15. William De Weese Cairns

of the many mathematicians who were involved with war work. He was 94 when he died, having been a charter member of the MAA since its founding in 1915.

Unlike Hilbert's other male colonists, **William De Weese Cairns** (1871–1955) spent his teaching career at a small college, so he made his mark more for service to the MAA than for original contributions, although he did publish a dozen papers in various journals. Will Cairns graduated from Ohio Wesleyan College in 1892, after which he taught physics in high school until 1896, when he enrolled at Harvard, earning a second bachelor's degree and a master's degree, each in one year. Cairns then taught in high school for another year before accepting a position as instructor and surveyor at Oberlin College. In 1905, he took a two-year study leave in Göttingen, earning his doctorate in 1907 under Hilbert for a dissertation on integral equations and the calculus of variations. Like Noble, Cairns was a charter member of the MAA. He was the MAA's first secretary-treasurer, and held the position from 1916 to 1942. In 1943 Earle Hedrick summarized Cairns's vital role within the MAA:[22]

> To these two men [H.E. Slaught and Cairns] far more than to any others the Association owes its success and its present strong position. … For many decades to come [Cairns's] high achievement, his constructive leadership, and his great devotion will be remembered, and will serve as an inspiration to younger men who desire to advance the cause of mathematics in America.

Will Cairns was elected president of the MAA for a two-year term in 1943 at age 72. Another MAA activist, Lester Ford, added a clear reference to the battles taking place across the Atlantic and Pacific Oceans at the time:[23]

> It is a fortunate circumstance for us that he will guide the destinies of our organization during the difficult days that lie ahead,

At the end of his term in 1944, the MAA bestowed upon him the unprecedented title of Honorary President for Life. Cairns had served as professor and chair of the Oberlin department from 1920 to 1939, yet retirement did not equal cessation of activity. First, he taught a course in meteorology at the University of New Mexico during World War II, and at age 77 he taught at Caltech.

David Clinton Gillespie (1877–1935) graduated from the University of Virginia in 1900, spent a year at Johns Hopkins, and then went abroad. He received his doctorate in 1906 for a dissertation on the calculus of variations. Upon his return to the US, David Gillespie joined the faculty at Cornell and remained there for the rest of his life, serving as department head 1932–1935. He published several papers, mostly in analysis, and served as associate editor of the *Annals* (1927–1929). Gillespie's attitude toward students bears repeating:[24]

> The careless or inattentive student frequently feared his scorn; but the earnest student found him a kindly and sympathetic friend. He was amazingly tolerant of ignorance, even of stupidity; but impatient with laziness or pretence [*sic*].

The two remaining figures are related to female mathematicians, a sure sign of the surge in the ranks of women who pursued graduate degrees in mathematics after having been prevented from such pursuits just a decade earlier. I regard **Arthur Robert Crathorne** (1873–1946) as an American even though he was born in England, because he came to the US as a young boy and received all of his education in this country. Arthur Crathorne was associated with the University of Illinois for almost his entire career, including his 1898 BS degree. Upon graduation, he taught at the University of Maine for two years, before moving to Wisconsin (1900–1904). He earned another bachelor's degree, from the University of Chicago, in 1902. After leaving Wisconsin, Crathorne traveled to Göttingen to work under Hilbert, completing his dissertation on the calculus of variations in 1907. The 59-page work was published in German the same year. Upon his return to the US, Crathorne joined the faculty at the University of Illinois, where he and **Henry Lewis Rietz** (1875–1943) formed a commercially successful authoring team of elementary textbooks. Crathorne also directed eight doctoral dissertations, the last two at age 68.

Charlotte Elvira Pengra-Crathorne (1875–1916) received her AB degree in mathematics at the University of Wisconsin in 1897. Upon graduation, Charlotte Pengra, like most highly educated women at that time, taught in high school for two years. She was awarded a fellowship at the Madison campus in 1899, and earned her PhD two years later for the dissertation on the conformal representation of plane curves, *"On functions connected with special Riemann surfaces, in particular those for which P equals 3, 4, and 5,"* directed by Linnaeus Dowling. Pengra's dissertation was apparently never published, yet she was the first woman to obtain a doctorate in mathematics at Wisconsin, the third overall. She probably met Arthur Crathorne when their paths intersected at Wisconsin during the year 1900–1901. After the two married in June 1904, they traveled to Göttingen. The young wife also attended classes, but her pursuit of a higher degree ended with the birth of the first child in that quaint, German, university town in November 1906. (Two others were born in the US.) But tragedy struck in 1916, when Charlotte Pengra-Crathorne contracted breast cancer and died at age 40.

Charles Haseman (1880–1931) is another Hilbert student with a link to a woman mathematician. Haseman graduated from the University of Indiana with an AB in 1903, joined the faculty two years later, and earned an AM the next year. He spent 1906–1907 on leave in Göttingen, where he wrote a 46-page dissertation on integral equations directed by Hilbert. Haseman returned to Indiana after that, but remained only two years before accepting a position at the University of Nevada at Reno, where he remained 1909–1930. He died the next year after a year's illness.

Haseman's family boasted impressive accomplishments. All nine children from this farm family were college educated, and five earned doctorates. The seventh of the nine was **Mary Gertrude Haseman** (1889–1979), who received an AB degree *cum laude* from the University of Indiana in 1910. Then she taught for a year at a two-year college before winning a fellowship to Bryn Mawr College, where she studied under department head Charlotte Angas Scott and James R. Conner from 1911 until 1915. The next year, Gertrude Haseman left Bryn Mawr without a degree to attend lectures by Frank Morley, the head of the department at Johns Hopkins since 1900. Haseman taught at a private school in Baltimore (1915–1917) while writing her dissertation. She completed that work, passed her final exam in 1916, and received her PhD at Bryn Mawr in 1917. The dissertation, "On knots, with a census of the amphericheirals with twelve crossings," has garnered acclaim as late as 1999.[25] After expressing her indebtedness to Scott and Conner for their helpful criticisms, Haseman wrote:[26]

> I am especially glad to have this opportunity of expressing to Professor Scott my sincere gratitude for her valuable help and unfailing encouragement during the writing of this dissertation as well as throughout my graduate course.

She presented her work at a meeting of the Royal Society of Edinburgh in June 1917 and an extension of it in November 1918. Both presentations were published in the *Transactions of the Royal Society of Edinburgh.*[27]

Gertrude Haseman taught at Harcum School, a women's junior college, in 1917–1918, and then in high school for another year back home in Linton, Indiana. She accepted an instructorship at the University of Illinois in 1920 and remained a colleague of Hilbert Colonists Edgar Townsend and Arthur Crathorne until late October 1927, when she resigned suddenly to become the first mathematics professor at a junior college (now the University of Bridgeport, CT). Yet she remained there only until the fall of 1928, when she accepted a professorship (and advisor to women) at Hartwick College in Oneonta, NY. Haseman also left that position after only one year, reputedly over a dispute with college administrators. She then moved to Columbia, MO, where her brother Leonard was head of the department of entomology at the University of Missouri. She returned to Linton in 1936 and lived there for the rest of her life, apparently without employment. Gertrude Haseman died at age 90. She is remembered today as one of the early contributors to knot theory and is listed prominently on a website devoted to the history of the subject.[28] Although her brother Charles did not attain such acclaim, the two formed the first brother-sister team to earn PhDs in mathematics in the US.

Summary. Table 7.13 summarizes the 13 members of the Hilbert Colony according to their undergraduate institution and year the bachelor's degree was awarded (columns 2 and 3), the institution attended immediately before departing for Göttingen (column 4), and the university generally associated with the person (column 5). This listing highlights the attraction of Hilbert for students from the Midwest, with only Noble coming from the West Coast, and Kellogg and Westfall from the East. Kellogg is the only person on the list from a "Big 3 university" (Chicago, Harvard, and Princeton).

Table 7.13. Hilbert's American students

Name	Bachelor's	Year	Before Göttingen	University
Bosworth	Wellesley	1868	Rhode Island	Rhode Island
Cairns	Ohio Wesleyan	1892	Harvard	Oberlin
Crathorne	Illinois	1898	Wisconsin	Illinois
Gillespie	Virginia	1900	Johns Hopkins	Cornell
Haseman	Indiana	1903	Indiana	Nevada
Hedrick	Michigan	1898	Harvard	Missouri, UCLA
Hurwitz	Missouri	1906	Harvard	Cornell
Kellogg	Princeton	1899	Princeton	Missouri, Harvard
Mason	Wisconsin	1898		Wisconsin
Noble	Berkeley	1889		Berkeley
Reid	Virginia Military	1899		Haverford
Townsend	Albion	1890	Illinois	Illinois
Westfall	Yale	1901	Yale	Missouri

The Hilbert Colony was slightly larger than the group that earned degrees under Felix Klein and much more numerous than any other group that obtained foreign degrees at any time. Another Göttingen mentor for Americans was Edmund Landau, who directed dissertations by Henry Blumberg, Dunham Jackson, and Aubrey J. Kempner. Other American students, who later became influential mathematicians, received doctorates from such German universities as Strasbourg (Raymond Archibald, Edward Huntington, and Louis Karpinski), Erlangen (William Osgood and Harry Tyler), and Bonn (Julian Coolidge and William Graustein). Outside Germany, American mathematicians earned doctorates from Vienna (James Pierpont) and Zurich (Saul Epsteen, Lulu Hofmann, and William Marshall).

In a 1904 *Bulletin* article, Harvard's Julian Coolidge, a descendant of Thomas Jefferson, extolled the advantages for American professors to study with certain Italian specialists. Regarding the need for American students to go abroad for advanced degrees, however, Coolidge proved to be a poor prognosticator. He wrote, "It is safe to predict that for a long time to come, this migration [to Europe] will continue."[29] It may have seemed safe to make such a prediction in 1904, but conditions improved so quickly in the US that few American students traveled abroad to obtain doctorates beyond 1910, and no other colony formed around a cluster point. David Hilbert did direct the dissertation of one more American, Haskell Curry, but that was not until 1929. The Hilbert Colony in America from 1899–1910 remains the last—and the largest—of those groups in the history of America.

Endnotes

Preface

1 Irving Kaplansky, "Nomination for Hyman Bass," *Notices Amer. Math. Soc.* **46** (1999), 931–932

Chapter 1

1 On p. 224 of V. Frederick Rickey, "How Columbus encountered America," *Mathematics Magazine* **65** (1992), 219–225.

2 On p. 24 of Ottomar Götz, "Regiomontanus," *Mathematical Intelligencer* **25** (3) (2003), 44–46.

3 My main source on Harriot is the book Robert Fox (ed.), *Thomas Harriot: An Elizabethan Man of Science*, Ashgate, Burlington, VT, 2000.

4 We might point out that 67 years earlier, Brother Juan Diez joined Hernando Cortés on an expedition to the Yucatan peninsula. He wrote a pamphlet in 1556 containing arithmetic and algebra.

5 Thomas Harriot, *A Briefe and True Report of the New-found Land of Virginia*, Dover Publications, 1972. This book can still be purchased at a very reasonable price at http://store.doverpublications.com/. A reprint of the 1590 edition cost only $14.95 in 2017.

6 On p. 47 of Muriel Seltman and Eddie Mizzi, "Thomas Harriot: Father of English algebra?," *Mathematical Intelligencer* **19** (1) (1997), 46–49.

7 Fox, *Elizabethan Man.* [Endnote 3]

8 Taken from p. 9 of Florian Cajori, *The Teaching and History of Mathematics in the United States*, Government Printing Office, Washington, DC, 1890.

9 *Ibid.*

10 *Idem*, p. 10.

11 On p. 296 of Henry Barnard, "Educational biography," *American Journal of Education* **1** (1855), 295–296.

12 As quoted on p. 347 of J.L. Coolidge, "Three hundred years of mathematics at Harvard," *Amer. Math. Monthly* **50** (1943), 347–356.

13 On p. 584 of Lao G. Simons, "Short stories in colonial geometry," *Osiris* **1** (1936), 584–605.

14 On p. 17 of Amelia Atlas, "Pride of the Indian College," *Harvard Magazine* (May-June 2011), 15–17.

15 Geraldine Brooks, *Caleb's Crossing*, Viking Penguin, New York, 2011.

16 On p. 21 of Cajori, *Teaching and History.* [Endnote 8.]

17 *Ibid.*

18 On p. 518 of Andrew Motte, *Sir Isaac Newton's Mathematical Principles of Natural Philosophy and his System of the World* (1729). This English translation was revised and appended with a historical and explanatory appendix by Florian Cajori (University of California Press, Berkeley, 1934).

19 As quoted on p. 125 of Frederick G. Kilgour, "The rise of scientific activity in New England," *The Yale Journal of Biology and Medicine* **22** (1949), 123–138. **Frederick Gridley Kilgour** (1914–2006), a Harvard graduate (class of 1935) and lecturer at Yale on the history of science, was founding director of the Online Computer Library Center (OCLC), an international computer library network and database that changed the way people use libraries. He was president and executive director of OCLC from 1967–1980.

20 As quoted on p. 32 of David Eugene Smith and Jekuthiel Ginsburg, *A History of Mathematics in America before 1900*, Open Court, Chicago, 1934; reprinted by Arno Press, 1980.

21 On p. 172 of Lyon G. Tyler, "Education in Colonial Virginia. Part IV. The higher education," *William and Mary Quarterly* **6** (1898), 171–187. This article is the fourth in a series of five on education in Virginia in the seventeenth and eighteenth centuries. **Lyon Gardiner Tyler** (1853–1935) and his family played important roles in the early stages of William and Mary. His grandfather was a graduate in the same class as Thomas Jefferson, and his father, also named John Tyler, became the tenth president of the US. Lyon Tyler was a William and Mary historian who became the college's president in 1888 after the college had been virtually closed for seven years. From then until he stepped down in 1919, Lyon Tyler increased the sizes of the faculty and student body appreciably, dramatically boosted the endowment, arranged for the college to become a public institution in 1906, and oversaw the enrollment of women in his last year in office. William and Mary's history department was named after him when his son pledged $5 million to the department in 2001.

22 *Ibid.*

23 As quoted in *idem*, p. 173.

24 As quoted on p. 223 of Lyon G. Tyler, "Education in Colonial Virginia. Part I. Poor children and orphans," *William and Mary Quarterly* **5** (1897), 219–223.

25 As quoted in *idem*, p. 222.

26 As quoted in *idem*, p. 219.

27 As quoted on p. 74 of Lyon G. Tyler, "Education in Colonial Virginia. Part III. Free Schools," *William and Mary Quarterly* **6** (1897), 71–85.

28 *Idem*, pp. 76–77.

29 On p. 5 of Lyon G. Tyler, "Education in Colonial Virginia. Part II. Private schools and tutors," *William and Mary Quarterly* **6** (1897), 1–6.

30 *Ibid.*

31 *Idem*, p. 6.

32 *Ibid.*

33 On p. 71 of Tyler, "Part III. Free Schools." [Endnote 27.]

34 *Idem*, p. 83.

35 As quoted in footnote 2 of *ibid*.

36 As quoted on p. 177 of Tyler, "Part IV. Higher education." [Endnote 21.]

37 H.R. Phalen, "The first professorship of mathematics in the colonies," *Amer. Math. Monthly* 53 (1946), 579–582. The mathematician Harold Romaine Phalen graduated from Tufts in 1911 and received his PhD at the University of Chicago 15 years later. He was at William and Mary 1940–1955. This paper is our main source for Lefevre.

38 As quoted in *idem*, p. 579.

39 As quoted in *idem*, p. 580.

40 H.R. Phalen states Cambridge in Endnote 37 but Oxford in Endnote 41. Cajori remains mute on this issue.

41 On p. 461 of H.R. Phalen, "Hugh Jones and octave computation," *Amer. Math. Monthly* 56 (1949), 461–465.

42 As quoted in *idem*, p. 462.

43 As quoted in *idem*, p. 464.

44 As quoted on p. 33 of Cajori, *Teaching and History*. [Endnote 8.]

45 On p. 582 of Phalen, "First professorship." [Endnote 37.]

46 As quoted on p. 505 of Herbert L. Ganter, "William Small, Jefferson's Beloved Teacher," *William and Mary Quarterly* (3rd Ser.) 4 (1947), 505–511. This paper contains much more information about Small, including his relationship with Benjamin Franklin when both were in England.

47 As quoted on p. 73 of Lyon G. Tyler, "Early presidents of William and Mary," *William and Mary College Quarterly Historical Papers* **1** (1892), 63–75.

48 On p. 222 of E. Alfred Jones, "Two professors of William and Mary College," *William and Mary Quarterly* **26** (1918), 221–231.

49 As quoted on p. 581 of Phalen, "First professorship." [Endnote 37.]

50 *Ibid.*

51 As cited on p. 184 of Tyler, "Higher education." [Endnote 21.]

52 On p. 4 of Louis C. Karpinski, *Bibliography of Mathematical Works Printed in America through 1850*, University of Michigan Press, Ann Arbor, MI, 1940.

53 Details are provided in *ibid*.

54 As quoted on p. 173 of Clifford K. Shipton, *New England Life in the Eighteenth Century: Representative Biographies from Sibley's Harvard Graduates*, Harvard University Press, Cambridge, MA, 1963. The chapter on Greenwood (170–182) is our major source of biographical facts.

55 *Ibid.*

56 *Idem*, p. 170.

57 *Ibid.*

58 *Idem*, p. 173.

59 As quoted in *idem*, p. 174.

60 As quoted in *idem*, p. 175.

61 Isaac Greenwood, *A Course of Philosophical Lectures, with a great Variety of Curious Experiments, Illustrating and Confirming Sir Isaac Newton's Laws of Matter and Motion*, Boston, 1726. The title page is shown in Fig. 33 on p. 41 of Karpinski, *Bibliography*. [Endnote 52.]

62 On p. 175 of Shipton, *New England Life*. [Endnote 54.]

63 On p. 38 of Todd Timmons, *Mathematics in Nineteenth-Century America: The Bowditch Generation*, Docent Press, Boston, 2013.

64 Isaac Greenwood, "An account of an aurora borealis seen in New-England on the 22d of October, 1730," *Philosophical Transactions* **37** (1731–1732), 55-69.

65 Isaac Greenwood, "A new method for composing a natural history of meteors," *Philosophical Transactions* **35** (1727–1728), 390-402; Isaac Greenwood, A brief account of some of the effects and properties of damps, *Philosophical Transactions* **36** (1729–1730), 184-191.

66 See the footnote on p. 63 of Lao Genevra Simons, "Algebra at Harvard College in 1730," *Amer. Math. Monthly* **32** (1925), 63–70.

67 As quoted on page 39 of Smith and Ginsburg, *America before 1900*. [Endnote 20.]

68 As quoted on p. 352 of Shipton, *New England Life*. [Endnote 54.] The chapter on Winthrop, pp. 349–373, is our major source of biographical facts.

69 As quoted on p. 179 of Shipton, *New England Life*. [Endnote 54.]

70 For more information on this work, see Shirley B. Gray and C. Edward Sandifer, "The Sumario Compendioso: The New World's first mathematics book," *The Mathematics Teacher* **94** (2001), 98–103. The complete title of the book is *Sumario Compendioso de las quentas de plata y oro que in los reynos del Piru son necessarias a los mercaderes y todo genero de tratantes. Los algunas reglas tocantes al Arithmetica* (Comprehensive summary of the counting of silver and gold, which, in the kingdoms of Peru, are necessary for merchants of all kinds of traders. The other rules touching on the Arithmetic.)

71 On p. 33 of Cajori, *Teaching and History*. [Endnote 8.]

72 *Idem*, p. 24.

73 On p. 156 of Frederick E. Brasch, "John Winthrop (1714–1779), America's first astronomer, and the science of his period," *Publications of the Astronomical Society of the Pacific* **28** (1916), 153–170.

74 On p. 356, Shipton, *New England Life*. [Endnote 54.]

75 *Idem*, p. 367.

76 *Idem*, p. 364.

77 As quoted in *idem*, p. 368.

78 As quoted in *idem*, p. 371.

79 Frederick G. Kilgour, "Professor John Winthrop's notes on sun spot observations (1739)," *Isis* **29** (1938), 355–361.

80 *Idem*, p. 358.

81 On p. 1152 of John Winthrop, "Concerning some natural curiosities in those parts, especially a very strange and very curiously contrived fish," *Philosophical Transactions* **5** (1670), 1151–1153.

82 Frank R. Freemon, "American colonial scientists who published in the 'Philosophical Transactions' of the Royal Society," *Notes and Records of the Royal Society of London* **39** (1985), 191–206.

83 On p. 358 of Kilgour, "Winthrop's notes on sun spot observations." [Endnote 79.]

84 On p. 349 of Coolidge, "Three hundred years." [Endnote 12.]

85 On p. 572 of John Winthrop, "Concerning the transit of Mercury over the sun, April 21, 1740, and of an eclipse of the moon, Dec. 21, 1740," *Philosophical Transactions* **42** (1742–1743), 572-578. We have adopted modern American spelling and capitalization.

86 *Idem*, p. 573.

87 *Idem*, p. 575.

88 *Idem*, p. 577.

89 On p. 505 of John Winthrop, "Observation of the transit of Mercury over the sun, October 25, 1743," *Philosophical Transactions* **59** (1769), 505–506.

90 *Idem*, p. 506.

91 On p. 52 of John Winthrop, "A letter from Dr. Franklyn, F.R.S., to the Astronomer Royal; Containing an observation of the transit of Mercury over the sun, November 9th 1769," *Philosophical Transactions* **61** (1771), 51–52.

92 On p. 1 of John Winthrop, "An account of the earthquake felt in New England, and the neighbouring parts of America, on the 18th of November 1755," *Philosophical Transactions* **50** (1757-1758), 1–18. We have adopted modern American spelling and capitalization.

93 On pp. 161–162 of Brasch, *America's first astronomer*. [Endnote 73.]

94 John Winthrop, "Observation of the transit of Venus, June 6, 1761, at St. John's, Newfound-Land," *Philosophical Transactions* **54** (1764), 279–283.

95 *Idem*, p. 283.

96 On p. 18 of William Smith, John Ewing, Owen Biddle, Hugh Williamson, Thomas Combe, and David Rittenhouse, "Apparent time of the contacts of the limbs of the sun and Venus; With other circumstances of most note, in the different European observations of the transit, June 3d,1769," *Transactions of the American Philosophical Society* **1** (1771), 12–20.

97 See pp. 165–168 of Brasch, *America's first astronomer*. [Endnote 73.]

98 John Winthrop, "Cogitata de Cometis," *Philosophical Transactions* **57** (1767), 132–154.

99 On pp. 57–58 of Cajori, *Teaching and History*. [Endnote 8.]

100 On p. 62 of Shipton, *New England Life*. [Endnote 54.]

101 *Idem*, p. 67.

102 As quoted on p. 29 of Cajori, *Teaching and History*. [Endnote 8.] Cajori misspelled Dummer.

103 On p. 62 of Leonard Tucker, "President Thomas Clap of Yale College: Another 'Founding Father' of American Science," *Isis* **52** (1961), 55–77.

104 *Idem*, p. 63.

105 As quoted on p. 473 of Frederick G. Kilgour, "Thomas Robie (1689–1729), colonial scientist and physician," *Isis* **3** (1939), 473–490.

106 On p. 46 of Thomas Clap, *An Essay on the Nature and Foundation of Moral Virtue and Obligation*, Yale, New Haven, 1765.

107 As quoted on p. 63 of Tucker, "President Clap." [Endnote 103.]

108 On p. 31 of Cajori, *Teaching and History*. [Endnote 8.]

109 *Ibid*.

110 On p. 249 of Robert Friend Rothschild, *Two Brides for Apollo: The Life of Samuel Williams (1743–1817)*, Indiana University Press, Bloomington, Indiana, 2009.

111 On p. 459 of Baxter Perry Smith, *The History of Dartmouth College*, Houghton, Osgood and Co., Boston, MA, 1878.

112 *Idem*, p. 58.

113 As quoted on p. 64 of Tucker, "President Clap." [Endnote 103.]

Chapter 2

1 As quoted on p. 15 of Edward C. Carter II, *"One Grand Pursuit": A Brief History of the American Philosophical Society's First 250 Years, 1743–1993*, American Philosophical Society, Philadelphia, 1993.

2 The term "conic sections" generally refers to the study of parabolas, ellipses, and hyperbolas, whereas the term "fluxions" refers to calculus in the style of Isaac Newton.

3 As quoted on p. 69 of Leonard Tucker, "President Thomas Clap of Yale College: Another 'Founding Father' of American Science," *Isis* **52** (1961), 55–77.

4 *Idem*, p. 56.

5 *Idem*, p. 70.

6 *Ibid.*

7 See p. 10, fn. 4 in Todd Timmons, *Mathematics in Nineteenth-Century America: The Bowditch Generation*, Docent Press, Boston, 2013.

8 On p. 244 of Brooke Hindle, "The Quaker background and science in colonial Philadelphia," *Isis* **46** (1955), 243–250.

9 As quoted on p. 12 of Carter II, *One Grand Pursuit.*" [Endnote 1.]

10 As quoted on p. 35 of "A brief history of the American Philosophical Society," *Year Book 1980*, American Philosophical Society, Philadelphia, 1981.

11 As quoted on pp. 13–14 of Carter, *One Grand Pursuit.* [Endnote 1.]

12 On p. 51 of Timmons, *Bowditch Generation.* [Endnote 7.]

13 Alexandra Oleson and Sanborn C. Brown (eds.), *The Pursuit of Knowledge in the Early American Republic*, Johns Hopkins University Press, Baltimore and London, 1976.

14 On p. 16 of John C. Greene, "Science, learning, and utility: Patterns of organization in the early American Republic," in *idem*, pp. 1–20

15 On p. 15 of Carter, *One Grand Pursuit.* [Endnote 1.] The quoted member was not named.

16 On p. 1 of Thomas Archibald and Louis Charbonneau, "Mathematics in Canada before 1945: A preliminary survey," in Peter Fillmore (ed.), *Mathematics in Canada, Volume I*, Canadian Mathematical Society, Ottawa, 1995, pp. 1–90.

17 M.W. Burke-Gaffney, "Boutet de Saint-Martin, Martin," available as of March 12, 2018 at the online version of the *Dictionary of Canadian Biography*: http://www.biographi.ca/en/bio/boutet_de_saint_martin_martin_1E.html

18 *Idem*, p. 3.

19 *Idem*, p. 4.

20 Most of the material on Franquelin is taken from the biography by M.W. Burke-Gaffney available as of March 12, 2018 at the online version of the *Dictionary of Canadian Biography*: http://www.biographi.ca/en/bio/franquelin_jean_baptiste_louis_2E.html

21 The major source on the science of David Rittenhouse, though not his mathematics, is Brooke Hindle, *David Rittenhouse*, Princeton University Press, Princeton, 1964. An analysis of Rittenhouse's mathematics is given in David E. Zitarelli, "David Rittenhouse: Modern mathematician," *Notices Amer. Math. Soc.* **62** (2015), 11–14.

22 On p. 531 of M.J. Babb, "The David Rittenhouse bicentenary," *Scientific Monthly* **35** (1932), 522–542. Babb's opinionated and piercing prose is reminiscent of his contemporary E.T. Bell. For instance, Babb wrote (on p. 538) that "Cadwallader closed Conway's lying mouth with a bullet."

23 On p. 553 of Silvio A. Bedini, "David Rittenhouse," *American National Biography* **18** (1999), 553–555.

24 W.C. Rufus, "David Rittenhouse as a mathematical disciple of Newton," *Scripta Math.* **8** (1941), 228–231.

25 David Rittenhouse, "An optical problem, proposed by Mr. Hopkinson," *Transactions of the Americal Philosophical Society* **2** (1786), 201–206.

26 On p. 553 of Bedini, "Rittenhouse." [Endnote 23.]

27 On p. 19 of M. Nevière et al, *Electromagnetic Resonances in Nonlinear Optics*, Gordon and Breach, Amsterdam, 2000.

28 David Rittenhouse, "A method of finding the sum of the several powers of the sines," *Transactions of the American Philosophical Society* **3** (1793), 155–156.

29 As quoted on p. 89 of Hindle, *David Rittenhouse.* [Endnote 21.]

30 On p. 396 of Owen Nulty, "Investigation of a theorem, proposed by Dr. Rittenhouse, respecting the summation of the several powers of the sines; With its application to the problem of a pendulum vibrating in circular arcs," *Transactions of the American Philosophical Society* (New Series) **1** (1818), 395–400.

31 David Rittenhouse, "Method of raising the common logarithm of any number immediately," *Transactions of the American Philosophical Society* **4** (1799), 69–71.

32 Frederick A. Homann, "David Rittenhouse: Logarithms and leisure," *Mathematics Magazine* **60** (1987), 15–20.

33 Brook Taylor, "A new method of computing logarithms," *Philosophical Transactions* **30** (1717), 618–622.

34 On p. 60 of Daniel Shanks, "A logarithm algorithm," *Mathematical Tables and Aids to Computation* **8** (1954), 60–64.

35 Donald Knuth, *The Art of Computer Programming*, Addison-Wesley, Reading, MA, 1968.

36 Paul C. Pasles, *Benjamin Franklin's Numbers: An Unsung Mathematical Odyssey*, Princeton University Press, Princeton, 2008. This is our primary source on Benjamin Franklin.

37 Michael Scanlan, "Who were the American postulate theorists?" *J. Symbolic Logic* **56** (1991), 981–1002.

38 As quoted on p. 8 of Pasles, *Franklin's Numbers*. [Endnote 36.]

39 As quoted in *idem*, p. 70.

40 *Idem*, p. 10.

41 As quoted in *idem*, p. 71.

42 *Idem*, pp. 71–72.

43 As quoted in *idem*, p. 72.

44 On p. 95 of Benjamin Franklin, *The Autobiography of Benjamin Franklin*, University of Pennsylvania Press, Philadelphia, 2011.

45 On p. 207 of Pasles, *Franklin's Numbers*. [Endnote 36.]

46 *Idem*, p. 134.

47 As quoted on p. 130 of J.A. Leo Lemay, *The Life of Benjamin Franklin, Volume 3: Soldier, Scientist, and Politician*, 1748–1757, University of Pennsylvania Press, Philadelphia, 2014.

48 On p. 44 of David Eugene Smith and Jekuthiel Ginsburg, *A History of Mathematics in America before 1900*, Open Court, Chicago, 1934; reprinted by Arno Press, 1980.

49 On pp. 101–102 of I. Bernard Cohen, *Science and the Founding Fathers: Science in the Political Thought of Thomas Jefferson, Benjamin Franklin, John Adams, and James Madison*, W.W. Norton, New York, 1995.

50 Frank J. Swetz, "Mathematical Treasure: Thomas Jefferson's Octagon," *Convergence* (April 2016): http://www.maa.org/press/periodicals/convergence/mathematical-treasure-thomas-jeffersons-octagon

51 Claudi Alsina and Roger B. Nelsen, "Proof without words: President Garfield and the Cauchy–Schwarz inequality," *Math. Magazine* **88** (2015), 144–145.

52 On p. 21 of John C. Greene, *American Science in the Age of Jefferson*, Iowa State University Press, Ames, Iowa, 1984.

53 Nerida F. Ellerton and M.A. (Ken) Clements, *Rewriting the History of School Mathematics in North America, 1607–1861: The Central Role of Cyphering Books*, Springer, New York, 2012.

54 As quoted on p. 14 of Theodore J. Crackel, V. Frederick Rickey, and Joel Silverberg, "Reassembling Humpty Dumpty again: Putting George Washington's cyphering manuscript back together again," *CSHPM/SCHPM Proceedings* **27** (2015). This paper is one of our two sources on the surveying of George Washington. The other is Theodore J. Crackel, V. Frederick Rickey, and Joel Silverberg, "George Washington's use of trigonometry and logarithms," *CSHPM/SCHPM Proceedings* **26** (2014), 98–115.

55 *Idem*, p. 10.

56 On p. 581 of H.R. Phalen, "The first professorship of mathematics in the colonies," *Amer. Math. Monthly* **53** (1946), 579–582.

57 Much of Washington's background in the mathematics of surveying is contained in Crackel, Rickey, Silverberg, "Trigonometry and logarithms." [Endnote 54.]

58 As stated on p. 7 of Crackel, Rickey, and Silverberg, "Humpty Dumpty." [Endnote 54.]

59 On p. 114 of Crackel, Rickey, and Silverberg, "Trigonometry and logarithms." [Endnote 54.]

60 *Idem*, p. 111.

61 As quoted on p. 15 of Crackel, Rickey, and Silverberg, "Humpty Dumpty." [Endnote 54.]

62 *Ibid*.

63 The most trustworthy account of his life and work is the book by Silvio A. Bedini, *The Life of Benjamin Banneker: The First African-American Man of Science*, 2nd ed., Maryland Historical Society, Baltimore, 1999.

64 *Idem*, p. 127.

65 As quoted in *idem*, p. 154.

66 As quoted in *idem*, p. 235.

67 Beatrice Lumpkin, "From Egypt to Benjamin Banneker: African origins of false position solutions," in Ronald Calinger (ed.), *Vita Mathematica*, Mathematical Association of America, Washington, DC, 1996, pp. 279–289. The problem is stated on p. 286.

68 On p. 148 of John Fauvel and Paulus Gerdes, "African slave and calculating prodigy: Bicentenary of the death of Thomas Fuller," *Historia Mathematica* **17** (1990), 141–151. This is our major source of information on Fuller. Benjamin Rush's letter is reproduced on page 148.

69 *Idem*, Note 4 on p. 149.

70 *Idem*, p. 149.

71 *Idem*, p. 146.

72 On p. 181 of David Kahn, *The Codebreakers: The Story of Secret Writing*, Scribner, New York, 1996. This is a revised edition of the 1967 original published by Macmillan.

73 As quoted in *idem.*, p. 184.

74 On p. 256 of Brooke Hindle, *The Pursuit of Science in Revolutionary America, 1735–1789*, University of North Carolina Press, Chapel Hill, 1956.

75 On p. 34 of Richard DeCesare, "Robert Patterson. American 'revolutionary' mathematician," *BSHM Bulletin* **31** (2016), 31–46.

76 Quoted in *idem*, p. 36.

77 *Idem*, p. 38.

78 *Idem*, p. 40.

79 See p. 107 of Philip S. Jones, "Some early American mathematicians," in Dalton Tarwater (ed.), *The Bicentennial Tribute to American Mathematics: 1776–1996*, Mathematical Association of America, Washington, DC, 1977, pp. 107–110.

80 Both quotations in this paragraph are taken from p. 263 of Hindle, *Pursuit*. [Endnote 74.]

81 *Idem*, p. 152.

82 Our primary source on the American Academy is Walter Muir Whitehill, "Early learned societies in Boston and vicinity," in Alexandra Oleson and Sanborn C. Brown (eds.), *The Pursuit of Knowledge in the Early American Republic*, Johns Hopkins University Press, Baltimore and London, 1976, pp. 151–173.

83 As quoted in *idem*, p. 153.

84 On p. viii of "Preface," *Memoirs of the American Academy of Arts and Sciences* **1** (1785), iii–xi.

85 On page 12 of John C. Greene, "Science, learning, and utility: Patterns of organization in the early American Republic," in Oleson and Brown (eds.), *Pursuit of Knowledge*, pp. 1–20. [Endnote 82.]

86 I am using here the three major terms adopted by Nathan Reingold in his paper "Definitions and speculations: The professionalization of science in America in the nineteenth century," in Oleson and Brown (eds.), *Pursuit of Knowledge*, pp. 33–69. [Endnote 82.]

87 On p. 66 of Florian Cajori, *The Teaching and History of Mathematics in the United States*, Government Printing Office, Washington, DC, 1890.

88 On p. 30 of Timmons, *Bowditch Generation*. [Endnote 7.]

89 As quoted on p. 45 of Cajori, The *Teaching and History*. [Endnote 87.]

90 As quoted on p. 28 of Timmons, *Bowditch Generation*. [Endnote 7.]

91 On p. 47 of Cajori, The *Teaching and History*. [Endnote 87.]

92 On p. 3 of Karen Parshall and David Rowe, *The Emergence of the American Mathematical Research Community, 1876–1900: J.J. Sylvester, Felix Klein, and E.H. Moore*, American Mathematical Society, Providence, RI, and London Mathematical Society, London, 1994.

Transition 1776: The Patriot

1 Robert Patterson, "Life of the author," in John Ewing, *A Plain Elementary and Practical System of Natural Experimental Philosophy*, Hopkins & Earle, Philadelphia, PA, 1809, p. xi.

2 *Idem*, pp. xi–xii.

3 *Idem*, p. xv.

Chapter 3

1 On p. viii of "Prefatory Note," *Transactions of the American Philosophical Society* **3** (1830), vii–viii.

2 On p. 12 of Todd Timmons, *Mathematics in Nineteenth-Century America: The Bowditch Generation*, Docent Press, Boston, 2013.

3 On p. 11 of Joe Albree, David C. Arney, and V. Frederick Rickey, *A Station Favorable to the Pursuits of Science: Primary Materials in the History of Mathematics at the United States Military Academy*, American Mathematical Society, Providence, 2000. This book's introduction provides an excellent source on the role that mathematics played at the Military Academy.

4 For information on some aspects of mathematics in this part of the country, see David E. Kullman, "Undergraduate mathematics in the old northwest, 1803–1873," in Amy Shell-Gellasch (ed.), *History of Undergraduate Mathematics in America*, United States Military Academy, West Point, NY, 2001, pp. 195–205.

5 As quoted on p. 13 of Albree, Arney, and Rickey, *Station Favorable*. [Endnote 3.]

6 On p. 34 of Karen Parshall and David Rowe, *The Emergence of the American Mathematical Research Community, 1876–1900: J.J. Sylvester, Felix Klein, and E.H. Moore*, American Mathematical Society, Providence, RI, and London Mathematical Society, London, 1994.

7 Reportedly, Partridge is one of only two cadets to be commissioned as a first lieutenant in the history of West Point. See p. 96 of Dick Jardine, "Alden Partridge: Educational innovator," in Amy Shell-Gellasch (ed.), *History of Undergraduate Mathematics in America*, United States Military Academy, West Point, NY, 2001, pp. 95–105.

8 Jardine, *ibid.*

9 Issues of *The Mathematical Correspondent* are available online at https://babel.hathitrust.org/cgi/ls?field1=ocr;q1=the%20mathematical%20correspondent;a=srchls

10 On p. 5 of Frederik Nebeker, *Astronomy and the Geophysical Tradition in the United States in the Nineteenth Century: A Guide to Manuscript Sources in the Library of the American Philosophical Society*, American Philosophical Society Library, Philadelphia, PA, 1991.

11 On p. 413 of S. Tyler, "Life of Lenhart the mathematician," *Biblical Repertory and Princeton Review* **13** (1841), 394–416.

12 On p. 95 of Florian Cajori, *The Teaching and History of Mathematics in the United States*, Government Printing Office, Washington, DC, 1890.

13 He wrote, " $\frac{n}{m} \times x^{\frac{n}{m}-1} \dot{x}$ represents the cotemporary fluxion of $x^{\frac{n}{m}}$."

14 On p. 132 of David S. Hart, "Historical sketch of American mathematical periodicals," *Analyst* **2** (1875), 131–138.

15 On p. 203 of Benjamin F. Finkel, "A history of American mathematical journals," *National Mathematics Magazine* **14** (1940), 197–210.

16 On p. 267 of Benjamin F. Finkel, "A history of American mathematical journals," *National Mathematics Magazine* **14** (1940), 261–270.

17 See p. 10 of George H. Daniels, *American Science in the Age of Jackson*, Columbia University Press, New York, 1968.

18 On p. 62 of Julian L. Coolidge, "Robert Adrain and the beginnings of American mathematics," *American Mathematical Monthly* **33** (1926), 61–76.

19 On p. 340 of Frank J. Swetz, "The mystery of Robert Adrain," *Math. Mag.* **81** (2008), 332–344.

20 A fuller account of the *Analyst; or Mathematical Museum* appeared on pp. 321–328 of Benjamin F. Finkel, "A history of American mathematical journals," *National Mathematics Magazine* **14** (1940), 317–328, and the article in Endnote 21.

21 On p. 500 of Brian Hayes, "Science on the farther shore," *American Scientist* **90** (2002), 499–502. Swetz provides an "outline of Adrain's derivation strategy" on pp. 337–339 of [Endnote 19].

22 On p. 34 of Mansfield Merriman, "On the history of the method of least squares," *Analyst* **4** (1877), 33–36. This is not the *Analyst* edited by Robert Adrain; this was *The Analyst* edited by Joel Hendricks 1874–1883.

23 See Benjamin F. Finkel, "A history of American mathematical journals," *National Mathematics Magazine* **14** (1940), 383–407.

24 Both quotations in this paragraph are taken from p. 502 of Hayes, "Farther shore." [Endnote 21.]

25 On p. 343 of Swetz, "Mystery." [Endnote 19.]

26 Further details about the *Analyst* appeared on pp. 404–407 of Finkel, "A history." [Endnote 23.]

27 See p. 465 of Benjamin F. Finkel, "A history of American mathematical journals," *New Mathematics Magazine* **14** (1940), 461–468.

28 Benjamin F. Finkel, "A history of American mathematical journals," *National Mathematics Magazine* **15** (1940), 27–34, and pp. 83–87 of the second part of that article, 83–96.

29 On pp. 134–135 of Hart, "Historical sketch." [Endnote 14.]

30 On p. 164 of Edward Hogan, "A proper spirit is abroad: Peirce, Sylvester, Ward, and American mathematics," 1829–1843, *Historia Mathematica* **18** (1991), 158–172.

31 These statistics are taken from my paper, David E. Zitarelli, "The bicentennial of American mathematics journals," *College Mathematics Journal* **36** (2005), 2–15. Another important source on the journal is Edward R. Hogan, "George Baron and the Mathematical Correspondent," *Historia Mathematica* **3** (1976), 403–415.

32 On p. 75 of Coolidge, "Beginnings." [Endnote 18.]

33 On p. 343 of Swetz, "Mystery." [Endnote 19.]

34 Todd Timmons, "A prosopographical analysis of the early American mathematics publication community," *Historia Mathematica* **31** (2004), 429–454.

35 As quoted on p. 76 of Cajori, *Teaching and History*. [Endnote 12.]

36 *Idem*, p. 77.

37 *Idem*, p. 76.

38 Kennard B. Bork, "Parker Cleaveland," in *American National Biography*, Vol. 5, Oxford University Press, Oxford, 1999, pp. 42–44. See also Edward H. Kraus, "A notable centennial in American mineralogy," *American Mineralogist* **23** (1938), 145–148.

39 On p. 75 of Timmons, *Bowditch Generation*. [Endnote 2.]

40 As quoted on p. 125 of Benjamin F. Finkel, "A history of American mathematical journals," *National Mathematics Magazine* **15** (1940), 121–128.

41 Details on this part of the *Mathematical Companion* are given on pp. 90–96 of Benjamin F. Finkel, "A history of American mathematical journals," *National Mathematics Magazine* **15** (1940), 83–96. The Prize Question appears on p. 96.

42 Endnote 40 is devoted to the second part of the *Mathematical Companion*. It reproduces all 27 problems.

43 The list of all published works by John D. Williams known up to 1892 is given in *Miscellaneous Notes and Queries* **10** (1892), 269–272. This source states that Williams published the *Mathematical Companion* 1828–1831.

44 As quoted on p. 127 of Endnote 40.

45 *Idem*, p. 56.

46 On p. 13 of Barbara L. Narendra, "Benjamin Silliman and the Peabody Museum," *Discovery* **14** (1979), 13–29.

47 On pp. 62–75 of Timmons, *Bowditch Generation*. [Endnote 2.]

48 *Idem*, p. 148.

49 On p. 9 of Joseph P. Bradley, "Memoir of Theodore Strong," 1790–1869, *Memoirs of the National Academy of Sciences* (1879), 8–28.

50 As quoted in *idem*, p. 10.

51 "The calculus question: differentials or fluxions, synthesis or analysis," pp. 151–161 in [Endnote 2].

52 On p. 79 of Lyman Beecher Stowe, *Saints, Sinners and Beechers*, Books for Libraries Press, Freeport, NY, 1934.

53 Timmons, *Prosopographical analysis*. [Endnote 34.]

54 On p. 73 of Timmons, *Bowditch Generation*. [Endnote 2.]

55 A.M. Fisher, "On maxima and minima of functions of two variable quantities," *Amer. J. Science & Arts* **5** (1822), 82–93.

56 Day was one of very few Americans who were born before the Revolutionary War and died after the Civil War.

57 On p. 125 of Helena Pycior, "British synthetic vs. French analytical styles of algebra in the early American republic," in David Rowe and John McCleary (eds.), *The History of Modern Mathematics, Volume 1: Ideas and their Reception*, Academic Press, Boston, 1989, 125–154.

58 On p. 200 of Louis C. Karpinski, *Bibliography of Mathematical Works Printed in America Through 1850*, University of Michigan Press, Ann Arbor, MI, 1940.

59 See fn. 46 on p. 199 of Timmons, *Bowditch Generation*. [Endnote 2.]

60 The historian Helena Pycior described this issue in detail on pp. 134–137 of "British synthetic vs. French analytical." [Endnote 57.]

61 *Idem*, p. 129.

62 On p. 215 of Timmons, *Bowditch Generation*. [Endnote 2.]

63 On p. 148 of Elizabeth Bancroft Schlesinger, "Two early Harvard wives: Eliza Farrar and Eliza Follen," *New England Quarterly* **38** (2) (June 1965), 147–167.

64 For additional information, see Amy Ackerberg-Hastings, "John Farrar: Forgotten figure of American mathematics," in J.J. Tattersall (ed.), *Proceedings of the Canadian Society for the History and Philosophy of Mathematics*, Vol. 11, University of Ottawa, Ottawa, 1998, pp. 63–68.

65 On p. 145 of Pycior, "British synthetic vs. French analytical." [Endnote 57.]

66 *Idem*, pp. 139–142.

67 *Idem*, p. 137.

68 See Albree, Arney, Rickey, *Station Favorable*. [Endnote 3.]

69 On p. 58 of Joe Albree, "Salem's Bowditch," *Mathematical Intelligencer* **14** (1) (1992), 58–60. This article presents a walking tour of Salem and shows Bowditch's home, now a Registered National Historic Landmark.

70 On p. 133 of Gerald L. Alexanderson, "Laplace and his American translator," *Bulletin Amer. Math. Soc.* **51** (2014), 131–135.

71 For more information on J.H. Moore, see Charles H. Cotter, "John Hamilton Moore and Nathaniel Bowditch," *Journal of Navigation* **30** (1977), 323–326.

72 See Timmons, *Bowditch Generation*. [Endnote 2].

73 On p. 3 of Nathaniel Bowditch, "Observations of the comet from 1807," *Memoirs of the American Academy of Arts and Sciences* **III** (Part 1) (1809), 1–17.

74 Nathaniel Bowditch, "On the motion of a pendulum suspended from two points," *Memoirs of the American Academy of Arts and Sciences* **III** (Part 2) (1815), 413–436.

75 On pp. 109–110 of Timmons, *Bowditch Generation*. [Endnote 2.]

76 As quoted on p. 122 of *idem*.

77 On p. 305 of Karpinski, *Bibliography*. [Endnote 58.]

78 In fn. 80 on p. 124 of Timmons, *Bowditch Generation*. [Endnote 2.]

79 *Idem*, p. 126.

80 On p. 194 of James Webb, *A Sense of Honor*, Naval Institute Press, Annapolis, MD, 1995.

81 On p. 166 of Ebenezer Baldwin, *Annals of Yale College, in New Haven, Connecticut, from its Foundation to the Year 1831*, Hezekiah Howe, New Haven, 1831.

82 On p. 151 of Cajori, *Teaching and History*. [Endnote 12.]

83 *Columbia Encyclopedia*, 6th ed., 2001.

84 For comments on the need for changes at Virginia as late as the 1940s, see Truman Botts, E.J. McShane in "1938–42: A personal recollection," *Notices of the Amer. Math. Soc.* **45** (1998), 686–687.

85 As quoted on p. 1 of Gilbert de B. Robinson, *The Mathematics Department in the University of Toronto 1827–1978*, University of Toronto Press, Toronto, Ontario, 1979. This author is grateful to Toronto historian Craig Fraser for loaning his copy of this invaluable source.

86 As quoted on p. 141 of John George Hodgins, *Documentary History of Education in Upper Canada, Volume 5*, Education Department of Ontario, Toronto, 1897.

87 On p. 41 of Parshall and Rowe, *Emergence*. [Endnote 6.]

88 On p. 231 of Walter F. Willcox, "Lemuel Shattuck, statist, founder of the American Statistical Association," *Journal Amer. Statistical Assoc.* **35** (part 2) (1940), 224–235.

89 On p. 298 of "Minutes of the first six meetings," *Journal Amer. Statistical Assoc.* **35** (1940), 298–302. The signatures of the founding fathers and charter members of the ASA appear on the page (unnumbered)

following this article. Portraits of the five founders appear on the subsequent two pages (also unnumbered).

90 *Idem*, p. 301.

91 The entire membership list from 1840 was reprinted in the *Journal Amer. Statistical Assoc.* **35** (1940), 305–308.

92 "By-Laws of the American Statistical Association," *Journal Amer. Statistical Assoc.* **35** (1940), 305.

93 On p. 9 of Walter F. Willcox, "Response," *Journal Amer. Statistical Assoc.* **42** (1947), 7–10.

94 The complete lists of presidents and secretaries of the ASA through 1989 are given on p. 69 of R.L. Mason, J.D. McKenzie, Jr., and S.J. Ruberg, "A brief history of the American Statistical Association, 1839–1989," *The American Statistician* **44** (1990), 68–73.

95 On p. 9 of Patti W. Hunter, "Drawing the boundaries: Mathematical statistics in twentieth-century America," *Historia Mathematica* **23** (1996), 7–30.

96 On p. 305 of Paul J. FitzPatrick, "Leading American statisticians in the nineteenth century," *Journal Amer. Statistical Assoc.* **52** (1957), 301–321. FitzPatrick was head of the economics department at Catholic University.

97 The circular is reprinted in *Journal Amer. Statistical Assoc.* **35** (1940), 303.

98 On pp. 39–41 of Carroll D. Wright, *The History and Growth of the United States Census*, Government Printing Office, Washington, DC, 1900.

99 Lemuel Shattuck, *A History of the Town of Concord; Middlesex County, Massachusetts, from its Earliest Settlement to 1832*, Rossell, Odiorne and Co., Boston, 1835.

100 As quoted on p. 303 of FitzPatrick, *Leading*. [Endnote 96.]

101 On p. 233 of Willcox, *Statist*. [Endnote 88.]

102 Lemuel Shattuck, *Report of a General Plan for the Promotion of Public and Personal Health*, Harvard University Press, Cambridge, 1948. [Facsimile Reprint]

103 Editorial, "Lemuel Shattuck (1793–1859): Prophet of American public health," *Amer. Journal of Public Health* **49** (1959), 676–677.

104 Warren Winkelstein, "Lemuel Shattuck: Architect of American public health," *Epidemiology* **19** (2008), 634.

105 On p. 310 of FitzPatrick, "Leading." [Endnote 96.]

106 As quoted in *ibid*.

107 *Idem*, p. 311.

108 *Idem*, p. 301.

109 Paul J. FitzPatrick, "Statistical societies in the United States in the nineteenth century," *The American Statistician* **11** (1957), 13–21.

110 *Idem*, p. 13.

111 *Idem*, p. 14.

112 On p. 690 of Paul J. FitzPatrick, "Leading American statisticians in the nineteenth century II," *Journal Amer. Statistical Assoc.* **53** (1958), 689–701.

113 As quoted in *idem*, p. 690.

114 This quotation and the next one appear on p. 14 of Paul J. Fitzpatrick, "Statistical works in early American statistics courses," *American Statistician* **10** (1956), 14–19.

115 On p. 17 of FitzPatrick, *Statistical societies*. [Endnote 109.]

Chapter 4

1 The authoritative source for calculus books published in America before 1910 is George M. Rosenstein, Jr., "The best method: American calculus textbooks of the nineteenth century," in Peter Duren (ed.), *A Century of Mathematics in America: Part III*, American Mathematical Society, Providence, RI, 1989, pp. 77–109. Much of our material in this and later sections is taken from this article.

2 On p. 332 of Frank J. Swetz, "The mystery of Robert Adrain," *Math. Mag.* **81** (2008), 332–344.

3 Section 5.4, pp. 151–161, in Todd Timmons, *Mathematics in Nineteenth-Century America: The Bowditch Generation*, Docent Press, Boston, 2013.

4 On p. 80 of Florian Cajori, *The Teaching and History of Mathematics in the United States*, Government Printing Office, Washington, DC, 1890.

5 The books by Ryan, Boucharlat, and Davies are all in the Artemas Martin Collection in the archives at American University. This collection is a wonderful resource for nineteenth-century books and papers.

6 See Table 1 on p. 84 of Rosenstein, "Best method." [Endnote 1.]

7 *Idem*, p. 85.

8 A modern, similarly long-lived text is *Calculus and Analytical Geometry* by **George Brinton Thomas** (1914–2006), whose first edition came out in 1952; its revision by **Ross Lee Finney** (1933–2000) was the 11th edition when it appeared in 1998; and its 12th edition was by Maurice D. Weir and Joel Hass in 2009. This text was still in use worldwide 60 years after it first appeared. Thomas was a longtime professor of mathematics at MIT. Finney was the son of a famous American composer.

9 On p. 269 of Dianzhou Zhang and Joseph W. Dauben, "Mathematical exchanges between the United States and China: A concise overview (1850–1950)," in Eberhard Knobloch and David E. Rowe, *The History of Modern Mathematics, Volume III: Images, Ideas, and Communities*, Academic Press, San Diego, 1994, pp. 263–297.

10 The English translation appears on *idem*, p. 268.

11 *Idem*, p. 267.

12 On p. 136 of David S. Hart, "Historical sketch of American mathematical periodicals," *Analyst* **2** (1875), 131–138.

13 A fairly complete account of *Mathematical Miscellany* is given in Benjamin F. Finkel, "A history of American mathematics journals," *National Mathematics Magazine* **15** (1941), 177–190 and 245–247.

14 As quoted on page 251 of Edward R. Hogan, "The *Mathematical Miscellany* (1836–1839)," *Historia Mathematica* **12** (1985), 245–257.

15 As quoted in *idem*, p. 246.

16 For more on Nancy Buttrick Root, see *idem*, pp. 249–250.

17 The tables of contents of the four volumes are listed on p. 113 of Deborah Kent, "The *Mathematical Miscellany* and *The Cambridge Miscellany of Mathematics*: Closely connected attempts to introduce research-level mathematics in America, 1836–1843," *Historia Mathematica* **35** (2008) 102–122.

18 On p. 251 of Hogan, *Mathematical Miscellany*. [Endnote 14.]

19 As quoted in *idem*.

20 On p. 108 of Kent, "Closely connected." [Endnote 17.]

21 *Ibid*.

22 On p. 337 of B. Osgood Peirce, "Joseph Lovering 1813–1892," *Biographical Memoirs National Acad. Sciences* **6** (1909), 329–344.

23 On p. 281 of David Eugene Smith, "Early American mathematical periodicals," *Scripta Mathematica* **1** (1932/1933), 277–285.

24 This currency conversion is based on the website **www.measuringworth.com**.

25 Most of my material on the *Cambridge Miscellany* is based on the paper in Endnote 18 as well as archival research carried out by Gino Pagano, an undergraduate in my course "The history of mathematics in America" during the spring 2004 semester. His report can be found online at **https://davidzitarelli. wordpress.com/part-2-new-republic-1800-1876/**.

26 Table 2 on p. 117 of Kent, "Closely connected," presents an analysis of those who submitted correct solutions. [Endnote 17.]

27 *Idem*, p. 111.

28 As quoted on p. 9 of Edward R. Hogan, *Of the Human Heart: A Biography of Benjamin Peirce*, Lehigh University Press, Bethlehem, PA, 2008. This is the most comprehensive account of the life and work of Peirce to date.

29 On p. 181 of Andrew P. Peabody, *Harvard Reminiscences*, Ticknor and Co., Boston, 1888.

30 On pp. 164–165 of Edward Hogan, "A proper spirit is abroad: Peirce, Sylvester, Ward, and American mathematics, 1829–1843," *Historia Mathematica* **18** (1991), 158–172.

31 This is not Henri Lebesgue (1875–1941), the founder of the Lebesgue integral and modern measure theory.

32 My coverage of the history of the AAAS is based on the paper by Sally Gregory Kohlstedt, "Savants and professionals: The American Association for the Advancement of Science," in Alexandra Oleson

and Sanborn C. Brown (eds.), *The Pursuit of Knowledge in the Early American Republic*, Johns Hopkins University Press, Baltimore and London, 1976, pp. 299–325.

33 On p. 8 of Walter R. Johnson, "Proceedings," *Proceedings of the American Academy of Arts and Sciences*, **1** (1849), 5–12.

34 Benjamin Silliman, *Proceedings of the American Academy of Arts and Sciences* **1** (1849), 55–60.

35 On p. 102 of *Proceedings of the American Academy of Arts and Sciences* **2** (1850).

36 *Idem*, pp. 444–445.

37 *Proceedings of the American Academy of Arts and Sciences* **1** (1849), 314.

38 *Idem*, p. 299.

39 *Idem*, Appendix II, p. 321.

40 As quoted in *idem*, p. 310.

41 Central High School was the second public high school in the country when it opened in 1838. The president was A.D. Bache. The initial, three-story building was in the shape of a "T," with the observatory above the third floor. Central was all-male until 1983.

42 As quoted on p. 19 of Hogan, *Human Heart*. [Endnote 28.]

43 As quoted in *idem*, p. 15.

44 As quoted in *idem*, p. 19.

45 As quoted on p. 18 of R.C. Archibald, "Biographical sketch," in I. Bernard Cohen (ed.), *Benjamin Peirce: "Father of Pure Mathematics" in America*, Arno Press, New York, 1980, pp. 8–30. This book contains four reminiscences of the work of Benjamin Peirce, a biographical sketch, and a reprint of "Linear associative algebras" from the *American Journal of Mathematics*. [Endnote 47.]

46 On pp. 2–3 of Charles W. Eliot, "Reminiscences of Peirce." [Endnote 45.]

47 On p. 97 of Benjamin Peirce, "Linear associative algebra," *American Journal of Mathematics* **4** (1881), 97–229.

48 As quoted on p. 6 of W.E. Byerly, "Reminiscence," in Cohen, Peirce: *"Father,"* pp. 5–7. [Endnote 45.]

49 This quotation is taken from the Preface [unpaginated] of *ibid*.

50 On p. 6 of W.E. Byerly, "Reminiscence." [Endnote 45.]

51 On p. 271 of George D. Birkhoff, "Fifty years of American mathematics," in R.C. Archibald (ed.), *Semicentennial Addresses of the American Mathematical Society*, Amer. Math. Soc., New York, 1938.

52 On p. 107 of Peirce, "Linear associative algebra." [Endnote 47.]

53 See the table in *idem*, p. 132.

54 *Idem*, p. 106.

55 *Idem*, p. 104.

56 As quoted on p. 280 of Jekuthiel Ginsburg, "A hitherto unpublished letter by Benjamin Peirce," *Scripta Mathematica* **2** (1934), 278–282.

57 On pp. 547–548 of Helena M. Pycior, "Benjamin Peirce's 'Linear Associative Algebra'," *Isis* **70** (1979), 537–551.

58 On p. 87 of H.E. Hawkes, "Estimate of Peirce's 'Linear Associative Algebra'," *Amer. Journal Math.* 24 (1902), 87–95

59 Herbert Edwin Hawkes, "On hypercomplex number systems," *Transactions Amer. Math. Soc.* **3** (1902), 312–330.

60 On p. 191 of Joseph Henry, "Memoir of Alexander Dallas Bache," *Memoirs of the National Academy of Science*, April 16, 1869, 181–212.

61 A recommended paper providing details for Bache's strategy for navigating the interrelationships between scientists and society is Hugh Richard Slotten, "The dilemmas of science in the United States: Alexander Dallas Bache and the US Coast Survey," *Isis* **84** (1993), 26–49.

62 As quoted on p. 35 of Karen Parshall and David Rowe, *The Emergence of the American Mathematical Research Community, 1876–1900: J.J. Sylvester, Felix Klein, and E.H. Moore*, American Mathematical Society, Providence, RI, and London Mathematical Society, 1994.

63 On p. 271 of Birkhoff, "Fifty years." [Endnote 51.]

64 *Ibid*.

65 Anonymous, "The Nautical Almanac Office," *Science* **3** (1884), 588.

66 For additional biographical information on Runkle, see H.W. Tyler, "Biography: John Daniel Runkle," *Amer. Math. Monthly* **10** (1903), 183–185.

67 On p. 279 of Cajori, *Teaching and History*. [Endnote 4.]

68 On pp. 136–137 of Hart, "Historical sketch." [Endnote 12.]

69 Anonymous, *Nautical Almanac Office*. [Endnote 65.]

70 *Ibid.*

71 On p. 301 of Benjamin F. Finkel, "A history of American mathematics journals," *National Mathematics Magazine* **15** (1941), 294–302 and 357–368.

72 On p. 318 of Samuel S. Greene, "Pliny Earle Chase," *Proceedings Amer. Antiquarian Soc.* **4** (1887), 316–321.

73 On p. 194 of Pliny Earle Chase, "Mathematical holocryptic cyphers," *Math. Monthly* **1** (March 1859), 194–196.

74 *Idem*, p. 195.

75 Pliny Earle Chase, "On the mathematical probability of accidental linguistic resemblances," *Proceedings of the APS* **13** (1865), 25–33.

76 On p. 321 of Greene, "Pliny Earle Chase." [Endnote 72.]

77 As quoted on p. 218 of David Kahn, *The Codebreakers: The Story of Secret Writing*, Scribner, New York, 1996. This is a revised edition of the 1967 original published by Macmillan.

78 Both quotations in this paragraph are taken from *idem*, p. 221.

79 For information on President Garfield's proof, see https://io9.gizmodo.com/james-garfield-was-the-only-u-s-president-to-prove-a-m-1037750658

80 On p. 356 of W.W. Campbell, "Biographical memoir of Edward Singleton Holden 1846–1914," *Biog. Memoirs Nat. Acad. Sci.* **8** (1919), 345–372.

81 As quoted on p. 30 of Roger Cooke and V. Frederick Rickey, "W.E. Story of Hopkins and Clark," in Duren, *Century*, pp. 29–76. [Endnote 1.]

82 Today only two subway stops on Boston's famous T (short for Massachusetts Transportation Authority, or MTA) separate Harvard and MIT.

83 On pp. 157–158 of Michael J. Crowe, *A History of Vector Analysis: The Evolution of the Idea of a Vectorial System*, University of Notre Dame Press, Notre Dame/London, 1967. This important work was republished by Dover Publications in 1985.

84 My major source is J.M. Colaw, "Biography: Dr. Joel E. Hendricks, AM," *Amer. Math. Monthly* **1** (1894), 64–67.

85 On p. 1 of Joel E. Hendricks, "Introductory remarks," *The Analyst* **1** (1874), 1–2.

86 *Ibid.*

87 *Analyst* **9** (4) (July 1882), 114–118.

88 W.W. Beman, "A brief account of the essential features of Grassmann's extensive algebra," *The Analyst* **8** (No. 3, May 1881), 96–97 and (No. 4, July 1881), 114–124. Beman was also an active participant with the problems section of *The Analyst*.

89 Thomas Craig, "Determination of a sphere which cuts five different spheres at the same angle," *The Analyst* **7** (No. 1, 1880), 13–16.

90 H.T. Eddy, "The application of the exponential polygon to the Hessian," *The Analyst* **4** (No. 3, 1877), 94. Eddy was also an active contributor with queries and problems in the journal.

91 Muir was offered the chair of the department at Stanford in 1890 but was lured by Cecil Rhodes, then Prime Minister of Cape Colony, to South Africa instead. His papers appeared in *The Analyst*, Vol. 8 (No. 6, 1881), 169–171 and Vol. 10 (No. 1, 1883), 8–9.

92 Stephen M. Stigler, "De Forest, Erastus Lyman," *American National Biography Online*, Feb. 2000: http://www.anb.org.libproxy.temple.edu/articles/13/13-00409.html. See also Stigler's outstanding paper, "Mathematical statistics in the early States," *Annals of Statistics* **6** (1978), 239–265; it was reprinted in Duren, *Century* [Endnote 1], pp. 537–564.

93 E.L. De Forest, "On adjustment formulas," *The Analyst* **4** (1877), 79–86 and 107–113.

94 R.S. Woodward, "On the actual and probable errors of interpolated values derived from numerical tables by means of first differences," *The Analyst* **9** (1882), 143–149 and 169–175.

95 On p. 249 of Raymond Garver, "The Analyst, 1874–1883," *Scripta Mathematica* **1** (1932), 247–251 and 322–326.

96 See p. 140 of E.L. De Forest, "On an unsymmetrical probability curve," *The Analyst* **9** (1882), 135–142. This paper was continued in three more parts: *The Analyst* **9** (1882), 161–168 and *The Analyst* **10** (1883), 1–7 and 67–74.

97 Chas. H. Kummell, "New investigation of the law of errors of observations," *The Analyst* **3** (1876), 132–140; 165-171.

98 Chas. H. Kummell, "Revision of proof of the formula for the error of observation," *The Analyst* **6** (1879), 80–81.

99 See pp. 326–330 of Asta Shomberg and James Tattersall, "Life and statistical legacy of Charles Hugo Kummell," *Mathematics Magazine* **86** (2013), 323–339.

100 Chas. H. Kummell, "An account of Cauchy's 'Calcul des residues'," *The Analyst* **6** (1879), 1–9, 41-46, and 173–176.

101 Chas. H. Kummell, "Reduction of observation equations which contain more than one observed quantity," *The Analyst* **6** (1879), 97–105; "Proof of some remarkable relations in the method of least squares," *The Analyst* **7** (1880), 84–88.

102 Chas. H. Kummell, "Some relations deduced from Euler's theorem on the curvature of surfaces," *The Analyst* **8** (1881), 93–95.

103 Chas. H. Kummell, "Remarks on Mr. Meech's article on elliptic functions" and "Evaluation of elliptic functions of the second and third species," *The Analyst* **5** (1878), 17–19 and 97–104.

104 Chas. H. Kummell, "Approximate multisection of an angle and hints for reducing the unavoidable error to the smallest amount," *The Analyst* **5** (1878), 172–174.

105 As quoted on p. 17 of Judy Green and Jeanne LaDuke, *Pioneering Women in American Mathematics*, American Mathematical Society, Providence, RI, and London Mathematical Society, London, 2009.

106 On p. 121 of Judy Green, "Christine Ladd Franklin (1847–1930)," in Louise S. Grinstein and Paul J. Campbell, *Women of Mathematics: A biobibliographic Sourcebook*, Greenwood Press, 1987, pp. 121–128.

107 On p. 84 of Beatrice Gormley, *Maria Mitchell: The Soul of an Astronomer*, Eerdmans Books for Young Readers, 2004.

108 On p. 52 of Christine Ladd, "Crelle's Journal," *The Analyst* **2** (1875), 51–52.

109 On p. 280 of Florian Cajori, *Teaching and History*. [Endnote 4.]

110 On p. 85 of "Solutions of problems in Number 2," *The Analyst* **3** (1876), 83–86.

111 Christine Ladd, "Note," *The Analyst* **4** (1877), 25, and "Determination of the locus of O," *The Analyst* **2** (1877), 47–48.

112 Christine Ladd, "Query," *The Analyst* **5** (1878), 64.

113 Christine Ladd, "Quaternions," *The Analyst* **4** (1877), 172–174.

114 Christine Ladd, "On some properties of four circles inscribed in one and circumscribed about another," *The Analyst* **5** (1878), 116–117, and "The polynomial theorem," *The Analyst* **5** (1878), 145–147.

115 Christine Ladd, "The nine-line conic," *The Analyst* **7** (1880), 147–149.

116 *The Analyst* **4** (1877), 160.

117 Thomas Craig, "Determination of a sphere which cuts five given spheres at the same angle," *The Analyst* **7** (1880), 13–16.

118 A.S. Hathaway, "A case of symbolic vs. operative expansion" and "A problem with solution," *The Analyst* **5** (1878), 38–39 and 86–87.

119 C.A. Van Velzer, "Answer to query" and "Solution of problem," *The Analyst* **9** (1882), 116–118 and 184–185.

120 J.E. Oliver, "Untitled," *The Analyst* **1** (1874), 29. Other sources in this paragraph are C.M. Woodward, "Tangency of hyperboloids of revolution," *The Analyst* **1** (1874), 106–107; H.T. Eddy, "The application of the exponential polygon to the Hessian,"*The Analyst* **2** (1875), 104–106; W.W. Johnson, "On the distribution of primes," *The Analyst* **2** (1875), 9–11; W.W. Beman, "Solution of a special case of problem 288, where the circles are tangent at the same point of AB," *The Analyst* **7** (1880), 115–116.

121 H.T. Eddy, "The application of the exponential polygon to the Hessian," *The Analyst* **4** (1877), 94.

122 W.W. Beman, "A brief account of the essential features of Grassmann's exterior algebra," *The Analyst* **8** (1881), 96–97 and 114–124.

123 Ormond Stone, "A quasi-proof of the arithmetic mean," *The Analyst* **7** (1882), 150–151.

124 Asaph Hall, "To change a series into a continued fraction," *Math. Monthly* **3** (1860), 262–268.

125 Asaph Hall, "On an experimental determination of π," *Messenger Math.* **2** (1873), 113–114.

126 Asaph Hall, "The Besselian function," *The Analyst* **1** (1874), 81–84.

127 Asaph Hall, "Approximate quadrature," *The Analyst* **3** (1876), 1–10.

128 This quotation, and the next, appear on p. 83 of Asaph Hall, "Notes on Gauss's 'Theoria motus'," *The Analyst* **8** (1881), 83–88.

129 *Idem*, p. 88.

130 I attended Pennsylvania Military College (PMC) for three years but left without a degree, so I never became an officer (or a gentleman). PMC became Weidner University in 1972.

131 The two quotations in this paragraph appear on pp. 54 and 105 of Crowe, *Vector Analysis*. [Endnote 83.]

132 On p. 137 of E.W. Hyde, "Mechanics by quaternions," *The Analyst* **7** (1880), 137–144 and 177–184; *The Analyst* **8** (1881), 17–24 and 49–55.

133 E.W. Hyde, "Proposition in transversals," *The Analyst* **5** (1878), 113–115; "Proof of a proposition in solid geometry," *The Analyst* **7** (1880), 157–158.

134 E.W. Hyde, "Foliate curves," *The Analyst* **2** (1875), 12–14; "Solution of Mr. Church's problem," *The Analyst* **2** (1875), 76–77; "Proof that no two different ellipses can be parallel," *The Analyst* **2** (1875), 144; "The section of a circular torus by a plane passing through the center and tangent at opposite sides," *The Analyst* **3** (1876), 78–79; "To fit together two or more quadrics so that their intersections shall be a plane," *The Analyst* **3** (1876), 97–99.

135 E.W. Hyde, "Limits of the prismoidal formula," *The Analyst* **3** (1876), 113–116.

136 M.A. Merriman, "A list of writings related to the method of least squares, with historical and critical notes," *Transactions of the Connecticut Acad. Arts and Sci.* **4** (1877), 151–232.

137 As quoted on p. 330 of Shomberg and Tattersall, "Life and statistical legacy." [Endnote 99.]

138 M.A. Merriman, "An elementary discussion of the principle of least squares," *Journal Franklin Inst.* **104** (1877), 270–274.

139 The remaining quotations in this paragraph appeared on p. 330 of Shomberg and Tattersall, "Life and statistical legacy." [Endnote 99.]

140 On p. 138 of Mansfield Merriman, "Note on the reactions of continuous beams," *The Analyst* **2** (1875), 138–140.

141 On p. 33 of Mansfield Merriman, "On the history of the method of least squares," *The Analyst* **4** (1877), 33–36.

142 On p. 140 of Mansfield Merriman, "On the history of the method of least squares," *The Analyst* **4** (1877), 140–143.

143 *Idem*, p. 143.

144 Daniel Kirkwood, "On the relative positions of the asteroidal orbits," *The Analyst* **1** (1874), 2–3.

145 Robert J. Aley, "Biography: Daniel Kirkwood," *Amer. Math. Monthly* **1** (1894), 140–149.

146 Daniel Kirkwood, "On the limit of planetary stability," *The Analyst* **8** (1881), 1–3.

147 Daniel Kirkwood, "Reminiscences of William Lenhart, Esq.," *The Analyst* **2** (1875), 181–182.

148 Daniel Kirkwood, "On the variation of the length of the day," *The Analyst* **7** (1880), 9–10.

149 Orson Pratt, Sen., "Six original problems," *The Analyst* **3** (1876), 186–187.

150 On p. 19 of Orson Pratt, Sen., "Solutions to the six original problems published in No. six, Vol. III," *The Analyst* **4** (1877), 15–19.

151 *The Analyst* **4** (1877), 55–56.

152 "Book notices," *The Analyst* **3** (1876), 12.

153 Artemas Martin, "Extraction of roots by logarithms," *The Analyst* **3** (1876), 172.

154 See the footnote on p. 80 of J.B. Mott, "Differentiation," *The Analyst* **3** (1876), 79–81.

155 On p. 159 of Joel Hendricks, "Announcement," *The Analyst* **6** (1883), 159–160.

156 Joel Hendricks, "Announcement," *The Analyst* **6** (1883), 166.

157 On p. 281 of Smith, "Early American mathematical periodicals." [Endnote 23.]

158 Benjamin F. Finkel, "A history of American mathematics journals," *National Mathematics Magazine* **15** (1941), 403–418, and **16** (1941), 64–68.

159 The bibliophile Keith Dennis might be regarded as a modern-day Artemas Martin. Anyone who has access to central New York State should arrange to see the "extension" (a trailer!) he added to his house near Cornell University to store his private collection of books.

160 The material on Martin was gleaned from Patricia R. Allaire and Antonella Cupillari, "Artemas Martin: An amateur mathematician of the nineteenth century," in *Combined Proceedings Sixth and Seventh Midwest Hist. Math. Conferences*, University of Wisconsin, La Crosse, 1999, pp. 21–32. The word "amateur" comes from the root "amare," to love. Allaire and Cupillari describe Martin as an amateur mathematician in this sense of the word and also for the idea of someone who performs for pleasure instead of personal gain.

161 As quoted on p. 360 of Steve Batterson, "The father of the father of American mathematics," *Notices Amer. Math. Soc.* **55** (2008) 352–363. This paper is the source of many of the facts cited here on H.A. Newton.

162 As quoted on p. 89 of Harold L. Dorwart, "Mathematics and Yale in the nineteen twenties," in Peter Duren (ed.), *A Century of Mathematics in America: Part II*, pp. 87–97.

163 See pp. 359–361 of Batterson, "Father of the father." [Endnote 161.]

164 *Idem*, p. 360.

165 As quoted in *idem*, p. 354.

166 As quoted in *idem*, p. 357.

167 As quoted in *idem*, p. 356.

Transition 1876: Story vs. Klein

1 As quoted on p. 31 of Roger Cooke and V. Frederick Rickey, "W.E. Story of Hopkins and Clark," in Peter Duren (ed.), *A Century of Mathematics in America*, American Mathematical Society, Providence, Vol. 3, 1989, pp. 29–76.

2 On p. 706 of Michael Atiyah and Don Zagier (coordinating eds.), "Friedrich Hirzebruch (1927–2012)," *Notices Amer. Math. Soc.* **61** (2014), 706–727.

3 On p. 352 of Steve Batterson, "The father of the father of American mathematics," *Notices Amer. Math. Soc.* **55** (2008) 352–363.

4 As quoted on p. 264 of Steve Batterson, "The contribution of John Parker Jr. to American mathematics," *Notices of the AMS* **58** (2011), 262–273.

5 As quoted in *idem*, p. 266.

6 On p. 607 of David E. Zitarelli, "Towering figures of American mathematics, 1890–1950," *American Mathematical Monthly* **108** (2001), 606–635.

7 *Ibid.*

Chapter 5

1 This currency conversion is based on the site www.measuringworth.com.

2 On p. 214 of *Dictionary of American Biography*, Vol. 9–10, Charles Scribner's Sons, New York.

3 As quoted on p. 3 of W. Norton Grubb and Marvin Lazerson, "Vocationalism in higher education: The triumph of the education gospel," *Journal Higher Education* **76** (2005), 1–25.

4 As quoted on p. 5 of Calvin C. Moore, *Mathematics at Berkeley: A History*, A. K. Peters, Wellesley, MA, 2006.

5 In the Preface to Albert E. Church, *Elements of Differential and Integral Calculus*, Wiley and Putnam, New York, 1842.

6 *Idem*, p. 34.

7 As quoted on p. 123 of Karen Parshall and David Rowe, *The Emergence of the American Mathematical Research Community, 1876–1900: J.J Sylvester, Felix Klein, and E.H. Moore*, American Mathematical Society, Providence, RI, and London Mathematical Society, London, 1994.

8 For a detailed analysis of Story's vast contributions to the history of mathematics in America, see Roger Cooke and V. Frederick Rickey, "W.E. Story of Hopkins and Clark," in Peter Duren (ed.), *A Century of Mathematics in America*, Vol. 3, 1989, pp. 29–76, American Mathematical Society, Providence, RI.

9 For more information about Peirce's contributions, see Carolyn Eisele, "The New Elements of mathematics by Charles S. Peirce," in J. Dalton Tarwater, John T. White, and John Miller (eds.), *Men and Institutions in American Mathematics*, Texas Tech Press, Lubbock, TX, 1976, pp. 111–121, and Carolyn Eisele, "Peirce, Charles Sanders," in Charles C. Gillispie (ed.), *Dictionary of Scientific Biography*, Charles Scribner's Sons, New York, 1970–1990, Vol. 10, pp. 482–488.

10 On p. 265 of Florian Cajori, *The Teaching and History of Mathematics in the United States*, Government Printing Office, Washington, DC, 1890.

11 On p. 235 of Karen Hunger Parshall, *James Joseph Sylvester: Jewish Mathematician in a Victorian World*, Johns Hopkins University Press, Baltimore, MD, 2006.

12 As quoted in fn. 56 in *idem*, p. 394.

13 As quoted in *idem*, p. 239.

14 On p. 483 of Judith Grabiner's review of the book in Endnote 11, *Bull. Amer. Math. Soc.* **44** (2007), 481–485.

15 As quoted on pp. 257–258 of Cajori, *Teaching and History*. [Endnote 10.]

16 Source: http://www.math.wisc.edu/oldhome/directories/alumni/1897.htm

17 Artemas Martin and J.B. Reynolds come to mind.

18 For a full account of the life and work of Ladd Franklin, see Judy Green, "Christine Ladd Franklin (1847–1930)," in Louise S. Grinstein and Paul J. Campbell (eds.), *Women of Mathematics: A Biobibliographic Sourcebook*, Greenwood Press, New York, 1987, pp. 121–128.

19 See Note 20 on p. 53 of Garrett Birkhoff, "Mathematics at Harvard, 1836–1944," in Peter Duren (ed.), *A Century of Mathematics in America*, American Mathematical Society, Providence, RI, Vol. 2, 1989, pp. 3–58.

20 As quoted on p. 255 of Parshall, *James Joseph Sylvester*. [Endnote 11.]

21 *Ibid.*

22 On p. 59 of John Woodbury, "Record of the Class, 1895–1900," in John Woodbury, *Harvard College: Class of 1880*, Wright & Potter, Boston, 1900, pp. 13–76.

23 "Death of Dr. George Stetson Ely," on p. 6 of *The Fredonia Center*, December 26, 1917.

24 On p. 258 of Parshall, *James Joseph Sylvester*. [Endnote 11.]

25 All three quotations in this paragraph are taken from p. 267 of Cajori, *Teaching and History*. [Endnote 10.]

26 Arthur S. Hathaway, "Some papers on the theory of numbers," *Amer. Journal Math.* **6** (1883-1884), 316–330.

27 See Chapter 21 of Eric Temple Bell, *Men of Mathematics*, Simon and Schuster, New York, 1937. The moniker "invariant twins" is likely to stick for posterity.

28 On pp. 68–69 of Daniel Coit Gilman, "Seventh annual report of the President of The Johns Hopkins University," 1882.

29 On p. 27 of "Notes on Sir William Thomson's recent lectures on molecular dynamics," *University Circulars* **35** (December 1884).

30 Footnote † on p. 113 of *University Circulars* **41** (July 1885).

31 As quoted on p. 113 of *University Circulars* **41** (July 1885).

32 *Ibid.*

33 As quoted on p. 118 of *University Circulars* **41** (July 1885).

34 On p. 110 of Simon Newcomb, "Abstract Science in America, 1776–1876," *North Amer. Review* **122** (1876), 88–123.

35 J.J. Sylvester, assisted by F. Franklin, "Tables of the generating functions and groundforms for the binary quantics of the first ten orders," *Amer. Journal Math.* **2** (1879), 223–251; J.J. Sylvester, assisted by F. Franklin, "Tables of the generating functions and groundforms for simultaneous binary quantics of the first four orders, taken two and two together," *Amer. Journal Math.* **2** (1879), 293–306.

36 K.H. Parshall and D.E. Rowe, *The Emergence of the American Mathematical Research Community, 1876–1900: J.J. Sylvester, Felix Klein, and E.H. Moore*, History of Mathematics, vol. 8, American Mathematical Society, Providence, and London Mathematical Society, London, 1994, as quoted on p. 129.

37 J.J. Sylvester, with insertions by F. Franklin, "A constructive theory of partitions, arranged in three acts, an interact and an exodion," *Amer. Journal Math.* **5** (1882), 251–330.

38 As quoted on p. 129 of Parshall and Rowe, *Emergence*. [Endnote 36.]

39 *Ibid.*

40 On p. 1 of Guillermo P. Curbera, *Mathematicians of the World, Unite! The International Congress of Mathematicians—A Human Endeavor*, A.K. Peters Ltd., Wellesley, MA, 2008.

41 A.B. Kempe, "On the geographical problem of the four colours," *Amer. Journal Math.* **2** (1879), 193–200.

42 Footnote 25 on p. 225 of Karen Hunger Parshall, "Mathematics and the politics of race: The case of William Claytor (PhD, University of Pennsylvania, 1933)," *Amer. Math. Monthly* **123** (2016), 214–240.

43 For a complete examination of the correspondence of J.J. Sylvester, see Karen Parshall, *James Joseph Sylvester: Life and Work in Letters*, Clarendon Press, Oxford, 1998.

44 On p. 223 of Parshall and Rowe, *Emergence*. [Endnote 36].

45 On p. 727 of Oswald Veblen, "Henry Burchard Fine—In Memoriam," *Bull. Amer. Math. Soc.* **35** (1929), 726–730.

46 See pp. 4–5 of Virginia Chaplin, "Princeton & mathematics: A notable record," *Princeton Alumni Weekly*, May 9, 1958, 1–13.

47 As quoted on p. 101 of Raymond C. Archibald, *A Semicentennial History of the American Mathematical Society, 1888–1938*, Amer. Math. Soc., New York, 1938.

48 For further details, see David E. Rowe, "Episodes in the Berlin-Göttingen rivalry, 1870–1930," *Mathematical Intelligencer* **22** (1) (Winter, 2000), 60–69.

49 On p. 210 of Parshall and Rowe, *Emergence*. [Endnote 36.]

50 On p. 353 of J.L Coolidge, "Three hundred years of mathematics at Harvard," *Amer. Math. Monthly* **50** (1943), 347–356.

51 *Ibid.*

52 For more on Osgood, see J.L. Coolidge, G.D. Birkhoff, and E.C. Kemble, "William Fogg Osgood," *Science* **98** (2549) (November 1943), 399–400. An analysis of Osgood's publication record is given in Peter Duren (ed.), *A Century of Mathematics in America: Part II*, American Mathematical Society, Providence, RI, and London Mathematical Society, London, 1989, pp. 79–85.

53 See *idem*, pp. 350–351, for a brief account of a competition that the Phi Beta Kappa chapter at Harvard conducted in which Boston Latin School bested such prestigious private academies as Exeter, Andover, and Hotchkiss. Although Boston public schools were superb at the time, Boston Latin was the only high-scoring public school.

54 On p. 353 of J.L. Brenner, "Student days—1930," *Amer. Math. Monthly* **86** (1979), 350–356.

55 W.F. Osgood, "On the existence of the Green's function for the most general simply connected plane region," *Trans. Amer. Math. Soc.* **1** (1900), 310–314.

56 For more on Bôcher, see W.F. Osgood, "The life and service of Maxime Bôcher," *Bull. Amer. Math. Soc.* **25** (1919), 337–350. A critical assessment of Osgood's work is given in G.D. Birkhoff, "The scientific work of Maxime Bôcher," *Bull. Amer. Math. Soc.* **25** (1919), 197–215.

57 On p. 353 of J.L Coolidge, "Three hundred years." [Endnote 50.]

58 On p. 294 of G.D. Birkhoff, "Fifty years of American mathematics," in *Semicentennial Addresses of the American Mathematical Society*, American Mathematical Society, New York, 1938, pp. 270–315.

59 On p. 350 of W.F. Osgood, "Life and service." [Endnote 56.]

60 On pp. 22–23 of Henry Seely White, "Autobiographical memoir of Henry Seely White," *Biographical Memoirs Nat. Acad. Sci.* **25** (1944), 16–33.

61 *Idem*, p. 22.

62 As quoted on p. 69 of G.W. Hill, "James Edward Oliver, 1829–1895," *Biographical Memoirs Nat. Acad. Sci.* (1896), 57–73.

63 On p. 13 of Thomas S. Fiske, The beginnings of the American Mathematical Society, Appendix B to "The Semicentennial Celebration," *Bulletin Amer. Math. Soc.* **45** (1939).

64 *Ibid.*

65 In 1897 the college became Columbia University and moved to the present Morningside Heights campus on West 116th St., where the AMS moved its headquarters.

66 On p. 34 of R.C. Archibald, *History AMS*. [Endnote 47.]

67 *Idem*, pp. 34–35.

68 On p. v of E.H. Moore, O. Bolza, H. Maschke, and H.S. White (eds.), *Mathematical Papers Read at the International Mathematical Congress Held in Conjunction with the World's Columbian Exposition, Chicago, 1893*, Macmillan and Co., New York, 1896.

69 See p. 42 of Duren, *A Century of Mathematics in America.* [Endnote 19.]

70 See R.C. Archibald (ed.), *Semicentennial Addresses of the American Mathematical Society*, New York, 1938, p. 113.

71 On p. 213 of Thomas S. Fiske, "Mathematical progress in America," *Science* **21** (Feb. 10, 1905), 209–215.

72 At the time this book went to press, the AMS had elected a third female president, Jill Pipher, whose term is 2019–2020.

73 On p. 61 of R.C. Archibald *Semicentennial.* [Endnote 70.]

Chapter 6

1 Florian Cajori, *The Teaching and History of Mathematics in the United States*, Government Printing Office, Washington, DC, 1890.

2 This currency conversion is based on the site **www.measuringworth.com.**

3 On p. 9 of Harry Press, "Diamond treasure found: Baseball scorebook sheds light on early games," *Sandstone & Tile* (Stanford Historical Society) **16** (Summer 1992), 8–10.

4 Jane Lathrop Stanford Papers, SC 033b, Box 2, Folder 18, Stanford University Archives, Stanford, Calif.

5 Douglas Houghton Campbell, "Memorial Resolution: Robert E. Allardice (1862–1928)," Stanford University, undated.

6 On pp. 18–19 Roxanne L. Nilan, "A prompt grasp of the situation: Palo Alto, Stanford, and the relief of San Francisco," *Sandstone & Tile* (Stanford Historical Society) **30** (2006), 17–27. http://histsoc. stanford.edu/pdfST/ST30no1.pdf)

7 On p. 11 of Frederick E. Basch, "Obituary: Rufus Lot Green," *Science* **77** (January 6, 1933), 11–12.

8 As quoted on p. 47 of Roger Cooke and V. Frederick Rickey, "W.E. Story of Hopkins and Clark," in Peter Duren (ed.), *A Century of Mathematics in America*, American Mathematical Society, Providence, RI, and London Mathematical Society, London, Vol. 3, 1989, pp. 29–76. This article is our primary source of information on Clark University and William Story's role in its mathematics department.

9 *Idem*, p. 48.

10 *Idem*, p. 55.

11 As quoted on pp. 200–201 of Karen Parshall, and David Rowe, *The Emergence of the American Mathematical Research Community, 1876–1900: J.J Sylvester, Felix Klein, and E.H. Moore*, American Mathematical Society, Providence, RI and London Mathematical Society, London, 1994.

12 On p. 24 of Cooke and Rickey, "Hopkins and Clark." [Endnote 8.]

13 *Ibid.*

14 For more details on the Perott–Kovalevskaya relationship, see Roger Cooke, "Joseph Perott: The documentary record of a scholar's life" [to appear].

15 The PhB is a "bachelor of philosophy" degree that involves some research.

16 On pp. 170–171 of Smith and Ginsburg, *A History of Mathematics in America before 1900*, Open Court Pub. Co., Chicago, 1934.

17 William E. Story and Louis N. Wilson (eds.), *Clark University 1889–1899: Decennial Celebration*, Clark University, Worcester, MA, 1899.

18 *Idem*, p. 68.

19 *Idem*, p. 67.

20 Atwater's presidency lasted 25 years, 1919–1944. For Clark mathematics, it was the Dark Ages. The Renaissance occurred with his successor Howard B. Jefferson, who began rebuilding the department in the late 1940s, though it was 1968 before the first PhD (since 1917) was awarded.

21 On pp. 182, 185, and 195 of Smith and Ginsburg, *History of Mathematics in America.* [Endnote 16.]

22 For additional information on Herbert Keppel, see Chapter 5 in Paul Ehrlich, *Mathematics at the University of Florida in Gainesville: The First 65 Years.* Available online at http://www.theral.net/index.htm.

23 On p. 64 of Story and Wilson, *Decennial.* [Endnote 17.]

24 *Ibid.*

25 *Idem*, pp. 64–65.

26 Further details about the storied life and career of Leona May Peirce can be found on pp. 263–264 of Judy Green and Jeanne LaDuke, *Pioneering Women in American Mathematics*, American Mathematical Society, Providence, and London Mathematical Society, London, 2009.

27 On p. 344 of Solomon Lefschetz, "Reminiscences of a mathematical immigrant in the United States," *American Mathematical Monthly* **77** (1970), 344–350.

28 On p. 273 of Phillip Griffiths, Donald Spencer, and George Whitehead, "Solomon Lefschetz," *Biographical Memoirs of the National Academy of Science*, 1992.

29 For further details on Wheeler's life and the role he played in the American mathematical community, see David L. Roberts, "Albert Harry Wheeler (1873–1950): A case study in the stratification of American mathematical activity," *Historia Mathematica* **23** (1996), 269–287.

30 On p. 62 of James Hilty, *Temple University: 125 Years of Service to Philadelphia, the Nation and the World*, Temple University Press, Philadelphia, PA, 2010.

31 On p. 191 of Smith and Ginsburg, *History of Mathematics in America*. [Endnote 16.]

32 For more information on John Hutchinson, see Virgil Snyder, "In Memoriam," *Bulletin Amer. Math. Soc.* **42** (1936), 164.

33 For more information on Ernest Skinner, see E.B. Van Vleck, "In Memoriam," *Bulletin AMS* **41** (1935), 593–594.

34 On pp. 141–143 of Gary G. Cochell, "The early history of the Cornell mathematics department: A case study in the emergence of the American mathematical research community," *Historia Mathematica* **25** (1998), 133–153.

35 On p. 135 of T.W. Goodspeed, *A History of the University of Chicago Founded by John D. Rockefeller: The First Quarter-Century*, University of Chicago Press, Chicago, 1916.

36 On p. 35 of Raymond C. Archibald, *A Semicentennial History of the American Mathematical Society, 1888–1938*, American Mathematical Society, New York, 1938.

37 Forewarning from personal experience: do not confuse Jacob William Albert Young with John Wesley Young. See p. 613 of my paper, "Towering figures in American mathematics, 1890–1950," *Amer. Math. Monthly* **108** (2001), 606–635.

38 On p. 276 of G.D. Birkhoff, "Fifty years of American mathematics," in Archibald, *Semicentennial History*, pp. 270–315. [Endnote 36.]

39 A fourth organization, AMATYC (The American Mathematical Association of Two-Year Colleges), was founded in 1974. It is devoted exclusively to providing a national forum for the improvement of mathematics instruction in the first two years of college.

40 Believed to be in the public domain from Library of Congress, Prints and Photographs Collections. http://www.flickr.com/photos/pingnews/486409173/

41 On p. 13 of T.F. Fiske, "Notes." *Bull. New York Math. Soc.* **2** (1892), 12–15.

42 On p. 259 of T.F. Fiske, "Notes." *Bull. New York Math. Soc.* **2** (1893), 259–260.

43 Moore, E.H., "Notes." *Bull. New York Math. Soc.* **2** (1893), 259.

44 On p. 110 of T.F. Fiske, "Notes," *Bull. New York Math. Soc.* **2** (1893), 109–111.

45 David Hilbert, "Mathematical problems," *Bull. Amer. Math. Soc.* **8** (1902), 437–479.

46 *Idem*, p. 472.

47 *Ibid.*

48 On p. 747 of T.R. Hollcroft, "The summer meeting in State College," *Bull. Amer. Math. Soc.* **43** (1937), 745–757.

49 Mary W. Newson, "Review of 'Thomas Jefferson and Mathematics'," *National Mathematics Magazine* **14** (1940), 492.

50 This assertion is based on Della Dumbaugh Fenster and Karen Hunger Parshall, "Women in the American mathematical research community: 1891–1906, in Eberhard Knobloch and David E. Rowe," *The History of Modern Mathematics, Volume III: Images, Ideas, and Communities*, Academic Press, 1994, pp. 229–261.

51 On p. 134 of Felix Klein, "The present state of mathematics," in E.H. Moore, O. Bolza, H. Maschke, and H.S. White (eds.), *Mathematical Papers Read at the International Mathematical Congress Held in Conjunction with the World's Columbian Exposition, Chicago, 1893*, Macmillan and Co., New York, 1896, pp. 133–135.

52 Philip J. Davis and David Mumford, "Henri's crystal ball," *Notices Amer. Math. Soc.* **55** (2008), 458–466.

53 H.W. Tyler, "The Mathematical Congress at Chicago," *Bull New York Math. Soc.* **3** (1893), 14–19.

54 On pp. 309–327 of Parshall and Rowe, *Emergence.* [Endnote 11.]

55 On p. 171 of Artemas Martin, "On fifth-power numbers whose sum is a fifth power," in E.H. Moore et al. (eds.), *Mathematical Congress*, pp. 168–174. [Endnote 51.]

56 Germans generally adopted the phrase "*unbegrenzte Möglichkeiten*," meaning a land of "unlimited possibilities." See p. 333 of Parshall and Rowe, *Emergence.* [Endnote 11].

57 Albert M. Sawin, "The algebraic solution of equations," *Annals Math.* **6** (1892), 169–177.

58 On p. 208 of E.H. Moore, "A doubly-infinite system of simple groups," in E.H. Moore et al (eds.), *Mathematical Congress*, pp. 208–242. [Endnote 51.]

59 *Idem*, p. 211.

60 For an historical account of the Wedderburn theorem and the role of Moore's theorem in it, see K.H. Parshall, "In pursuit of the finite division algebra theorem and beyond: Joseph H. M. Wedderburn, Leonard E. Dickson, and Oswald Veblen," *Arch. Internat. Hist. Sci.* **32** (1983), 223–349.

61 On p. 1 of E.H. Moore, "Introduction to a form of general analysis," in *The New Haven Colloquium*, Yale University Press, New Haven, 1910, pp. 1–150.

62 On p. 273 of G.D. Birkhoff, "Fifty years of American mathematics," in Archibald, *Semicentennial*, pp. 270–315. [Endnote 36.]

63 On p. 192 of H.E. Slaught, "Eliakim Hastings Moore: An appreciation," *Amer. Math. Monthly* **40** (1933), 191–195.

64 Eduard Riecke, Ernst Schering, Woldemar Voigt, Felix Klein, Issai Schur, and Heinrich Weber, "The teaching of mathematics at Göttingen," *Bulletin New York Math. Soc.* **3** (1893), 80–88.

65 On p. xii of E.H. Moore et al. (eds.), *Mathematical Congress.* [Endnote 51.]

66 *Ibid.*

67 *Ibid.*

68 *Idem*, p. v.

69 On p. 328 of Henry B. Fine, "Papers of the mathematical congress," *Bull. Amer. Math. Soc.* **2** (1896), 327–329.

70 The book is titled *The Evanston Colloquium: Lectures on Mathematics.* In addition to the lectures, it contains an appendix with a translation by Harry Tyler of a historical sketch "The development of mathematics at the German universities" that Felix Klein had written that year for the work *Die deutschen Universitäten.*

71 On p. 22 of T.F Fiske, "Notes," *Bull. New York Math. Soc.* **2** (1893), 22–25.

72 The lectures in the book vary slightly from those that were delivered. On p. viii of the Preface, Alexander Ziwet wrote, "The only change made consists in obliterating the conversational form of the frequent questions and discussions by means of which Professor Klein understands so well to enliven his discourse."

73 For a detailed analysis of the mathematical contents of these lectures, see pp. 333–354 of Parshall and Rowe, *Emergence.* [Endnote 11.]

74 On p. 95 of F. N. Cole, "The October meeting of the American Mathematical Society," *Bull. Amer. Math. Soc.* **6** (1899), 95–103.

75 On p. 1 of "Evansville man will apply for patent on invention," *Enterprise* (Evansville, Wisconsin), November 2, 1906.

76 On p. 260 of L.C. Karpinski, J.W. Bradshaw, T.H. Hildebrandt, and Peter Field, "Alexander Ziwet—In Memoriam," *Bull. Amer. Math. Soc.* **35** (1929), 259–260.

77 As quoted on p. 213 of Parshall and Rowe, *Emergence.* [Endnote 11.]

78 A popular T-shirt today pronounces, "Ithaca is gorges."

79 On p. 22 of T.F. Fiske, "Notes." *Bull. New York Math. Soc.* **2** (1893), 22–25.

80 *Ibid.*

81 Anonymous, "Notes," *Bull. Amer. Math. Soc.* **3** (1896), 92–93.

82 Felix Klein, "On the stability of a sleeping top," *Bull. Amer. Math. Soc.* **3** (1897), 129–132.

83 On pp. 360–361 of Parshall and Rowe, *Emergence.* [Endnote 11.]

84 For further information on colloquia, see Chapter 6 of Archibald, *Semicentennial History* [Endnote 36], pp. 66–73, for the period up to 1938, and Everett Pitcher, *A History of the Second Fifty Years: American Mathematical Society 1839–1988*, pp. 38-44, for the next 50 years.

85 Eugen Netto, "Zur Theorie der Tripelsysteme," *Mathematische Annalen* **43** (1893), 143–152.

86 On p. 73 of E.H. Moore, "A doubly infinite system of simple groups," *Bulletin of the NYMS* **3** (1893), 73–78.

87 E. Hastings Moore, "Concerning the abstract groups of order $k!$ and $(k!/2)$ holoedrically isomorphic with the symmetric and the alternating substitution-groups on k letters," *Proceedings of the London Mathematical Society* **28** (1896), 357–366.

88 In the penultimate footnote on p. 39 of E.H. Moore, "Concerning Jordan's linear groups," *Bull. Amer. Math. Soc.* **2** (1895), 33–43.

89 On p. 72 of E.H. Moore, "On crinkly curves," *Trans. Amer. Math. Soc.* **1** (1900), 72–90.

90 On p. 191 of H.E. Slaught, "Eliakim Hastings Moore: An appreciation," *Amer. Math. Monthly* **40** (1933), 191–195.

91 For further details, see pp. 398–399 of Parshall and Rowe, *Emergence*. [Endnote 11.]

92 For an account of Dickson's family, see Della D. Fenster, "Leonard Eugene Dickson (1874-1954): An American legacy in mathematics," *Mathematical Intelligencer* **21** (4) (1999), 54–59.

93 On p. 34 of Garrett Birkhoff, "Some leaders in American mathematics: 1891–1941," in J. Dalton Tarwater, John T. White, and John Miller (eds.), *Men and Institutions in American Mathematics*, Texas Tech Press, Lubbock, TX, 1976, pp. 25–78.

94 Leonard Eugene Dickson, "The analytic representation of substitutions on a power of a prime number of letters with a discussion of the linear group," *Annals Math.* **11** (1897), 65–120 and 161–183.

95 On p. 324 of Della D. Fenster, "American initiatives toward internationalization: The case of Leonard Dickson," in Karen Hunger Parshall and Adrian C. Rice (eds.), *Mathematics Unbound: The Evolution of an International Mathematical Research Community, 1800–1945*, American Mathematical Society, Providence, and London Mathematical Society, London, 2002, pp. 311–333.

96 On p. 9 of Karen Hunger Parshall, "A study in group theory: Leonard Eugene Dickson's Linear Groups," *Mathematical Intelligencer* **13** (1991), 7–11.

97 On p. 58 of Della D. Fenster, "Leonard Eugene Dickson (1874–1954): An American legacy in mathematics," *Mathematical Intelligencer* **21** (4) (1999), 54–59.

98 On p. 143 of Saunders Mac Lane, "Mathematics at the University of Chicago: A brief history," in Peter Duren (ed.), *A Century of Mathematics in America*, Vol. 3, American Mathematical Society, Providence, RI, and London Mathematical Society, London, 1989, pp. 127–154.

99 Della D. Fenster, "What makes a student the best?" *Amer. Math. Monthly* [to appear].

100 Burton Wadsworth Jones should not be confused with F. Burton Jones, who earned his Texas PhD in 1935 under R.L. Moore. Burton W. Jones directed the dissertations of three notable mathematicians at Cornell: Irving Reiner (1924–1986; University of Illinois), Mary Dolciani (1923–1985; Hunter College), and William LeVeque (1923–2007 and AMS Executive Director 1977–1988).

101 Mildred Sanderson, "Generalization in the theory of numbers and theory of linear groups," *Annals Math* **13** (1911), 36–39.

102 Mildred Sanderson, "Formal modular invariants with application to binary modular covariants," *Trans. Amer. Math. Soc.* **14** (1913), 489–500.

103 On p. 22 of E.T. Bell, "Fifty years of algebra in America," 1888–1938, in *Semicentennial Addresses of the American Mathematical Society*, American Mathematical Society, New York, 1938, pp. 1–34.

104 A point is said to be of the "first kind" if no geodesics passing through it contain conjugate points; otherwise the point is of the "second kind."

Chapter 7

1 As quoted on p. 67 of R.C. Archibald, *A Semicentennial History of the American Mathematical Society, 1888–1938*, American Mathematical Society, New York, 1938.

2 On p. 278 of "Notes," *Bulletin Amer. Math. Soc.* **2** (1896) 276–279.

3 On p. 49 of Thomas S. Fiske, "The Buffalo Colloquium," *Bulletin Amer. Math. Soc.* **3** (1896) 49–59.

4 R.C. Archibald, *Semicentennial*, p. 61. [Endnote 1.]

5 Thomas S. Fiske, "The Buffalo Colloquium," *Bulletin Amer. Math. Soc.* **3** (1896), 49–59.

6 On pp. 51–52 of *idem*.

7 On p. 59 of *idem*.

8 On p. 409 of "Notes," *Bulletin Amer. Math. Soc.* **4** (1898), 409–410.

9 On p. 58 of H.S. White, "The Cambridge Colloquium," *Bulletin Amer. Math. Soc.* **5** (1898), 57–58.

10 On p. 113 of F.N. Cole, "The October meeting of the American Mathematical Society," *Bulletin Amer. Math. Soc.* **7** (1900), 113–120.

11 On p. 23 of Edward Kasner, "The Ithaca Colloquium," *Bulletin Amer. Math. Soc.* **9** (1901), 22–25.

12 See p. 68 of R.C. Archibald, *Semicentennial*. [Endnote 1.]

13 On p. 272 of "Notes," *Bulletin Amer. Math. Soc.* **9** (1903), 271–275.

14 On p. 222 of F. N. Cole, "The tenth annual meeting of the American Mathematical Society," *Bulletin Amer. Math. Soc.* **10** (1904), 221–229.

15 All currency conversions in this book are based on the site **www.measuringworth.com**.

16 On p. 71 of Virgil Snyder, "The New Haven colloquium," *Bulletin Amer. Math. Soc.* **13** (1906), 71–74.

17 See Chapter 6, "Colloquium Lectures and Colloquium Publications," in Archibald, *Semicentennial*, for further details on these prestigious lectures. [Endnote 1.]

18 On p. 164 of "Notes," *Bulletin Amer. Math. Soc.* **3** (1897), 164–166.

19 Both quotations in this paragraph appear on p. 1 of Thomas S. Fiske, "The summer meeting of the American Mathematical Society," *Bulletin Amer. Math. Soc.* **1** (1894), 1–6.

20 Thomas S. Fiske, "The second summer meeting of the American Mathematical Society," *Bulletin Amer. Math. Soc.* **2** (1895), 1–7.

21 On p. 1 of F. N. Cole, "The third summer meeting of the American Mathematical Society," *Bulletin Amer. Math. Soc.* **3** (1896), 1–9.

22 Due to confusion with the name of the International Congress of Mathematicians, the Canadian Mathematical Congress was changed to the Canadian Mathematical Society in 1979.

23 See footnote 13 on p. 65 of Sloan Evans Despeaux, "International mathematical contributions to the British scientific journals, 1800–1900," in Karen Hunger Parshall and Adrian C. Rice (eds.), *Mathematics Unbound: The Evolution of an International Mathematical Research Community, 1800–1945*, American Mathematical Society, Providence, and London Mathematical Society, London, 2002, pp. 61–87.

24 See pp. 38–43 of E. Pitcher, *A History of the Second Fifty-Years: American Mathematical Society, 1939–1988*, American Mathematical Society, Providence, 1988.

25 F.N. Cole, "The fourth summer meeting of the American Mathematical Society," *Bulletin Amer. Math. Soc.* **4** (1897), 1–11.

26 On p. 16 of Elaine McKinnon Riehm and Frances Hoffman, *Turbulent Times in Mathematics: The Life of J.C. Fields and the History of the Fields Medal*, American Mathematical Society, Providence RI, and London Mathematical Society, London, 2011.

27 On p. 13 of Gilbert de B. Robinson, *The Mathematics Department in the University of Toronto 1827–1978*, University of Toronto Press, Toronto, Ontario, 1979.

28 AMS Archives, Box 1, Folder 4 (May 3, 1924).

29 On p. 75 of Archibald, *Semicentennial*. [Endnote 1.]

30 On p. 199 of "Notes," *Bulletin Amer. Math. Soc.* **3** (1897), 199–200.

31 On p. 258 of "Notes," *Bulletin Amer. Math. Soc.* **3** (April 1897), 257–260.

32 As quoted on p. 9 of Archibald, *Semicentennial*. [Endnote 1.]

33 As quoted in *idem*, p. 56.

34 *Ibid.*

35 As quoted in *idem*, p. 57.

36 *Ibid.*

37 As quoted in *idem.*, p. 58.

38 On p. 15 of Thomas S. Fiske, "The beginnings of the American Mathematical Society," Appendix B to "The Semicentennial Celebration," *Bulletin Amer. Math. Soc.* **45** (1939), 12–15. Reprinted in Peter Duren (ed.), *A Century of Mathematics in America*, Vol. 1, American Mathematical Society, Providence, RI, and London Mathematical Society, London, 1988, pp. 13–17.

39 As quoted on p. 60 of Archibald, *Semicentennial*. [Endnote 1.]

40 The home base for the *Annals of Mathematics* moved to Princeton in 1911 and has been published there since then.

41 Details on Pell's fascinating life are contained in the book by R. Pipes, *The Degaev Affair: A Terror and Treason in Tsarist Russia*, Yale University Press, New Haven, 2003.

42 On pp. 369–370 of Gaston Darboux, "Sophus Lie," *Bulletin Amer. Math. Soc.* 5 (1899), 367–370. This article was translated by Edgar Odell Lovett from the original French version that appeared in the February 27, 1899, issue of *Comptes rendus*. Lovett was one of Lie's most famous American students.

43 *Idem*, p. 368.

44 All three quotations in this paragraph appear on pp. 191–192 of G.A. Miller, "Some reminiscences in regard to Sophus Lie," *Amer. Math. Monthly* 6 (1899), 191–193.

45 On p. 97 of George Bruce Halsted, "Biography: Sophus Lie," *Amer. Math. Monthly* 6 (1899), 97–101.

46 This quotation, and the next, appear on p. 236 of Karen Parshall, and David Rowe, *The Emergence of the American Mathematical Research Community, 1876–1900: J. J. Sylvester, Felix Klein, and E.H. Moore*, American Mathematical Society, Providence, and London Mathematical Society, London, 1994.

47 James M. Page, "On the primitive groups of transformations in space of four dimensions" *Amer. Journal Math.* 10 (1888), 293–346.

48 On p. 451 of L.E. Dickson, "Note on Page's 'Ordinary Differential Equations'," *Bulletin Amer. Math. Soc.* 5 (1899), 451–455.

49 See the final footnote on p. 353 of Edgar Odell Lovett, "Page's 'Differential Equations'," *Bulletin Amer. Math. Soc.* 4 (1898), 349–353.

50 Edgar O. Lovett, "The theory of perturbations and Lie's theory of contact transformations," *Quarterly Journal of Pure and Applied Mathematics* 30 (1898), 47–149.

51 For more information on Bouton, see the memorial article prepared by William Osgood, Julian Coolidge, and George Chase, published in *Bulletin Amer. Math. Soc.* 28 (1922), 123–124.

52 On p. 460 of "Notes," *Bulletin Amer. Math. Soc.* 5 (1899), 458–461.

53 On p. 35 of Charles L. Bouton, "Nim, a game with a complete mathematical theory," *Annals of Mathematics* 3 (1901–1902), 35–39.

54 On p. 235 of Haley Dozier and John Perry, "Androids armed with poisoned chocolate squares: Ideal Nim and its relatives," *Mathematical Magazine* 89 (2016), 235–250.

55 David A. Rothrock, "Point invariants for the finite continuous groups of the plane," *Amer. Math. Monthly* 5 (1898), 249–264.

56 On p. 146 of Gary G. Cochell, "The early history of the Cornell mathematics department: A case study in the emergence of the American mathematical community," *Historia Mathematica* 25 (1998), 133–153.

57 On p. 365 of "Notes," *Bulletin Amer. Math. Soc.* 7 (1901), 362–369.

58 H.F. Blichfeldt, "On a certain class of groups of transformation in three dimensional space," *Amer. J. Math.* 22 (1900), 113–120.

59 On p. 184 of E.T. Bell, "Hans Frederik Blichfeldt 1873–1945," *Biographical Memoirs of the National Academy of Sciences*, 1951, 180–189. This article contains a bibliography of Blichfeldt's works.

60 On p. 8 of E.P. Lane, "Ernest Julius Wilczynski—In Memoriam," *Bulletin Amer. Math. Soc.* 39 (1933), 7–14. This article contains a bibliography of Wilczynski's works.

61 *Idem*, p. 7.

62 Both quotations in this paragraph appear on p. 204 of Jeremy J. Gray, "Languages for mathematics and the language of mathematics in the world of nations," in Parshall and Rice, *Unbound*, pp. 201–228. [Endnote 23.]

63 A modern account of the languages used in scientific writing is Michael D. Gordin, *Scientific Babel: How Science Was Done Before and After Global English*, University of Chicago Press, Chicago and London, 2015.

64 The list of Pierpont's students is given on p. 482 of Øystein Ore, "James Pierpont—In Memoriam," *Bulletin Amer. Math. Soc.* 45 (1939), 481–486. This article contains a bibliography of Pierpont's works. A more recent study underscored the difficulty of assigning doctoral advisors at Yale by revealing two women PhDs who earned their degrees under Pierpont; see footnote 37 on p. 53 of Judy Green and Jeanne LaDuke, *Pioneering Women in American Mathematics*, American Mathematical Society, Providence, RI, London Mathematical Society, London, 2009.

65 On p. 482 of Ore, "In Memoriam." [*Idem*.]

66 The two quotations in this paragraph appear on pp. 225–226 of James Pierpont, "Mathematical instruction in France," *Bulletin Amer. Math. Soc.* **6** (1900), 225–249.

67 All three quotations in this paragraph appear on p. 9 of J.L. Coolidge, "The opportunities for mathematical study in Italy," *Bulletin Amer. Math. Soc.* **11** (1904), 9–17.

68 Richardson's method missed some figures, including John Westfall.

69 Our account is based on H.R. Brahana, "George Abram Miller (1863–1951)," *Biographical Memoirs. National Academy of Sciences* **30** (1957), 257–312.

70 On p. 55 of Green and LaDuke, *Pioneering Women*. [Endnote 64.]

71 As quoted on p. 142 of Walter Isaacson (ed.), *A Benjamin Franklin Reader*, Simon & Schuster, New York and London, 2003.

72 The primary reference for the history of the mathematics department at the University of Pennsylvania is the pamphlet *Centennial Celebration* that accompanied its first 100 years on October 30, 1999.

73 *Idem*, p. 13.

74 Most of this material on the development of mathematics at Cornell in the nineteenth century is drawn from Cochell, Early history. [Endnote 56.]

75 As quoted on p. 63 of G.W. Hill, "Memoir of James Edward Oliver, 1829–1895," *Biographical Memoirs* **4** (1902), National Academy of Sciences, Washington DC, 57–74. This is our major source of information on Oliver.

76 *Ibid.*

77 *Idem*, p. 65.

78 As quoted on p. 213 of Parshall and Rowe, *Emergence*. [Endnote 46.]

79 On p. 69 of Hill, Memoir Oliver. [Endnote 75.]

80 On p. 545 of G.B. Halsted, "James Edward Oliver," *Science* **1** (20) (May 17, 1895), 544–545.

81 On p. 213 of Parshall and Rowe, *Emergence*. [Endnote 46.]

82 As quoted on p. 146 of Cochell, "The early history." [Endnote 56.]

83 On p. 557 of "Notes," *Bulletin Amer. Math. Soc.* **4** (1898), 554–558.

84 The two main sources on Snyder are pp. 218–223 of Archibald *Semicentennial* [Endnote 1], and A.B. Coble, "Virgil Snyder 1869–1950," *Bulletin Amer. Math. Soc.* **56** (1950), 468–471.

85 On p. 218 of Parshall and Rowe, *Emergence*. [Endnote 46.]

86 On pp. 130–131 of Charles F. Thwing, *The College Woman*, Taylor & Baker, New York, 1894.

87 On p. 31 of Margaret Rossiter, *Women Scientists in America*, Johns Hopkins University Press, Baltimore, 1982.

88 As quoted on p. 41 of Green and LaDuke, *Pioneering Women*. [Endnote 64.]

89 See *idem.*, especially p. 248, for additional information on Ida Metcalf. She is also cited for taking in an eight-year-old ward who later became very successful.

90 On pp. 97–98 of Annie L. MacKinnon, "Concomitant binary forms in terms of the roots," *Annals Math.* **9** (1894–1895), 95-157.

91 *Idem*, p. 157.

92 Annie L. MacKinnon, "Concomitant binary forms in terms of the roots," *Annals Math.* **12** (1898–1899), 95-109.

93 The enrollment in Klein's number theory course was 13, including four women, three of whom were the American women mentioned here. One of the men was the Johns Hopkins graduate J.C. Fields, for whom the Fields Medal is named.

94 It seems relevant here to point out that no Canadian university granted a PhD to a woman until 1930, so Baxter was forced to go to the US for graduate study.

95 See Green and LaDuke, *Pioneering Women*, for additional information on these women, pp. 174–175 for Annie MacKinnon Fitch, and pp. 201–202 for Agnes Baxter Hill. [Endnote 64.]

96 As quoted on p. 506 of Susan E. Kelly and Sarah A. Rozner, "Winifred Haring (Edgerton) Merrill: 'She opened the door,' " *Notices Amer. Math. Soc.* **59** (2012), 504–512. An analysis of this dissertation is given on pp. 506–508.

97 "Girls fight school fire," *New York Times*, February 19, 1924, p. 3.

98 Ernest William Brown, "Mathematics," pp. 52–55 in *Alumnæ, Graduate School, Yale University, 1894–1920*, Yale University, New Haven, 1920.

99 Both quotations in this paragraph appear on p. 72 of Karen Hunger Parshall, "Training women in mathematical research: The first fifty years of Bryn Mawr College," *Math. Intelligencer* **37** (2) 2015, 71–83.

100 *Idem*, p. 73.

101 Quoted on p. 194 of Patricia Clark Kenschaft, "Charlotte Angas Scott (1858–1931)," in Louise S. Grinstein and Paul J. Campbell (editors), *Women of Mathematics: A Biobibliographic Sourcebook*, Greenwood Press, New York, 1987, pp. 193–203. This is our main source for information on Scott.

102 Mary W. Gray was elected vice president of the AMS in 1976. Julia Robinson was the first woman president of the Society, serving in this capacity 1983–1984.

103 For details on Gentry's life, including correcting the record on the date of her degree, see p. 180 of Green and LaDuke, *Pioneering Women*, pp. 180–181. [Endnote 64.]

104 Felix Klein, "The arithmetizing of mathematics," *Bulletin Amer. Math. Soc.* **2** (1896), 241–249.

105 On p. 72 of "Notes," *Bulletin Amer. Math. Soc.* **28** (1922), 72–82.

106 Isabel Maddison, "Two books on elementary geometry," *Bulletin Amer. Math. Soc.* **3** (1897), 253–255.

107 Isabel Maddison, "Shorter notices," *Bulletin Amer. Math. Soc.* **4** (1898), 234–235.

108 Isabel Maddison, "Shorter notices," *Bulletin Amer. Math. Soc.* **6** (1899), 113–115.

109 Isabel Maddison, "Note on the history of the map-coloring problem," *Bulletin Amer. Math. Soc.* **3** (1897), 257.

110 On p. 90 of Patricia Clark Kenschaft, *Change is Possible: Stories of Women and Minorities in Mathematics*, American Mathematical Society, Providence, RI, 2005.

111 *Idem*, p. 92.

112 As quoted on p. 19 of Paul J. FitzPatrick, "Statistical societies in the United States in the nineteenth century," *The Amer. Statistician* **11** (1957), 13–21.

113 Additional details are provided in the paper by Paul J. FitzPatrick, "The early teaching of statistics in American colleges and universities," *The Amer. Statistician* **9** (1955), 12–18.

114 On p. 313 of Paul J. FitzPatrick, "Leading American statisticians in the nineteenth century," *Journal of the Amer. Statistical Assoc.* **52** (1957), 301–321.

115 As stated on the website http://www.wharton.upenn.edu/about/wharton-history.cfm.

116 For more information on the career and publications of Falkner, see F. Leslie Hayford, "Roland Post Falkner, 1866–1940," *Journal Amer. Stat. Assoc.* **36** (1941), 543–545.

117 For more information on the course and textbook, see pp. 16–17 of Paul J. FitzPatrick, "Statistical works in early American statistics courses," *The Amer. Statistician* **10** (1956), 14–19.

118 As stated on p. 15 of FitzPatrick, "Early teaching." [Endnote 113.]

119 As quoted on p. 17 of FitzPatrick, "Statistical works." [Endnote 117.]

120 The first two quotations in this paragraph appear on p. 220 of Theodore M. Porter, *The Rise of Statistical Thinking 1820–1900*, Princeton University Press, Princeton, NJ, 1986. Third quotation appears on pp. 219–227.

121 *Ibid.*

122 As quoted on p. 16 of FitzPatrick, "Early teaching." [Endnote 113.]

123 *Ibid.*

124 *Ibid.*

125 On p. 17 of Carroll D. Wright, "Statistics in Colleges," *Publications Amer. Economic Assoc*, **3** (1888), 5–28.

126 Sloan Evans Despeaux, "International mathematical contributions to British scientific journals, 1800–1900," in Parshall and Rice, *Mathematics Unbound*, pp. 61–87. [Endnote 23.]

Transition 1900: Hilbert's American Colony

1 On p. 48 of Constance Reid, *Hilbert*, Springer-Verlag, New York–Berlin–Heidelberg, 1970.

2 On p. 184 of E.T. Bell, "Hans Frederik Blichfeldt 1873–1945," *Biographical Memoirs Nat. Acad. Sci.* **25** (1951), 180–189.

3 Reid, Constance, *Hilbert*, Springer-Verlag, New York–Berlin, 1970, pp. 93-94. Oliver Kellogg is the only other member of the Hilbert Colony in America mentioned in this well-received biography.

4 On p. 228 of George Bruce Halsted, "Supplementary report on non-Euclidean geometry," *Amer. Math. Monthly* **8** (1901), 216–230.

5 On p. 410 of W.B. Ford, "Obituary: Earle Raymond Hedrick," *Amer. Math. Monthly* **50** (1943), 409–411.

6 On p. 504 of Joseph D. Zund, "Oliver Dimon Kellogg," *American National Biography* **12** (1999), 503–504.

7 Patti Wilger Hunter, "Max Mason," *American National Biography* online (2000): http://www.anb.org.libproxy.temple.edu/articles/13/13-01059.html

8 As quoted on p. 208 of Warren Weaver, "Max Mason 1877–1961," *Biographical Memoirs Nat. Acad. Sci.* **37** (1963), 205–236.

9 As quoted on p. 69 of R.C. Archibald, *A Semicentennial History of the American Mathematical Society, 1888–1938*, American Mathematical Society, New York, 1938.

10 Hunter, "Max Mason." [Endnote 7.]

11 On p. 274 of George Bruce Halsted, "Book review," *Amer. Math. Monthly* **9** (1902), 274–275. See also George Bruce Halsted, "Book review," *Science* **16** (1902), 307–308.

12 B.F. Finkel, "Book review," *Amer. Math. Monthly* **16** (1909), 78; Hans H. Dalaker, "Book review," *Amer. Math. Monthly* **24** (1917), 228–229; and Albert W. Raab, "Book review," *Amer. Math. Monthly* **36** (1929), 330–332.

13 On p. 201 of Archibald, "Semicentennial." [Endnote 9.]

14 *Idem*, pp. 22 and 24.

15 On p. 100 of Mark Kac, *Enigmas of Chance: An Autobiography*, Harper & Row, New York, 1985.

16 Griffith C. Evans, Thomas Buck, and Hans Lewy, "In memoriam: Charles Albert Noble (1867–1962)," University of California at Berkeley, 1963.

17 On p. 209 of Weaver, "Max Mason." [Endnote 8.]

18 Formally, Stringham was not the chair of the department, as that term did not come into use at Berkeley until 1920, when it was applied to Mellen Haskell, who was appointed to the position upon Stringham's death in 1907. Both Stringham and Haskell had studied under Felix Klein, the former at Leipzig and the latter in Göttingen.

19 Gilman Hall is named for Daniel Coit Gilman, the initial president of the University of California at Berkeley who, a few years later, brought J.J. Sylvester to Johns Hopkins University.

20 On p. 224 of Archibald, *Semicentennial*. [Endnote 9.]

21 D.R. Curtiss, "Book review," *National Math. Magazine* **15** (1940), 158.

22 Earl [*sic*] Raymond Hedrick, "Our retiring secretary-treasurer," *Amer. Math. Monthly* **50** (1943), 1.

23 Lester R. Ford, "Our retiring secretary-treasurer," *Amer. Math. Monthly* **50** (1943), 1.

24 On p. 299 of W.A. Hurwitz, "David Clinton Gillespie—In Memoriam," *Bulletin Amer. Math. Soc.* **42** (1936), 298–299.

25 Jósef H. Przytyki, "Little and Haseman—early American tabulators of knots," lecture at the AMS Special Session on "The Development of topology in the Americas," Austin, TX, 8–10, 1999.

26 Mary Gertrude Haseman, "On knots, with a census of the amphericheirals with twelve crossings," *Transactions Royal Soc. Edinburgh* **52** (Part I) (1917–1918), 235-255.

27 See also Mary Gertrude Haseman, "Amphericheiral knots," *Transactions Royal Soc. Edinburgh* **52** (Part III) (1919–1920), 597-602.

28 The website http://www.maths.ed.ac.uk/~aar/knots/ is maintained by Jozef Przytycki and Andrew Ranicki.

29 On p. 9 of J.L. Coolidge, "The opportunities for mathematical study in Italy," *Bull. Amer. Math. Soc.* **11** (1904), 9–17.

Bibliography

[1] D. J. Albers and G. L. Alexanderson (eds.), *Fascinating mathematical people: Interviews and memoirs*, Princeton University Press, Princeton, NJ, 2011. MR2866911

[2] D. J. Albers, G. L. Alexanderson, and C. Reid, *International mathematical congresses: An illustrated history, 1893–1986*, Springer-Verlag, New York, 1987. MR874336

[3] S. F. Kennedy, D. J. Albers, G. L. Alexanderson, D. Dumbaugh, F. A. Farris, D. B. Haunsperger, and P. Zorn (eds.), *A century of advancing mathematics*, Mathematical Association of America, Washington, DC, 2015. MR3381845

[4] N. E. Albert, A^3 *& his algebra: How a boy from Chicago's west side became a force in American mathematics*, iUniverse, Inc., Lincoln, NE, 2005. With a preface by Erwin Kleinfeld. MR2147379

[5] J. Albree, D. C. Arney, and V. F. Rickey, *A station favorable to the pursuits of science: Primary materials in the history of mathematics at the United States Military Academy*, History of Mathematics, vol. 18, American Mathematical Society, Providence, RI; London Mathematical Society, Cambridge, 2000. MR1730857

[6] R. C. Archibald, *A correction*, Bull. Amer. Math. Soc. **42** (1936), no. 3, 170, DOI 10.1090/S0002-9904-1936-06267-1. MR1563264

[7] R. C. Archibald, *History of the American Mathematical Society, 1888–1938*, Bull. Amer. Math. Soc. **45** (1939), no. 1, 31–46, DOI 10.1090/S0002-9904-1939-06908-5. MR1563903

[8] Aubin, David and Catherine Goldstein (editors), "The War of Guns and Mathematics," *Mathematical Practices and Communities in France and Its Western Allies around World War I*, American Mathematical Society, Providence RI, 2014.

[9] C. E. Aull and R. Lowen (eds.), *Handbook of the history of general topology. Vol. 1*, Kluwer Academic Publishers, Dordrecht, 1997. MR1619254

[10] S. Batterson, *American mathematics 1890–1913: catching up to Europe*, MAA Spectrum, Mathematical Association of America, Washington, DC, 2017. MR3675486

[11] S. Batterson, *Pursuit of genius: Flexner, Einstein, and the early faculty at the Institute for Advanced Study*, A K Peters, Ltd., Wellesley, MA, 2006. MR2248799

[12] S. Batterson, *Stephen Smale: the mathematician who broke the dimension barrier*, American Mathematical Society, Providence, RI, 2000. MR1737026

[13] S.A. Bedini, *The Life of Benjamin Banneker: The First African-American Man of Science*, (second edition), Maryland Historical Society, Baltimore, 1999.

[14] E. T. Bell, *Men of mathematics*, Simon & Schuster, New York, 1937. MR3728303

[15] E. T. Bell, *The Development of Mathematics*, McGraw-Hill Book Company, Inc., New York, 1945. MR0016318

[16] N. Bowditch, *American Practical Navigator*, Government Printing Office, Washington DC, 1802.

[17] F. E. Browder (ed.), *Mathematical developments arising from Hilbert problems*, Proceedings of Symposia in Pure Mathematics, Vol. XXVIII, American Mathematical Society, Providence, RI, 1976. MR0419125

[18] F. Cajori, *The Teaching and History of Mathematics in the United States*, Government Printing Office, Washington DC, 1890.

[19] J. Carlson, A. Jaffe, and A. Wiles (eds.), *The Millennium Prize Problems*, Clay Mathematics Institute, Cambridge, MA, and American Mathematical Society, Providence, RI, 2006. MR2246251

[20] *Astronomical papyri from Oxyrhynchus Vol. I, II*, Memoirs of the American Philosophical Society, vol. 233, American Philosophical Society, Philadelphia, PA, 1999. P. Oxy. 4133–4300a; Edited, with translations from the Greek and commentaries by Alexander Jones. MR1757948

[21] B.A. Case (ed.), *A Century of Mathematics Meetings*, American Mathematical Society, Providence RI, 1996.

[22] I. B. Cohen (ed.), *Benjamin Peirce: "Father of pure mathematics" in America: Three Centuries of Science in America*, Arno Press, New York, 1980. MR628145

[23] I.B. Cohen, *Science and the Founding Fathers: Science in the Political Thought of Thomas Jefferson, Benjamin Franklin, John Adams, and James Madison*, W. W. Norton, New York, 1995.

[24] M. J. Crowe, *A history of vector analysis. The evolution of the idea of a vectorial system*, University of Notre Dame Press, Notre Dame, Ind.-London, 1967. MR0229496

[25] G. P. Curbera, *Mathematicians of the world, unite!: The International Congress of Mathematicians— a human endeavor*, A K Peters, Ltd., Wellesley, MA, 2009. With a foreword by Lennart Carleson. MR2499757

[26] G.H. Daniels, *American Science in the Age of Jackson*, Columbia University Press, New York, 1968.

[27] P. Duren et al. (eds.), *A Century of Mathematics in America: Part I*, American Mathematical Society, Providence RI, 1988.

[28] P. Duren et al. (eds.), *A Century of Mathematics in America: Part II*, American Mathematical Society, Providence RI, 1989.

[29] P. Duren et al. (eds.), *A Century of Mathematics in America: Part III*, American Mathematical Society, Providence RI, 1989.

[30] M. A. Clements and N. F. Ellerton, *Thomas Jefferson and his decimals 1775–1810: neglected years in the history of U. S. school mathematics*, Springer, Cham, 2015. With a foreword by Douglas L. Wilson. MR3288726

[31] H. W. Eves, *Mathematical circles revisited: A second collection of mathematical stories and anecdotes*, The Prindle, Weber & Schmidt Series in Mathematics, PWS-KENT Publishing Co., Boston, MA, 1971. MR958118

[32] J. Ewing (ed.), *A century of mathematics: Through the eyes of the Monthly*, Mathematical Association of America, Washington, DC, 1994. MR1272018

[33] P. Fillmore (ed.), *Mathematics in Canada: Les mathematiques au Canada, Volume/Tome 1*, Canadian Mathematical Society/Society mathematiques du Canada, Ottawa, 1995.

[34] B.F. Finkel, *A Mathematical Solution Book*, Kibler & Co., Springfield, MO, 1888.

[35] R. Fox (ed.), *Thomas Harriot: An Elizabethan Man of Science*, Ashgate, Burlington, VT, 2000.

[36] D.R. Francis, *The Universal Exposition of 1904*, Louisiana Purchase Exposition Company, St. Louis, 1913.

[37] C. C. Gillispie, *Eloge: Charles Scribner, Jr., 13 July 1921–11 November 1995*, Isis **88** (1997), no. 2, 302–303, DOI 10.1086/383694. MR1472607

[38] H. H. Goldstine, *The computer from Pascal to von Neumann*, Princeton University Press, Princeton, NJ, 1993. Reprint of the 1972 original. MR1271142

[39] T.W. Goodspeed, *A History of the University of Chicago Founded by John D. Rockefeller: The First Quarter-Century*, University of Chicago Press, Chicago, 1916.

[40] J. R. Goodstein, *The Volterra chronicles: The life and times of an extraordinary mathematician 1860–1940*, History of Mathematics, vol. 31, American Mathematical Society, Providence, RI; London Mathematical Society, London, 2007. MR2287463

[41] M.D. Gordin, *Scientific Babel: How Science Was Done Before and After Global English*, University of Chicago Press, Chicago/London, 2015.

[42] I. Grattan-Guinness, *The search for mathematical roots, 1870–1940: Logics, set theories and the foundations of mathematics from Cantor through Russell to Gödel*, Princeton Paperbacks, Princeton University Press, Princeton, NJ, 2000. MR1807717

[43] J.J. Gray and K.H. Parshall (eds.), *History of Algebra in the 19th and 20th Centuries*, American Mathematical Society and London Mathematical Society, Providence RI and London, 2007.

[44] J. Green and J. LaDuke, *Women in the American mathematical community: the pre-1940 Ph.D.s*, Math. Intelligencer **9** (1987), no. 1, 11–23, DOI 10.1007/BF03023568. MR869536

[45] D.A. Grier, *When Computers Were Human*, Princeton University Press, Princeton and Oxford, 2005.

[46] L. S. Grinstein and P. J. Campbell (eds.), *Women of mathematics: A biobibliographic sourcebook*, Greenwood Press, Westport, CT, 1987. MR911481

[47] T. Harriot, *A Briefe and True Report of the New-found Land of Virginia*, Dover Publications, Garden City NJ, 1972.

[48] S. J. Heims, *John von Neumann and Norbert Wiener: From mathematics to the technologies of life and death*, MIT Press, Cambridge, Mass.-London, 1980. MR592236

[49] J. Hilty, *Temple University: 125 Years of Service to Philadelphia, the Nation and the World*, Temple University Press, Philadelphia PA, 2010.

[50] B. Hindle, *David Rittenhouse*, Princeton University Press, Princeton, 1964.

[51] B. Hindle, *The Pursuit of Science in Revolutionary America, 1735–1789*, University of North Carolina Press, Chapel Hill, 1956.

[52] E. R. Hogan, *Of the human heart: A biography of Benjamin Peirce*, Lehigh University Press, Bethlehem, PA, 2008. MR2542268

[53] W. Isaacson (ed.), *A Benjamin Franklin Reader*, Simon & Schuster, New York/London, 2003.

[54] I. James, *Remarkable mathematicians: From Euler to von Neumann*, MAA Spectrum, Mathematical Association of America, Washington, DC; Cambridge University Press, Cambridge, 2002. MR1964301

[55] D. Jerison and D. W. Stroock, *Norbert Wiener*, The Legacy of Norbert Wiener: A Centennial Symposium (Cambridge, MA, 1994), Proc. Sympos. Pure Math., vol. 60, Amer. Math. Soc., Providence, RI, 1997, pp. 3–19, DOI 10.1090/pspum/060/1460269. MR1460269

[56] N. L. Johnson and S. Kotz (eds.), *Leading personalities in statistical sciences: From the seventeenth century to the present*, Wiley Series in Probability and Statistics: Probability and Statistics, John Wiley & Sons, Inc., New York, 1997. A Wiley-Interscience Publication. MR1469759

[57] P.S. Jones (ed.), *A History of Mathematics Education in the United States and Canada*, National Council of Teachers, Washington, DC, 1970.

[58] M. Kac, *Enigmas of chance: An autobiography*, Alfred P. Sloan Foundation, Harper & Row, Publishers, New York, 1985. MR837421

[59] D. Kahn, *The Codebreakers: The Story of Secret Writing*, Scribner, New York, 1996.

[60] L. C. Karpinski, *Bibliography of Mathematical Works Printed in America through 1850*, University of Michigan Press, Ann Arbor, Mich., 1940. MR0001733

[61] P. C. Kenschaft, *Change is possible: Stories of women and minorities in mathematics*, American Mathematical Society, Providence, RI, 2005. MR2167140

[62] P. A. Kidwell, A. Ackerberg-Hastings, and D. L. Roberts, *Tools of American mathematics teaching, 1800–2000*, Johns Hopkins Studies in the History of Mathematics, Smithsonian Institution, Washington, DC; Johns Hopkins University Press, Baltimore, MD, 2008. MR2426819

[63] M. Kline, *Mathematical thought from ancient to modern times*, Oxford University Press, New York, 1972. MR0472307

[64] E. Knobloch and D. E. Rowe (eds.), *The history of modern mathematics. Vol. III: Images, ideas, and communities*, Academic Press, Inc., Boston, MA, 1994. MR1282498

[65] O. Lehto, *Mathematics without borders: A history of the International Mathematical Union*, Springer-Verlag, New York, 1998. MR1488698

[66] J.A.L. Lemay, *The Life of Benjamin Franklin, Volume 3: Soldier, Scientist, and Politician, 1748–1757*, University of Pennsylvania Press, Philadelphia, 2014.

[67] P. R. Masani, *Norbert Wiener: 1894–1964*, Vita Mathematica, vol. 5, Birkhäuser Verlag, Basel, 1990. MR1032520

[68] K.O. May (ed.), *The Mathematical Association of America: Its First Fifty Years*, Mathematical Association of America, Washington, DC, 1972.

[69] C. C. Moore, *Mathematics at Berkeley: a history*, A K Peters, Ltd., Wellesley, MA, 2007. MR2289685

[70] E.H. Moore, O. Bolza, H. Maschke, and H. S. White (eds.), *Mathematical Papers Read at the International Congress Held in Connection with the World's Columbian Exposition Chicago 1893*, McMillan & Company, New York, 1896.

[71] S.E. Morison, *The Development of Harvard University, 1869–1929*, Harvard University Press, Cambridge, 1930.

[72] S. Nadis and S.-T. Yau, *A history in sum: 150 years of mathematics at Harvard (1825–1975)*, Harvard University Press, Cambridge, MA, 2013. MR3100544

[73] S. Nasar, *A beautiful mind*, Simon & Schuster, New York, 1998. MR1631630

[74] F. Nebeker, *Astronomy and the Geophysical Tradition in the United States in the Nineteenth Century: A Guide to Manuscript Sources in the Library of the American Philosophical Society*, American Philosophical Society Library, Philadelphia, PA, 1991.

[75] J.R. Newman (ed.), *The World of Mathematics: A small library of the literature of mathematics from A'hmosé the Scribe to Albert Einstein, presented with commentaries and notes*, 4 vols., Simon and Schuster, New York, 1956.

[76] C. O'Connor, *Beginning a Great Work: Washington University in St. Louis, 1853-2003*, Washington University Press, St. Louis, 2003.

[77] A. Oleson and S.C. Brown (eds.), *The Pursuit of Knowledge in the Early American Republic*, Johns Hopkins University Press, Baltimore MD and London, 1976.

[78] K. H. Parshall, *James Joseph Sylvester: Jewish mathematician in a Victorian world*, Johns Hopkins University Press, Baltimore, MD, 2006. MR2216541

[79] K. H. Parshall, *James Joseph Sylvester: Life and work in letters*, The Clarendon Press, Oxford University Press, New York, 1998. MR1674190

[80] K.H. Parshall and A.C. Rice (eds.), *Mathematics Unbound: The Evolution of an International Mathematical Research Community, 1800–1945*, American Mathematical Society and London Mathematical Society, Providence RI and London, 2002.

[81] K. H. Parshall and D. E. Rowe, *The emergence of the American mathematical research community, 1876–1900: J. J. Sylvester, Felix Klein, and E. H. Moore*, History of Mathematics, vol. 8, American Mathematical Society, Providence, RI; London Mathematical Society, London, 1994. MR1290994

[82] P. C. Pasles, *Benjamin Franklin's numbers: An unsung mathematical odyssey*, Princeton University Press, Princeton, NJ, 2008. MR2355634

[83] A.P. Peabody, *Harvard Reminiscences*, Ticknor and Co., Boston, 1888.

[84] E. R. Phillips (ed.), *Studies in the history of mathematics*, MAA Studies in Mathematics, vol. 26, Mathematical Association of America, Washington, DC, 1987. MR913096

[85] R. Pipes, *The Degaev Affair: A Terror and Treason in Tsarist Russia*, Yale University Press, New Haven, 2003.

[86] E. Pitcher, *A History of the Second 50 Years: American Mathematical Society 1939–1988*, American Mathematical Society, Providence RI, 1988.

[87] D. R. Porter, *William Fogg Osgood at Harvard: Agent of a transformation of mathematics in the United States*, Docent Press, Boston, MA, 2013. MR3135333

[88] T. M. Porter, *The Rise of Statistical Thinking 1820–1900*, Princeton University Press, Princeton NJ, 1986.

[89] C. Reid, *Courant*, Springer-Verlag, New York, 1996.

[90] C. Reid, *Hilbert*, Springer-Verlag, New York, 1970.

[91] C. Reid, *Neyman—from life*, Springer-Verlag, New York, 1982. MR680939

[92] E. M. Riehm and F. Hoffman, *Turbulent times in mathematics: The life of J. C. Fields and the history of the Fields Medal*, American Mathematical Society, Providence, RI; Fields Institute for Research in Mathematical Sciences, Toronto, ON, 2011. MR2850575

[93] D. L. Roberts, *American mathematicians as educators, 1893–1923: Historical roots of the "math wars"*, Docent Press, Boston, MA, 2012. MR3136004

[94] G. de B. Robinson, *The Mathematics Department in the University of Toronto 1827–1978*, University of Toronto Press, Toronto, Ontario, 1979.

[95] H.J. Rogers, *Congress of Arts and Science: Universal Exposition, St. Louis, 1904*, Houghton, Mifflin and Company, Boston and New York, 1905.

[96] M.W. Rossiter, *Women Scientists in America: Struggles and Strategies to 1940*, Johns Hopkins University Press, Baltimore MD, 1982.

[97] D. Rowe and J. McCleary (editors), *The History of Modern Mathematics, Volume 1: Ideas and their Reception*, Academic Press, Boston, 1989.

[98] D. G. Saari (ed.), *The way it was: Mathematics from the early years of the Bulletin*, American Mathematical Society, Providence, RI, 2003. MR2049152

[99] S. L. Segal, *Mathematicians under the Nazis*, Princeton University Press, Princeton, NJ, 2003. MR1991149

[100] R. E. O'Malley Jr., *Recountings: conversations with MIT mathematicians [book review of MR2516491]*, SIAM Rev. **51** (2009), no. 4, 802. MR2573949

[101] A. Shell-Gellasch (ed.), *History of Undergraduate Mathematics in America*, United States Military Academy, West Point NY, 2001.

[102] A. Shell-Gellasch, *In service to mathematics: The life and work of Mina Rees*, Docent Press, Boston, MA, 2011. MR2883650

[103] C.K. Shipton, *New England Life in the Eighteenth Century: Representative Biographies from Sibley's Harvard Graduates*, Harvard University Press, Cambridge, MA, 1963.

[104] R. Siegmund-Schultze, *Rockefeller and the internationalization of mathematics between the two world wars: Documents and studies for the social history of mathematics in the 20th century*, Science Networks. Historical Studies, vol. 25, Birkhäuser Verlag, Basel, 2001. MR1826906

[105] D.E. Smith, *A History of Modern Mathematics*, John Wiley, New York, 1906.

[106] D.E. Smith and J. Ginsburg, *A History of Mathematics in America Before 1900*, Mathematical Association of America, Chicago, 1934.

[107] D. E. Smith and J. Ginsburg, *A history of mathematics in America before 1900*, Three Centuries of Science in America, Arno Press, New York, 1980. Reprint of the 1934 original. MR749578

[108] F.F. Stephens, *A History of the University of Missouri*, University of Missouri Press, Columbia, MO, 1962.

[109] D. J. Struik, *A concise history of mathematics*, 4th ed., Dover Publications, Inc., New York, 1987. MR919604

[110] D. J. Struik, *Yankee Science in the Making*, Little, Brown and Company, Boston, Mass., 1948. MR0029828

[111] D. Tarwater (ed.), *The bicentennial tribute to American mathematics: 1776–1976*, The Mathematical Association of America, Washington, D.C., 1977. Papers presented at the Fifty-Ninth Annual Meeting of the Mathematical Association of America, held in San Antonio, Tex., January 1976. MR0497411

[112] D. Tarwater, J.T. White, and J.D. Miller (eds.), *Men and Institutions in American Mathematics*, Texas Tech Press, Lubbock, 1976.

[113] C.F. Thwing, *The College Woman*, Taylor & Baker, New York, 1894.

[114] T. Timmons, *Mathematics in nineteenth-century America—the Bowditch generation*, Docent Press, Boston, MA, 2013. MR3157131

[115] D.R. Traylor, *Creative Teaching: Heritage of R. L. Moore*, University of Houston Press, Houston, 1972.

[116] L.R. Veysey, *The Emergence of the American University*, University of Chicago Press, Chicago, 1965.

[117] J. Webb, *A Sense of Honor*, Naval Institute Press, Annapolis, MD, 1995.

[118] N. Wiener, *Ex-prodigy. My childhood and youth*, Simon and Schuster, New York, 1953. MR0057817

[119] N. Wiener, *I Am a Mathematician*, Doubleday, Garden City New York, 1956.

[120] C.D. Wright, *The History and Growth of the United States Census*, Government Printing Office, Washington DC, 1900.

[121] D.E. Zitarelli, *EPADEL: A Semisesquicentennial History, 1926-2000*, Raymond-Reese Book Co., Elkins Park PA, 2001.

Index

Abel, Niels Henrik, 133
Academy of Natural Sciences, 179
acoustics, 69, 117, 143, 188, 392
Adams, Daniel, 99
    Scholar's Arithmetic, 99
Adams, Henry Carter, 410, 412
Adams, John, 16, 39, 74, 95, 96
Adams, John Quincy, 96
Adrain, Robert, 111, **123**, 121–130, 216
Agricultural, Mining, and Mechanical Arts
    College (AMMAC), 241
Albert, Abraham Adrian, 351
Albion College, 310
d'Alembert, Jean le Rond, 96
Alexander, James, **58**, 58
Alexander, Stephen, 180
algebra, 2, 7, 8, 14, 27, 30, 34, 39, 79, 98, 117,
    131, 135, 138, 139, 188, 221, 241, 322, 392
    analytic style, 137, 139, 150
    synthetic style, 135, 137, 139, 150
Allardice, Robert Edgar, 297
Allen, Reginald Bryant, 306
American Academy, *see* American Academy of
    Arts and Sciences
American Academy for the Advancement of
    Science (AAAS), 179–182
American Academy of Arts and Sciences, 9, 54,
    89, 95–97, 112, 129, 133, 142, 153, 162, 165,
    172
American Civil War, 25, 98, 109, 162, 165, 193,
    196–199, 402, 439
American Geographical and Statistical Society
    (AGSS), 161–163
American Journal of Mathematics, 2, 131, 200,
    202, 249, 253, 264–271, 275, 276, 367, 369
American Journal of Science and the Arts
    (*AJSA*), 128, 129, 131–134, 137, 152
American Mathematical Monthly, 3, 26, 121,
    128, 165, 215, 349, 369, 424
American Mathematical Society (AMS), xv, 200,
    207, 212, 236, 239, 271, 273, 275, 283, 285,
    287–295, 357
American Philosophical Society (APS), 9, 54,
    57–62, 96

American Revolutionary War, 9, 12, 16, 54, 74,
    91, 100, 101, 111, 165, 439
American Statistical Association (ASA), 112,
    153–161, 411
Amherst College, 257
Amity College, 314
AMS Colloquium Lecture Series, 358–361, 381
The Analyst, 125, 165, 195, 199, 202–217, 255,
    260, 290, 368
Analyst; or Mathematical Museum, 93, 122–126,
    203, 214, 216, 288
Andrews, Grace, 418
Andrews, Robert, 29
Annals of Mathematics, 202, 368, 369
Annals of Statistics, 444
Annapolis, US Naval Academy, 146–147, 177,
    223, 230, 241, 303
arithmetic, 18, 27, 30, 34, 98, 131, 139, 241
Artis analyticae Praxis, 14
Association of American Geologists and
    Naturalists, 180
Association of Collegiate Alumnae (ACA), 395
Astronomische Nachrichten, 174
astronomy, 7, 8, 11, 37, 39, 51, 79, 97, 101, 117,
    147, 149, 182–184, 188, 241, 326, 392
Atwood, Wallace W., 307

Bache, Alexander Dallas, 189
Bacon, Clara, 418
Baker, Alfred, 363
Bancroft, George, 139, 230
Banneker, Benjamin, 9, **83**, 82–88
    *Almanack and Ephemeris for the Year of our
    Lord, 1792*, 85
Bard College, 402
Barnard College, 399, 402, 403
Barney, Ida, 401
Barnum, Charlotte Cynthia, 330, 400, 401
Baron, George, 112, 118, 216
Barron, William Amherst, 113
Bartram, John, 58
Baxter, Agnes Sime, 391, 397, 398
Beman, Wooster Woodruff, 331
Benner, Henry, 309, 315